Atmospheric Data Analysis

Roger Daley
Canadian Climate Centre

Published by the Press Syndicate of the University of Cambridge
The Pitt Building, Trumpington Street, Cambridge CB2 1RP
40 West 20th Street, New York, NY 10011-4211, USA
10 Stamford Road, Oakleigh, Melbourne 3166, Australia

First published 1991
First paperback edition 1993

Printed in the United States of America

Library of Congress Cataloging-in-Publication Data is available.

A catolog record for this book is available from the British Library.

ISBN 0-521-38215-7 hardback
ISBN 0-521-45825-0 paperback

Contents

Preface

The demand for atmospheric and environmental information is escalating rapidly, requiring the design and deployment of increasingly exotic and comprehensive observing systems. Atmospheric data analysis is the diagnosis of the complete, physically consistent four-dimensional atmospheric structure from this vast, asynchronous, heterogeneous data base.

Analyses of the atmospheric state are widely used in forecasting, in observational and phenomenological studies, and in formulating and verifying theoretical models of atmospheric behavior. They are also valuable for climatological studies, oceanography, and other disciplines of the terrestrial environment. In their most popular form, animated full-color atmospheric analyses are the mainstay of the television weather show. Although atmospheric analyses are widely used, the process of atmospheric analysis is not as widely understood as it should be. In large part, this is due to the paucity of pedagogical material in the field. This book, *Atmospheric Data Analysis*, evolved from a series of discussions with Tony Hollingsworth in England during a sabbatical year in 1983. Both of us proceeded to develop pedagogical material on atmospheric analysis, which formed the basis for a course at Colorado State University in 1984. Among the student critiques of the lecturer was one perceptive comment that a textbook would have substantially improved the course. I could not disagree and set to work immediately.

Atmospheric Data Analysis covers the complete field of atmospheric objective analysis or data assimilation. It is designed to be self-contained and thus includes some topics from atmospheric dynamics and statistics. The emphasis is on the theoretical foundations of the subject, and most of the developments are analytic. However, practical aspects and examples are introduced where appropriate. The book is intended for graduate or advanced undergraduate students with no prior knowledge of the subject. The level of mathematics is at the advanced undergraduate level, and background material on less familiar mathematical concepts is provided. Some prior knowledge of undergraduate atmospheric dynamics would be helpful but is not absolutely necessary. Problems of varying levels of difficulty are included after each chapter.

The graduate or undergraduate curricula could use this book for a self-contained course or could include it in courses on numerical weather prediction, climate modeling, or atmospheric and climate dynamics. The material is also applicable in oceanography and related disciplines.

The emphasis in early chapters is on spatial analysis, and the treatment is largely mathematical or statistical. Gradually, more atmospheric physics, in the form of multivariate constraints, is introduced into the discussion, and there is an increasing emphasis on temporal aspects. The last chapter introduces the student to the most recent developments in the field.

This book could not have been written without the support and cooperation of a number of people on three continents. Tony Hollingsworth provided the initial inspiration, gave me many useful ideas and endless encouragement, and reviewed large segments of the text. Andrew Lorenc and David Williamson reviewed the final draft of the manuscript. Steve Cohn, Ron Errico, Akira Kasahara, John Lewis, Hersh Mitchell, Tom Schlatter, J. J. Stephens, and Clive Temperton each reviewed substantial portions of the manuscript and provided me with much constructive criticism. Other portions of the manuscript were reviewed by Andrew Bennett, John Derber, Tzvi Gal-Chen, Carl Hane, Peter Lynch, Kikuro Miyakoda, Phil Rasch, Yoshi Sasaki, Bob Seaman, Adrian Simmons, and Grace Wahba. Helpful ideas and encouragement came from Wayman Baker, T. N. Krishnamurti, Norm McFarlane, I. M. Navon, Wayne Schubert, Duane Stevens, Phillip Thompson, Joe Tribbia, and Warren Washington.

I am extremely indebted to Pearl Burke, who persevered in typing the manuscript on an increasingly cantankerous word processor. Other typing was done by Lynda Smith, Una Ellis, and Valerie Moore. Original graphics were drawn by Tom Chivers, Bob Donohue, and Brian Taylor of the Atmospheric Environment Service of Canada (AES) and by the graphics department of the National Center for Atmospheric Research (NCAR). I would also like to thank NCAR, AES, and my immediate superiors, Akira Kasahara, Jim McCulloch, and Kirk Dawson, for their support over the years. My editors at Cambridge University Press, Peter-John Leone and Judi Steinig, provided much-needed advice and support for a first-time author.

Finally, I am particularly grateful for the patience and forbearance of my wife, Lucia, during the course of this project and for her help in proofreading the final text.

Partial list of symbols

a earth radius, Fourier coefficient, arbitrary constant

b Fourier coefficient, arbitrary constant

c expansion coefficient, weight (Section 4.1), constant

d residual, derivative

$\underline{d}$ observation increment vector, observation vector (Section 13.1)

e 2.71828

$\underline{e}$ eigenvector

f arbitrary function, Coriolis parameter

g spectrally transformed variable, gravitational constant

h height, equivalent depth, arbitrary basis function

i $\sqrt{-1}$, index

$\mathbf{i}$ unit vector

j index

$\mathbf{j}$ unit vector

k two-dimensional wavenumber, index

$\mathbf{k}$ unit vector

l longitudinal velocity component, norm, index

m x wavenumber, index

n y wavenumber, normal direction, index

p probability, element of $\underline{P}$, arbitrary function, index

q specific humidity, element of $\underline{\underline{Q}}$, arbitrary function, index

$\underline{q}$ column vector (Section 5.1)

r absolute distance, random noise component

$\mathbf{r}$ position vector

s complex frequency, arbitrary state variable, index

$\underline{s}$ column vector of dependent variables

t transverse velocity component, time

u eastward (x) or radial velocity component

v	northward (y) or tangential velocity component
$\mathbf{v}$	horizontal velocity vector
w	specified a priori weight, vertical velocity component
x	eastward space coordinate, normal mode amplitude
$\times$	vector product
y	northward space coordinate, slow mode amplitude
$\underline{y}$	eigenvector (Section 4.5)
z	vertical space coordinate, fast mode amplitude
$\underline{z}$	arbitrary column vector
A, $\underline{\underline{A}}$	analyzed value (Section 4.2), analysis error covariance matrix
B, $\underline{\underline{B}}$	background value (Section 4.2), background error covariance matrix
C	covariance (Section 5.1), interaction coefficient (Section 10.3), condensation (Section 13.6), curve on complex plane (Section 13.4)
$\underline{\underline{C}}$	weight matrix (Section 3.5), covariance matrix (Appendix D)
C_{P}	specific heat at constant pressure
D	divergence, damping parameter
E	eastward gravity mode (Section 9.4)
E	error, evaporation
$\underline{\underline{E}}$	matrix of eigenvectors
F	slack function, arbitrary function
$\mathbf{F}$	friction force
$\underline{\underline{F}}$	forward interpolation error covariance matrix
G	inertia–gravity mode, arbitrary function
$\underline{\underline{G}}$	Gram matrix
H	forward interpolation operator (Section 13.1)
$\underline{\underline{H}}$	matrix of basis functions evaluated on network (Section 2.3), forward interpolation matrix (Section 13.1)
I, $\underline{\underline{I}}$	functional, identity matrix
J	functional, Bessel function, summation limit
K	Kelvin mode (Section 9.4)
K	shallow water parameter (Section 6.4), frictional coefficient, summation limit
$\underline{\underline{K}}$	Kalman–Bucy gain matrix
L	Laplace transform, characteristic scale, summation limit
$\underline{\underline{L}}$	linearized f-plane shallow water equations matrix
L_{H}, L_{Z}, L_{R}	horizontal and vertical characteristic scales, Rossby radius
L_{B}, L_{O}	background and observation error characteristic scales
M	ellipticity measure, summation index
$\underline{\underline{M}}$	diagonal matrix (Section 3.5), linear model matrix (Section 13.2)
N	Brunt–Väisälä frequency, summation index
$\underline{\underline{N}}$	spline matrix (Section 2.7), Jacobian matrix (Section 10.7)
O, $\underline{\underline{O}}$	observed value, observation error covariance matrix
P, $\underline{\underline{P}}$	pressure, prediction error covariance matrix

Q	potential vorticity, diabatic heating, summation limit
$\underline{\underline{Q}}$	spline matrix (Section 2.7), diagonal SCM matrix (Section 3.7), KB system error matrix (Section 13.3)
R	Rossby mode (Section 9.4)
R	radius of influence, spectral response, gas constant, symbolic right-hand side
R_0	Rossby number
$\underline{\underline{R}}$	$\underline{\underline{Q}} + \underline{\underline{F}}$
S, $\underline{S}$	horizontal domain, second-derivative vector (Section 2.7)
$\underline{\underline{S}}$	symmetric SCM matrix (Section 3.5), model matrix (Section 12.3)
T	transpose
T	true value (Section 4.2), temperature (Section 7.3)
$\underline{\underline{T}}$	tri-diagonal matrix
U	constant zonal wind component
V	variance
W	westward gravity mode
W	a posteriori weight
$\underline{\underline{W}}$	a posteriori weight matrix
$\underline{Y}$	vector of slow mode amplitudes
Z	vertical structure function
$\underline{Z}$	vector of fast mode amplitudes
α	spectral transform of a posteriori weights, inertia–gravity wave frequency, direction cosine, Rossby projection angle
$\underline{\underline{\alpha}}$	a posteriori weight matrix
β	latitudinal derivative of Coriolis parameter, nondimensional data insertion frequency (Section 12.3)
γ	constant
Γ	static stability
δ	Dirac delta function, Kronecker delta, variational operator
∂	partial differential operator
$\boldsymbol{\nabla}$, $\boldsymbol{\nabla}\cdot$	gradient and divergence operators
$\boldsymbol{\nabla}\times$, ∇^2	curl and Laplacian operators
ε	error, small parameter (Section 7.3), spherical harmonic parameter (Appendix A)
$\boldsymbol{\zeta}$	relative vorticity vector
ζ	$\mathbf{k}\cdot\boldsymbol{\zeta}$
κ	R/C_P
η	mean value, arbitrary function
θ, $\underline{\underline{\theta}}$	potential temperature, Hessian matrix
λ	eigenvalue, Lagrange multiplier, longitude
$\underline{\underline{\Lambda}}$	diagonal matrix of eigenvalues
μ	sin(latitude), geostrophic coupling parameter, eigenvalue, x wave-number, index

ν	eigenvalue, divergent coupling parameter, y wavenumber, index
π	3.14159
Π, $\underline{\underline{\Pi}}$	vertical pressure scale, slow mode projection matrix
ρ	correlation, density, spectral radius
$\underline{\rho}$, $\underline{\underline{\rho}}$	correlation vector, matrix
σ	standard deviation, frequency
Σ	summation
τ, $\boldsymbol{\tau}$	time scale, tangential unit vector
ϕ	tangential coordinate, latitude
Φ, $\tilde{\Phi}$	geopotential, horizontally averaged geopotential
χ	velocity potential
ψ	streamfunction
ω	normalized a posteriori weight, vertical velocity (P coordinates), frequency
$\boldsymbol{\Omega}$	earth angular rotation vector
Ω	$\mathbf{k}\cdot\boldsymbol{\Omega}$, forward interpolation operator
$\underline{\underline{\Omega}}$	forward interpolation matrix

1

Introduction

One of the worst snowstorms in living memory struck the U.S. mid-Atlantic states during 18–19 February 1979. Total snowfall amounts exceeded 0.5 meter (m), and accompanying high winds created drifts of up to 2.5 m. Airports and roads were closed from Atlanta to New York City, and a state of emergency was declared in many areas. This legendary storm, often referred to as the "Presidents' Day snowstorm," was notable not only for its severity but also because it was exceptionally poorly forecast.

Media dissemination of weather forecasts to the public is the final step in a complex process. The first step is to collect all the atmospheric observations from the entire globe for a given time. Second, these observations are diagnosed or analyzed to produce a regular, coherent spatial representation of the atmosphere at that time. Third, this analysis becomes the initial condition for the time integration of a numerical weather prediction model based on the governing differential equations of the atmosphere. Finally, the numerical prediction is used by a human forecaster as the basis for the public forecast.

Not surprisingly, the Presidents' Day snowstorm has been widely studied. In one of these studies, Hollingsworth, Lorenc, Tracton, Arpe, Cats, Uppala, and Kallberg (1985) examined the sensitivity of the numerical forecast to changes in the initial conditions. In midlatitudes, errors tend to propagate eastward in the prevailing westerly flow. The investigators discovered that the predicted evolution of the Presidents' Day snowstorm in the western Atlantic was extremely sensitive to *small errors in the initial analysis in the northwestern Pacific four days earlier.* In other words, a small localized error in the initial analysis affected the forecast for locations far removed in space and time.

Clearly, accurate analyses of the state of the atmosphere are indispensable for forecasting. They are also invaluable for phenomenological and climatological studies and for the formulation and testing of theoretical models of atmospheric behavior. In their most popular form, animated full-color analyses are the mainstay of the TV weather show.

In the last century, only observations from local surface stations were available, and these were analyzed using subjective manual techniques. Over the years, as the demand for atmospheric and environmental information grew, increasingly exotic and comprehensive observing systems were deployed for sampling atmospheric parameters. Analysis techniques also had to evolve to provide this vast and heterogeneous real-time data base with spatial and temporal continuity and internal physical consistency.

Analysis problems that are equally complex occur in other scientific disciplines. What distinguishes the atmospheric analysis problem from other problems are the physical properties of the medium, the spatial and temporal characteristics of the observing system, and the applications toward which the analysis is directed. In this chapter we examine all of these characteristics, beginning with the most fundamental, the physical properties of the atmosphere itself.

1.1 Atmospheric characteristics

The important components of the earth climate system are

the atmosphere,
the hydrosphere (oceans),
the cryosphere (polar ice fields and sea ice, continental snow cover),
the lithosphere (earth's surface including hydrology and volcanism), and
the biosphere (vegetative cover and oceanic flora and fauna).

Each of these components is coupled to the others (Peixoto and Oort 1984). The atmosphere and hydrosphere are strongly coupled through the exchange of energy, momentum, and matter on many space and time scales at the air/sea interface. Changes in continental snow cover affect the surface albedo, and variations in the extent of sea ice affect the heat exchange between ocean and atmosphere. The lakes, rivers, and ground water of the lithosphere are essential elements of the terrestrial branch of the hydrological cycle and are connected to the atmospheric branch by evaporation and precipitation. The vegetative cover of the biosphere affects the surface roughness, albedo, evaporation, precipitation, and moisture capacity of the soil.

The earth climate system displays phenomena on a variety of space and time scales (Figure 1.1).

The atmosphere is the most variable component of the earth climate system in both space and time. The atmosphere system functions to store and distribute the heat received from the sun. Although there is net radiative cooling in the free atmosphere, the earth's surface has a radiative surplus, which is maximum in the tropics and is positive at all latitudes except the winter polar latitudes. This horizontal and vertical radiative distribution forces a circulation in the atmosphere that transports heat upward and poleward.

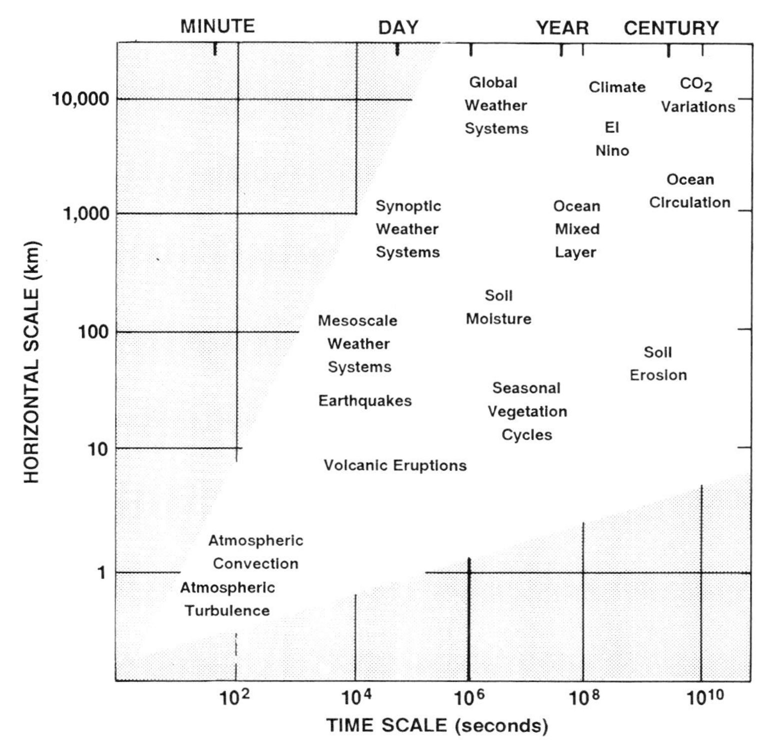

Figure 1.1 Space and time scales of phenomena of the earth climate system. Note that both scales are logarithmic. Phenomena with very long time scales are not included.

In the topics, the surface air is warm and moist, and the pressure at the surface of the earth is relatively low. As the heated air rises and cools, much of its water vapor condenses out as rain. At high levels, the air travels poleward and sinks at about 30 degrees north and south, creating belts of high pressure. This descending air is heated by (adiabatic) compression. The circulation in this convective cell is completed by an equatorward current at low level, which flows naturally from high to low pressure. The low-level equatorward flow from both hemispheres results in the Intertropical Convergence Zone. The Hadley cell, as this simple toroidal circulation is called, is a direct thermal cell.

As the earth rotates, moving fluid is deflected to the right in the northern hemisphere and to the left in the southern hemisphere (Coriolis force). At low levels in the tropics and subtropics, the Coriolis force deflects equatorward flowing air to the west giving rise to the easterly trade winds. In the same way, the air flowing poleward at high levels must develop a strong westerly component to conserve its absolute angular momentum. Because this westerly current is unstable to small perturbations, the flow at high latitudes is characterized by large-amplitude waves or eddies, which account for the major portion of the extratropical poleward heat transport. The eddies can vary in scale from planetary waves (which are due mainly to orography or thermal contrast between continent and ocean), through synoptic disturbances or cyclones (which extract energy from the westerly flow through instability mechanisms), right down to individual convective elements. The cyclones have life cycles of about one week and travel along well-defined cyclone tracks that are dictated by the more slowly moving planetary waves. The cyclones have a large rotational wind component and a small divergent wind component. The divergent component, even though small, is important because low-level convergence and upper-level divergence imply upward vertical motion, condensation, and precipitation.

This book concentrates primarily on atmospheric phenomena of global and synoptic space and time scales (see Figure 1.1): space scales from approximately 100 to 40,000 km (the earth circumference) and time scales from a few hours to about one month. However, we do not neglect atmospheric phenomena with different space and time scales and other components of the earth climate system. The shorter time and space scales of mesoscale weather systems and atmospheric convection are discussed in Section 13.7. Time scales longer than one month are considered in Section 1.6. The hydrosphere is discussed in Section 13.8 and the lithosphere in Section 13.6.

In the analysis of the atmospheric state, two aspects are of primary importance: the physical laws that govern the atmospheric circulation and the spatial and temporal spectra of atmospheric phenomena. The physical laws indicate how it might be possible to determine one variable from another, for example, analyzing wind from temperature observations. Knowledge of the spectra of atmospheric variables can be used to determine an acceptable spacing between observations. Spectra are also useful in designing analysis algorithms that successfully separate observational noise from the signal (the true values).

The governing equations of the atmosphere can be written in terms of the independent variables (three spatial variables and time) and the dependent variables (such as mass, temperature, the three components of motion, humidity, chemical species, and cloudwater). In general, these equations are nonlinear partial differential equations of which the most important are the equations of motion, the first law of thermodynamics, and the mass and humidity conservation equations.

Atmospheric phenomena can be characterized by their aspect ratio L_Z/L_H, where L_H is a characteristic horizontal scale and L_Z a characteristic vertical scale. A reasonable value for L_Z is about 10 km; so in the global and synoptic scales, $L_Z \ll L_H$. Under these conditions, the atmospheric flow is, to a very high degree of approximation, in hydrostatic balance, and vertical motions are much weaker than horizontal motions. In mid and high latitudes, the synoptic and to some extent the global scales exhibit a substantial degree of geostrophic balance. That is, the Coriolis forces are approximately balanced by the pressure gradient forces.

Space/time spectra of atmospheric flow are shown in Figure 1.2 for time scales of days and horizontal space scales of 10^3 kms. The contours are isolines of 500 millibar (mb) geopotential variance and are spaced logarithmically. It is evident

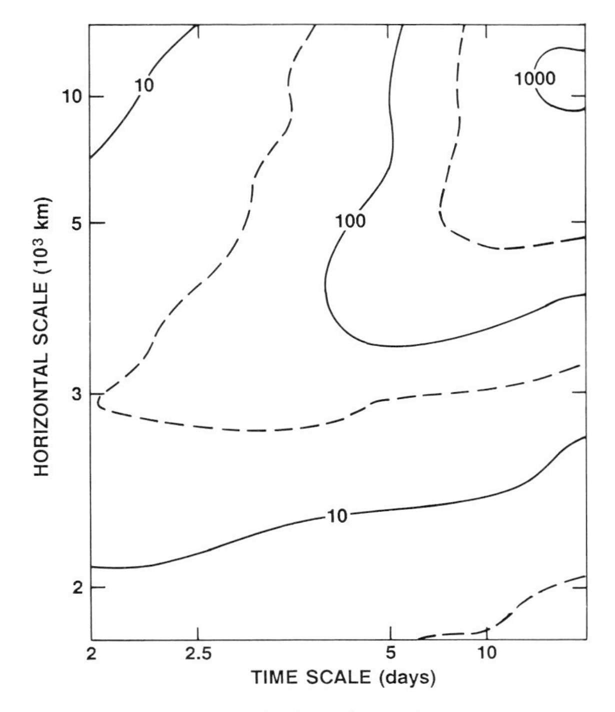

Figure 1.2 Space/time spectrum of atmospheric variance. (From Pratt, *J. Atmos. Sci.* **36**: 1681, 1979. The American Meteorological Society.)

that the bulk of the variance is at relatively larger time and space scales and that shorter time scales are associated with shorter space scales.

The temporal spectrum of atmospheric kinetic energy shown in Figure 1.3 is logarithmic in time and linear in kinetic energy. This diagram is somewhat schematic and so does not show units for the kinetic energy. Four obvious energy peaks appear. The peaks at one day and one year are the diurnal and annual cycles, respectively. The peak that occurs between one day and one month is associated with the baroclinic eddies or cyclones in the midlatitude westerlies; the peak at about one minute is associated with atmospheric turbulence and convection. A relative minimum occurs at time scales of about one hour.

A spatial spectrum of vertically integrated atmospheric kinetic energy at horizontal scales of 40,000 to 200 km is shown in Figure 1.4 as a log/log plot. The global wavenumber k is defined such that wavenumber 1 has a wavelength equal to the earth's circumference. Curve S (stationary) indicates all variance associated with time scales longer than about one month, and curve T (transient) indicates all variance associated with the remaining time scales. As would be expected from Figure 1.2, curve S has most of the variance at very long waves. The transient curve T has a maximum at wavenumbers 8–10 (several thousand kilometers) associated with the baroclinic eddies.

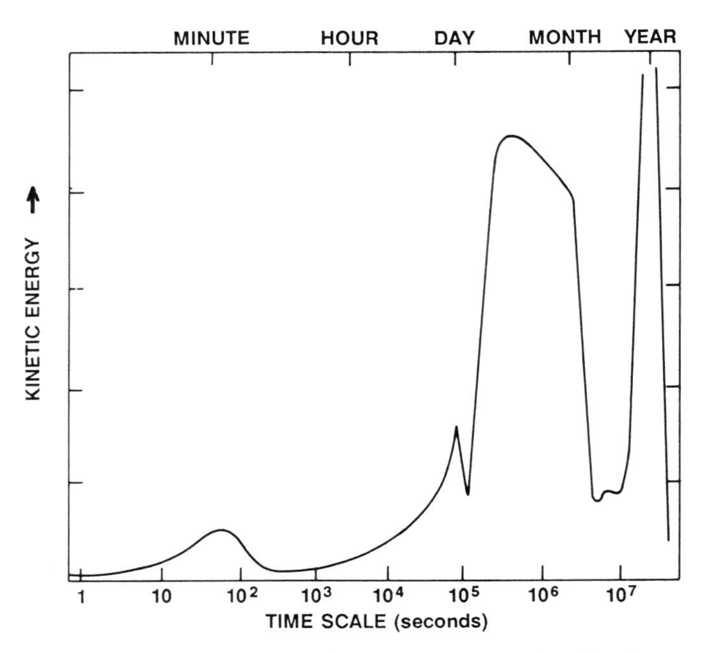

Figure 1.3 Temporal spectrum of atmospheric kinetic energy. (After Vinnichenko 1970)

Atmospheric turbulence theories suggest that inertial subranges should exist for which the kinetic energy spectra obey a $k^{-5/3}$ power law for a three-dimensional fluid and a k^{-3} power law for a two-dimensional fluid. The small aspect ratio in the synoptic and global scales suggests a largely two-dimensional flow, which should become increasingly three-dimensional as k increases. The slopes of the $k^{-5/3}$ and k^{-3} power laws are indicated on Figure 1.4. The observed spectra are not very reliable for the shorter scales (dashed line) but seem to fall somewhere between $k^{-5/3}$ and k^{-3}.

As noted earlier, the atmospheric circulation is governed by physical laws. The formulation of these governing laws in the nineteenth and twentieth centuries put meteorology on a scientific basis and led to an important conceptual breakthrough.

1.2 The ultimate problem in meteorology

The main impetus for the study of the atmospheric circulation has always been people's desire to forecast the weather (predict future atmospheric states). Bjerknes (1911) referred to this as the ultimate problem in meteorology and outlined an approach for tackling it. According to this approach, two conditions must be satisfied to successfully predict future atmospheric states:

I The present state of the atmosphere must be characterized as accurately as possible.

II The intrinsic laws, according to which the subsequent states develop out of the preceding ones, must be known.

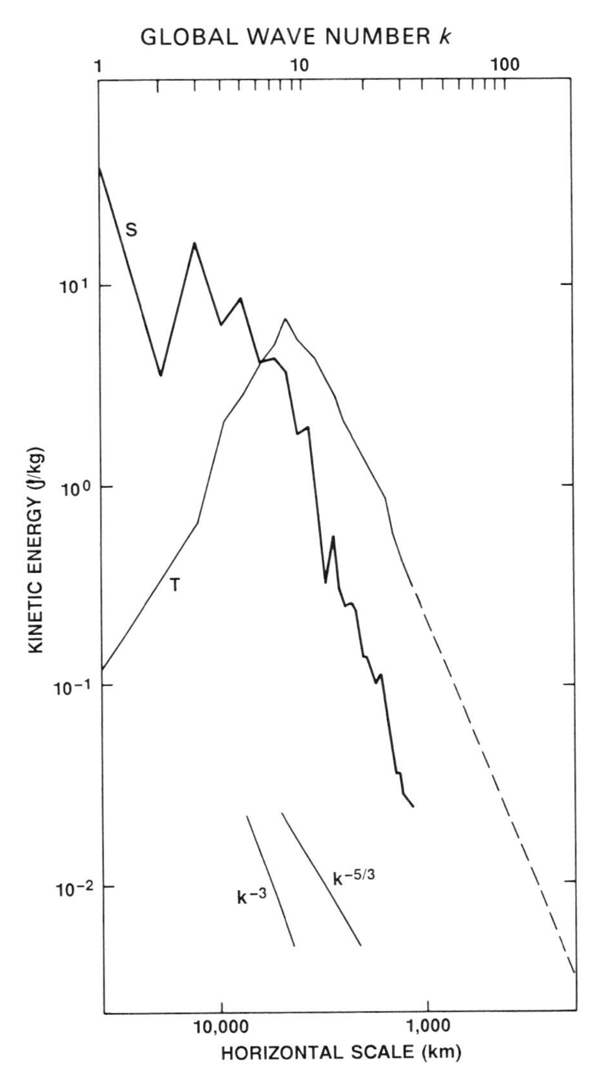

Figure 1.4 Spatial spectrum of atmospheric kinetic energy. (Curves S and T are from Boer and Shepherd, *J. Atmos. Sci.* **40**: 164, 1983. The American Meteorological Society, with dashed portion of T curve inferred from Brown and Robinson, *J. Atmos. Sci.* **36**: 270, 1979. The American Meteorological Society.)

Conditions I and II define weather prediction as an initial-value problem. In today's terminology, Bjerknes' approach would be called deterministic because future states of the atmosphere are assumed to be completely determined from the present state. For the practical implementation of the initial-value approach, Bjerknes outlined a program that was subdivided into three partial problems or components:

1 The observation component

2 The diagnostic or analysis component
3 The prognostic component

Components 1 and 2 are related to the characterization of the present state (condition I); component 3 is related to condition II.

Component 1 requires an observation network distributed throughout the atmosphere. At the observing points, the dependent variables (mass, wind, etc.) are measured. The distances between observation stations and the time between observations should be sufficiently small to adequately resolve the space and time scales of the phenomena of interest. Bjerknes also suggested that all observations be taken at approximately the same time (the principle of simultaneity), but this restriction is no longer felt to be practical.

In the diagnostic component (2), the observations are analyzed in a way that produces regular spatial representations of the dependent variables at fixed times. Such representations could consist of regularly spaced points on a grid or coefficients of a functional expansion. The density of the gridpoints or the number of terms in the functional expansion should be sufficient to adequately represent each of the dependent variables. The diagnostic function should consider not only the observations themselves but also the intrinsic relations between different dependent variables (mass and wind, for example).

In the prognostic component (3), the governing equations of the atmosphere are used to predict future states from the present state. The governing equations are highly implicit and have to be rewritten so that they can be used to *integrate* or *march* the dependent variables forward in time from their initial values. Bjerknes' era lacked the resources and knowledge to carry out his program successfully, but now all three components are performed routinely.

We now consider all three functions in more detail in order to establish the context in which the diagnostic component (the subject of this book) is exercised.

1.3 The observing system

> An existing meteorological station in the British Isles has been either an outgrowth from an astronomical or magnetic observatory, or it has adjoined the house of an enthusiast who lived there for reasons unconnected with meteorology, or it has been pushed out to the confines of the islands to grasp as much weather as possible, or it has been placed in charge of coastguards because they are on duty at night, or it has been set on a mountain to test the upper air. Excellent practical reasons all these, but it is remarkable that the properties of the atmosphere, which are expressed by its dynamical equations and its equation of continuity, appear to have no influence on the selection. (Richardson 1922, p. 217)

Richardson's remarks, written in 1922, indicate quite clearly his frustration with the British meteorological observing system of the time. His complaint was essentially of the heterogeneity of the observing system, which was designed for human convenience rather than sampling the atmosphere in a manner consistent with underlying physical principles. As will be seen in the following historical review, the heterogeneity of the observing system has its roots deep in history (Khrgian 1970).

The present-day global meteorological observing system has been evolving for 300 years. The basic meteorological instruments (thermometer, barometer, hygrometer, and anemometer) had been invented by the middle of the eighteenth century. Meteorological observatories such as the Paris Observatory commenced taking regular observations in the late 1600s. Short-lived meteorological observing networks were set up in Britain (Royal Society, 1724–1735) and Russia (Great Northern Expedition, 1730–1745). The observations collected by these networks were dignosed long after the fact due to the poor communications of the times. The Palatine Academy of Sciences and Letters in Mannheim, Germany organized the first international observing network, in which regular observations were collected during the 1780s and 1790s from as far away as the Ural Mountains and Cambridge, Massachusetts.

The idea of a global real-time observing system first became a reality in the nineteenth century. A number of technological and organizational developments occurred after 1800 that made a global network possible. On the technical side, meteorological instrumentation substantially improved. More important, however, was the invention of the telegraph, which allowed meteorological observations to be communicated rapidly.

On 14 November 1854, a severe storm in the Black Sea destroyed the French fleet at Balaklava. Because the storm had been observed the previous day over the Mediterranean, the French government asked Urbain LeVerrier, director of the Paris Astronomical Observatory, to study the circumstances surrounding this phenomenon. He obtained weather observations from around Europe and was able to trace the path of the storm. In 1855 he presented to Napoleon III a plan for a great meteorological network, designed to warn mariners of impending storms. Thus was born the first permanent observing network and the first national weather service. Other nations quickly followed suit.

Meteorological phenomena, of course, transcend national boundaries, and it was realized that the observation network would require extensive international cooperation. In particular, observing practices, units, and observation times had to be standardized. Increasing international cooperation and consultation took place throughout the latter half of the nineteenth century, but the International Meteorological Conference of Vienna in 1873 placed this cooperation on a formal diplomatic basis, and a permanent international committee was established shortly thereafter. One of its main responsibilities was the standardization of meteorological observations. This was no mean task; even the general adoption of metric units by English-speaking meteorologists took another 75 years.

Thus, by the beginning of the twentieth century, a global real-time observing system was in place. It was primarily a surface network, mainly confined to land and distributed rather arbitrarily, as noted by Richardson. Sporadic attempts to sound the free atmosphere with balloons, kites, and aircraft occurred in the early years of this century, but not until the second world war was an international upper air network established. This network employed the balloon-borne radiosonde, which measured directly the temperature, pressure, and humidity and transmitted encoded observations to nearby ground stations by radio. The radiosonde balloons could be tracked by theodolite or radar to yield wind measurements.

By 1950, the upper air network provided good coverage over land, but there were still large oceanic areas from which observations were rarely received. In fact, Spilhaus (1951) reported that over large areas of the southern oceans less than 500 meteorological observations (of any kind) had been made in half a century. In the later 1960s, satellite-borne radiometers were developed and launched. With the deployment of these instruments came the hope of obtaining uniform observational coverage over the globe and thus finally filling in the large oceanic data voids. The invention of the electronic computer in the late 1940s made possible the real-time processing of all these new data.

In 1979, the nations of the world organized the Global Weather Experiment (GWE) to observe the atmosphere as systematically as possible using the most advanced technology then available. In addition to the operational observing network, many special observing systems were deployed, such as drifting buoys to measure surface variables over the oceans, dropwindsondes ejected from aircraft, and constant-level balloons to observe the upper circulation. The experiment ran from December 1978 to November 1979 and provided a comprehensive global meteorological data set.

The international meteorological observing system today is known as the World Weather Watch (WWW). It is supervised by the World Meteorological Organization (WMO), which is a United Nations agency. The WWW has three components. The Global Observing System (GOS) consists of the basic surface and radiosonde networks, aircraft and satellite systems run by the national meteorological services. The Global Telecommunications System (GTS) consists of telecommunication facilities and arrangements for rapid transmission of observations and processed information. The Global Data Processing System (GDPS) has world meteorological centers in Melbourne, Moscow, and Washington D.C. plus national meteorological centers that collect, store, process, archive, and disseminate the observational data in real time.

Weather observations from the surface and radiosonde networks are processed in the following way. At the observing station, the observer processes the raw information, making appropriate corrections for local effects and checking the internal consistency of the observations. The observations are then coded in a form suitable for transmission, using an international code called SYNOP (for land surface observations), SHIP (for sea surface observations), and TEMP (for upper air observations). This coded information is transmitted to regional centers and then to national and world meteorological centers. The information from all stations is collected, decoded, checked for errors, archived, and disseminated. Unfortunately, this process takes time: a complete set of global observations is not usually available until several hours after observation time.

Surface and radiosonde observations are taken regularly at *synoptic* times. These are 00 and 12 GMT (Greenwich mean time) for radiosondes and 00, 03, . . ., 21, 24

GMT for surface observations. Satellite-borne observing systems transmit their information directly to ground receiving stations from which it is forwarded to special centers where the information is decoded and processed on computers. Satellite observations are taken continuously and are called *asynoptic*.

This historical survey illustrates several aspects of the observing network that are still true today. First, it is a multipurpose network based on international cooperation and thus there is a substantial political component. Second, rapid communications are of the utmost importance. Third, because it is based on many types of instruments and is controlled by many nations, there is a great variation in coverage and quality. In fact, Richardson's comment still has some validity.

It is important to distinguish between the international global observing network and the local short-duration observing networks that are set up for field experiments. In the latter case, the station sites, sampling frequency, and so forth are chosen with respect to the phenomena to be studied. The field network is designed to maximize the signal extracted, subject only to the constraint of a given financial expenditure. Consequently, network design is very important for field experiments but plays a smaller role in the international global observing network, being used primarily in the deployment of new observing systems. Some of the analysis tools, to be developed in later chapters, can be used in network design, but, apart from a brief discussion in Appendix H, we do not pursue this subject.

For the various observation systems and instruments, the important characteristics are the dependent variables measured (such as temperature and wind), the error characteristics of the measurement, and the spatial and temporal distribution.

We divide meteorological instruments into three classes:

Class 1. Instruments that make in situ measurements at points. Here *point* does not have the usual mathematical definition; we simply assume that the instrument occupies a much smaller volume than the phenomenon to be sampled. This class includes the conventional surface instruments (e.g., mercury barometer, thermometers, anemometers, and hygrometers) and the radiosonde instruments for the measurement of temperature, pressure, humidity, chemical constituents, aerosols, and so on.

Class 2. Instruments that sample an area (on the earth's surface) or a volume (of the atmosphere) remotely. These instruments take advantage of the transmission properties of the earth's atmosphere with respect to different wavelengths of the electromagnetic spectrum. They can be ground based, aircraft based, or mounted on orbital platforms and are either active or passive. In an active system, electromagnetic pulses are transmitted through the atmosphere, and the reflected or scattered signal is processed to yield an indirect estimate of some atmospheric or surface property. Examples are radar for measuring precipitation and winds (via Doppler shift) and profilers and lidars for measuring water vapor, aerosols, and winds (via Doppler shift). In a passive system, properties of the atmosphere or earth's surface are inferred from radiation emitted, scattered, and/or reflected in the visible, infrared, or microwave wavelengths. Properties such as temperature, water vapor, sea surface temperature, and chemical constituents can be obtained by inverting the measured radiances.

Class 3. Instruments that calculate wind velocities from Lagrangian trajectories. In this case a physical target is followed remotely, and velocities are calculated from observed displacements at fixed times. Examples are radiosonde balloons tracked by theodolite or radar, constant-level balloons tracked with satellite navigation systems, and cloud elements tracked visually (or using pattern recognition techniques) from geostationary satellites.

The error characteristics of these instruments are not straightforward, unfortunately, and some preliminary discussion is in order. Each meteorological instrument is accurate to within a certain tolerance called the *instrument error*. This error is a function of the instrument design and the ambient conditions in which it must operate. For example, the mercury barometer for surface (not mean sea level) pressure measurements has an expected instrument error of about 0.25 mb for a single reading due to ambient temperature and wind effects (Atmospheric Environment Service of Canada 1954).

A more nebulous type of error, the *error of representativeness*, is more difficult to quantify, though an attempt has been made by Petersen and Middleton (1963). In qualitative terms this error can be explained as follows. The spatial spectrum of Figure 1.4 indicates variance at all spatial scales in the atmosphere, with generally less variance at smaller scales. The observation network, however, has a finite spacing between observation stations. Consider a network with average spacing L_{N} between stations; it is reasonable to suppose (and is shown quantitatively in Appendix H) that phenomena with scales much greater (smaller) than L_{N} will be sampled very well (poorly) by this network. For example, consider the observation of thunderstorms or tornadoes with a characteristic scale of 10 km or less in an observing network with $L_{\mathrm{N}} = 1000$ km. If the thunderstorm lies between stations, it will not be "seen" by the network at all. On the other hand, if it lies directly over one of the stations, it may be misrepresented as a much larger-scale phenomenon. Thus, the error of representativeness is a measure of the error caused by the misrepresentation of all scales smaller than L_{N}.

The error of representativeness is different in instruments of Classes 1 and 2. Suppose $s(\mathbf{r})$ is some atmospheric property (the true signal) and $\mathbf{r}$ represents the one-, two-, or three-dimensional spatial coordinate. Define $\tilde{s}(\mathbf{r}_k)$ to be an estimate of the signal at the kth observation station by an instrument of Classes 1 or 2. Petersen (1968) suggests the following simplified representation for $\tilde{s}(\mathbf{r}_k)$:

$$\tilde{s}(\mathbf{r}_k) = \int_{\mathbf{r}} s(\mathbf{r}) y(\mathbf{r}, \mathbf{r}_k)\, d\mathbf{r} + \varepsilon(\mathbf{r}_k) \qquad (1.3.1)$$

where $\varepsilon(\mathbf{r}_k)$ is the instrument error at $\mathbf{r}_k$. In an instrument of Class 1, $y(\mathbf{r}, \mathbf{r}_k) = \delta(\mathbf{r} - \mathbf{r}_k)$, the Dirac delta function. The function $\delta(\mathbf{r} - \mathbf{r}_k)$ is defined to be infinite at $\mathbf{r} = \mathbf{r}_k$ and equal to zero elsewhere, and has the property

$$\int_{\mathbf{r}} \delta(\mathbf{r} - \mathbf{r}_k) s(\mathbf{r}) \, d\mathbf{r} = s(\mathbf{r}_k) \tag{1.3.2}$$

For Class 2 instruments, $y(\mathbf{r}, \mathbf{r}_k)$ is an *aperture function* appropriate to the instrument. The estimate $\tilde{s}(\mathbf{r}_k)$ contains both instrument error and error of representativeness for Class 1 and 2 instruments. However, for Class 2 instruments, the weighted area averaging implied by (1.3.1) tends to filter smaller-scale components and reduce the error of representativeness.

Observation errors can be temporally or spatially correlated (Appendix D). Some observing systems use a first guess or background field (Section 1.5) to obtain the observation. In these systems, the background and observation errors can be correlated.

For each observing system, representative coverage maps are shown in Figures 1.5 and 1.6. These maps show the location of various types of data received at the European Centre for Medium Range Forecasting for the period 0900–1500 GMT 2 January 1989. In accord with normal practice, the estimated observation errors given throughout this book consist of both instrument errors and errors of representativeness. Estimated observation errors are given in Bengtsson (1975) and Shaw et al. (1987).

Surface stations

Surface stations (see Figure 1.5) are distributed irregularly over land and along well-traveled shipping routes. The surface pressure, temperature, wind components, and relative humidity are observed at least every 3 hours. The surface pressure is then reduced to mean sea level pressure using algorithms that take into account the temperature and the topography. The errors are 0.5 mb for the surface pressure, 1 K for the temperature, 3 $\mathrm{ms^{-1}}$ for the wind, and 10 percent for the relative humidity. The errors in calculating mean sea level pressure can be much larger in regions of high terrain; also, wind errors are larger from moving ships.

Radiosondes and pilot balloons

Radiosonde launching stations (see Figure 1.5) are distributed irregularly over land and are located on some remote islands. The density is particularly high in Europe and China and much lower in uninhabited areas. There are also weather ships and special container ships that launch radiosondes automatically. The variables observed are the temperature, pressure, humidity, and wind components every 12 hours. Only the wind is observed by pilot balloons. The errors are 1 K in temperature, 3–5 $\mathrm{ms^{-1}}$ in the wind, and 10 percent for the relative humidity. The errors are largely random, but they can be vertically correlated, particularly in determining the geopotential hydrostatically from the temperature. Wind errors tend to be larger at higher elevations.

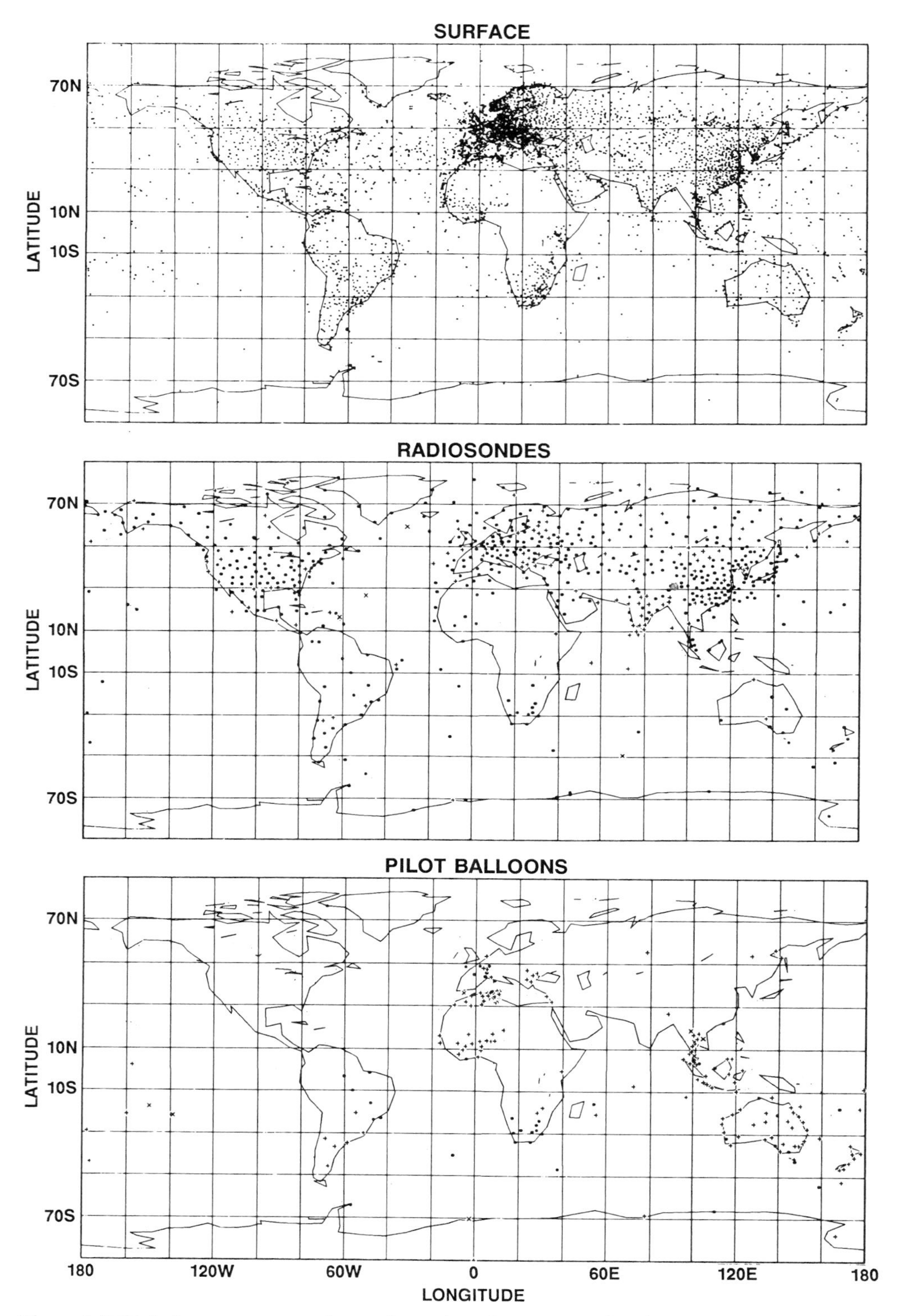

Figure 1.5 Global coverage map for various observing systems for the 6-hour period from 0900–1500 GMT 2 January 1989. (European Centre for Medium Range Forecasting.)

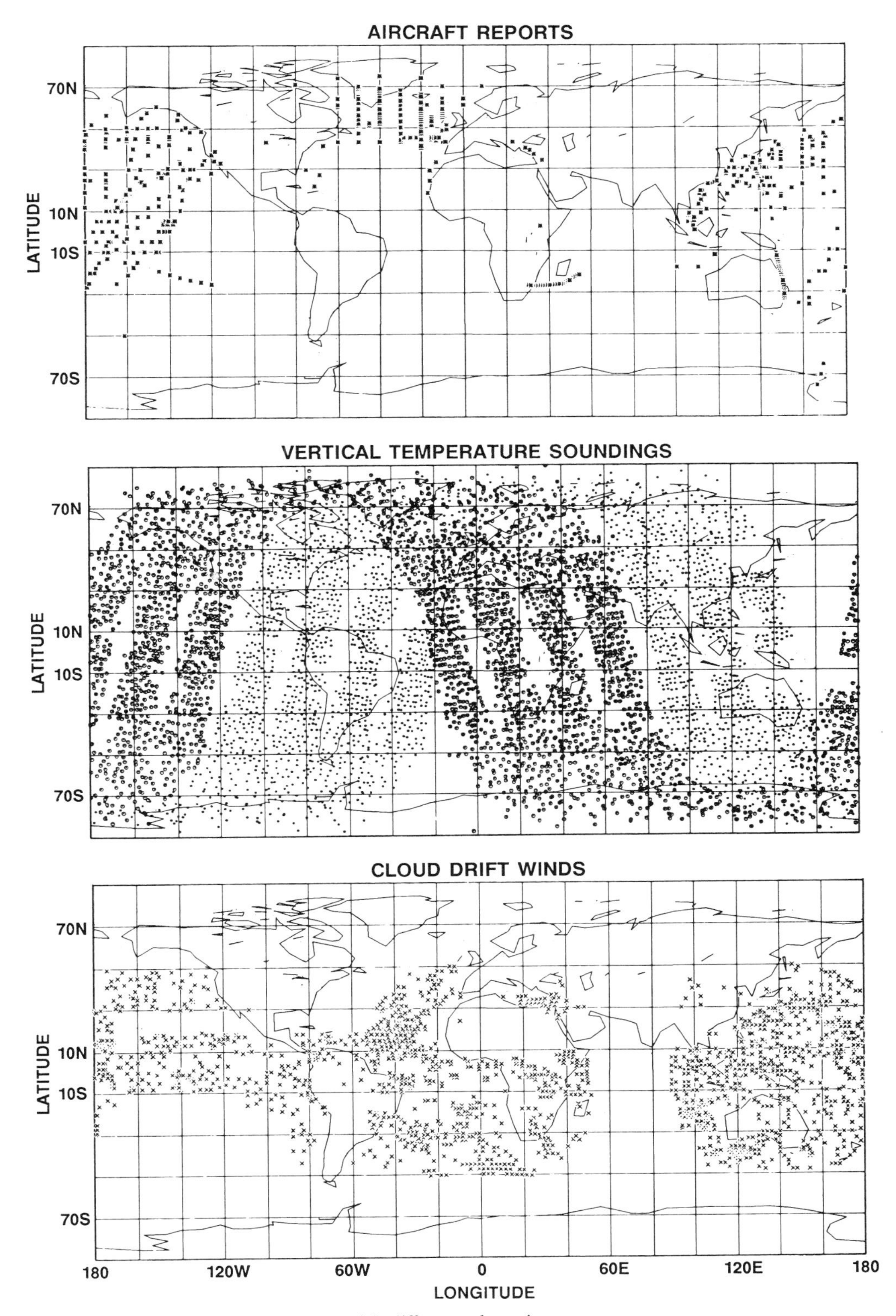

Figure 1.6 Same as Figure 1.5 but with different observing systems.

Aircraft reports

Aircraft reports (see Figure 1.6) come mainly from commercial aircraft on well-traveled routes (which are clearly visible) at about 200 mb. Data collection uses both manual procedures (AIREPS) and automatic procedures (ASDAR). Winds and temperatures are sampled asynoptically. Wind errors are random with root mean square error of 3–5 ms^{-1} which tends to increase with wind velocity.

Vertical temperature soundings

Vertical soundings of air temperature are made by polar-orbiting satellites that contain radiometers. Radiances are observed on several spectral channels and then converted to temperature using sophisticated inversion algorithms (Deepak 1977; Isaacs, Hoffman, and Kaplan 1986). The coverage is global but asynoptic. The descending and ascending orbits of two satellites are visible in Figure 1.6. The observations represent areas rather than points and have a horizontal resolution of about 250 km. The vertical resolution is coarse, about 200 mb. The errors are 2–3 K, but the error is highly correlated in the horizontal. These observations are more reliable in clear air than under cloudy conditions.

Cloud drift winds

Observations of cloud drift winds – high-level winds (200 mb) and low-level winds (850 mb) – are obtained from sequences of cloud photographs taken from geosynchronous satellites (see Figure 1.6). From the photographic sequences, individual cloud elements can be tracked manually, yielding estimates of wind speed and direction. This technique has a number of difficulties, particularly in determining the altitude of the cloud element being followed and its speed, because clouds do not necessarily move at the speed of the wind. The observations are asynoptic and represent areas rather than points. They are considered most valuable in the tropics and have an error of 3 ms^{-1} at low levels and 8 ms^{-1} at high levels. Wind direction can be determined reasonably accurately, but magnitudes are less accurate. Although this was originally a manual technique, it is now done automatically using correlation techniques.

Other observing systems

Other observing systems (drifting buoys, dropwindsondes, and constant-level balloons) have also been deployed. Orbiting active microwave scatterometers have been used to measure surface winds over the ocean. Ground-based wind profilers (microwave instruments using the backscatter from turbulence or aerosols) are now being deployed.

The number of observations available is a strong function of the time of day. Usually, only asynoptic observations are available and then at relatively few locations at any particular instant. At 00, 03, 06, 09, 12, . . ., GMT, there are surface observations; and at 00 and 12 GMT, there are also radiosondes. Thus, the most complete data coverage occurs at 00 and 12 GMT.

The future will bring increasing use of space-based instruments of Class 2. In addition to improved instruments for observing the traditional meteorological variables such as wind and temperature, there will be more emphasis on measuring other properties of the earth climate system (Report of Earth System Science Committee, NASA, Washington, January 1988). Among the many proposed instruments are the following:

Orbiting lidars that use the Doppler shift in the backscatter from aerosols to measure wind velocities

Orbiting radars to measure precipitation

Various types of orbiting radiometers for measuring components of the radiation budget, concentrations of radiatively and chemically active trace gases, land surface properties (albedo, vegetative cover, surface wetness), ocean variables (sea surface temperature, sea ice extent and drift, sea level, chlorophyll), and geophysical variables.

1.4 Subjective analysis

In the first half of the nineteenth century, a number of attempts were made to represent meteorological observations in a coherent fashion on charts or maps. These early diagnostic efforts had two motivations. First, it was hoped that the governing laws of the atmospheric flow could be deduced from these charts. Second, it was felt that the diagnostic charts of the past and present atmospheric states would help in prognosis of future states. Not much progress was made, however, until the invention of the synoptic chart.

This invention can be largely attributed to LeVerrier and to Admiral Robert Fitzroy. When LeVerrier attempted to diagnose the severe storm of 1854 (see previous section), he obtained observations from around Europe for the period before and during the storm. From these observations, he constructed a series of synoptic charts, each of which displayed the European meteorological observations at a fixed time on a geographical background. The whole sequence of synoptic charts provided a time series from which the development and movement of the storm could be determined.

The synoptic chart quickly became accepted, and by 1860 Admiral Fitzroy (then head of the British Meteorological Department and previously captain of the *Beagle* during Darwin's famous voyage) was preparing synoptic charts in real time and using them as a basis for prognosis. Figure 1.7 is an example of an early synoptic chart. It is taken from the book of Loomis (1885) and shows isobars and wind directions in a clearly recognizable form.

Synoptic charts were prepared entirely by hand. The meteorological observations were first plotted on the maps at their appropriate locations using a special code. Then the meteorologist drew isobars (lines of constant pressure), isotherms (lines of constant temperature), lines of constant dewpoint spread, streamlines (lines tangent

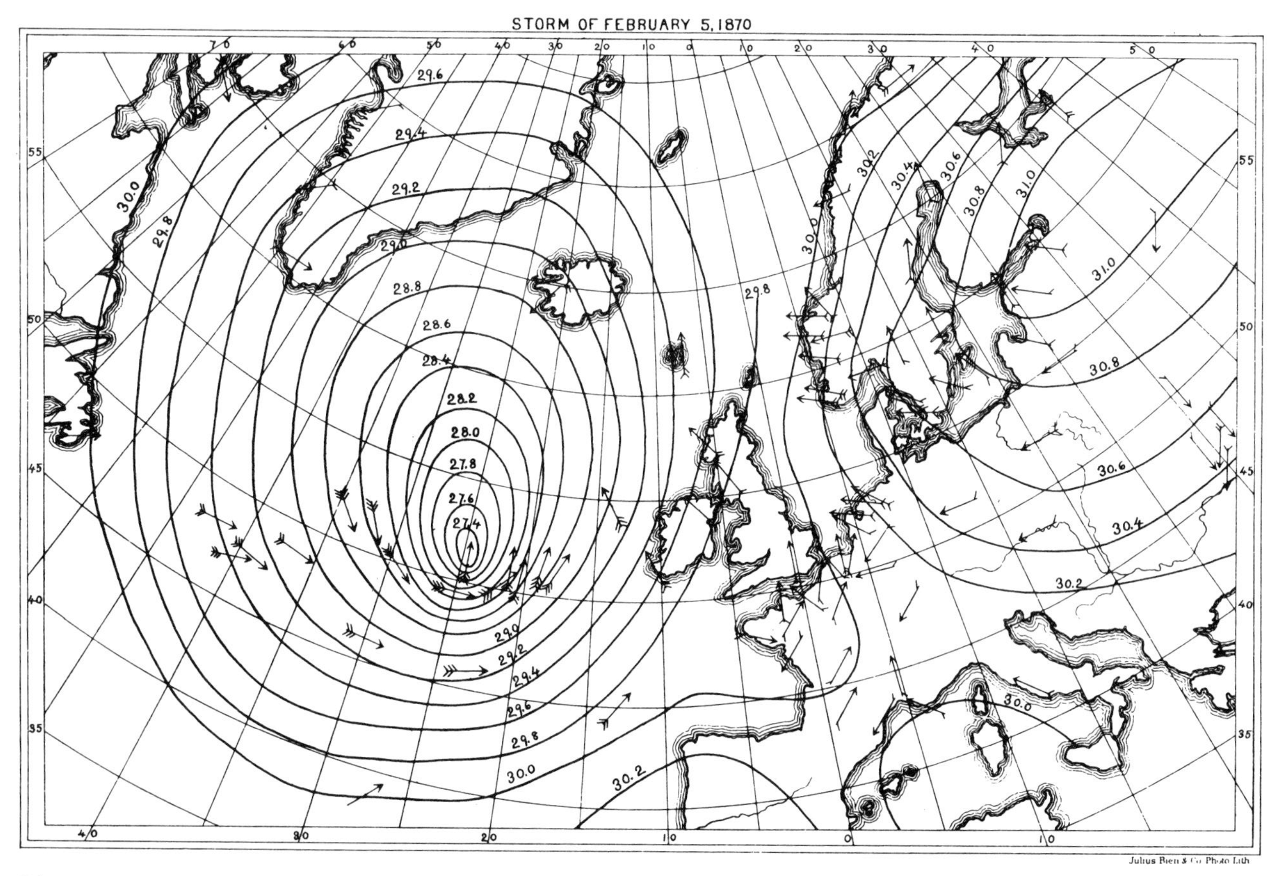

Figure 1.7 Early synoptic chart. (After Loomis 1885)

to the flow), isotachs (lines of constant windspeed), and a variety of other isolines. The analysis shown in Figure 1.7 is a surface isobaric chart, but a number of other charts could be prepared. Upper air charts on constant height, constant pressure, or constant potential temperature (isentropic) surfaces, and vertical–horizontal or vertical–time cross sections were also drawn.

Graphical techniques (Bjerknes 1911) were developed that allowed calculation and manipulation of complex quantities directly from the charts. Addition, subtraction, multiplication, and division of scalars; addition of vectors; differentiation and integration of vectors and scalars; divergence and curl operations; and line integration could all be done graphically. Saucier (1955) showed how these techniques could be used to calculate gradient and thermal winds, horizontal temperature advection, horizontal deformation, divergence and vorticity, vertical motion, surface pressure tendency, and a number of other derived quantities. Fjortoft (1952) derived a technique for integrating the barotropic vorticity equation (see Section 7.8) using graphical techniques.

Synoptic charts were prepared by analysts. Because the resulting diagnosis or analysis relied extensively on their judgments, these procedures are now called *subjective analysis*. This work was labor intensive, and the weather services of the larger nations devoted substantial resources to subjective analysis. Weather centers employed numerous meteorologists, plotters, clerks, and others in the preparation of synoptic charts. This activity reached its maximum during and shortly after the second world war. Vederman (1949) described the activities of the U.S. Weather Bureau–Air Force–Navy Analysis Center of that time.

Synoptic charts were found to be useful for predicting future atmospheric states. Weather forecasters extrapolated from present and past synoptic charts to produce prognostic charts of expected weather a day or two in the future. Progress was slow, and very little increase in forecast skill was achieved in the period 1860–1920. During the first world war, Bjerknes and his collaborators in Bergen developed a useful conceptual model of the atmosphere that was widely used by forecasters. The use of this Norwegian Frontal Model, together with a steadily expanding upper air network, resulted in some increase in predictive accuracy. However, the increase in predictive skill during the era 1860–1950 was slight (Reed 1977).

1.5 Objective analysis: first attempts

Bjerknes had conceived of weather prediction as an initial-value problem in which future atmospheric states could be prognosed using the governing equations. This was first attempted by Richardson (1922). His governing equations were discretized (written in finite-difference form) and integrated forward in time from the initial state using numerical procedures. The domain of the forecast (shown in Figure 1.8) was Europe. Richardson's procedure required that at the initial time the pressure be defined at the regular array of points marked P and the wind velocities (momenta actually) be defined at the points marked M. The observing stations were located at

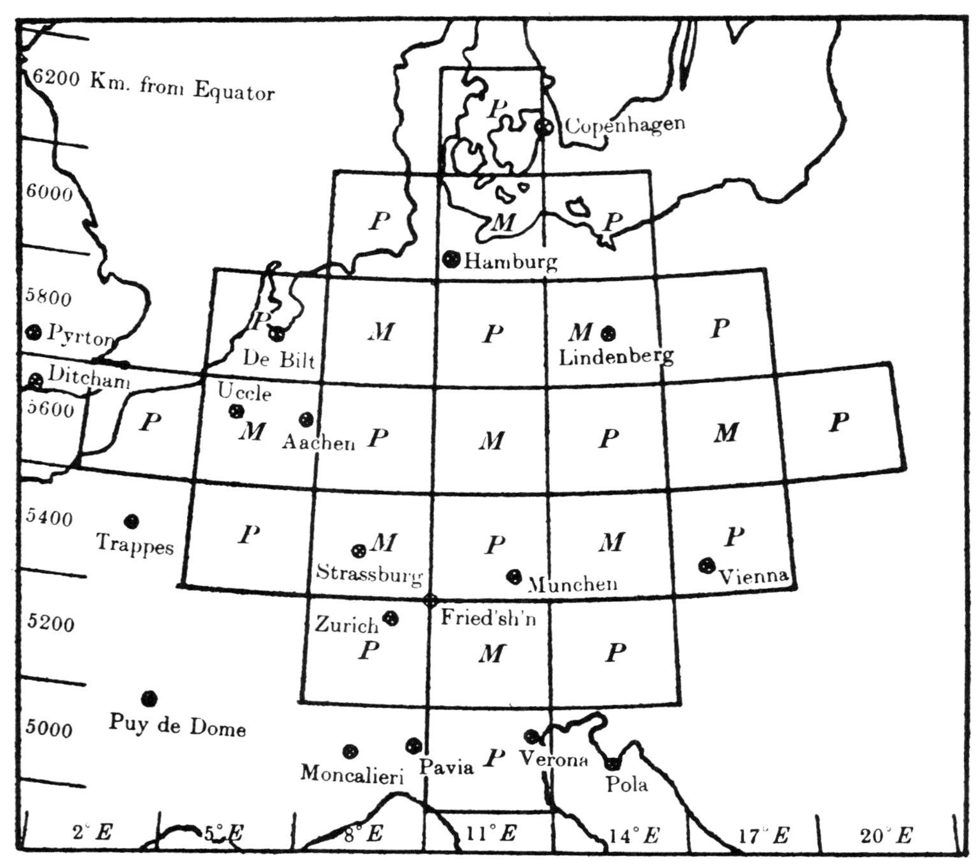

Figure 1.8 Forecast grid and observation stations for Richardson's experiment. (After Richardson 1922)

the named locations indicated by a circled cross. The diagnostic problem in Richardson's experiment was to obtain the values of the pressure and momentum at the points P and M from the irregularly spaced observing stations. Richardson's solution to this problem was to analyze the observations subjectively and then read off (digitize) the appropriate values at the points P and M. After this was done, the prognostic portion of the experiment could proceed. In the spirit of the quotation that prefaces Section 1.3, Richardson suggested facetiously that the diagnostic problem could be avoided by relocating the observing stations at the grid points (P and M). Unfortunately, this pioneering effort in numerical weather prediction failed for several reasons, which are discussed in Platzman (1967). It is interesting to note that one of the reasons for failure was that the concept of initialization (to be discussed in Section 1.6 and Chapter 6) was not understood at the time.

In 1950, a second effort was made to predict the weather using the ideas of Bjerknes and Richardson. The calculations were performed at the Princeton Institute for

Advanced Study on the first multipurpose electronic digital computer (ENIAC), using a modified form of the atmospheric equations (the barotropic vorticity equation). This time the experiment was successful (see Charney, Fjortoft, and Von Neumann 1950; Platzman 1979). As in Richardson's experiment, the initial values of the dependent variables were required on a regular grid. The investigators used subjective analysis followed by digitization but found it to be extremely time consuming; in fact, it took much longer than the forecast. A small group was set up at the Institute for Advanced Study to find a more elegant and efficient procedure for diagnosing the initial conditions.

What was needed was an automatic procedure to estimate the atmospheric dependent variables on a regular two- or three-dimensional grid using the data available from the irregularly spaced observation network. The procedure had to be robust enough to work without human intervention and without consuming an inordinate amount of computer time. Such procedures have come to be known as *objective analysis* procedures. They were called "objective" because they did not rely on the judgment of a human analyst. However, the objectivity of objective analyses is largely a fiction because analyses of an event produced by two different algorithms may differ to approximately the same extent as the subjective analyses produced by different forecasters. Every analysis algorithm embodies a mathematical or statistical model of the field structure or process, and the degree of success depends on the artfulness of the model choice.

The first serious attempt at objective analysis was that of Panofsky (1949). His scheme used a polynomial expansion to fit all the observation points in a small area of the analysis domain that included several analysis gridpoints. The coefficients of the expansion were determined by a least square fit, and the smoothness of the objective analysis could be controlled by the number of coefficients in the expansion. The observations were weighted according to their presumed accuracies, but the weights were specified in an ad hoc manner. An attempt was made to incorporate a dynamic constraint (the geostrophic relation) into the procedure in a way that would couple the wind and mass fields. The fitting of polynomials over areas results in analysis discontinuities between the areas fitted. To remedy this problem, Panofsky suggested using spherical harmonic functions to fit observations over the entire globe.

An objectively analyzed 700-mb geopotential field for 25 March 1947 produced by Panofsky is shown in panel (a) of Figure 1.9. In panel (b) is a subjectively analyzed version of the same observations. Comparisons between objectively and subjectively analyzed fields were commonplace in the early years because of lack of faith in the objective analysis procedures.

An important step forward was taken by Gilchrist and Cressman (1954), who also used a polynomial expansion, but the fit was local rather than over an area. Thus, polynomials were fitted to all the observations in a local region surrounding each individual gridpoint. This region has come to be known as the *region of influence*. The observation weights were determined by the best fit to subjective analyses, and the geostrophic constraint was incorporated as in Panofsky's scheme. Gilchrist and

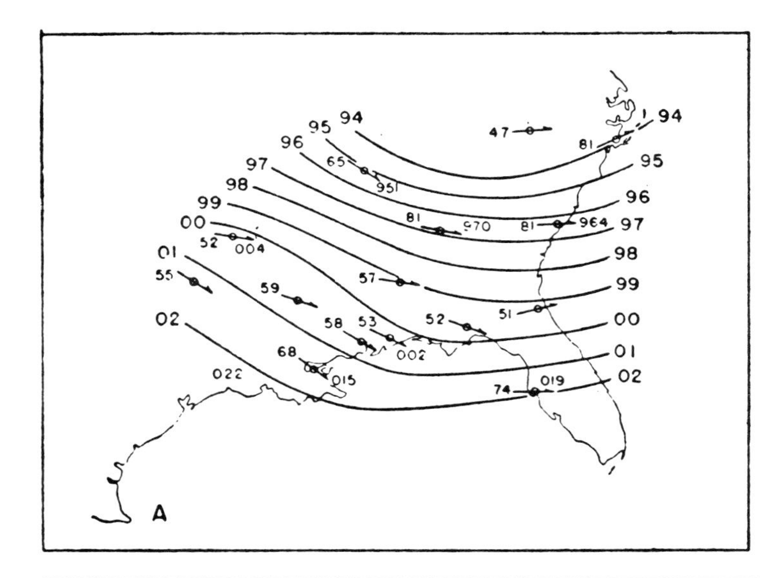

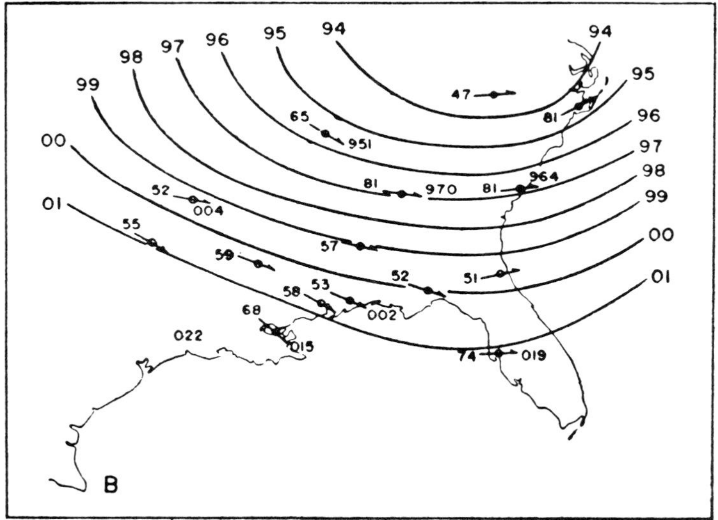

Figure 1.9 (a) Objectively analyzed and (b) subjectively analyzed versions of the 700 mb geopotential field for 25 March 1947. (From Panofsky, *J. Appl. Meteor.* **6**: 386, 1949. The American Meteorological Society.)

Cressman also made two important suggestions that were incorporated in subsequent work – that the data quality could be checked automatically rather than manually and that the analysis might be improved if a preliminary estimate of the analysis could be obtained from a previous numerical forecast. This previous estimate of the analysis is now called the *background field*, *first guess field*, or *prior estimate*.

The next advance was by Bergthorsson and Doos (1955) who devised an analysis method that eventually developed into the method of successive corrections. This

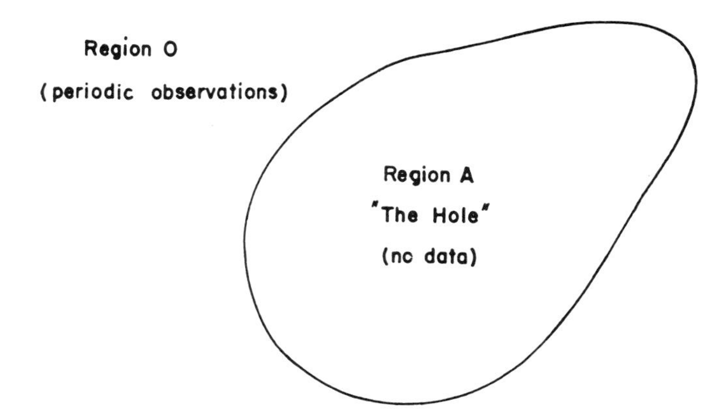

Figure 1.10 Schematic diagram of persistent data voids (holes) and observation-rich (O) regions. (After Thompson 1961)

procedure did not directly analyze the observations. Rather, a background field (obtained from climatology or a previous forecast) was subtracted from each observation to produce *observation increments*. These observation increments were then analyzed to produce *analysis increments*, which were then added onto the background field to produce the final analysis. The analysis increments at each individual analysis gridpoint were weighted linear combinations of the observation increments in the surrounding region of influence. The observation weights were inversely proportional to the distance between the observation location and the analysis gridpoint. Bergthorsson and Doos attempted to derive optimal weights statistically using a large data base and thus anticipated the later development of statistical interpolation procedures. Automatic data checking was used for the first time even though the rejection of observations as erroneous was done manually.

Another important concept was introduced by Thompson (1961). The global observing system contains some fairly substantial data voids or "holes" from which new observations are rarely received. (The situation was much worse in the 1950s and 1960s.) Surrounding these holes are regions from which relatively dense observations are received regularly (called the "O" region by Thompson). This situation is pictured in Figure 1.10. Now, suppose an objective analysis is made from the observations in the O region and some sort of guess for the hole. Presumably the analysis would be good in the O region and poor in the hole. Then suppose a numerical forecast model is integrated from this analysis to produce a forecast that is valid at the next observation time (6 or 12 hours later). At the next observation time, we would have new observations from the O region and model-predicted data for the hole. An analysis at this new time would be based on genuine observations from the O region and a new and (presumably) better approximation (the first guess) to conditions in the hole. In other words, information has propagated from the O region into the hole. The prediction would then be advanced to the next observation time, and the process repeated, producing successively better analyses for the holes.

By 1960, objective analysis procedures were operational in real time at the larger national weather centers. Cressman (1959) describes the U.S. system, which was based on the scheme of Bergthorsson and Doos (1955). Thus, it used a background field and analyzed observation increments. It used distance-dependent weighting for all observations within the radius of influence of each analysis gridpoint. Preliminary data processing and quality control were done automatically.

1.6 The data assimilation cycle

Progress in objective analysis continued after 1960. The use of numerical forecast models to produce background fields became almost universal. A powerful technique, statistical interpolation (Eliassen 1954; Gandin 1963), gained acceptance in the 1970s. For a time, the diagnostic component of Section 1.2 was adequately handled by objective analysis techniques alone. This changed in the early 1970s with the introduction of numerical forecast models based on the primitive equations (7.3.1–4).

The primitive equation models used a more general form of the governing equations and required a much more carefully diagnosed initial state than hitherto. Unlike the filtered models that had been used in the early years of numerical weather prediction (see Section 6.2), the primitive equation models permitted inertia–gravity oscillations with very short-time scales. If the initial state did not satisfy an appropriate balance between the mass and wind fields, large-amplitude spurious inertia–gravity waves would be excited in these models. Consequently, *balancing* or *initialization* procedures were introduced.

The term *data assimilation cycle* was coined to describe the increasingly complex diagnostic component. It can be thought of as having four subcomponents:

1 Quality control (data checking)
2 Objective analysis
3 Initialization
4 Short forecast to prepare next background field

This book focuses primarily on the second and third subcomponents (objective analysis in Chapters 2–5 and initialization in Chapters 6–11). The other two subcomponents are very important, however, and it is appropriate to say a few words about them here.

The discussion of the observing system in Section 1.3 suggests that the data base provided by the global observing system contains observations of many types and qualities. Quality-control algorithms are designed to reject or modify bad data. The data from the observation system contains errors that can be classified into two types:

Natural error: instrument error
error of representativeness
Gross error: improperly calibrated instruments
incorrect registration of observations
incorrect coding of observations
telecommunication errors

These errors can be either random or spatially or temporally correlated with each other or with the synoptic situation, and there can be systematic biases (Appendix D). For example, although the errors in satellite radiometers are thought to be uncorrelated, temperature soundings obtained by inverting the radiances can be horizontally and vertically correlated. Radiosondes are sometimes incorrectly shielded from the sun and thus yield systematic temperature errors.

The calibration in remote or space-based instruments can often drift, resulting in biases. The calibration errors in space-based instruments can sometimes be corrected by comparison with on-board standards or ground-based in situ measurements. Hollingsworth et al. (1986) have demonstrated that the data assimilation system itself can be used to detect improperly calibrated instruments.

A number of quality-control checks on the observations are routinely performed. First, checks are done for coding errors and correct station location (ship reports over land would be rejected here). Next, gross checks are done on the physical reasonableness of the observations. The remaining quality control consists of checks that rely on some common information and therefore some redundancy between observations. In other words, the observation is checked against its neighbors, and spatial and temporal consistency is demanded. Dynamic relations such as the hydrostatic or geostrophic relation can be used to check geopotential against temperature and mass against wind. The observations can also be checked against the background field. Quality control has become more and more sophisticated, using the objective analysis framework itself to make the process more internally consistent. Recent advances include the Bayesian approach of Lorenc and Hammon (1988) and the complex quality control method of Gandin (1988). Analysis intercomparison experiments have shown that the analysis can be very sensitive to quality-control decisions (Hollingsworth et al. 1985).

The numerical forecast model used to prepare the background field at the next observation time is often referred to as the *assimilation model*. It is usually a high-resolution numerical model based on the primitive equations (7.3.1–4) and containing sophisticated parametrizations of various physical processes such as convection, radiation, the hydrological cycle, surface biological processes, planetary boundary layer turbulence, and air/sea interaction. Such assimilation models are not different in principle from numerical models used for daily, weekly, and monthly forecasting or even for long-time climate simulations (Haltiner and Williams 1980; Washington and Parkinson 1986 and references cited therein). The climate of a numerical model is obtained by time averaging a long simulation or ensemble averaging a number of realizations. An assimilation model should include the physical parametrizations necessary to ensure that *if it is not updated with new observations*, the model climate will approximate the "true" climate. This ensures that in persistent data voids the background fields produced by the assimilating model remain physically plausible.

A schematic outline of a data assimilation cycle is shown in Figure 1.11. This type of data assimilation is called intermittent. Every 6 hours all the observations within

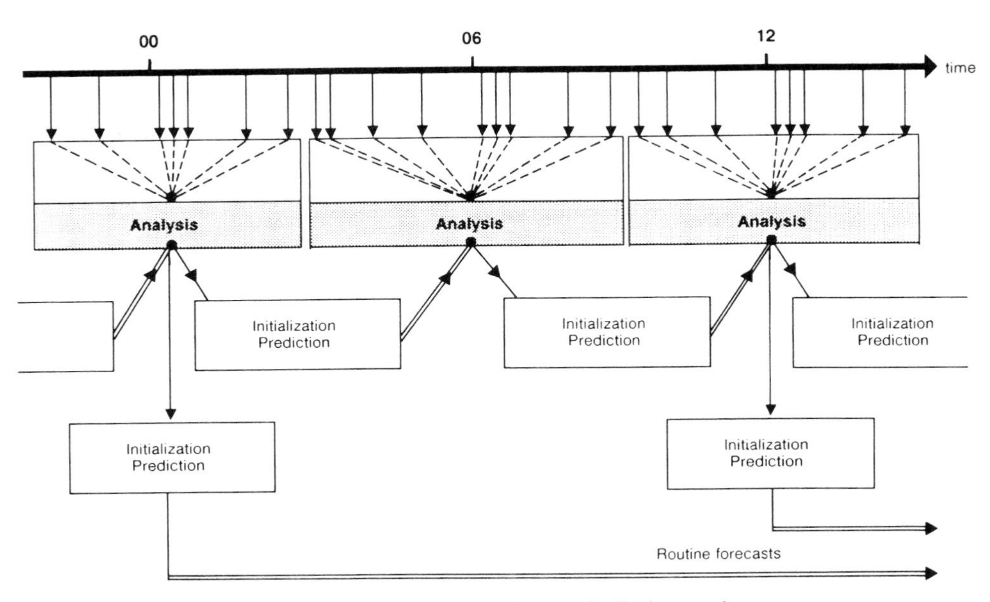

Figure 1.11 Schematic diagram of intermittent data assimilation cycle.

± 3 hours of the midpoint of the time interval are collected and quality checked. Background estimates of the state variables at the analysis gridpoints are obtained by integrating the assimilating model, using the objective analysis from 6 hours previously as the initial conditions. Background estimates of the state variables at the observation stations are obtained using standard mathematical interpolation (Appendix E) of the background estimates on the regular, homogeneous three-dimensional analysis grid. This process is called *forward interpolation*. The background estimates at the observation stations are subtracted from the observations to produce *observation increments* (also known as *observed-minus-background differences* or the *innovation vector*). *Analysis increments* (also known as *analysis-minus-background differences* or the *correction vector*) on the objective analysis grid are obtained by objective analysis of the observation increments. The analyzed values are simply the sum of the analysis increments and the background estimates at the analysis gridpoints.

The initialization step is then performed, and the forecast model is run from this analyzed/initialized state to produce a 6 hour forecast. This 6 hour forecast is then used as the background field for the next analysis. The "routine forecasts" shown at the bottom of Figure 1.11 are the longer forecasts of the prediction component and are the major raisons d'etre for the whole process.

Figure 1.12 shows an example of a 6 hour forecast (background), observations, observation increments, analysis increments, and the analysis. This example (after Daley 1985) is a latitude–longitude plot of the 200 mb wind field in the tropical western Pacific during the GWE. Wind speed is indicated by the length of the arrow in the box at the lower right of each plot. Note the change of scale in the two

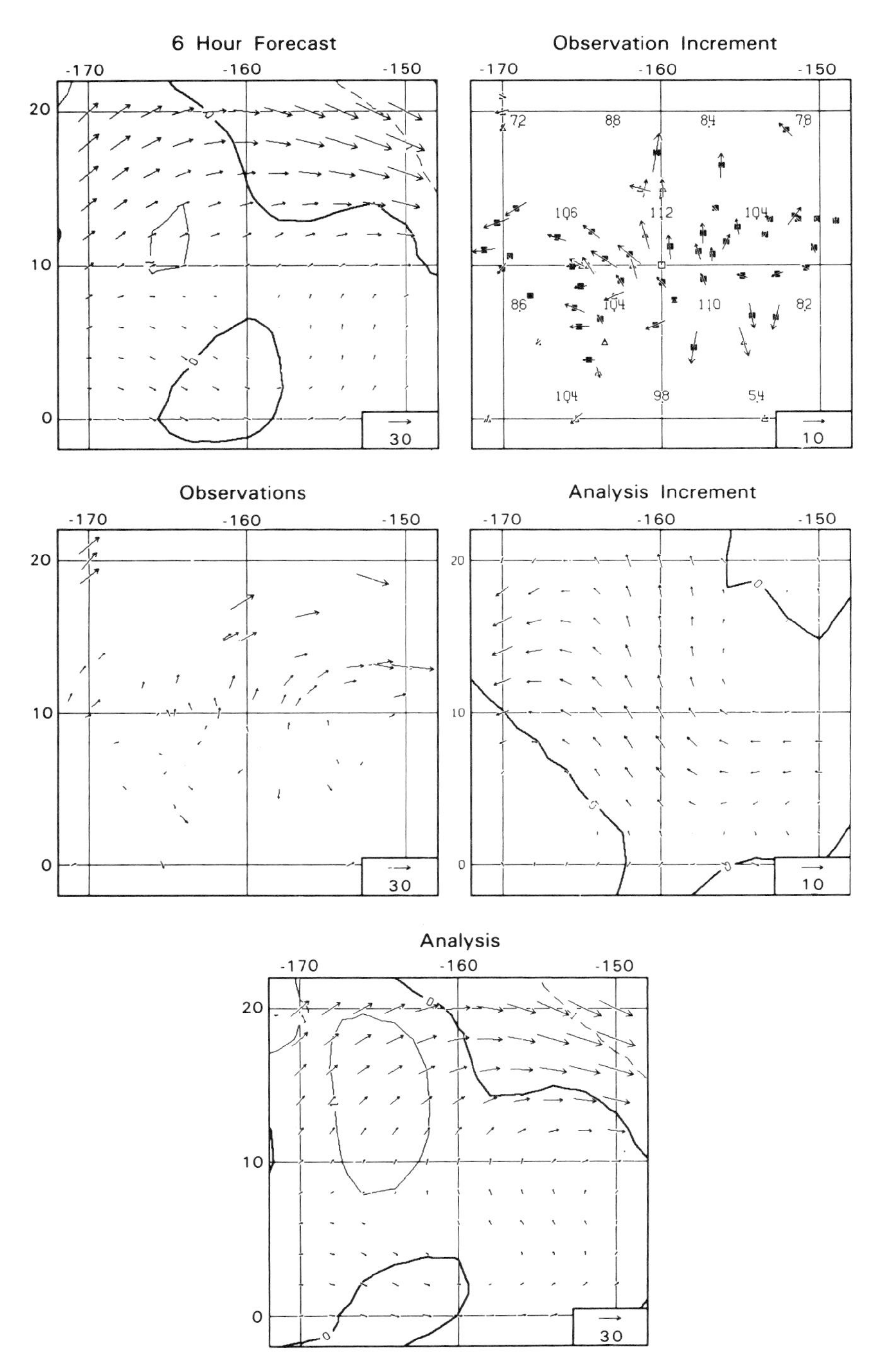

Figure 1.12 Illustration of data analysis of observation increments for the 200 mb wind field. (From Daley, *Mon. Wea. Rev.* **113**: 1066, 1985. The American Meteorological Society.)

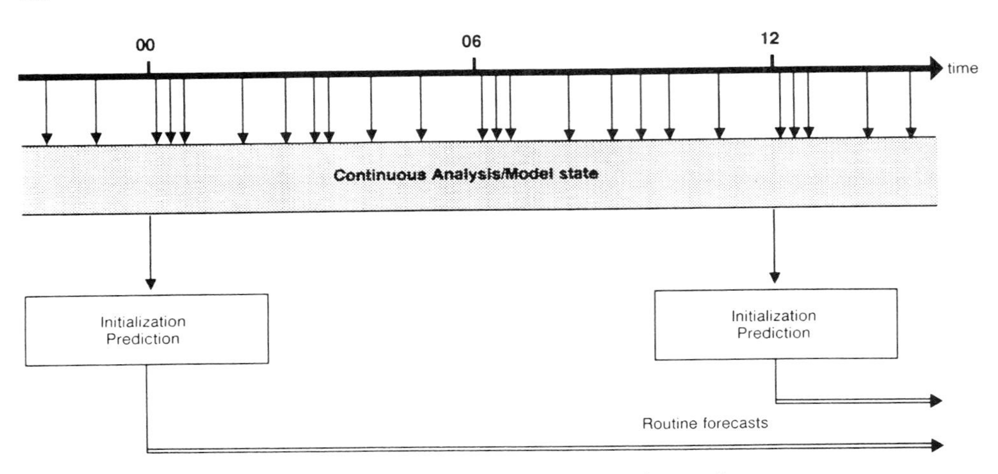

Figure 1.13 Schematic diagram of continuous data assimilation cycle.

increment fields. The observation and observation increment fields are irregularly distributed, but the remaining fields are on the regular analysis grid. It is apparent from this figure that the background field does provide a reasonably accurate estimate of the analysis, and that the observation and analysis increments provide only a rather minor adjustment to the background field.

This result can be formalized as follows: Suppose at time t, there exists an initialized analysis of the atmosphere, $s(t)$. A numerical model is then integrated from $s(t)$ to produce a forecast at $t+6$. Denote the forecast change in the atmospheric state during this 6 hour integration as $\Delta_F s$, where F indicates forecast. The 6 hour forecast is used as a background field for the objective analysis of the observations at $t+6$. The objective analysis produces a change in the estimate of the atmospheric state $\Delta_O s$. Finally, an initialization is performed, which also produces a small change $\Delta_I s$. After these processes, a new analyzed/initialized state $s(t+6)$ has been obtained. Then

$$s(t+6) = s(t) + \Delta_F s + \Delta_O s + \Delta_I s \tag{1.6.1}$$

Hollingsworth et al. (1986) argue that in a successful data assimilation cycle,

$$\Delta_F s > \Delta_O s > \Delta_I s \tag{1.6.2}$$

Condition (1.6.2) does seem to hold generally for present-day data assimilation systems, at least in data-rich areas, because both the 6 hour forecast and the observations are much closer to the "truth" than the objective analysis at time t. Thus, despite the fact that the observations are slightly more accurate than the 6 hour forecast, most of the adjustment toward the "true" values has already occurred in the forecast step before the new observations are introduced.

Intermittent data assimilation as described in Figure 1.11 was the most widely used assimilation process of the late 1980s. It was not, however, universal. *Continuous data assimilation* or more strictly *continuous forward data assimilation*, depicted in Figure 1.13, was also favored by some people. Figure 1.13 is in the same format as

Figure 1.11 and can be readily compared with it. In continuous data assimilation, the observations are assimilated at the same rate that they are observed but somewhat behind real time to allow for data to be communicated and processed. The atmospheric state simulated by the numerical model is continually adjusted to fit the new observations. A routine forecast can be initiated at any time during the cycle.

The historical development outlined in the previous sections has concentrated on the preparation of objective analyses as initial states for the prediction of future atmospheric states. This always has been, and still is, the primary motivation for analysis. It has not, however, been the only motivation.

Objective analyses are invaluable in observational studies of the atmospheric general circulation. Many important components of the atmospheric energy, heat, moisture, and chemical constituent budgets, such as boundary fluxes, generation, dissipation and conversion terms can be calculated from objective analyses. However, dangers exist in using objective analyses for this purpose. In some cases, diagnosed values of general circulation quantities may reflect the implicit asssumptions in the assimilating model or the objective analysis algorithm itself. For example, the spatial kinetic energy spectrum, shown in Figure 1.4 (for $k < 30$), was derived from series of objective analyses. The dependency of the observed kinetic energy on k may, to some extent, depend on the assumptions built into the analysis algorithm.

The systematic depiction of atmospheric phenomena by objective analyses provide a basis for conjectures or simplified mathematical models of the phenomena that eventually lead to increased theoretical understanding.

In recent years, interest in phenomena of much longer time scales has increased. An example is the anthropogenic increase of CO_2, which is expected to affect the global climate on time scales of decades to centuries. Traditionally, such long time scale phenomena have been studied by the examination of time series from individual observing stations with long and stable records. However, to determine long-term trends on global and regional scales, series of objective analyses are extremely useful (Knox et al. 1988). Modern optical storage systems have made it possible to store up to 20 years of global analyses on a single compact disc, a very attractive possibility for a climatologist (Mass et al. 1987).

Unfortunately, the use of objective analyses for detecting subtle long-term climate changes has some serious pitfalls. False signals may be detected because of changes in the data base or changes in the operational objective analysis algorithm over the years (Trenberth and Olsen 1988). There have, in fact, been proposals to reanalyze the last 30 years of observations, using the latest data assimilation techniques, in order to produce a climate history with the maximum degree of temporal consistency (Bengtsson and Shukla 1988).

1.7 Spatial analysis

The purpose of atmospheric data analysis is the characterization of the state of the atmosphere in a manner that is physically consistent and spatially and temporally coherent. The raw material for the process consists of the countless individual observations, irregularly distributed in space and time and often erroneous and conflicting. The final product is a regular space/time representation of the atmospheric state in which it is hoped that the maximum signal has been extracted from the observations and that the effect of observational noise has been minimized (Appendix D).

The temporal and spatial dimensions of the problem are equally important and are given equal weight in this book. Following historical precedent, the spatial aspects will be discussed first, with temporal aspects being delayed until Chapter 6. In Chapters 2–5, three types of spatial analysis techniques will be considered; function fitting in Chapter 2, successive corrections in Chapter 3, and statistical interpolation in Chapters 4 and 5. More advanced techniques are discussed in Chapter 13.

At first sight, spatial analysis (or spatial objective analysis) seems to be little more than classical interpolation. After all, the observations are distributed irregularly and estimates of the atmospheric variables are required at regular intervals on a grid. If the observations were perfect, then classical interpolation techniques such as those of Appendix E would suffice. But the observations are not simply evaluations of an analytic function, and the spatial analysis of atmospheric variables is much more than classical interpolation. First, the observations have errors that can be temporally and spatially correlated with each other or with the signal. Second, the atmospheric state variables are related to each other by the governing laws; they are not independent. Third, there are large data voids where no direct observations exist but where other useful information can be exploited. Fourth, there is usually available a good prior estimate (or background) of the analysis obtained via numerical forecast from a previous analysis.

Spatial analysis can be defined formally as follows:

> *Spatial analysis* is the estimation by numerical algorithm of atmospheric state variables on a three-dimensional regular grid (or as coefficients of a functional expansion) from observations available at irregularly distributed locations.

An ideal spatial analysis algorithm would filter the noise inherent in the observations without appreciable alteration of the true field spectrum. Moreover, the algorithm would be designed to ensure that the final analysis satisfied the governing laws of the atmosphere. Thus, in addition to the interpolation aspects of spatial analysis, there are filtering aspects. An ideal spatial analysis algorithm would filter observational noise, variance associated with scales too small to be properly resolved by the network, and fluctuations that do not satisfy the governing laws.

Traditionally, most spatial analysis procedures used in the atmospheric sciences have been linear in the following sense. Suppose $\tilde{s}(\mathbf{r}_k)$, $1 \leq k \leq K$, defines a set of observations and/or background estimates of s at the locations $\mathbf{r}_k$. Here $\mathbf{r}$ is the spatial coordinate. Define $s_A(\mathbf{r}_i)$ as the analyzed value of s at the ith analysis point $\mathbf{r}_i$. Then linear spatial analysis procedures can be written in the following simple form:

$$s_A(\mathbf{r}_i) = \sum_{k=1}^{K} W_{ik}\tilde{s}(\mathbf{r}_k) \tag{1.7.1}$$

The W_{ik} will be referred to here as the a posteriori weights and will always be written with uppercase letters. Equation (1.7.1) is linear when the W_{ik} are independent of the actual observed and background values $\tilde{s}(\mathbf{r}_k)$, but depend only on their positions and accuracies.

In recent years there has been increasing interest in nonlinear analysis procedures. In this situation, one can still write the analysis equation in the form (1.7.1), but the W_{ik} will then be a function of $\tilde{s}(\mathbf{r}_k)$. Formally, the a posteriori weights can then be written as

$$W_{ik} = \frac{\partial s_{\mathrm{A}}(\mathbf{r}_i)}{\partial \tilde{s}(\mathbf{r}_k)} \tag{1.7.2}$$

This form indicates how the a posteriori weights can be determined numerically for procedures that are tedious to analyze: A small perturbation is made to the kth value, $\tilde{s}(\mathbf{r}_k)$, and the change in the analysis field at the ith analysis point evaluated.

Chapters 2–5 will be entirely concerned with linear analysis procedures; but there will be brief discussions of the nonlinear analysis problem in Chapter 13.

2

Function fitting

The objective analysis of meteorological fields was first seriously discussed at the Princeton Institute for Advanced Study by an informal group consisting of John von Neumann, Jule Charney, Joseph Smagorinsky, George Platzman, and others in the late 1940s and early 1950s. The analysis method that seemed most promising for the primitive computers of the time was polynomial fitting, and two forms of this method were studied: regional polynomial fitting (Panofsky 1949) and local quadratic fitting (Gilchrist and Cressman 1954).

Both of these procedures are examples of a spatial analysis method that will be referred to here as function fitting. The analysis is presumed to be expanded in a finite series of ordered mathematical basis functions with unknown expansion coefficients. This series is evaluated at the observation stations, and the squared difference between the observations and the evaluated series is minimized. This least squares minimization leads to a linear relation for the unknown expansion coefficients, which can then be determined by inverting a matrix and the spatial analysis obtained by evaluating the series at the analysis gridpoints.

The early function-fitting techniques of Panofsky, Cressman, and Gilchrist were quickly superseded by the successive-correction method (Chapter 3) and later by statistical interpolation (Chapters 4 and 5). However, function fitting has been used sporadically in meteorological analysis over the years, and advanced methods are under active investigation at the present time.

The principle of least squares is the basis for both function fitting and statistical interpolation (Chapters 4 and 5). Before discussing this important principle, however, we illustrate a simple function-fitting algorithm inspired by the historic study of Gilchrist and Cressman (1954).

2.1 Local polynomial fitting

Figure 2.1 illustrates an array of 16 analysis gridpoints whose absolute spatial locations are denoted $\mathbf{r}_i$. Around each gridpoint is drawn a circle called the radius

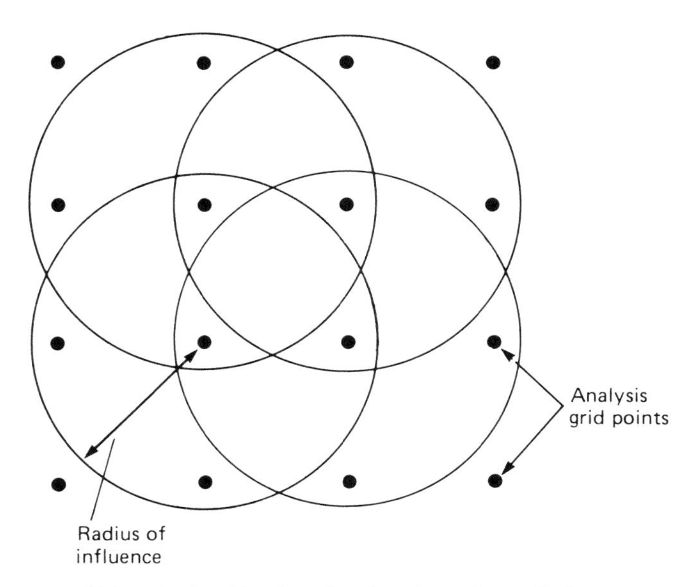

Figure 2.1 An array of 16 analysis gridpoints showing the regions of influence of each gridpoint.

of influence (R_i), and the area circumscribed is called the region of influence. In local fitting, it is assumed that the analysis at the ith gridpoint is influenced by all the observations that lie within the region of influence of that particular gridpoint.

Consider the spatial analysis of the dependent variable f. Define a local coordinate system (x, y) at the ith gridpoint such that $x = 0$, $y = 0$ at the gridpoint itself. Consider observations $f_O(x_k, y_k)$ (which usually contain errors) surrounding the ith gridpoint such that $x_k^2 + y_k^2 \leq R_i^2$, and suppose that there are K_i such observations.

Define the analyzed value of f within the ith region of influence as $f_A(x, y)$ and assume that it can be represented by a two-dimensional polynomial expansion of the form

$$f_A(x, y) = \sum_m \sum_n c_{mn} x^m y^n, \qquad (m + n \leq M), \qquad (m, n \geq 0) \tag{2.1.1}$$

where the c_{mn} are the (as yet undetermined) real expansion coefficients. Note that at the analysis gridpoint itself $x = y = 0$ and $f_A(0, 0) = c_{00}$. Form the following quadratic expression

$$I = \frac{1}{2} \sum_{k=1}^{K_i} \left[\sum_m \sum_n c_{mn} x_k^m y_k^n - f_O(x_k, y_k) \right]^2 \tag{2.1.2}$$

by formally evaluating $f_A(x, y)$ at each of the observation stations $1 \leq k \leq K_i$ in the ith region of influence. I in (2.1.2) is minimized by differentiating it with respect to each of the coefficients c_{mn} in turn and setting the results equal to zero:

$$\frac{\partial I}{\partial c_{mn}} = \sum_{k=1}^{K_i} x_k^m y_k^n \left[\sum_\mu \sum_\nu c_{\mu\nu} x_k^\mu y_k^\nu - f_O(x_k, y_k) \right] = 0$$

or

$$\sum_{\mu} \sum_{\nu} c_{\mu\nu} \sum_{k=1}^{K_i} x_k^{m+\mu} y_k^{n+\nu} = \sum_{k=1}^{K_i} x_k^m y_k^n f_{\mathrm{O}}(x_k, y_k), \qquad \mu + \nu \leq M \tag{2.1.3}$$

for all $m + n \leq M$. μ and ν are dummy indices.

In the study of Gilchrist and Cressman (1954), $M = 2$ and (2.1.1) was a quadratic polynomial. Define the operator

$$\overline{(\ \ )} = \frac{1}{K_i} \sum_{k=1}^{K_i} (\ \ )$$

Then the six expansion coefficients c_{00}, c_{10}, c_{01}, c_{20}, c_{02}, and c_{11} are given by the matrix relation

$$\begin{pmatrix} 1 & \overline{x_k} & \overline{y_k} & \overline{x_k^2} & \overline{y_k^2} & \overline{x_k y_k} \\ \overline{x_k} & \overline{x_k^2} & \overline{x_k y_k} & \overline{x_k^3} & \overline{x_k y_k^2} & \overline{x_k^2 y_k} \\ \overline{y_k} & \overline{x_k y_k} & \overline{y_k^2} & \overline{x_k^2 y_k} & \overline{y_k^3} & \overline{x_k y_k^2} \\ \overline{x_k^2} & \overline{x_k^3} & \overline{x_k^2 y_k} & \overline{x_k^4} & \overline{x_k^2 y_k^2} & \overline{x_k^3 y_k} \\ \overline{y_k^2} & \overline{x_k y_k^2} & \overline{y_k^3} & \overline{x_k^2 y_k^2} & \overline{y_k^4} & \overline{x_k y_k^3} \\ \overline{x_k y_k} & \overline{x_k^2 y_k} & \overline{x_k y_k^2} & \overline{x_k^3 y_k} & \overline{x_k y_k^3} & \overline{x_k^2 y_k^2} \end{pmatrix} \begin{vmatrix} c_{00} \\ c_{10} \\ c_{01} \\ c_{20} \\ c_{02} \\ c_{11} \end{vmatrix} = \begin{vmatrix} \overline{f_{\mathrm{O}}(x_k, y_k)} \\ \overline{x_k f_{\mathrm{O}}(x_k, y_k)} \\ \overline{y_k f_{\mathrm{O}}(x_k, y_k)} \\ \overline{x_k^2 f_{\mathrm{O}}(x_k, y_k)} \\ \overline{y_k^2 f_{\mathrm{O}}(x_k, y_k)} \\ \overline{x_k y_k f_{\mathrm{O}}(x_k, y_k)} \end{vmatrix} \tag{2.1.4}$$

where the elements of the matrix on the left-hand side and of the column vector on the right-hand side are all known.

The c_{mn} can be determined by inverting the 6×6 left-hand matrix (provided it is nonsingular). The analyzed value at the ith gridpoint is equal to c_{00}, as noted following (2.1.1). The process is repeated for each of the analysis gridpoints $\mathbf{r}_i$.

This simple illustration considers univariate analysis (a single dependent variable) and a circular region of influence. In actual fact, Gilchrist and Cressman's (1954) algorithm differed somewhat because it was multivariate (included observations of two or more dependent variables) and had a square region of influence. The most important step in this illustration is the least squares minimization (2.1.2–3), which is examined in more detail in the next section.

2.2 Least squares estimation

> If the astronomical observations and other quantities on which the computation of orbits is based were absolutely correct, the elements also, whether deduced from three or four observations, would be strictly accurate (so far indeed as the motion is supposed to take place exactly according to the laws of Kepler) and, therefore, if other observations were used, they might be confirmed, but not corrected. But since our measurements and observations are nothing more than approximations to the truth, the same must be true of all calculations resting upon them, and the highest aim of all computation made concerning concrete phenomena must be to approximate, as nearly as practicable, to the truth. But this can be accomplished in no other way than by a suitable combination of more observations than the number absolutely requisite for the determination of the

unknown quantities. This problem can only be properly undertaken when an approximate knowledge of the orbit has been already attained, which is afterward to be corrected so as to satisfy all of the observations in the most accurate manner possibly. (Gauss 1809, tr. 1963)

The earliest stimulus for the development of estimation theory was provided by astronomical studies in which planet and comet motion was inferred from telescopic measurement data. The motions of these bodies can be completely characterized by six parameters, and the problem that was considered was that of estimating the values of these parameters from the measurement data. To solve this problem, the method of least squares was invented by Gauss in 1795.

The least squares method is discussed in detail in Whittaker and Robinson (1924) and Tarantola (1987), but it can be simply illustrated as follows. Suppose there exist N observations $s_1, s_2, \ldots, s_N$ of the variable s. Assume that these observations were taken with different types of instruments and that the observation error associated with each measurement is given by $\varepsilon_n = s_n - s$. Assume that the observation errors are *random*, *unbiased*, and *normally distributed* [Equation (D6) of Appendix D]. Then the probability that the error of the nth observation lies between ε_n and $\varepsilon_n + d\varepsilon_n$ is

$$p(\varepsilon_n) = \frac{1}{\sigma_n\sqrt{2\pi}} \exp\left[-\frac{\varepsilon_n^2}{2\sigma_n^2}\right] \tag{2.2.1}$$

where

$$\sigma_n^2 = \langle (s_n - s)^2 \rangle = \langle \varepsilon_n^2 \rangle = \int_{-\infty}^{\infty} \varepsilon_n^2 p(\varepsilon_n)\, d\varepsilon_n, \qquad \langle \varepsilon_n \rangle = 0$$

and $\langle\ \rangle$ is the expectation operator defined in Appendix D.

Suppose there is only one observation, $N = 1$. Then the most probable value of s is that value for which the probability $p(\varepsilon_1)$ is a maximum. From (2.2.1), this occurs at $\varepsilon_1 = 0$, or $s = s_1$. Now, when there are N observations, the joint probability that ε_1 lies between ε_1 and $\varepsilon_1 + d\varepsilon_1$, ε_2 lies between ε_2 and $\varepsilon_2 + d\varepsilon_2, \ldots, \varepsilon_N$ lies between ε_N and $\varepsilon_N + d\varepsilon_N$ is the product of all the individual probabilities (2.2.1). Thus,

$$\begin{aligned} p(\varepsilon_1, \ldots, \varepsilon_N) &= p(\varepsilon_1)p(\varepsilon_2)\cdots p(\varepsilon_N) \\ &= \prod_{n=1}^{N} \frac{1}{\sigma_n\sqrt{2\pi}} \exp\left(-\frac{\varepsilon_n^2}{2\sigma_n^2}\right) = \left[\prod_{n=1}^{N} \frac{1}{\sigma_n\sqrt{2\pi}}\right] \exp\left[-\sum_{n=1}^{N} \frac{(s_n - s)^2}{2\sigma_n^2}\right] \end{aligned}$$

where $\prod$ is the product operator. In this case, the most probable value of s is that value for which the probability $p(\varepsilon_1, \ldots, \varepsilon_N)$ is a maximum. This obviously occurs when the summation inside the preceding exponential is a minimum. The most probable value, denoted s_a, is often called the *maximum likelihood estimate* of s, and it must minimize

$$I = \frac{1}{2}\sum_{n=1}^{N} \sigma_n^{-2}(s_\mathrm{a} - s_n)^2 = \frac{(s_\mathrm{a} - s_1)^2}{2\sigma_1^2} + \frac{(s_\mathrm{a} - s_2)^2}{2\sigma_2^2} + \cdots + \frac{(s_\mathrm{a} - s_N)^2}{2\sigma_N^2} \tag{2.2.2}$$

Equation (2.2.2) can be minimized by differentiating with respect to s_a and setting the result to zero,

$$\sum_{n=1}^{N} \sigma_n^{-2}(s_a - s_n) = \frac{s_a - s_1}{\sigma_1^2} + \frac{s_a - s_2}{\sigma_2^2} + \cdots + \frac{s_a - s_N}{\sigma_N^2} = 0$$

or

$$s_a = \frac{\sum_{n=1}^{N} \sigma_n^{-2} s_n}{\sum_{n=1}^{N} \sigma_n^{-2}} \tag{2.2.3}$$

In other words, the most probable value (or maximum likelihood estimate) of s is given by a weighted average of the observations, where the weights are inversely proportional to the expected observation error variances of each observation. Note that (2.2.3) is a linear estimate and that it is also unbiased as $\langle s_n \rangle = \langle s \rangle, 1 \le n \le N$.

Define the error of the estimate as $\varepsilon_a = s_a - s$. Then the expected error variance of the estimate s_a is

$$\langle \varepsilon_a^2 \rangle = \left\langle \left[\frac{\sum_{n=1}^{N} \sigma_n^{-2}(s_n - s)}{\sum_{n=1}^{N} \sigma_n^{-2}} \right]^2 \right\rangle = \left[\sum_{n=1}^{N} \sigma_n^{-2} \right]^{-1} \tag{2.2.4}$$

as $\langle \varepsilon_n \varepsilon_\eta \rangle = 0$ for $n \ne \eta$, because ε_n is random.

In the case in which all observations were taken with the same type of instrument, $\sigma_n^2 = \sigma^2$, $1 \le n \le N$,

$$s_a = \frac{1}{N} \sum_{n=1}^{N} s_n \qquad \text{and} \qquad \langle \varepsilon_a^2 \rangle = \frac{\sigma^2}{N} \tag{2.2.5}$$

Here the estimate of s is the simple average of the observations, and the expected error variance is directly proportional to the observed error variance and inversely proportional to the number of observations.

The estimate s_a is unbiased. When the observations are biased, $\langle \varepsilon_n \rangle \ne 0$, and (2.2.3) implies s_a is biased. A minor modification of (2.2.2) produces an unbiased estimate. Presuming that the instrument biases $\langle \varepsilon_n \rangle$ are known, replace s_n in (2.2.2) by $s_n - \langle \varepsilon_n \rangle$ and redefine σ_n^2 as $\langle \varepsilon_n^2 \rangle - \langle \varepsilon_n \rangle^2$. Minimization with respect to s_a yields

$$s_a = \frac{\sum_{n=1}^{N} \sigma_n^{-2}(s_n - \langle \varepsilon_n \rangle)}{\sum_{n=1}^{N} \sigma_n^{-2}}$$

Subtraction of the true value s from both sides and application of the expectation operator gives $\langle \varepsilon_a \rangle = 0$. In most of the discussion of the next four chapters, it will be assumed that the observations are unbiased, because if they are not and the bias is known, then its removal is straightforward.

Equation (2.2.2) can be rewritten as

$$I = \sum_{n=1}^{N} w_n d_n^2 \tag{2.2.6}$$

where $w_n = 0.5\sigma_n^{-2}$ and $d_n = s - s_n$. The w_n is the Gauss precision modulus or weight. Since w_n is specified, it will be called the *a priori weight* and denoted with lowercase letters to distinguish it from the a posteriori weights of (1.7.1). The d_n is the *residual* of the nth observation. It should be noted that the least squares estimate obtained by minimizing (2.2.6) is a maximum likelihood estimate only if the observation errors are normally distributed and $w_n = 0.5\sigma_n^{-2}$.

The special case $N = 2$ arises repeatedly in this book. For future reference, we now rewrite this case using a special notation. Denote the two observations of s as s_o and s_b with corresponding error variances σ_o^2 and σ_b^2, respectively. Then (2.2.2–4) can be rewritten as

$$I = \frac{(s_a - s_o)^2}{2\sigma_o^2} + \frac{(s_a - s_b)^2}{2\sigma_b^2}$$

$$s_a = \frac{\sigma_o^{-2} s_o + \sigma_b^{-2} s_b}{\sigma_o^{-2} + \sigma_b^{-2}} = \frac{\sigma_b^2 s_o + \sigma_o^2 s_b}{\sigma_o^2 + \sigma_b^2} = s_b + \frac{\sigma_b^2}{\sigma_o^2 + \sigma_b^2}[s_o - s_b]$$

$$\langle \varepsilon_a^2 \rangle = \sigma_b^2 - \frac{\sigma_b^4}{\sigma_o^2 + \sigma_b^2} = \frac{\sigma_b^2 \sigma_o^2}{\sigma_o^2 + \sigma_b^2} = (\sigma_o^{-2} + \sigma_b^{-2})^{-1} \tag{2.2.7}$$

The simple example (2.2.2–7) defined a *scalar* or zero-dimensional application of the principle of least squares. When the variable has spatial and/or temporal dependence, then a *vector* version of the principle of least squares can be applied. Spatial dependence is considered here and temporal dependence in Sections 8.6 and 13.2.

Consider first the case $N = 1$. At a fixed time, define a dependent or state variable $f(\mathbf{r})$, where $\mathbf{r} = (x, y, z)$ is a three-dimensional spatial coordinate. Define $f_O(\mathbf{r}_k)$ to be an observation of f at the observation station $\mathbf{r}_k$, with expected observation error variance $\langle \varepsilon_O^2(\mathbf{r}_k) \rangle$. Suppose there are K such observation stations and the observation errors are normally distributed, unbiased, and spatially uncorrelated $\langle \varepsilon_O(\mathbf{r}_k)\varepsilon_O(\mathbf{r}_l) \rangle = 0$, $l \neq k$. Designate the analyzed field of f as $f_A(\mathbf{r})$.

In this case, there is only a single observation $f_O(\mathbf{r}_k)$ at each observation station $\mathbf{r}_k$. Introduce the quadratic form

$$I = \sum_{k=1}^{K} w_k d_k^2 = \frac{1}{2} \sum_{k=1}^{K} \langle \varepsilon_O^2(\mathbf{r}_k) \rangle^{-1} [f_O(\mathbf{r}_k) - f_A(\mathbf{r}_k)]^2 \tag{2.2.8}$$

Minimization of (2.2.8) with respect to each of the unknown analysis values $f_A(\mathbf{r}_k)$ leads to the trivial solution

$$f_A(\mathbf{r}_k) = f_O(\mathbf{r}_k), \qquad 1 \leq k \leq K$$

Because there is only a single observation at each observation state, the most probable value of f at $\mathbf{r}_k$ is equal to that observation, consistent with (2.2.3). In the case of perfect observations (no observation error), this is the correct solution. Usually, quadratic forms such as (2.2.8) are minimized subject to some explicit or implicit constraint, and the solution is not trivial. In Section 2.1, for example, the objective analysis was constrained to have the polynomial representation (2.1.1). In Sections 2.6 and 2.7, other types of constraints are introduced.

The quadratic form (2.2.8) can be written in matrix form as

$$I = [\underline{f}_A - \underline{f}_O]^T \underline{\underline{w}} [\underline{f}_A - \underline{f}_O]$$

or

$$I = 0.5[\underline{f}_A - \underline{f}_O]^T \underline{\underline{O}}^{-1} [\underline{f}_A - \underline{f}_O] \tag{2.2.9}$$

where $\underline{f}_A$ and $\underline{f}_O$ are column vectors of length K of the analyzed values $f_A(\mathbf{r}_k)$ and observed values $f_O(\mathbf{r}_k)$, respectively. The $\underline{\underline{w}}$ is the $K \times K$ diagonal matrix with elements w_k, T indicates matrix transpose, and $\underline{\underline{O}}$ is the diagonal $K \times K$ matrix with elements $\langle \varepsilon_O^2(\mathbf{r}_k) \rangle$. The convention of double underlining for matrices and single underlining for column vectors is used throughout this text.

If the observation errors are spatially correlated $\langle \varepsilon_O(\mathbf{r}_k)\varepsilon_O(\mathbf{r}_l) \rangle \neq 0$, then $\underline{\underline{O}}$ is a full covariance matrix (see Appendix D). The form of (2.2.9) does not change in this case, but clearly $\underline{\underline{O}}$ must be nonsingular.

Equation (2.2.9) is the vector equivalent of (2.2.2) for the case $N = 1$. For future comparison with results in Chapters 4 and 5, the vector case for $N = 2$ is also derived here. That is, Equation (2.2.9) is generalized to include background estimates of the function f at the observing stations. Thus, suppose there exist background estimates $f_B(\mathbf{r}_k)$, $1 \leq k \leq K$. Here the subscript B indicates background. With each background estimate is an associated background error $\varepsilon_B(\mathbf{r}_k)$. Assume that background and observation errors are unbiased, random, normally distributed, and spatially correlated, but not with each other:

$$\begin{aligned} &\langle \varepsilon_O(\mathbf{r}_k)\varepsilon_O(\mathbf{r}_l) \rangle \neq 0, \qquad \langle \varepsilon_B(\mathbf{r}_k)\varepsilon_B(\mathbf{r}_l) \rangle \neq 0 \\ &\langle \varepsilon_O(\mathbf{r}_k)\varepsilon_B(\mathbf{r}_l) \rangle = 0, \qquad \text{for all } k,\ l \end{aligned} \tag{2.2.10}$$

Then the vector generalization of (2.2.2) for $N = 2$ is

$$I = 0.5\{[\underline{f}_A - \underline{f}_O]^T \underline{\underline{O}}^{-1} [\underline{f}_A - \underline{f}_O] + [\underline{f}_A - \underline{f}_B]^T \underline{\underline{B}}^{-1} [\underline{f}_A - \underline{f}_B]\} \tag{2.2.11}$$

where $\underline{\underline{B}}$ is the background error covariance matrix with elements $\langle \varepsilon_B(\mathbf{r}_k)\varepsilon_B(\mathbf{r}_l) \rangle$. $\underline{f}_A$, $\underline{f}_O$, and $\underline{f}_B$ are column vectors of the analyzed, observed, and background values at the observation stations. The properties of the error covariance matrices $\underline{\underline{B}}$ and $\underline{\underline{O}}$ are discussed in Appendix D and Chapter 4, but for now it is enough to note that they must both be nonsingular. By analogy with (2.2.8), Equation (2.2.11) can be written as

$$\begin{aligned} I = 0.5 \sum_{k=1}^{K} \sum_{l=1}^{K} \{&[f_A(\mathbf{r}_k) - f_O(\mathbf{r}_k)][f_A(\mathbf{r}_l) - f_O(\mathbf{r}_l)]\tilde{o}_{kl} \\ &+ [f_A(\mathbf{r}_k) - f_B(\mathbf{r}_k)][f_A(\mathbf{r}_l) - f_B(\mathbf{r}_l)]\tilde{b}_{kl}\} \end{aligned} \tag{2.2.12}$$

where $\tilde{o}_{kl}$ and $\tilde{b}_{kl}$ are elements of the matrices $\underline{\underline{O}}^{-1}$ and $\underline{\underline{B}}^{-1}$, respectively. If the background and observation errors were spatially uncorrelated, then $\underline{\underline{B}}$ and $\underline{\underline{O}}$ would be diagonal and only the terms for $l = k$ would be retained in (2.2.12). Differentiating (2.2.12) with respect to each of the $f_A(\mathbf{r}_k)$, $1 \le k \le K$, and setting the result to zero gives

$$\frac{\partial I}{\partial f_A(\mathbf{r}_k)} = 0 = \sum_{l=1}^{K} \{[f_A(\mathbf{r}_l) - f_O(\mathbf{r}_l)]\tilde{o}_{kl} + [f_A(\mathbf{r}_l) - f_B(\mathbf{r}_l)]\tilde{b}_{kl}\}, \qquad 1 \le k \le K$$

or

$$\underline{\underline{B}}^{-1}[\underline{f}_A - \underline{f}_B] + \underline{\underline{O}}^{-1}[\underline{f}_A - \underline{f}_O] = 0 \tag{2.2.13}$$

Rearrangement of (2.2.13) gives

$$\underline{f}_A = [\underline{\underline{B}}^{-1} + \underline{\underline{O}}^{-1}]^{-1}[\underline{\underline{B}}^{-1}\underline{f}_B + \underline{\underline{O}}^{-1}\underline{f}_O] \tag{2.2.14}$$

$$\underline{f}_A - \underline{f}_B = \underline{\underline{B}}[\underline{\underline{B}} + \underline{\underline{O}}]^{-1}[\underline{f}_O - \underline{f}_B] \tag{2.2.15}$$

Equation (2.2.14) is a generalization of (2.2.3) for the case $N = 2$, and thus $\underline{f}_A$ is the column vector of maximum likelihood estimates of f at the observation stations. Following the terminology of Chapters 1, $\underline{f}_A - \underline{f}_B$ is the column vector of analysis increments (at the observation stations), and $\underline{f}_O - \underline{f}_B$ is the column vector of observation increments. Note that (2.2.15) is not, in itself, an objective analysis algorithm because it provides no mechanism for determining f_A at any location that is not an observation station.

It is also possible to determine the expected error variance of the analyzed values $\underline{f}_A$. Define $\underline{f}_T$ to be the column vector of true values $f_T(\mathbf{r}_k)$. Subtract $\underline{f}_T$ from both sides of (2.2.14),

$$\underline{f}_A - \underline{f}_T = [\underline{\underline{B}}^{-1} + \underline{\underline{O}}^{-1}]^{-1}[\underline{\underline{B}}^{-1}(\underline{f}_B - \underline{f}_T) + \underline{\underline{O}}^{-1}(\underline{f}_O - \underline{f}_T)] \tag{2.2.16}$$

Define $\varepsilon_A(\mathbf{r}_k) = f_A(\mathbf{r}_k) - f_T(\mathbf{r}_k)$ to be the analysis error at observation station k, and $\underline{\varepsilon}_A$ to be the column vector with elements $\varepsilon_A(\mathbf{r}_k)$ and $\underline{\varepsilon}_O$ with elements $\varepsilon_O(\mathbf{r}_k)$. Then, (2.2.16) can be written as

$$\underline{\varepsilon}_A = [\underline{\underline{B}}^{-1} + \underline{\underline{O}}^{-1}]^{-1}[\underline{\underline{B}}^{-1}\underline{\varepsilon}_B + \underline{\underline{O}}^{-1}\underline{\varepsilon}_O] \tag{2.2.17}$$

Right multiply both sides of (2.2.17) by $\underline{\varepsilon}_A^T$ (where superscript T stands for transpose) and take expectation values. As $\langle \underline{\varepsilon}_B \underline{\varepsilon}_O^T \rangle = \langle \underline{\varepsilon}_O \underline{\varepsilon}_B^T \rangle = 0$,

$$\langle \underline{\varepsilon}_A \underline{\varepsilon}_A^T \rangle = [\underline{\underline{B}}^{-1} + \underline{\underline{O}}^{-1}]^{-1}[\underline{\underline{B}}^{-1}\langle \underline{\varepsilon}_B \underline{\varepsilon}_B^T \rangle \underline{\underline{B}}^{-1} + \underline{\underline{O}}^{-1}\langle \underline{\varepsilon}_O \underline{\varepsilon}_O^T \rangle \underline{\underline{O}}^{-1}][\underline{\underline{B}}^{-1} + \underline{\underline{O}}^{-1}]^{-1}$$

But

$$\langle \underline{\varepsilon}_B \underline{\varepsilon}_B^T \rangle = \underline{\underline{B}} \qquad \text{and} \qquad \langle \underline{\varepsilon}_O \underline{\varepsilon}_O^T \rangle = \underline{\underline{O}}$$

so

$$\langle \underline{\varepsilon}_A \underline{\varepsilon}_A^T \rangle = [\underline{\underline{B}}^{-1} + \underline{\underline{O}}^{-1}]^{-1} \tag{2.2.18}$$

Define the analysis error covariance matrix $\underline{\underline{A}}$ with elements $\langle \varepsilon_A(\mathbf{r}_k)\varepsilon_A(\mathbf{r}_l) \rangle$. Then

$$\underline{\underline{A}} = [\underline{\underline{B}}^{-1} + \underline{\underline{O}}^{-1}]^{-1}$$

or

$$\underset{=}{A} = \underset{=}{B}[\underset{=}{B} + \underset{=}{O}]^{-1}\underset{=}{O} = \underset{=}{B} - \underset{=}{B}[\underset{=}{B} + \underset{=}{O}]^{-1}\underset{=}{B} \tag{2.2.19}$$

Equations (2.2.11, 2.2.15, and 2.2.19) are the vector equivalents of the scalar equations (2.2.7), and there is an obvious similarity in the functional form, with constants being replaced by matrices. The elements along the main diagonal of $\underset{=}{B}$ and $\underset{=}{O}$ are the expected background and observation error variances at the observation stations. In the same way, the main diagonal elements of $\underset{=}{A}$ are the expected analysis error variances at the observation stations. Equations (2.2.15 and 2.2.19) can be derived more elegantly using the vector methods of Section 5.1.

Minimization with respect to quadratic forms such as (2.2.2) or (2.2.11) is referred to mathematically as minimization in the l_2-norm sense, with the subscript "2" referring to the power 2 in (2.2.2). Minimization in the l_2 sense leads to linear analysis equations (Section 1.7). However, it should be mentioned that minimization with respect to other norms is also possible. Thus an l_q-norm form corresponding to (2.2.2) might be,

$$I_q = \frac{1}{q} \sum_{n=1}^{N} \frac{|s_a - s_n|^q}{\sigma_n^q} \tag{2.2.20}$$

where | | indicates absolute value. Tarantola (1987) discusses minimization using l_q-norm criteria, for $1 \leq q \leq \infty$. Minimization of (2.2.20) would be appropriate if the error probability distribution were not normal (2.2.1) but of the form

$$p(\varepsilon_n) = c_q \exp\left[\frac{-|\varepsilon_n|^q}{q\sigma_n^q}\right] \tag{2.2.21}$$

where c_q is a constant determined from the requirement that the integral of $p(\varepsilon_n)$ between $\pm\infty$ must equal one. The cases $q = 1$ and $q = \infty$ are particularly important. When $q = 1$, the error distribution (2.2.21) is "long-tailed" in that there is a much higher probability of large errors than in (2.2.1). Results obtained using a minimum l_1-norm criterion are less sensitive to large errors than in the l_2-norm case. This makes the l_1-norm attractive for the quality control of data in a number of disciplines (Barrodale 1968).

Claerbout and Muir (1973) compared the l_1- and l_2-norms using the following analogy. "When a traveller reaches a fork in the road, the l_1-norm tells him to take one way or the other, but the l_2-norm instructs him to head off into the bushes. Likewise, a hunter when seeing two birds in the sky, might not choose to shoot at the mid-point between them." In other words, for the case $N = 2$, discussed in (2.2.7), the value s_a produced by an l_1-norm minimization would usually be close to either s_o or s_b, whereas with an l_2-norm approach it might be close to neither.

The l_2-norm procedures are emphasized in this book because they lead to simple linear forms that are relatively easy to implement and because the normal error distribution (2.2.1) is thought to be appropriate for many meteorological variables (provided grossly erroneous observations have been rejected by quality-control mechanisms). This background material on the theory of least squares estimation will be useful in Chapters 4 and 5, but a more immediate application is in developing function-fitting algorithms.

2.3 The Gram matrix

Section 2.1 gave a simple demonstration of function fitting. In that illustration, the function (2.1.1) was fitted to all the observations in a region of influence surrounding a given analysis gridpoint, and the functional fit was evaluated only at the gridpoint itself (i.e., at the centre of the data cluster). This type of fitting is referred to as *local fitting*.

In another type of function fitting, called *global fitting*, a single function is fitted to *all observations in the analysis domain*, and the analysis anywhere in the domain can be obtained by evaluating this function. A particularly important example of global fitting (see Section 2.6) occurs when the analysis domain is the entire earth's atmosphere.

The least squares minimization theory of the previous section is now used to explore function-fitting algorithms in more detail. Suppose we want to analyze a dependent variable $f(\mathbf{r})$ where $\mathbf{r}$ is a one-, two- or three-dimensional spatial coordinate. Assume that the analyzed field $f_A(\mathbf{r})$ can be represented by a finite series of ordered basis functions $h_0(\mathbf{r}), \ldots, h_m(\mathbf{r}), \ldots, h_M(\mathbf{r})$, with (as yet unknown) expansion coefficients c_m, $0 \leq m \leq M$,

$$f_A(\mathbf{r}) = \sum_{m=0}^{M} c_m h_m(\mathbf{r}) \tag{2.3.1}$$

The notation adopted in (2.3.1) is purposely compact and must be used with caution if the expansion is two or three dimensional. In fact, there are two summation indices in two dimensions (as in Section 2.1) and three in three dimensions. The index m in (2.3.1) is assumed to be a multidimensional index that combines the individual indices for each dimension. For example, in Section 2.1, the six expansion coefficients c_{00}, $c_{10}, c_{01}, c_{20}, c_{02}$, and c_{11} would be relabeled c_0, c_1, c_2, c_3, c_4, and c_5. The corresponding basis functions would be denoted $h_0 = 1, h_1 = x, h_2 = y, h_3 = x^2, h_4 = y^2$, and $h_5 = xy$.

Assume that there are K observations $f_O(\mathbf{r}_k)$ over the domain and that we want a global fit to these observations. This can be achieved by minimizing the following quadratic form:

$$I = \sum_{k=1}^{K} w_k d_k^2 = \sum_{k=1}^{K} w_k \left[\sum_{m=0}^{M} c_m h_m(\mathbf{r}_k) - f_O(\mathbf{r}_k) \right]^2 \tag{2.3.2}$$

where $w_k = 0.5\langle \varepsilon_O^2(\mathbf{r}_k)\rangle^{-1}$ and the observation errors are random, normal, unbiased, and spatially uncorrelated.

Equation (2.3.2) can be minimized by differentiating with respect to each coefficient

c_m and setting $\partial I/\partial c_m = 0$ for each m. This gives

$$\sum_{\mu=0}^{M} c_\mu \sum_{k=1}^{K} w_k h_m(\mathbf{r}_k) h_\mu(\mathbf{r}_k) = \sum_{k=1}^{K} w_k f_O(\mathbf{r}_k) h_m(\mathbf{r}_k), \ 0 \le m \le M, \tag{2.3.3}$$

where μ is a dummy index as before. Equations (2.3.3) are the *normal equations.*

Equation (2.3.1) evaluated at the observation stations $\mathbf{r}_k$ can be written in matrix form:

$$\underline{f}_A = \underline{\underline{H}}\underline{c} \tag{2.3.4}$$

where $\underline{f}_A$ is the column vector of length K whose elements are $f_A(\mathbf{r}_k)$; $\underline{c}$ is the column vector of length $M+1$ of expansion coefficients c_m; $\underline{\underline{H}}$ is the $K \times (M+1)$ rectangular matrix whose elements $h_{km} = h_m(\mathbf{r}_k)$. Then the normal equations (2.3.3) can be written as

$$\underline{\underline{G}}\underline{c} = \underline{\underline{H}}^T \underline{\underline{O}}^{-1} \underline{\underline{H}}\underline{c} = \underline{\underline{H}}^T \underline{\underline{O}}^{-1} \underline{f}_O \tag{2.3.5}$$

where $\underline{\underline{O}}$ is the diagonal $K \times K$ square matrix whose elements are $\langle \varepsilon_O^2(\mathbf{r}_k) \rangle$ and $\underline{f}_O$ is the column vector of length K whose elements are $f_O(\mathbf{r}_k)$. The matrix $\underline{\underline{G}} = \underline{\underline{H}}^T \underline{\underline{O}}^{-1} \underline{\underline{H}}$ is a $(M+1) \times (M+1)$ square matrix called the *Gram* matrix. The matrix on the left-hand side of (2.1.4) is such a matrix. $\underline{\underline{G}}$ is a square matrix (unlike $\underline{\underline{H}}$) and is, in principle, invertible:

$$\underline{c} = (\underline{\underline{H}}^T \underline{\underline{O}}^{-1} \underline{\underline{H}})^{-1} \underline{\underline{H}}^T \underline{\underline{O}}^{-1} \underline{f}_O \tag{2.3.6}$$

The analyzed field $f_O(\mathbf{r})$ can then be obtained at arbitrary locations in the domain by inserting (2.3.6) into (2.3.1). In global fitting, unlike local fitting, all the coefficients, c_m, $0 \le m \le M$ are required.

Because there are K observations, up to K coefficients can be determined from (2.3.6). In other words, $M \times 1 \le K$. In the special case $K = M + 1$, $\underline{\underline{H}}$ is a square matrix, and the normal equations (2.3.3) are said to be *fully determined*. In general, $M + 1 \le K$ and the normal equations are said to be *overdetermined*; that is, there are more independent observations than there are coefficients to be determined.

In the fully determined case, (2.3.5) becomes $\underline{\underline{H}}\underline{c} = \underline{f}_O$, and the specified weights are irrelevant. We illustrate this case using one-dimensional polynomial basis functions:

$$h_0(x) = 1,\ h_1(x) = x, \ldots, h_M(x) = x^M \tag{2.3.7}$$

Then, for the observation points x_k, $1 \le k \le K$, the matrix $\underline{\underline{H}}$ becomes

$$\underline{\underline{H}} = \begin{pmatrix} 1 & x_1 & x_1^2 & \cdots & x_1^M \\ \vdots & & & & \\ 1 & x_k & x_k^2 & \cdots & x_k^M \end{pmatrix} \tag{2.3.8}$$

In the special case $K = M + 1 = 2$, (2.3.6) becomes

$$\begin{vmatrix} c_0 \\ c_1 \end{vmatrix} = \begin{pmatrix} 1 & x_1 \\ 1 & x_2 \end{pmatrix}^{-1} \begin{vmatrix} f_O(x_1) \\ f_O(x_2) \end{vmatrix} \tag{2.3.9}$$

or

$$c_0 = \frac{x_2 f_O(x_1) - x_1 f_O(x_2)}{x_2 - x_1}, \qquad c_1 = \frac{f_O(x_2) - f_O(x_1)}{x_2 - x_1}$$

The analyzed field $f(x)$ from (2.3.4) becomes

$$f_A(x) = \frac{x_2 - x}{x_2 - x_1} f_O(x_1) + \frac{x - x_1}{x_2 - x_1} f_O(x_2) \tag{2.3.10}$$

which will be recognized as the Lagrange interpolation formula [Equation (E1) in Appendix E] for $K = 2$. In this fully determined case, the analysis $f_A(x)$ is required to exactly fit the observation stations x_k, $1 \leq k \leq K$. In fact, the residuals $d_k = f_A(x_k) - f_O(x_k)$ are identically equal to zero in this case. Thus, fully determined least squares minimization using polynomial basis functions is equivalent to Lagrange interpolation. The choice $M + 1 = K$ is appropriate when the observations are error free. Usually, the observations $f_O(x_k)$ are not error free, and a fully determined fit is not only inappropriate, it can lead to serious problems, as we demonstrate in the next section.

The Gram matrix $\underline{\underline{G}}$ has special properties. Usually, $\underline{\underline{G}}$ is a full matrix and must be inverted using the standard techniques of linear algebra. In the case for which the observation errors are spatially uncorrelated, the elements $g_{m\mu}$ of the Gram matrix may be written as

$$g_{m\mu} = \sum_{k=1}^{K} w_k h_m(\mathbf{r}_k) h_\mu(\mathbf{r}_k) \tag{2.3.11}$$

From (2.3.11), it is apparent that $\underline{\underline{G}}$ is real and symmetric. An arbitrary $N \times N$ matrix $\underline{\underline{Q}}$ is said to be positive semidefinite if (see Fadeev and Fadeeva 1963), for any arbitrary column vector $\underline{z}$ with elements z_n, $1 \leq n \leq N$,

$$\underline{z}^T \underline{\underline{Q}} \underline{z} \geq 0 \qquad \text{provided not all } z_n \text{ are equal to zero)} \tag{2.3.12}$$

When (2.3.12) is strictly positive, $\underline{\underline{Q}}$ is said to be positive definite. It is straightforward to show that the Gram matrix is positive semidefinite:

$$\begin{aligned} \underline{z}^T \underline{\underline{G}} \underline{z} &= \sum_{m=0}^{M} \sum_{\mu=0}^{M} g_{m\mu} z_m z_\mu = \sum_{k=1}^{K} w_k \sum_{m=0}^{M} \sum_{\mu=0}^{M} h_m(\mathbf{r}_k) h_\mu(\mathbf{r}_k) z_m z_\mu \\ &= \frac{1}{2} \sum_{k=1}^{K} \langle \varepsilon_O^2(\mathbf{r}_k) \rangle^{-1} \left[\sum_{m=0}^{M} h_m(\mathbf{r}_k) z_m \right]^2 \geq 0 \end{aligned} \tag{2.3.13}$$

Consider now the eigenstructure of $\underline{\underline{G}}$. If $\underline{e}$ is an eigenvector of $\underline{\underline{G}}$, then

$$\underline{\underline{G}} \underline{e} = \lambda \underline{e} \tag{2.3.14}$$

where λ is the eigenvalue. Two eigenvectors $\underline{e}_m$ and $\underline{e}_\mu$ of a matrix are said to be orthogonal if $\underline{e}_m^T \underline{e}_\mu = \delta_{m\mu}$ and $\delta_{m\mu}$ is the Kronecker delta function (Here we have assumed that the eigenvectors have been normalized so that $\underline{e}_m^T \underline{e}_m = 1$). Because $\underline{\underline{G}}$ is real and symmetric, its eigenvectors are orthogonal and its eigenvalues are real. If λ

is an eigenvalue of the positive semidefinite matrix $\underline{\underline{G}}$ and $\underline{e}$ is the corresponding eigenvector, then

$$\frac{\underline{e}^{T}\underline{\underline{G}}\underline{e}}{\underline{e}^{T}\underline{e}} = \lambda \geq 0 \tag{2.3.15}$$

because both numerator and denominator are positive. Thus, the eigenvalues of $\underline{\underline{G}}$ are real and nonnegative (positive if $\underline{\underline{G}}$ is positive definite).

The Gram matrix can always be inverted if all of its eigenvalues are positive. However, if one or more of its eigenvalues is vanishingly small, it can be numerically singular. Under what circumstances does this happen? If two of the basis functions $h_m(\mathbf{r})$, $h_\mu(\mathbf{r})$ are similar but not identical, then two rows of the Gram matrix will become similar, and the matrix can be numerically singular. This can also happen if one of the rows can be approximately represented by a linear combination of the other rows.

A matrix that is difficult to invert is called ill-conditioned, and the matrix inverse is not very accurate. A measure of the accuracy of the matrix inverse is given by the condition number defined by

$$N_c = \frac{\lambda_1}{\lambda_s} \tag{2.3.16}$$

where λ_1 is the largest eigenvalue of $\underline{\underline{G}}$, and λ_s is the smallest. A high condition number implies a λ_s that is very close to zero and near linear dependency of the rows (and columns) of the matrix. As shown by Dixon et al. (1972), the relative error in the coefficients $\underline{c}$ of (2.3.6) is directly proportional to the condition number of the Gram matrix $\underline{\underline{G}}$. Thus, if $\underline{\underline{G}}^{-1}$ is not very accurate, then the resulting objective analyses are not very satisfactory. Unfortunately, the matrices encountered in function fitting often have high condition numbers. For the polynomial basis functions (2.3.7), the condition number is high except when M is small. As discussed in Dorny (1975), $M = 7$ is about the practical limit for polynomial fitting.

Even when $\underline{\underline{G}}$ is well-conditioned, it may be difficult to invert in practice. The order of the Gram matrix is equal to the number of degrees of freedom in the expansion. There is a practical limit to the sizes of full matrices that can be inverted because of the finite memory size of the computer and because the computation time for matrix inversion increases as the cube of the order. This is not a problem in local fitting, where M is generally small, but it is a serious limitation in global fitting.

In principle, both of these problems can be circumvented by a Gram–Schmidt orthogonalization process. If the elements $g_{m\mu}$, $m \neq \mu$, defined in (2.3.11) can be made equal to zero, then the Gram matrix is diagonal and trivial to invert for any value of M. The off-diagonal elements of $\underline{\underline{G}}$ vanish if the basis functions are orthogonal to each other over the observation network $\mathbf{r}_k$, $1 \leq k \leq K$, with respect to the prescribed weights w_k. Orthogonal basis functions $p_m(\mathbf{r})$ can be constructed from nonorthogonal basis functions $h_m(\mathbf{r})$ as follows. Define an inner product

$$[p, q] = \sum_{k=1}^{K} w_k p(\mathbf{r}_k) q(\mathbf{r}_k) \tag{2.3.17}$$

Then (2.3.3) can be written as

$$\sum_{\mu=0}^{M} c_\mu [h_m, h_\mu] = [h_m, f_O] \tag{2.3.18}$$

and $[h_m, h_\mu]$ is not usually equal to zero when $m \neq \mu$. A new set of basis functions $p_m(\mathbf{r})$, $0 \leq m \leq M$, can be constructed:

$$p_0(\mathbf{r}) = h_0(\mathbf{r}),$$

$$p_1(\mathbf{r}) = h_1(\mathbf{r}) - \frac{[p_0, h_1]}{[p_0, p_0]} p_0(\mathbf{r})$$

$$p_m(\mathbf{r}) = h_m(\mathbf{r}) - \sum_{\mu=0}^{m-1} \frac{[p_\mu, h_m]}{[p_\mu, p_\mu]} p_\mu(\mathbf{r}) \tag{2.3.19}$$

Equation (2.3.18) can be rewritten in terms of the new basis functions $p_m(\mathbf{r})$ with different expansion coefficients $\tilde{c}_m$. Thus,

$$\sum_{\mu=0}^{M} \tilde{c}_\mu [p_m, p_\mu] = [p_m, f_O] \quad \text{or} \quad \tilde{c}_m = [p_m, p_m]^{-1} [p_m, f_O] \tag{2.3.20}$$

from (2.3.19). The Gram–Schmidt process (2.3.19) for a discrete network is completely analogous to classical Gram–Schmidt orthogonalization for continuous variables.

The process (2.3.19) can be laborious if M is very large. Also, as pointed out by Dorny (1975), the $p_m(\mathbf{r})$ must be constructed very carefully if the $h_m(\mathbf{r})$ are almost linearly dependent. The Gram–Schmidt procedure is attractive for a time-invariant observation network $\mathbf{r}_k$ that will be used many times. For then, the procedure (2.3.19) can be done once and for all because it does not depend on the observations, only on their locations and expected observation errors. The Gram–Schmidt procedure has been applied to polynomial fitting for meteorological spatial analysis by Dixon et al. (1972).

2.4 Underfitting, overfitting, and other problems

The previous section discussed two technical problems in the inversion of the Gram matrix: possible ill-conditioning (particularly for polynomial basis functions) and the order of the matrix being too large to invert (for global fitting). Unfortunately, there are other problems, and accurate inversion of the Gram matrix does not necessarily guarantee a satisfactory objective analysis. This section examines some of the remaining problems of function-fitting algorithms, with particular emphasis on underfitting and overfitting.

Problems of underfitting and overfitting are examined using trigonometric basis functions. The one-dimensional case will be considered; extension to two and three dimensions is straightforward. Consider the analysis domain $-\pi \leq x \leq \pi$, with periodic boundary conditions $f(-\pi) = f(\pi)$. Assume the analysis $f_A(x)$ is expanded in trigonometric basis functions as follows:

$$f_A(x) = \frac{a_0}{2} + \sum_{m=1}^{M} [a_m \cos mx + b_m \sin mx] \tag{2.4.1}$$

The subsequent developments are considerably simplified if complex Fourier notation is used. Define $c_m = 0.5(a_m - ib_m)$, where $i = \sqrt{-1}$. Because $f_A(x)$ is real, c_m is conjugate symmetric, $c_{-m} = c_m^*$, where (*) indicates complex conjugation. Then

$$f_A(x) = \sum_{m=-M}^{M} c_m e^{imx} \tag{2.4.2}$$

The spatial scale of the trigonometric basis functions decreases as m increases. Now consider the observations $f_O(x_k)$, $1 \leq k \leq K$, and define a quadratic form similar to (2.3.2):

$$I = \sum_{k=1}^{K} w_k d_k^2 = \sum_{k=1}^{K} w_k \left[\sum_{m=-M}^{M} c_m e^{imx_k} - f_O(x_k) \right]^2 \tag{2.4.3}$$

Minimizing (2.4.3) with respect to each of the 2M + 1 coefficients c_m gives the normal equations,

$$\sum_{\mu=-M}^{M} c_\mu \sum_{k=1}^{K} w_k e^{i(\mu - m)x_k} = \sum_{k=1}^{K} w_k f_O(x_k) e^{-imx_k} \tag{2.4.4}$$

where μ is a dummy index as before and we have made use of the fact that c_m is conjugate symmetric. The Gram matrix in (2.4.4) can be thought of as a $(2M + 1) \times (2M + 1)$ real, symmetric, positive semidefinite matrix whose elements can be determined from the real form of the trigonometric expansion (2.4.1).

A close correspondence exists between (2.4.4) and a truncated Fourier expansion for the case in which the specified weight functions w_k are all equal. If the sums $\sum_{k=1}^{K}$ ( ) in (2.4.3–4) are replaced by integrals $\int_{-\pi}^{\pi}$ ( ) dx, then (2.4.4) becomes

$$c_m = \frac{1}{2\pi} \int_{-\pi}^{\pi} f_O(x) e^{-imx} \, dx \tag{2.4.5}$$

In deriving (2.4.5), we have used the identity

$$\int_{-\pi}^{\pi} e^{i(\mu - m)x} \, dx = 2\pi \delta_{m\mu} \tag{2.4.6}$$

Note that (2.4.5) is the classical expression for the coefficients of a Fourier series.

Figure 2.2 illustrates a least squares fit using trigonometric basis functions. Over

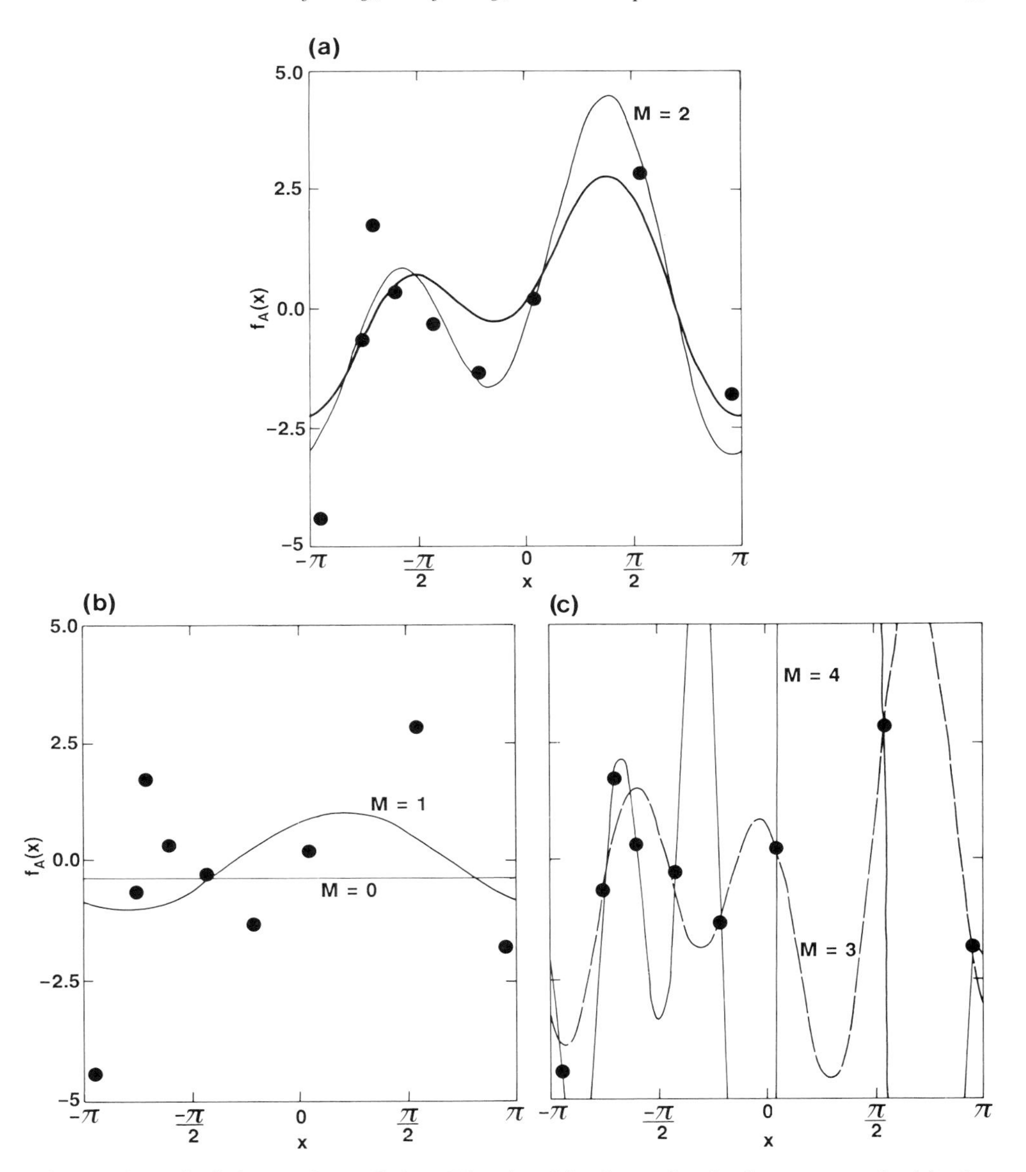

Figure 2.2 Underfitting and overfitting. The signal is shown by the heavy curve in (a); the observations (including error) are shown by solid dots. Fits for $M = 0, 1, 2, 3$, and 4 are shown in (a), (b), and (c).

the interval $-\pi \leq x \leq \pi$, the signal or truth is specified to be

$$f_T(x) = 0.5 + 1.2\cos(x) + 0.8\sin(x) - 1.3\cos(2x) + 0.6\sin(2x) \qquad (2.4.7)$$

where subscript T stands for the truth. The signal is shown by the dark solid line in Figure 2.2(a).

All the observations $f_O(x_k)$ are assumed to have the same expected observation error variance and were created by adding to $f_T(x_k)$ random errors drawn from a normal distribution (Appendix D) with mean zero and standard deviation equal to 1. There are nine observations, indicated by the dots in Figure 2.2(a). The observation location is shown by its position on the x axis (abscissa), and the values of the observation $f_O(x_k)$ by its value on the ordinate. Note that the observations are not distributed uniformly; they are relatively dense for $x < 0$ and relatively sparse for $x > 0$.

The signal (2.4.7) has 5 degrees of freedom $-2 \leq m \leq 2$. Figure 2.2(a) shows the analysis of nine observations using the fitting algorithm – [(2.4.4) and (2.4.2 with all w_k equal)] – for $M = 2$ (lightweight curve). From (a), we see that the analysis errors are largest when the actual observation error is large or when the observation density is low, in accord with intuition.

Now consider the case $M < 2$. That is, the correct basis functions are used for the analysis, but there are fewer degrees of freedom than in the signal. This is called *underfitting* and is illustrated in Figure 2.2(b). The signal is also given by (2.4.7), and the observation values and locations are indicated as in (a). Two cases of underfitting are shown: $M = 0$ and $M = 1$. The analysis for $M = 0$ is a constant and is a very poor fit to the signal. The case $M = 1$ (3 degrees of freedom) is a better fit but not as good as for $M = 2$.

Another case is $M > 2$, for which there are more degrees of freedom in the analysis than in the signal. This is called *overfitting* and is illustrated in Figure 2.2(c). Two cases are shown: $M = 3$ (7 degrees of freedom) and $M = 4$ (9 degrees of freedom). Note that for $M = 4$, there are exactly the same number of degrees of freedom in the expansion as there are observations, giving a fully determined fit. Every observation is fitted exactly for $M = 4$. The result is an extremely poor analysis in the regions with sparse observations. It is clear from (c) why an overdetermined fit is preferable to a fully determined fit.

In actual practice, of course, the signal is never known. However, the spectrum of the signal (Figure 1.4, for example) might be known, at least approximately. The optimum value of the truncation limit M would be chosen on the basis of knowledge of the spectrum and the characteristics of the observing system. There is usually a trade-off between choosing M too small and too large. If M is too small, the analysis will be smooth and will not fit the observations very well. If M is too large, the analysis will fit the observations well but is likely to be totally unacceptable in data voids. Overfitting is extremely dangerous.

Overfitting and underfitting can occur with any set of basis functions. Polynomial fitting has additional problems. All polynomials tend to $\pm\infty$ at large distances, and extrapolation is very risky, as noted in Appendix E. Local polynomial fitting is more reliable than global polynomial fitting, but there are difficulties when the data density is changing rapidly (at the edge of a data void).

Polynomial fitting has still another problem. Suppose one wants to calculate spatial derivatives of the analyzed field – for example, to calculate the geostrophic wind (7.4.5) from a geopotential analysis or the vorticity from a streamfunction analysis

(Section 6.4). First derivatives of the analyzed field are required in the first instance and second derivatives in the second instance.

The jth spatial derivative (in one dimension) of trigonometric and polynomial expansions is given, respectively, by

$$\frac{d^j f_A(x)}{dx^j} = \sum_{m=-M}^{M} c_m (im)^j e^{imx} \quad \text{and} \quad \frac{d^j f_A(x)}{dx^j} = \sum_{m=0}^{M-j} c_{m+j} \frac{(m+j)!}{m!} x^m \quad (2.4.8)$$

where (!) indicates factorial. The higher derivatives of an analysis based on a trigonometric expansion give more weight to the smaller-scale basis functions ($|m|$ large). Thus, the spatial derivatives of the analysis have more small-scale variance and are finer grained (less smooth) than the analysis itself. This is in accord with intuition and experience.

In the case of a polynomial expansion, the maximum order of the polynomial is reduced by spatial differentiation; and when $j > M$, the jth derivative of the analysis is reduced to zero. Thus, the derived fields obtained by differentiation of an analysis based on polynomial basis functions have fewer local maxima and minima than the analysis itself. Such derived fields lack detail and are, in general, smoother than the analyzed field. The derived fields, in this case, are unrealistic.

2.5 The a posteriori analysis weights

We can obtain considerable insight into any spatial analysis procedure by examining the a posteriori analysis weights defined in (1.7.1). Examination of the a posteriori weights is one way to compare the characteristics of various objective analysis algorithms. Another way to compare the spectral responses of different algorithms is discussed in Chapter 3.

An expression for the a posteriori weights (assuming that the observation errors are spatially uncorrelated) is obtained as follows. Define $\tilde{g}_{m\mu}$ to be an arbitrary element of $\underline{\underline{G}}^{-1}$, the inverse of the Gram matrix. Then, from (2.3.3),

$$c_m = \sum_{\mu=0}^{M} \sum_{k=1}^{K} w_k \tilde{g}_{m\mu} h_\mu(\mathbf{r}_k) f_O(\mathbf{r}_k), \qquad 0 \le m \le M \qquad (2.5.1)$$

At an analysis point $\mathbf{r}_i$, the analyzed value is

$$f_A(\mathbf{r}_i) = \sum_{m=0}^{M} c_m h_m(\mathbf{r}_i) = \sum_{k=1}^{K} W_{ik} f_O(\mathbf{r}_k)$$

where

$$W_{ik} = w_k \sum_{m=0}^{M} \sum_{\mu=0}^{M} \tilde{g}_{m\mu} h_m(\mathbf{r}_i) h_\mu(\mathbf{r}_k) \qquad (2.5.2)$$

is the a posteriori weight given to the observation $f_O(\mathbf{r}_k)$ in the analysis at $\mathbf{r}_i$.

Because the $\tilde{g}_{m\mu}$ are the elements of the inverse of the Gram matrix, which is usually a full matrix, we cannot get a simple analytic expression for W_{ik} except in very special circumstances, such as the following. Define the equally spaced one-dimensional observation network over the analysis domain $-\pi \le x \le \pi$:

$$x_k = -\pi + k\,\Delta x \qquad \text{where} \qquad \Delta x = \frac{2\pi}{K}, \qquad 1 \le k \le K \tag{2.5.3}$$

The mathematics is slightly simplified if K is chosen to be odd. Over this network, the normal equations for one-dimensional trigonometric basis functions (2.4.4), when all the specified weights w_k are equal, become

$$c_m = \frac{1}{2\pi} \sum_{k=1}^{K} f_{\text{O}}(x_k) \exp(-imx_k)\,\Delta x, \qquad -M \le m \le M \tag{2.5.4}$$

The off-diagonal elements in (2.5.4) vanish because of the following identity (Jenkins and Watt 1968), which is valid provided that $(2m+1)$ and $(2\mu+1)$ are both less than or equal to K:

$$\sum_{k=1}^{K} \exp[i(\mu - m)(-\pi + k\,\Delta x)]\,\Delta x = 2\pi\delta_{m\mu} \tag{2.5.5}$$

Substitution of (2.5.4) into

$$f_{\text{A}}(x_i) = \sum_{m=-M}^{M} c_m e^{imx_i} = \sum_{k=1}^{K} W_{ik} f_{\text{O}}(x_k)$$

yields

$$W_{ik} = \frac{\Delta x}{2\pi} \sum_{m=-M}^{M} e^{im(x_i - x_k)} = \frac{\Delta x}{\pi}\left[0.5 + \sum_{m=1}^{M} \cos m(x_k - x_i)\right] \tag{2.5.6}$$

The a posteriori weights W_{ik} indicate the weight that is given to an observation displaced by a distance $x_k - x_i$ from the analysis gridpoint. A number of properties of W_{ik} can be obtained by examining (2.5.6). The a posteriori weight is symmetric in the univariate case, $W_{ik} = W_{ki}$. The application of (2.5.5) shows that the sum of the weights is equal to one, $\sum_{k=1}^{K} W_{ik} = 1$. It is also simple to show that $W_{ik} \le W_{ii}$, where W_{ii} is the weight given to an observation if it occurs at the analysis gridpoint itself. In a fully determined fit, $W_{ik} = \delta_{ik}$ (see Exercise 2.7). For fixed M, $\lim_{K\to\infty} W_{ik} = 0$; so, as the number of observations increases (and the observation spacing Δx decreases), the weight given to an individual observation decreases.

$W_{ik}/\Delta x$ is purely a function of the displacement between observation stations and the analysis gridpoint and does not depend on K. From (2.5.6), $\lim_{M\to\infty} W_{ii}/\Delta x = \infty$. In Figure 2.3 we have plotted the function $W_{ik}/\Delta x$ (ordinate) for the equally spaced observation network using trigonometric basis functions (2.5.6). The abscissa is the distance between observation and analysis point, $x_k - x_i$, in the analysis domain $-\pi$ to π. Two cases are shown: $M = 2$ (5 degrees of freedom) and $M = 10$ (21 degrees of freedom).

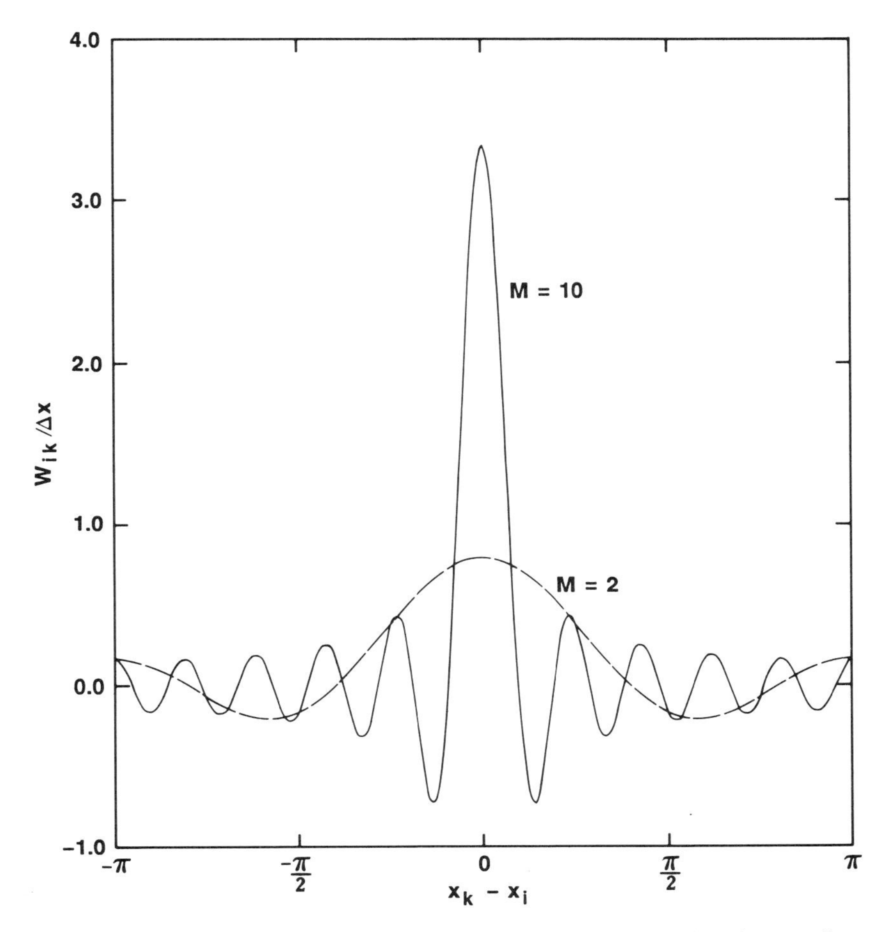

Figure 2.3 The a posteriori weight $W_{ik}/\Delta x$ as a function of $x_k - x_i$ for the equally spaced one-dimensional network in the univariate case.

In the case $M = 2$, relatively large weights are given to observations that lie considerable distances from the analysis gridpoint. For $M = 10$, the a posteriori weight tends to be small except in the immediate neighborhood of the analysis gridpoint. Specifying $M = 2$ produces a relatively large-scale analysis, which implies a relatively broad a posteriori weight function. Higher values of M imply increasingly localized a posteriori weight functions. Note that the a posteriori weight functions can be negative.

Now, the equally spaced case is rather special: A more realistic situation is shown in Figure 2.4. Again trigonometric basis functions (2.4.2) are used with $M = 4$ (9 degrees of freedom) over the analysis domain $-\pi$ to π. The observation network, this time, is irregular, and the 13 observation points x_k are indicated by large solid dots. It is evident that the observation network density is high for $x_k < 0$ and low for $x_k > 0$. The expected observation error variance $\langle \varepsilon_O^2(x_k) \rangle$ is assumed to be the

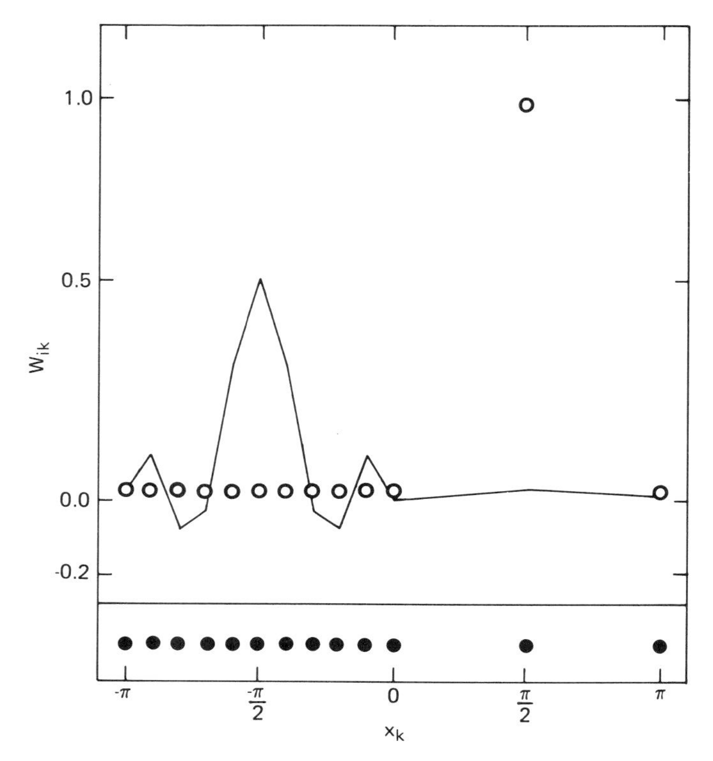

Figure 2.4 The a posteriori weight W_{ik} as a function of x_k for an irregular network with analysis gridpoints at $x_i = -\pi/2$ (solid curve) and $x_i = \pi/2$ (open circles). The observation locations are indicated by the solid dots.

same at all observation stations. The values of W_{ik} have been determined numerically for this case for two analysis points x_i. The first analysis point $x_i = -\pi/2$ happens to coincide with the sixth observation station from the left. The weight function W_{ik}, in this case, is shown by the continuous curve. The second analysis point chosen $x_i = \pi/2$ happens to coincide with the twelfth observation station from the left. The weights in this case are indicated by the open circles.

It can be seen that for the same value of M, the weight distribution differs markedly from location to location when the observation network is irregular. Where the network is relatively dense ($x_k < 0$), the a posteriori weight distribution is similar to that in Figure 2.3. On the other hand, where the observation density is low ($x_k > 0$), the analysis algorithm puts almost all the weight on the local observations.

2.6 Multivariate function fitting

One of the major themes of this book is the necessity for objective analyses to be consistent with the governing equations of the atmosphere. The dependent or state variables are coupled through the governing equations and therefore cannot be analyzed in isolation. The equations that relate the dependent variables are called dynamic or physical constraints. (Mathematical constraints are introduced in the next section.) When several dependent variables are analyzed simultaneously, with implicit or explicit dynamic constraints, the process is referred to as multivariate objective analysis.

A series of increasingly complex dynamic constraints will be developed in Chapters 6–10; but for present purposes, we consider only very simple constraints. Simple, linear dynamic constraints can be accommodated easily into the function-fitting formalism. In fact, multivariate function fitting goes back to the time of Panofsky (1949).

The geostrophic relation is a simple linear constraint that is derived in most elementary meteorology textbooks (discussed in detail in Chapter 6). In cartesian coordinates on an f-plane, it is

$$u = -\frac{1}{f_0}\frac{\partial \Phi}{\partial y}, \qquad v = \frac{1}{f_0}\frac{\partial \Phi}{\partial x} \tag{2.6.1}$$

where u and v are the eastward and northward velocity components, Φ the geopotential field, x and y are the eastward and northward spatial coordinates, and f_0 is the (constant) Coriolis parameter. Equation (2.6.1) is a low-order approximation to the governing equations with considerable validity in the extratropics, but it is completely inappropriate in the tropics.

Consider the simultaneous analysis of geopotential $\Phi_A(x, y)$ and horizontal wind components $u_A(x, y)$ and $v_A(x, y)$. Assume that the analyzed values are expanded in a truncated series of ordered two-dimensional basis functions:

$$\Phi_A(x, y) = \sum_{mn} c_{mn} h^{\Phi}_{mn}(x, y) \qquad u_A(x, y) = \sum_{mn} c_{mn} h^{u}_{mn}(x, y)$$

$$v_A(x, y) = \sum_{mn} c_{mn} h^{v}_{mn}(x, y) \tag{2.6.2}$$

where the h^{Φ}_{mn}, h^{u}_{mn}, and h^{v}_{mn} are basis functions corresponding to Φ, u, and v, respectively. Note that the expansion coefficients in each of the three expansions are the same. The dynamic constraint (2.6.1) is imposed implicitly by demanding common expansion coefficients for each of the three variables and a relationship between the basis functions. For example, suppose the basis functions for the expansion Φ_A were two-dimensional trigonometric functions. Then the geostrophic constraint (2.6.1) would imply

$$h^{\Phi}_{mn} = e^{i(mx+ny)}, \qquad h^{u}_{mn} = -\frac{in}{f_0} e^{i(mx+ny)}, \qquad h^{v}_{mn} = \frac{im}{f_0} e^{i(mx+ny)} \tag{2.6.3}$$

Let us now introduce the more compact notation of (2.3.1). Define $\mathbf{r} = (x, y)$ to be the two-dimensional spatial coordinate, and write the analyzed values Φ_A, u_A, and

v_A in (2.6.2) as

$$\Phi_A(\mathbf{r}) = \sum_m c_m h_m^{\Phi}(\mathbf{r}), \qquad u_A(\mathbf{r}) = \sum_m c_m h_m^u(\mathbf{r}), \qquad v_A(\mathbf{r}) = \sum_m c_m h_m^v(\mathbf{r}) \tag{2.6.4}$$

where m is now a two-dimensional index. Suppose there are observations $\Phi_O(\mathbf{r}_k)$, $1 \le k \le K$, at the observation stations $\mathbf{r}_k$ and $u_O(\mathbf{r}_l)$, $v_O(\mathbf{r}_l)$, $1 \le l \le L$, at the observation stations $\mathbf{r}_l$. Define $\varepsilon_\Phi(\mathbf{r}_k)$ and $\varepsilon_v(\mathbf{r}_l)$ to be the observation errors of geopotential and wind components, respectively, and assume that

$$\langle \varepsilon_\Phi(\mathbf{r}_q)\varepsilon_\Phi(\mathbf{r}_p)\rangle = \langle \varepsilon_v(\mathbf{r}_q)\varepsilon_v(\mathbf{r}_p)\rangle = 0 \qquad \text{for } p \ne q$$

and

$$\langle \varepsilon_\Phi(\mathbf{r}_p)\varepsilon_v(\mathbf{r}_q)\rangle = 0 \qquad \text{for all } p, q$$

Thus, the observation error is spatially uncorrelated, and there is no correlation between observation errors of wind and geopotential. Then the following quadratic form can be defined as

$$\begin{aligned} I = &\sum_{k=1}^{K} w_\Phi(\mathbf{r}_k)\left[\sum_m c_m h_m^{\Phi}(\mathbf{r}_k) - \Phi_O(\mathbf{r}_k)\right]^2 \\ &+ \sum_{l=1}^{L} w_v(\mathbf{r}_l)\left\{\left[\sum_m c_m h_m^u(\mathbf{r}_l) - u_O(\mathbf{r}_l)\right]^2 + \left[\sum_m c_m h_m^v(\mathbf{r}_l) - v_O(\mathbf{r}_l)\right]^2\right\} \end{aligned} \tag{2.6.5}$$

where $w_\Phi(\mathbf{r}_k) = 0.5/\langle \varepsilon_\Phi^2(\mathbf{r}_k)\rangle$ and $w_v(\mathbf{r}_l) = 0.5/\langle \varepsilon_v^2(\mathbf{r}_l)\rangle$.

Minimizing (2.6.5) with respect to each of the expansion coefficients c_m gives

$$\begin{aligned} &\sum_\mu c_\mu \left\{\sum_{k=1}^{K} w_\Phi(\mathbf{r}_k) h_m^{\Phi}(\mathbf{r}_k) h_\mu^{\Phi}(\mathbf{r}_k) + \sum_{l=1}^{L} w_v(\mathbf{r}_l)[h_m^u(\mathbf{r}_l)h_\mu^u(\mathbf{r}_l) + h_m^v(\mathbf{r}_l)h_\mu^v(\mathbf{r}_l)]\right\} \\ &\quad = \sum_{k=1}^{K} w_\Phi(\mathbf{r}_k)\Phi_O(\mathbf{r}_k)h_m^{\Phi}(\mathbf{r}_k) + \sum_{l=1}^{L} w_v(\mathbf{r}_l)[u_O(\mathbf{r}_l)h_m^u(\mathbf{r}_l) + v_O(\mathbf{r}_l)h_m^v(\mathbf{r}_l)] \end{aligned} \tag{2.6.6}$$

where μ is a dummy index as before. The analyzed values Φ_A, u_A, and v_A identically satisfy the imposed constraint (2.6.1) because the winds and geopotential are fitted simultaneously; the basis functions satisfy the relation; and the expansion coefficients are common. In the terminology to be introduced in Chapter 8, (2.6.1) is a strong constraint on the multivariate analysis because it is exactly satisfied.

Equation (2.6.6) is the multivariate equivalent of the univariate normal equations (2.3.3). The normal equations (2.6.6) can also be written in matrix form corresponding to (2.3.5). The Gram matrix in this case would have the same order as in the corresponding univariate matrix and, except for special networks, would be a full matrix. Over irregular observing networks, the multivariate fitting algorithm (2.6.4) and (2.6.6) has similar problems to those discussed in Sections 2.3 and 2.4: ill-conditioning, matrices too large to invert, unacceptable analyses in data voids, and so on.

During the 1970s, a multivariate function-fitting algorithm developed by Flattery (1971) was used operationally at the National Meteorological Center in Washington. The analysis domain was the whole earth's atmosphere and the fit was global in the sense of Section 2.3. The analyzed geopotential and wind were assumed to be expanded in a truncated series of Hough functions.

Hough functions are the eigenfunctions of a linearized form of the governing equations on a sphere, known as the Laplace tidal equations. They will be discussed in detail in Chapter 9, but for the present, the following simplified description will suffice. Each Hough mode is a function of latitude ϕ and longitude λ and has three components – a zonal (eastward) wind component, a meridional (northward) wind component, and a geopotential component. The Hough modes have a distinct horizontal scale and an associated eigenfrequency; they are orthogonal over the sphere in the continuous case. The Hough modes are often divided into two classes. Low-frequency Rossby–Hough modes tend to satisfy the geostrophic relation (2.6.1) in the extratropics whereas the higher frequency Hough modes correspond to inertia–gravity waves (see Chapter 6).

The frequency properties of Hough modes can be exploited as an implicit dynamical constraint. If the geopotential and wind are expanded in a series of Rossby–Hough modes only, the resulting objective analysis will satisfy a more general form of the geostrophic relation (2.6.1) that has some validity over the whole globe and not just in the extratropics. The constraints implied by a Rossby–Hough expansion are still linear, however.

In principle, a multivariate fitting algorithm using Rossby–Hough functions can be based on (2.6.4) and (2.6.6) with $\mathbf{r} = (\lambda, \Phi)$ and with the basis functions $h_m^\phi(\mathbf{r})$, $h_m^u(\mathbf{r})$, and $h_m^v(\mathbf{r})$ being the components of the Rossby–Hough functions. Some practical difficulties exist, however, because the Gram matrix is too large to invert. Flattery (1971) circumvented this problem by first locally interpolating the observations to a uniform global analysis grid. The Rossby–Hough functions are orthogonal on this special grid – [in the same way that trigonometric functions are orthogonal on the network (2.5.3)] – and the off-diagonal elements of the Gram matrix vanish. Therefore, (2.6.6) collapsed to a form similar to (2.5.4) and could be easily solved. However, Flattery's procedure does not strictly minimize (2.6.5) because of the preliminary interpolation of the raw observations to the special global grid. Flattery's approach also fits the geopotential and winds separately; so the expansion coefficients are not common, and the resulting objective analyses does not exactly satisfy the imposed dynamic constraint. More recent attempts to fit global atmospheric observations to Hough functions have been made by Halberstam and Tung (1984).

Multivariate function fitting can be examined further in the very simple one-dimensional context of the previous section. Suppose the analysis domain is $-\pi \leq x \leq \pi$ and the imposed dynamic constraint on the objective analysis is

$$\frac{d\Phi_A}{dx} = f_0 v_A \tag{2.6.7}$$

Assume Φ_A and v_A are expanded as

$$\Phi_A(x) = \sum_{m=-M}^{M} c_m e^{imx} \qquad \text{and} \qquad v_A(x) = \sum_{m=-M}^{M} \frac{im}{f_0} c_m e^{imx} \tag{2.6.8}$$

satisfying (2.6.7).

Consider observations $\Phi_O(x_k)$ and $v_O(x_k)$, $1 \le k \le K$, with expected observation error variances $E_\Phi^2 = \langle \varepsilon_\Phi^2(x_k) \rangle$ and $E_v^2 = \langle \varepsilon_v^2(x_k) \rangle$, which are independent of k. The normal equations (from 2.6.6) are

$$\sum_{\mu=-M}^{M} c_m[1 + m\mu\gamma] \sum_{k=1}^{K} e^{i(\mu-m)x_k} = \sum_{k=1}^{K} e^{-imx_k}[\Phi_O(x_k) - imf_0\gamma v_O(x_k)] \tag{2.6.9}$$

where $\gamma = E_\Phi^2/f_0^2 E_v^2$. In the derivation of (2.6.9), $w_\Phi(x_k) = 0.5E_\Phi^{-2}$, $w_v(x_k) = 0.5E_v^{-2}$, and the normal equations have been multiplied by E_Φ^2.

Over the special network (2.5.3), the off-diagonal terms of the Gram matrix vanish and

$$c_m = q_m \sum_{k=1}^{K} e^{-imx_k}[\Phi_O(x_k) - imf_0\gamma v_O(x_k)] \tag{2.6.10}$$

where $q_m = (1 + m^2\gamma)^{-1}$. At an analysis gridpoint x_i, $\Phi_A(x_i)$ and $v_A(x_i)$ can be written as

$$\Phi_A(x_i) = \sum_{k=1}^{K} W_{\Phi\Phi}(x_k - x_i)\Phi_O(x_k) + \sum_{k=1}^{K} W_{\Phi v}(x_k - x_i)v_O(x_k)$$

$$v_A(x_i) = \sum_{k=1}^{K} W_{v\Phi}(x_k - x_i)\Phi_O(x_k) + \sum_{k=1}^{K} W_{vv}(x_k - x_i)v_O(x_k) \tag{2.6.11}$$

where

$$W_{\Phi\Phi}(x_k - x_i) = \frac{\Delta x}{\pi}\left[\frac{1}{2} + \sum_{m=1}^{M} q_m \cos m(x_k - x_i)\right]$$

$$W_{\Phi v}(x_k - x_i) = -\frac{\Delta x\, \gamma f_0}{\pi} \sum_{m=1}^{M} m q_m \sin m(x_k - x_i)$$

$$W_{v\Phi}(x_k - x_i) = \frac{\Delta x}{f_0 \pi} \sum_{m=1}^{M} m q_m \sin m(x_k - x_i)$$

$$W_{vv}(x_k - x_i) = \frac{\Delta x\, \gamma}{\pi} \sum_{m=1}^{M} m^2 q_m \cos m(x_k - x_i)$$

are the multivariate a posteriori weight functions corresponding to (2.5.6) in the univariate case. Δx is defined in (2.5.3) and γ has dimensions of length squared.

Figure 2.5(a) shows $W_{\Phi\Phi}(x_k - x_i)/\Delta x$ and (b) shows $W_{\Phi v}(x_k - x_i)/f_0\, \Delta x$ as functions of $x_k - x_i$, in the same format as Figure 2.3. All plots are for $M = 4$ (9 degrees of freedom). In (a) and (b), the three curves correspond to $\gamma = 0$, 0.2, and 100. $\gamma = 0$

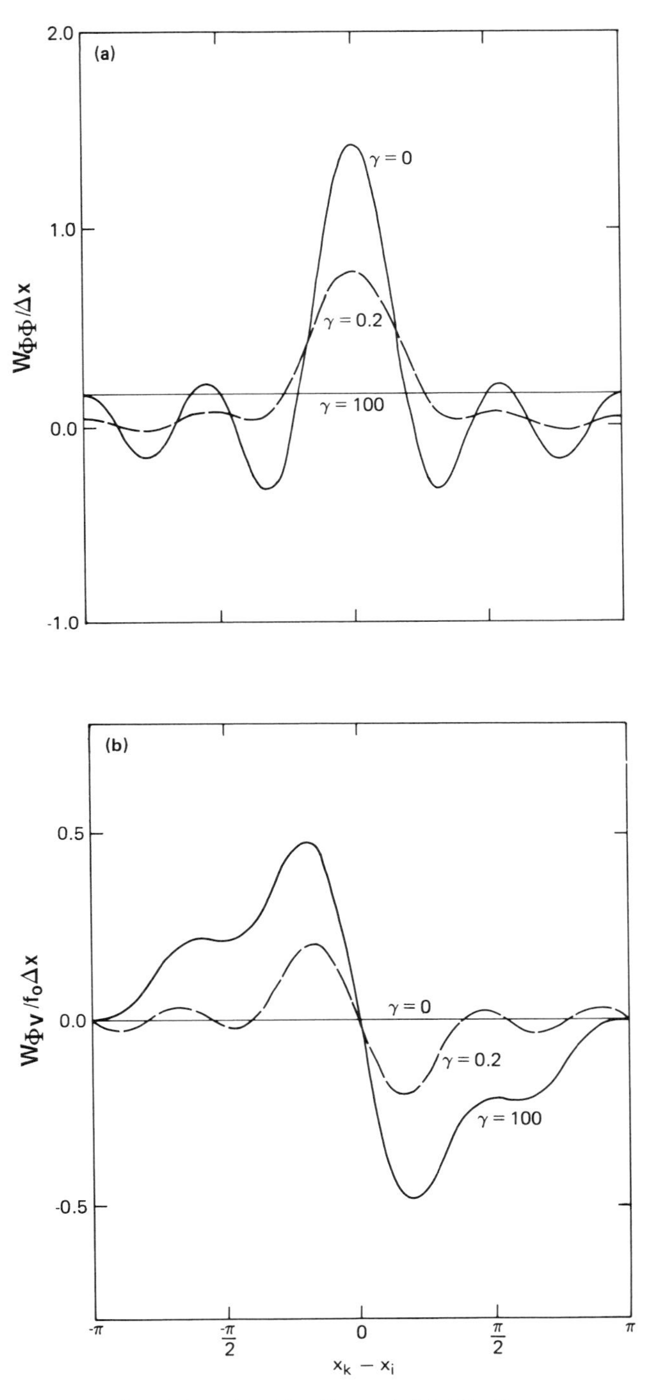

Figure 2.5 The a posteriori weights: (a) $W_{\Phi\Phi}/\Delta x$ and (b) $W_{\Phi v}/f_0\,\Delta x$ as a function of $x_k - x_i$ for three values of γ in the multivariate case on the equally spaced network.

implies highly inaccurate or nonexistent wind observations, and $\gamma = 100$ implies highly inaccurate or nonexistent geopotential observations.

In Figure 2.5(a), we see that as γ is increased, less and less weight is given to the geopotential observations $\Phi_O(x_k)$ in the analysis of the geopotential. Moreover, the weight given to $\Phi_O(x_k)$ becomes increasingly large scale in nature (similar to the weights for small values of M in Figure 2.3). In the limit $\gamma \to \infty$, $W_{\Phi\Phi}(x_k - x_i)/\Delta x$ approaches a constant, indicating that only the constant coefficient c_0 is derived from the geopotential observations. Thus, the values of c_m for large $|m|$ must be derived from the wind observations.

This is consistent with $W_{\Phi v}(x_k - x_i)/f_0\,\Delta x$ plotted in (b). Note that when $x_k = x_i$, $W_{\Phi v}(x_k - x_i) = 0$ for all γ. This indicates that a wind observation that happens to coincide with the analysis gridpoint is given no weight. Thus, in the analysis of $\Phi_A(x_i)$, the largest weights are given to wind observations that are somewhat displaced from the analysis gridpoint. In Figure 2.5(b), it is apparent that negative/positive values of $v_O(x_k)$ for x_k greater/less than x_i will increase $\Phi_O(x_i)$. This is in accord with intuition.

$W_{v\Phi}(x_k - x_i)$ and $W_{vv}(x_k - x_i)$ are not plotted, but their behavior can be inferred from Figure 2.5. The a posteriori weights derived in this section can be compared with a posteriori weights derived for other algorithms such as multivariate statistical interpolation in Chapter 5.

2.7 Constrained minimization: penalty functions

In recent years, function-fitting algorithms have been rarely used in atmospheric objective analysis. An important exception, however, is constrained minimization or spline-fitting algorithms, which have been actively investigated by mathematicians and statisticians for some years and are now discussed in the meteorological literature as well. In these techniques, quadratic forms such as (2.2.9) are minimized subject to imposed mathematical constraints. These algorithms generally contain several user-supplied parameters that should be chosen on the basis of physical characteristics (the spectrum, in particular) of the state variable to be analyzed. Consequently, the imposed formal mathematical constraints are implicitly physical constraints as well.

The technique is first illustrated using one-dimensional trigonometric functions on the domain $x_a \le x \le x_b$. The analyzed field $f_A(x)$ is represented as in (2.4.2),

$$f_A(x) = \sum_{m=-M}^{M} c_m e^{imx} \tag{2.7.1}$$

Define observations $f_O(x_k)$, $1 \le k \le K$, with expected observation errors $\langle \varepsilon_O^2(x_k) \rangle$ which are normally distributed and spatially uncorrelated. Define the following quadratic form:

$$I(\gamma, p) = I_1(f_A) + \gamma J_p(f_A) \tag{2.7.2}$$

where

$$I_1(f_A) = \sum_{k=1}^{K} w_k[f_A(x_k) - f_O(x_k)]^2, \qquad J_p(f_A) = \int_{x_a}^{x_b} \left[\frac{d^p}{dx^p} f_A(x)\right]^2 dx$$

and $w_k = 0.5/\langle \varepsilon_O^2(x_k)\rangle$. Also, $f_A(x_k)$ is given by (2.7.1), and γ and p are user-specified positive real constants (p integer).

$I_1(f_A)$ is a measure of the closeness with which $f_A(x)$ fits the observations $f_O(x)$ and is often referred to as the *cost function*. $J_p(f_A)$ is an imposed mathematical constraint known as a *penalty function*. As will be demonstrated shortly, $J_p(f_A)$ is a measure of the smoothness of the objective analysis $f_A(x)$.

Now choose $x_a = -\pi$ and $x_b = \pi$, and minimize (2.7.2) with respect to the coefficients c_m in (2.7.1), which yields the normal equations

$$\sum_{\mu=-M}^{M} c_\mu \sum_{k=1}^{K} w_k e^{i(\mu-m)x_k} + 2\pi\gamma m^{2p} c_m = \sum_{k=1}^{K} w_k f_O(x_k) e^{-imx_k} \qquad (2.7.3)$$

using (2.4.6).

The Gram matrix in (2.7.3) is similar to that of (2.4.4) except that it has the additional term $2\pi\gamma m^{2p}$ on the main diagonal. This extra term makes the Gram matrix more diagonally dominant. In the nonconstrained case ($\gamma = 0$), the Gram matrix can be ill-conditioned if two of the basis functions (and, thus, two of the rows) are similar. The addition of a term $2\pi\gamma m^{2p}$ ($\gamma > 0$) to the main diagonal of the Gram matrix makes these two rows less similar, increases the smallest eigenvalue λ_s [see (2.3.16)], and improves the conditioning of the matrix.

Now consider the observation network (2.5.3) and all w_k equal to w. The normal equations (2.7.3) for this special network can be written as

$$c_m = \frac{1}{1 + \alpha m^{2p}} \frac{1}{K} \sum_{k=1}^{K} f_O(x_k) e^{-imx_k}, \qquad -M \le m \le M \qquad (2.7.4)$$

and $\alpha = 2\pi\gamma/Kw$.

Equation (2.7.4) becomes equal to (2.5.4) when $\gamma = 0$. The expansion coefficient c_m is reduced by a factor of $(1 + \alpha m^{2p})^{-1}$ with respect to the value in (2.5.4). The addition of the term $J_p(f_A)$ acts like a high-wavenumber filter (also known as a low-pass filter) in reducing the amplitude of the large m (small spatial scale) components of the analysis $f_A(x)$. The amount of filtering is controlled by the specification of γ and p; γ controls the trade-off between the penalty function $J_p(f_A)$, which is a measure of the smoothness of the analysis, and the cost function $I_1(f_A)$, which is a measure of the fidelity of the analysis to the observations; p controls the spectrum of $f_A(x)$, and large values of p imply more effective filtering.

Figure 2.6 illustrates the effect of a penalty function for one-dimensional trigonometric functions over the periodic domain $-\pi \le x \le \pi$. The diagram is in the same format as Figure 2.2, with 10 observations indicated by solid dots. Equation (2.7.3) is applied to these observations with $M = 3$ (7 degrees of freedom), $p = 2$, and $w_k = 1$,

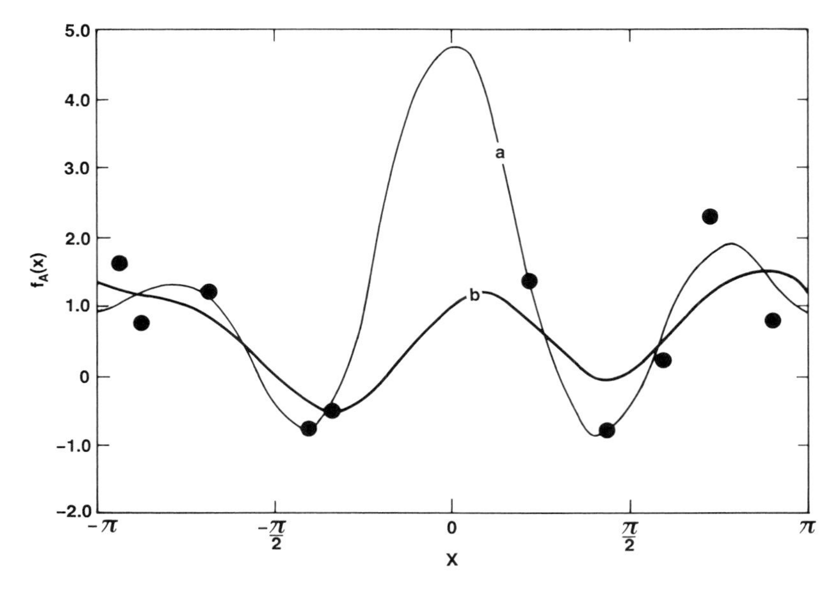

Figure 2.6 The effect of a penalty function on trigonometric fitting. Curve a is unconstrained ($\gamma = 0$), and curve b is constrained ($\gamma = 0.01$).

$1 \leq k \leq 10$. Two curves are shown: $\gamma = 0$ (curve a) and $\gamma = 0.01$ (curve b). As expected from Section 2.4, there is enormous overshooting in the data void near $x = 0$ for the unconstrained case ($\gamma = 0$). The addition of the penalty function has a relative minor effect in the data-rich areas but is very effective in reducing the overshooting in the data void.

One of the desirable properties of any objective analysis algorithm, as noted in Section 1.7, is the filtering of noise (primarily small-scale and nonphysical noise) from the objective analysis. It can be seen that the constrained minimization algorithm (2.7.2) is an effective small-scale filter. Equation (2.7.4) demonstrates the filtering properties (or spectral response) of the algorithm in (2.7.2) in a simple manner. In Chapters 3–5, the spectral responses of other objective analysis algorithms will be discussed in more detail.

In Appendix E, Equation (E9) suggests that cubic-spline functions have a minimum curvature property; that is, they minimize

$$\int_{x_a}^{x_b} [f''_A(x)]^2 \, dx \tag{2.7.5}$$

This observation was the basis of an important constrained minimization algorithm developed by Reinsch (1967). Consider the minimization of $I(\gamma, p)$ in (2.7.2) with $p = 2$, over the periodic one-dimensional domain, $x_a \leq x \leq x_b$:

$$I(\gamma, 2) = \sum_{k=1}^{K} w_k[f_A(x_k) - f_O(x_k)]^2 + \gamma J_2(f_A) \tag{2.7.6}$$

where $J_2(f_A) = \int_{x_a}^{x_b} [f''_A(x)]^2 \, dx$.

Denote $f''_A(x_k) = S_k$. Define $f''_A(x)$ by Equation (E3), and evaluate the penalty function of (2.7.6). After integration by parts, the result is

$$J_2(f_A) = \frac{1}{6}\sum_{k=1}^{K} \Delta_k[S_{k-1}^2 + S_k^2 + (S_{k-1} + S_k)^2] = \underline{S}^T\underline{\underline{T}}\underline{S} \tag{2.7.7}$$

where $\Delta_k = x_k - x_{k-1}$. $\underline{S}$ is the column vector of length K of unknown values of the second derivative S_k of the objective analysis evaluated at the observation stations. $\underline{\underline{T}}$ is the symmetric, positive definite, tri-diagonal matrix defined in Equation (E8). The positive definiteness of $\underline{\underline{T}}$ can be seen from (2.3.13) and (2.7.7).

Equation (2.7.6) can be written in matrix form:

$$I(\gamma, 2) = 0.5[\underline{f}_A - \underline{f}_O]^T\underline{\underline{O}}^{-1}[\underline{f}_O - \underline{f}_A] + \gamma\underline{S}^T\underline{\underline{T}}\underline{S} \tag{2.7.8}$$

where $\underline{f}_A$ and $\underline{f}_O$ are column vectors of length K of values $f_A(x_k)$ and $f_O(x_k)$. The observation error $\varepsilon_O(x_k)$, $1 \le k \le K$, is assumed to be spatially uncorrelated with expected variance $\langle\varepsilon_O^2(x_k)\rangle$. $\underline{\underline{O}}$ is the diagonal $K \times K$ observation error covariance matrix, and the a priori weights of (2.7.6) are $w_k = 0.5/\langle\varepsilon_O^2(x_k)\rangle$. In (2.7.8), both $\underline{S}$ and $\underline{f}_A$ are unknown at this point.

In the spline interpolation discussed in Appendix E, the approximation for $f_A(x)$ passed exactly through the observations, so $\underline{f}_A = \underline{f}_O$. This is not true in the case of (2.7.8). However, the relation (E8) between $\underline{S}$ and $\underline{f}_A$ is still valid because it is a consequence of requiring continuity in the first and second derivatives of $f_A(x)$. Consequently,

$$\underline{\underline{T}}\underline{S} = \underline{\underline{Q}}^T\underline{f}_A \tag{2.7.9}$$

where $\underline{\underline{Q}}$ is defined in Appendix E. It might be noted that $\underline{\underline{Q}}$ is actually a square symmetric matrix in the periodic case and, hence, the transpose notation is redundant. However, in the nonperiodic case, $\underline{\underline{Q}}$ is a rectangular matrix, so the transpose notation is retained. Introducing (2.7.9) into (2.7.8) gives

$$I(\gamma, 2) = 0.5\{[\underline{f}_A - \underline{f}_O]^T\underline{\underline{O}}^{-1}[\underline{f}_A - \underline{f}_O] + \underline{f}_A^T\underline{\underline{N}}^{-1}\underline{f}_A\} \tag{2.7.10}$$

where $\underline{\underline{N}}^{-1} = 2\gamma\underline{\underline{Q}}\underline{\underline{T}}^{-1}\underline{\underline{Q}}^T$.

Minimization with respect to $\underline{f}_A$ following Section 2.2 yields

$$\underline{\underline{O}}^{-1}[\underline{f}_A - \underline{f}_O] + \underline{\underline{N}}^{-1}\underline{f}_A = 0 \qquad \text{or} \qquad \underline{f}_A = \underline{\underline{N}}[\underline{\underline{N}} + \underline{\underline{O}}]^{-1}\underline{f}_O \tag{2.7.11}$$

Because $\underline{\underline{T}}$ is a symmetric, positive definite matrix, its eigenvalues are real and positive. This implies $\underline{\underline{T}}^{-1}$ is positive definite because its eigenvalues are the reciprocals of the eigenvalues of $\underline{\underline{T}}$. Consequently, $\underline{\underline{Q}}\underline{\underline{T}}^{-1}\underline{\underline{Q}}^T$ and $\underline{\underline{N}}$ are symmetric, positive definite (and consequently nonsingular). The functional form (2.7.11) might be compared with that of (2.2.15).

More practically, we can write (2.7.11) as

$$[\underline{\underline{T}} + 2\gamma\underline{\underline{Q}}^T\underline{\underline{O}}\underline{\underline{Q}}]\underline{S} = \underline{\underline{Q}}^T\underline{f}_O \tag{2.7.12}$$

Generally, (2.7.12) is solved for $\underline{S}$, and the objective analysis at a gridpoint $f_A(x_i)$ is obtained from (E3). The matrix on the left-hand side of (2.7.12) is a sparse matrix with nonzero elements only on the main diagonal and the two subdiagonals on either side of it. It can be inverted simply and efficiently.

The Reinsch algorithm (2.7.8–12) can be generalized to the two-dimensional case following Duchon (1976) and Wahba and Wendelberger (1981). Define a two-dimensional analogue to (2.7.2):

$$I(\gamma, p) = \sum_{k=1}^{K} [f_{\mathrm{A}}(x_k, y_k) - f_{\mathrm{O}}(x_k, y_k)]^2 + \gamma J_p(f_{\mathrm{A}}) \tag{2.7.13}$$

where

$$J_p(f_{\mathrm{A}}) = \int_x \int_y \sum_{q=0}^{p} \binom{p}{q} \left[\frac{\partial^p f_{\mathrm{A}}}{\partial x^q \, \partial y^{p-q}} \right]^2 dx\, dy$$

is the penalty function. Here $\binom{p}{q}$ defines the binomial coefficients. When $p = 2$, the penalty function becomes

$$J_2(f_{\mathrm{A}}) = \int_x \int_y \left[\left(\frac{\partial^2 f_{\mathrm{A}}}{\partial x^2} \right)^2 + 2\left(\frac{\partial^2 f_{\mathrm{A}}}{\partial x\, \partial y} \right)^2 + \left(\frac{\partial^2 f_{\mathrm{A}}}{\partial y^2} \right)^2 \right] dx\, dy \tag{2.7.14}$$

Minimization of (2.7.13) with $p = 2$, leads to thin-plate smoothing splines, the natural generalization of the Reinsch algorithm. The spectral filtering characteristics of (2.7.13) are given by a generalization of (2.7.4), $(1 + \alpha k^{2p})^{-1}$, where k is a two-dimensional wavenumber. Equation (2.7.13) can be generalized to three or more dimensions. As discussed by Wahba and Wendelberger (1981), if d is the number of dimensions, then $2p - d > 0$. Thus, for the two-dimensional case, $p \geq 2$. Wahba (1982) has also extended (2.7.13) to the spherical case. In (2.7.13), both γ and p must be specified. Methods for their optimal specification are discussed in Craven and Wahba (1979). Constrained minimization will also be approached from the point of view of weak variational constraints and the Euler–Lagrange equations in Chapter 8.

Spline-fitting algorithms are ultimately based on a minimization principle, as are the statistical interpolation algorithms of Chapters 4 and 5. The objective analysis technique to be discussed in the next chapter is not.

Exercises

2.1 The column vector $\underline{f}_{\mathrm{A}} - \underline{f}_{\mathrm{O}}$ (the difference between the analyzed and observed values at the observation locations) is called the residual. Show that the residual is orthogonal to each of the basis functions $h_m(\mathbf{r}_k)$, $0 \leq m \leq M$, over the network $\mathbf{r}_k$, $1 \leq k \leq K$, with respect to the a priori weights.

2.2 Write a quadratic form corresponding to (2.3.2) for the case of spatially correlated observation error. Show that minimizing this expression leads to (2.3.6).

2.3 Suppose the column vectors of observed, analyzed, and true values $\underline{f}_{\mathrm{O}}$, $\underline{f}_{\mathrm{A}}$, and $\underline{f}_{\mathrm{T}}$ defined at the observation locations satisfy

$$\underline{f}_{\mathrm{O}} = \underline{f}_{\mathrm{T}} + \underline{\varepsilon}_{\mathrm{O}}, \qquad \underline{f}_{\mathrm{A}} = \underline{\underline{H}}\,\underline{c}, \qquad \text{and} \qquad \underline{f}_{\mathrm{T}} = \underline{\underline{H}}\,\underline{\tilde{c}}$$

where $\underline{\underline{H}}$ is the matrix defined in (2.3.4), $\underline{c}$ and $\underline{\tilde{c}}$ are the analyzed and true expansion coefficients, and $\underline{\varepsilon}_O$ is the column vector of observation errors with observation error covariance $\underline{\underline{O}}$. (See, for example, Figure 2.2a.) Define $\underline{\varepsilon}_c = \underline{c} - \underline{\tilde{c}}$ and $\underline{\varepsilon}_A = \underline{f}_A - \underline{f}_T$ to be the column vectors of coefficient errors and analysis errors (at the observation locations), respectively.

(a) Show that if $\langle \underline{\varepsilon}_O \rangle = 0$, then $\langle \underline{\varepsilon}_c \rangle = 0$ and $\langle \underline{\varepsilon}_A \rangle = 0$.

(b) Show that the coefficient error covariance and analysis error covariance matrices are given by

$$\langle \underline{\varepsilon}_c \underline{\varepsilon}_c^T \rangle = [\underline{\underline{H}}^T \underline{\underline{O}}^{-1} \underline{\underline{H}}]^{-1} \qquad \text{and} \qquad \langle \underline{\varepsilon}_A \underline{\varepsilon}_A^T \rangle = \underline{\underline{H}}[\underline{\underline{H}}^T \underline{\underline{O}}^{-1} \underline{\underline{H}}]^{-1} \underline{\underline{H}}^T$$

2.4 Consider an observation network with uncorrelated observation errors with expected observation error variance equal to E_O^2 at the observation locations. Suppose there are two basis functions $h_0(x) = 1$ and $h_1(x) = x$. Consider two cases:
(a) Two observation locations at x_1 and x_2
(b) Three observation locations at x_1, x_2, and x_3
Using the expressions derived in Exercise 2.3, find the coefficient error covariance matrix for the two cases. What happens when all observations coincide? Show that the expected analysis error variances at the observation locations are equal to E_O^2 in case a and are less than or equal to E_O^2 in case b.

2.5 Consider an observation network with two observation stations x_1 and x_2 with uncorrelated observation error whose expected observation error variance is equal to 1 at x_1 and x_2. Assume the basis functions are the same as in Exercise 2.4. Find an expression for the eigenvalues of the Gram matrix $\underline{\underline{G}} = \underline{\underline{H}}^T \underline{\underline{H}}$ in this case. Show that the eigenvalues are real and nonnegative and that the condition number becomes infinite as $x_2 \to x_1$.

2.6 Consider the observation network with three observation stations at x_1, x_2, and x_3. Assume the observation errors are uncorrelated with expected observation error variance $E_O^2(x_1) = E_O^2(x_3) = 1$ and $E_O^2(x_2) = \frac{1}{2}$. Assume basis functions $h_0(x) = 1$ and $h_1(x) = x$. Construct a new set of orthogonal basis functions using the Gram–Schmidt procedure and generate a Gram matrix for the new set.

2.7 For the equally spaced network (2.5.3) show that the a posteriori weights given in Equation (2.5.6) satisfy the following:

(a) $$\sum_{k=1}^{K} W_{ik} = 1$$

(b) $$W_{ik} \le W_{ii} \quad \text{and} \quad 0 \le W_{ii} \le 1$$

(c) $$W_{ik} = \delta_{ik} \qquad \text{for a fully determined fit}$$

3

The method of successive corrections

The polynomial-fitting techniques developed in the 1950s proved to be unsatisfactory for meteorological analysis. There were a number of difficulties, as noted in Sections 2.3 and 2.4, and the inversion of the Gram matrices was computationally expensive on the computers of the time.

In 1955, Bergthorsson and Doos published "Numerical Weather Map Analysis," a paper that outlined a completely different method for meteorological spatial analysis. In Figure 2.3, the a posteriori weight functions for univariate function fitting reach a maximum at the analysis gridpoint and then fall off with increasing distance. In the Bergthorsson/Doos procedure, the a posteriori weights are *specified a priori* to be monotonically decreasing functions of the distance between observing station and analysis gridpoint. Because the a posteriori weights are specified in advance, the matrix to be inverted is diagonal, and consequently the computational demands are minimal.

The original formulation of Bergthorsson and Doos specified a single iteration or pass through the observations. In subsequent formulations, the objective analysis procedure was usually applied iteratively by making successive corrections to a background or first-guess field. Thus, the procedure is now known as the *method of successive corrections* or *SCM*. The original Bergthorsson/Doos weights were determined empirically using statistics on analysis error; thus, to some extent, Bergthorsson and Doos's ideas anticipated the later development of statistical interpolation (Chapters 4 and 5).

The basic formulation for a single correction is now introduced, followed by a discussion of the iteration cycle. At that point, we include two sections on the spectral properties of the technique. This material will be useful in discussing the convergence properties of the SCM method in Section 3.5 and will also provide important background material for several sections in Chapters 4 and 5.

3.1 Basic formulation for a single correction

Versions of the SCM algorithm have been used both operationally and for observational studies from the mid-1950s to the present. Most variants of the technique, though not all (Section 3.6), require a background field. The introduction of the background field was the most valuable of a number of new concepts introduced by Bergthorsson and Doos (1955).

The background field was briefly discussed in Sections 1.5, 1.6, and 2.2. We presume that an approximate representation (either coefficients of a functional expansion or values on a regular analysis grid) of the dependent variable to be analyzed is available a priori. The background field can be obtained from climatology, a short (3–12 hour) forecast of the assimilating model from a previous objective analysis, or some optimum blend of the two. A climatological background is usually the average of a long sequence of observations of the state variable at a particular spatial location for a given season of the year. If the background field is represented by a functional expansion, a background estimate can be generated at any location in the analysis domain by evaluating the functions. If the background field is given on a dense, regular grid, background estimates at arbitrary locations can be obtained by forward interpolation using classical interpolation techniques (Appendix E) with minimal error.

In the time of Bergthorsson and Doos, the background field was chosen to be a blend of forecast and climatology. Forecast models were not very accurate, and climatology was given considerably more weight than it would be today. Climatology is still useful, however, in regions of low data density such as the Southern Ocean or the stratosphere, where forecasts are infrequently updated with new observations and may drift far from reality.

We now illustrate the first correction or iteration of the SCM procedure. The treatment is based on the work of Bergthorsson and Doos though the notation has been completely changed.

Consider the analysis of the dependent variable $f(\mathbf{r})$, where $\mathbf{r}$ defines the spatial location in a one-, two-, or three-dimensional domain. Define the ith analysis gridpoint $\mathbf{r}_i$ and kth observing station location $\mathbf{r}_k$, as in Chapter 2. Denote the observed value and background estimates at the observing station as $f_O(\mathbf{r}_k)$ and $f_B(\mathbf{r}_k)$, respectively. Now define $f_A(\mathbf{r}_i)$ and $f_B(\mathbf{r}_i)$ to be the analyzed and background values, respectively, at the analysis gridpoint. We assume that the background estimates are available everywhere in the domain, as noted earlier.

At the analysis gridpoint and observation station, the background estimates contain errors denoted $\varepsilon_B(\mathbf{r}_i)$ and $\varepsilon_B(\mathbf{r}_k)$. In the same way, the observation error is denoted $\varepsilon_O(\mathbf{r}_k)$. (The background and observation errors are discussed in a cursory fashion in this chapter; a more detailed treatment is postponed until Chapter 4.) Assume that the background error is homogeneous (Appendix D) and that consequently the expected background error variance $E_B^2 = \langle \varepsilon_B^2(\mathbf{r}) \rangle$ is independent of location. Assume that the observation error is spatially uncorrelated and that consequently $\langle \varepsilon_O(\mathbf{r}_k)\varepsilon_O(\mathbf{r}_l)\rangle = 0$ for two separate observation stations k and l. The expected observation error variance at station k is denoted $E_O^2(k) = \langle \varepsilon_O^2(\mathbf{r}_k)\rangle$ and is assumed to be a function of instrument type only. It is also assumed that the observation error is not correlated with the background error.

Consider first the case of a single observation at $\mathbf{r}_k$. Bergthorsson and Doos (1955) considered two estimates of f at the gridpoint $\mathbf{r}_i$:

$$f_B(\mathbf{r}_i) \qquad \text{and} \qquad f_B(\mathbf{r}_i) + [f_O(\mathbf{r}_k) - f_B(\mathbf{r}_k)]$$

The first estimate is the background value at the gridpoint, and the second assumes that the difference between the observation $f_O(\mathbf{r}_k)$ and the background estimate $f_B(\mathbf{r}_k)$ is constant along the line joining $\mathbf{r}_k$ and $\mathbf{r}_i$. Bergthorsson and Doos attempted to weight these estimates in an optimal way to produce the analyzed value $f_A(\mathbf{r}_i)$. The result was

$$f_A(\mathbf{r}_i) = \frac{E_B^{-2} f_B(\mathbf{r}_i) + E_O^{-2}(k) w(\mathbf{r}_k - \mathbf{r}_i)[f_B(\mathbf{r}_i) + f_O(\mathbf{r}_k) - f_B(\mathbf{r}_k)]}{E_B^{-2} + E_O^{-2}(k) w(\mathbf{r}_k - \mathbf{r}_i)} \tag{3.1.1}$$

where $w(\mathbf{r}_k - \mathbf{r}_i) = 1$ when $\mathbf{r}_k = \mathbf{r}_i$ and $w(\mathbf{r}_k - \mathbf{r}_i) \to 0$ as $|\mathbf{r}_k - \mathbf{r}_i| \to \infty$. Thus, when $|\mathbf{r}_k - \mathbf{r}_i|$ is large, $f_A(\mathbf{r}_i)$ becomes equal to the background value $f_B(\mathbf{r}_i)$. If $\mathbf{r}_k = \mathbf{r}_i$, then (3.1.1) becomes

$$f_A(\mathbf{r}_i) = \frac{E_B^{-2} f_B(\mathbf{r}_i) + E_O^{-2} f_O(\mathbf{r}_i)}{E_B^{-2} + E_O^{-2}} \tag{3.1.2}$$

where E_O^2 is the expected observation error variance at $\mathbf{r}_i$. Equation (3.1.2) is consistent with (2.2.7). A more useful form of (3.1.1) is

$$f_A(\mathbf{r}_i) - f_B(\mathbf{r}_i) = \frac{E_O^{-2}(k) w(\mathbf{r}_k - \mathbf{r}_i)}{E_B^{-2} + E_O^{-2}(k) w(\mathbf{r}_k - \mathbf{r}_i)} [f_O(\mathbf{r}_k) - f_B(\mathbf{r}_k)] \tag{3.1.3}$$

where $[f_O(\mathbf{r}_k) - f_B(\mathbf{r}_k)]$ is the observation increment and $[f_A(\mathbf{r}_i) - f_B(\mathbf{r}_i)]$ is the analysis increment defined in Section 1.6. The $w(\mathbf{r}_k - \mathbf{r}_i)$ is indicated with lowercase w, as in Chapter 2, because it is specified a priori. An a posteriori weight W_{ik} consistent with the definition in Section 2.5 can also be defined for (3.1.3):

$$f_A(\mathbf{r}_i) - f_B(\mathbf{r}_i) = W_{ik}[f_O(\mathbf{r}_k) - f_B(\mathbf{r}_k)] \tag{3.1.4}$$

where

$$W_{ik} = \frac{E_B^2 w(\mathbf{r}_k - \mathbf{r}_i)}{E_B^2 w(\mathbf{r}_k - \mathbf{r}_i) + E_O^2(k)}$$

Bergthorsson and Doos estimated the weights W_{ik} for the 500 mb geopotential field by statistical regression with respect to a series of subjectively analyzed synoptic charts. They found that $w(\mathbf{r}_k - \mathbf{r}_i)$ was essentially a function only of the relative displacement between $\mathbf{r}_k$ and $\mathbf{r}_i$; it was not a function of either their absolute positions or their relative orientations with respect to each other. In two dimensions,

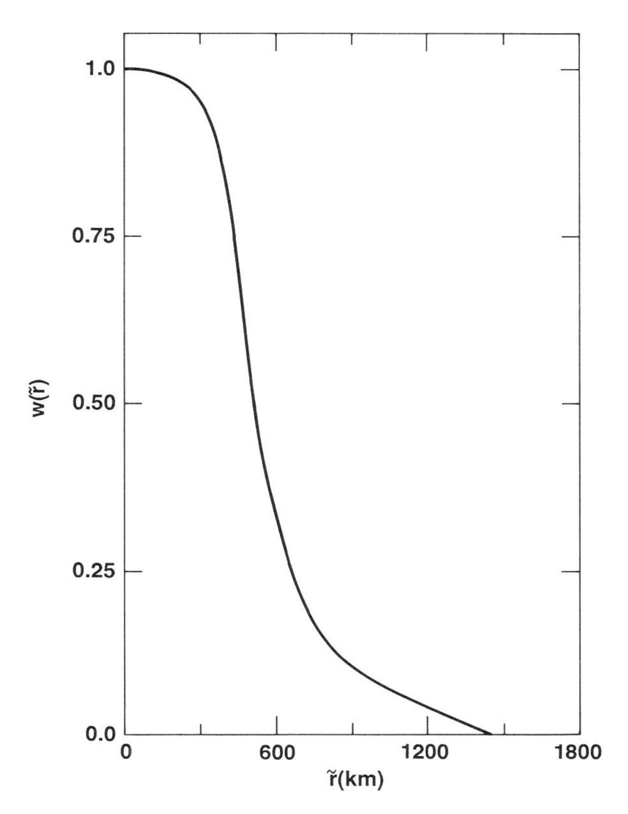

Figure 3.1 The weight function $w(\tilde{r})$ determined by Bergthorsson and Doos. (After Bergthorsson and Doos 1955)

$w(\mathbf{r}_k - \mathbf{r}_i) = w(\tilde{r})$, where

$$\tilde{r}^2 = (x_k - x_i)^2 + (y_k - y_i)^2 \tag{3.1.5}$$

The ($\tilde{}$) notation indicates that $\tilde{r}$ is the magnitude of the relative distance between observation station and analysis gridpoint. Because radiosonde observations were used in this study, $E_{\mathrm{O}}^2(k)$ was equal to a constant E_{O}^2 for all observations.

The function $w(\tilde{r})$ as a function of $\tilde{r}$ (in km) is plotted in Figure 3.1. Consistent with Figure 2.3, this curve shows that $w(\tilde{r})$ is a maximum at $\mathbf{r}_k = \mathbf{r}_i$ and falls off to near zero at about 1500 km. Figure 3.1 is consistent with intuition, which suggests that observations closer to the analysis gridpoint should be given more weight than those further away. Note, however, that $w(\tilde{r})$ in Figure 3.1 is always positive, unlike the curves of Figure 2.3. Bergthorsson and Doos found that for the 500 mb geopotential field, $E_{\mathrm{O}}^2/E_{\mathrm{B}}^2 = \frac{1}{9} \approx 0.11$; in other words, the observations were much more accurate than the background estimates.

The Bergthorsson/Doos scheme is a local scheme in the sense of Section 2.3, and only observations that lie within the radius of influence (Figure 2.1) of the gridpoint $\mathbf{r}_i$ are allowed to influence the objective analysis $f_{\mathrm{A}}(\mathbf{r}_i)$. Suppose K_i such observations are made with different types of instruments. Then, (3.1.3) becomes

$$f_A(\mathbf{r}_i) - f_B(\mathbf{r}_i) = \frac{\sum_{k=1}^{K_i} E_O^{-2}(k)w(\tilde{r})[f_O(\mathbf{r}_k) - f_B(\mathbf{r}_k)]}{\sum_{k=1}^{K_i} E_O^{-2}(k)w(\tilde{r}) + E_B^{-2}} \tag{3.1.6}$$

The special case in which all the observation stations are co-located with the analysis gridpoint is considered in Exercise 3.1.

If there is a single instrument type, then (3.1.6) can be written as

$$f_A(\mathbf{r}_i) - f_B(\mathbf{r}_i) = \frac{\sum_{k=1}^{K_i} w(\tilde{r})[f_O(\mathbf{r}_k) - f_B(\mathbf{r}_k)]}{\sum_{k=1}^{K_i} w(\tilde{r}) + \varepsilon_O^2} \tag{3.1.7}$$

where $\varepsilon_O^2 = E_O^2/E_B^2$ is the constant expected observation error variance normalized by the constant expected background error variance.

One variant of SCM (Cressman 1959) that has been widely used, drops the term E_B^{-2} from (3.1.6). In the case of a single observation at the analysis gridpoint, the Cressman algorithm produces $f_A(\mathbf{r}_i) = f_O(\mathbf{r}_i)$, rather than (3.1.2), which is appropriate only if the observations are perfect or there is no background estimate.

Although Bergthorsson and Doos (1955) derived the weight $w(\tilde{r})$ statistically from data, subsequent formulations of the SCM algorithm specified functional forms of the weights. For example, Cressman (1959) adopted the following form:

$$w(\tilde{r}) = \begin{cases} \dfrac{R^2 - \tilde{r}^2}{R^2 + \tilde{r}^2}, & \tilde{r} \le R \\ 0, & \tilde{r} > R \end{cases} \tag{3.1.8}$$

$w(\tilde{r})$ is equal to 1 at $\tilde{r} = 0$ and falls to zero at $\tilde{r} = R$.

Another form of $w(\tilde{r})$ was developed by Sasaki (1960) and Barnes (1964),

$$w(\tilde{r}) = \exp\left(\frac{-\tilde{r}^2}{2R^2}\right) \tag{3.1.9}$$

and is plotted in Figure 3.3(a), curve 2. In this case, $w(0) = 1$ and $w(\tilde{r}) > 0$ for all $\tilde{r}$. Choosing $w(\mathbf{r}) > 0$ guarantees that the denominator of (3.1.7) is finite, even if $\varepsilon_O^2 = 0$.

The single correction algorithm described in this section has four properties:

$$\sum_{k=1}^{K_i} W_{ik} \le 1 \tag{3.1.10}$$

$$W_{ik} \ge 0, \qquad 1 \le k \le K_i \tag{3.1.11}$$

$$|f_A(\mathbf{r}_i) - f_B(\mathbf{r}_i)| \le \text{maximum}|f_O(\mathbf{r}_k) - f_B(\mathbf{r}_k)| \tag{3.1.12}$$

$$W_{ik} \quad \text{is independent of the relative separation of observations} \tag{3.1.13}$$

Property (3.1.10) follows from (3.1.6), property (3.1.11) from (3.1.8–9), and property (3.1.12) from (3.1.10–11). Property (3.1.12) implies that the correction algorithm is not capable of any meaningful extrapolation. Property (3.1.13) follows because W_{ik} depends only on the distance of the observation from the analysis gridpoint. Consider two observations at an equal distance $\tilde{r}$ from the analysis gridpoint. Imagine two situations, one in which the observation stations are almost co-located and a second in which the two observation stations are on opposite sides of the analysis gridpoint. When the two observations are almost co-located, the second observation provides almost no new information about the state variable that is not already available from the first observation. When the two observations are far apart, the second observation provides important new information not available from the first observation. However, the single correction algorithm does not distinguish between these two situations and ignores the correlations between the observations. Thus, observations occurring in areas of high data density are given too much weight relative to observations in areas of low data density.

The four properties (3.1.10–13) of the single correction algorithm are very restrictive and do not apply in the powerful statistical interpolation algorithms of Chapters 4 and 5. As will be seen in the next section, these properties also do not apply when the correction algorithm is iterated.

3.2 The iteration cycle

The Bergthorsson/Doos procedure was really a forerunner of the method of successive corrections. The original procedure required only a single iteration or scan of the observations, and consequently there was only a single correction of the background field. Later work suggested that multiple iterations might lead to a better analysis. Assume that the background error is homogeneous and that all observations within the radius of influence of the analysis gridpoint $\mathbf{r}_i$ are made with the same type of instrument with spatially uncorrelated observation errors. Then, (3.1.7) can be written as

$$f_{\mathrm{A}}^{1}(\mathbf{r}_i) = f_{\mathrm{B}}(\mathbf{r}_i) + \underline{\mathrm{W}}_i^{\mathrm{T}}[\underline{\mathrm{f}}_{\mathrm{O}} - \underline{\mathrm{f}}_{\mathrm{B}}] \tag{3.2.1}$$

where $\underline{\mathrm{f}}_{\mathrm{O}}$ and $\underline{\mathrm{f}}_{\mathrm{B}}$ are column vectors of length K_i with elements $f_{\mathrm{O}}(\mathbf{r}_k)$ and $f_{\mathrm{B}}(\mathbf{r}_k)$ of observation and background estimates at observation sites within the radius of influence. The superscript 1 on $f_{\mathrm{A}}(\mathbf{r}_i)$ indicates that this is the first iterate. $\underline{\mathrm{W}}_i$ is a column vector of length K_i of a posteriori weights

$$W_{ik} = \frac{w(r_{ik})}{\sum_{k=1}^{K_i} w(r_{ik}) + \varepsilon_{\mathrm{O}}^2} \tag{3.2.2}$$

where $r_{ik}^2 = (x_k - x_i)^2 + (y_k - y_i)^2$ is the same as $\tilde{r}^2$ in (3.1.7). T stands for matrix transpose. The next iteration of the procedure is defined to be

$$f_{\mathrm{A}}^{2}(\mathbf{r}_i) = f_{\mathrm{A}}^{1}(\mathbf{r}_i) + \underline{\mathrm{W}}_i^{\mathrm{T}}[\underline{\mathrm{f}}_{\mathrm{O}} - \underline{\mathrm{f}}_{\mathrm{A}}^{1}] \tag{3.2.3}$$

Here $\underline{\mathrm{f}}_{\mathrm{A}}^{1}$ is the column vector of estimates at the observation stations r_k obtained by forward interpolation of the gridpoint values $f_{\mathrm{A}}^{1}(\mathbf{r}_i)$. The general iterative step is

$$f_{\mathrm{A}}^{j+1}(\mathbf{r}_i) = f_{\mathrm{A}}^{j}(\mathbf{r}_i) + \underline{\mathrm{W}}_i^{\mathrm{T}}[\underline{\mathrm{f}}_{\mathrm{O}} - \underline{\mathrm{f}}_{\mathrm{A}}^{j}] \tag{3.2.4}$$

for which (3.2.3) corresponds to $j=1$ and (3.2.1) corresponds to $j=0$ if $f_A^0(\mathbf{r}_i)=f_B(\mathbf{r}_i)$ and $\underline{f}_A^0=\underline{f}_B$.

Consider an important special case. Suppose there are K observations, and the analyzed value at each analysis gridpoint is a function of all K observations. Because each analyzed value is a function of the same set of observations, this is a global rather than a local analysis. Further suppose that the set of observation stations and the set of analysis gridpoints coincide; that is, the analysis gridpoints are $\mathbf{r}_k$, $1\le k\le K$. Then (3.2.4) becomes

$$f_A^{j+1}(\mathbf{r}_k)=f_A^j(\mathbf{r}_k)+\underline{W}_k^T[\underline{f}_O-\underline{f}_A^j] \tag{3.2.5}$$

In this case, no forward interpolation is required. If we consider the objective analysis at all observation stations, (3.2.5) becomes

$$\underline{f}_A^{j+1}=\underline{f}_A^j+\underline{\underline{W}}^T[\underline{f}_O-\underline{f}_A^j] \tag{3.2.6}$$

where $\underline{f}_A^j$ is the jth iterate of the objective analysis at the observation stations $\mathbf{r}_k$. $\underline{\underline{W}}^T$ is the $K\times K$ real, square matrix of a posteriori weights. The transpose notation is retained because the weight matrix is not, in general, symmetric.

The iteration process is illustrated in Figure 3.2. The domain is periodic, $-\pi\le x\le\pi$, and there are 20 equally spaced observation stations $x_k=-\pi+k\pi/10$. The observed values are indicated by the solid dots in (a). The SCM algorithm is applied to these observations using Cressman's weights (3.1.8) with $R=\pi/4$. This implies that there are five observations between $\pm R$. It is important to distinguish between the case that has no background field ($\varepsilon_O^2=0$) and the case that has a background field ($\varepsilon_O^2\ne 0$) but is equal to zero everywhere. In this illustration, $\varepsilon_O^2=0.25$, but the background field is zero everywhere. The analysis is performed at the observation stations, using the SCM algorithm (3.2.6). The results for iterations $j=0$, 1, and 2 are shown in (a). It is clearly evident that the analyzed values at the observation stations converge toward the observed values.

The a posteriori weights $\underline{\underline{W}}^T$ given by (3.2.2) are really only the a posteriori weights for a single iteration. The true a posteriori weights at the observation stations after $j+1$ corrections can be determined from (3.2.6):

$$\underline{f}_A^{j+1}-\underline{f}_O=[\underline{\underline{I}}-\underline{\underline{W}}^T][\underline{f}_A^j-\underline{f}_O]=[\underline{\underline{I}}-\underline{\underline{W}}^T]^{j+1}[\underline{f}_A^0-\underline{f}_O] \tag{3.2.7}$$

or

$$\underline{f}_A^{j+1}-\underline{f}_B=\underline{\underline{W}}^T(j+1)[\underline{f}_O-\underline{f}_B],\qquad\text{where}\qquad \underline{\underline{W}}^T(j)=\underline{\underline{I}}-[\underline{\underline{I}}-\underline{\underline{W}}^T]^j \tag{3.2.8}$$

and $\underline{f}_A^0=\underline{f}_B$. Here $\underline{\underline{W}}^T(j)$ is the a posteriori weight matrix for the jth correction. These weights are plotted in Figure 3.2(b) for the same case as in (a), for $j=1$, 2, and 5 at the observation station $x_k=0$.

When $j=1$, the weights have the form given by (3.1.8) and (3.2.2). As the iteration number increases, the a posteriori weight $W_{kk}(j)$ increases, while $W_{lk}(j)$, $l\ne k$, may become negative like the weights in Figure 2.3. Because the a posteriori weights may become negative, property (3.1.11) is true only for the first iteration. Note also that

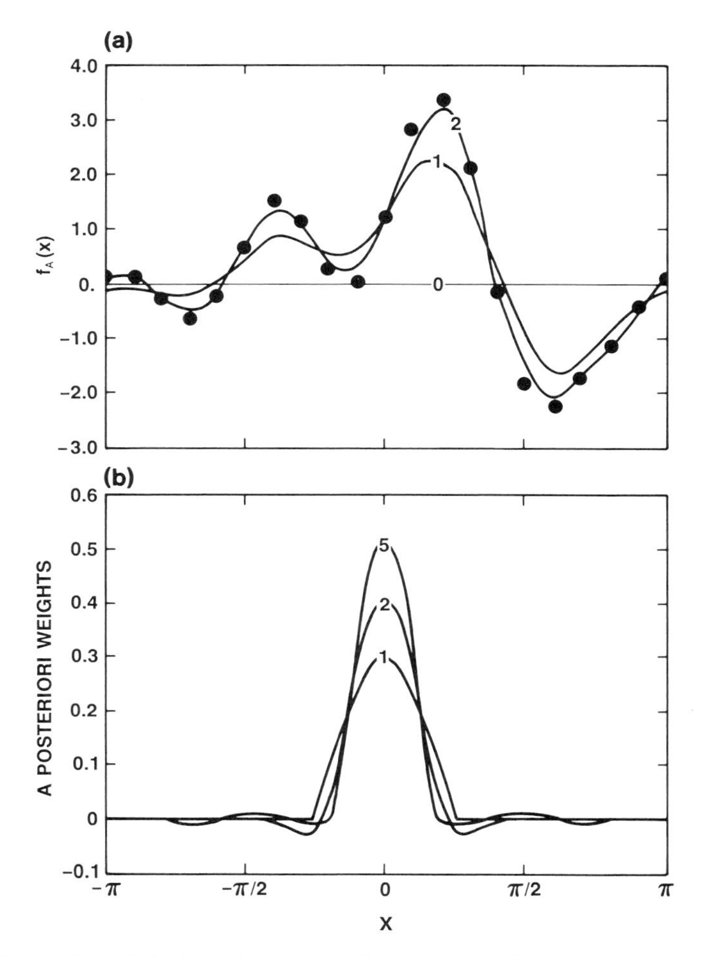

Figure 3.2 Illustration of the iteration process for the SCM algorithm for 20 equally spaced observations on a one-dimensional periodic domain: (a) shows the analyzed values as a function of iteration number; (b) shows the a posteriori weights.

for $j > 1$, the weights $\underline{\underline{W}}^{\mathrm{T}}(j)$ can become nonzero for $|x_k - x_l| > R$. For this regular, periodic network $\underline{\underline{W}}^{\mathrm{T}}$ happens to be symmetric, but this is not generally so.

Exercise 3.2 demonstrates that property (3.1.13) also does not apply when $j > 1$. In fact, none of the properties (3.1.10–13) necessarily hold when $j > 1$.

The convergence of the iteration cycle (3.2.6) is discussed in Section 3.5. Before proceeding, however, we need to discuss the spectral properties of the weight function $w(r)$.

3.3 Spectral response: infinite continuous networks

The primary goal of the SCM (or any other objective analysis) algorithm is to maximize the signal extracted while suppressing as much observational noise as possible. Atmospheric spectra (typified by Figure 1.4) generally show decreasing variance with increasing wavenumber; that is, the atmospheric signal contains a preponderance of its variance at the larger scales (see also Appendix H). On the other hand, the noise associated with observational error is predominantly small scale though this is not always the case; sometimes observation error is highly spatially correlated and contains considerable large-scale variance (see Section 4.8).

When the observational error is not spatially correlated, an objective analysis algorithm should attempt to extract the large-scale signal while suppressing the small-scale observational noise. Thus, the objective analysis algorithm acts as a *filter* as well as an *interpolator*.

The filtering properties of any objective analysis algorithm can be demonstrated by examining the spectral response of the algorithm. The spectral response of the SCM algorithm is particularly easy to obtain, and the methodology developed here will be used later for the univariate and multivariate statistical interpolation algorithms of Chapters 4 and 5.

We first examine the spectral response of the algorithm for a rather special observation network. All real observation networks cover finite domains, and the observations are discrete. An important limiting case is an infinitely dense observation network on an infinite domain. This network is not very realistic; but it provides a good introduction to the spectral response of objective analysis algorithms on the more realistic networks to be found in the following section. The single correction algorithm on one-dimensional and two-dimensional domains is considered in the following discussion. The spectral response of the SCM algorithm itself will be discussed in Section 3.5.

Consider first the domain $-\infty \leq x \leq \infty$ and continuous observations $f_{\mathrm{O}}(x)$ of the signal $f(x)$. Define $f_{\mathrm{A}}(x)$ and $f_{\mathrm{B}}(x)$ as the continuous analyzed and background fields. A continuous analogue of (3.1.7) is given by

$$f_{\mathrm{A}}(x) = f_{\mathrm{B}}(x) + \frac{1}{2\pi L}\int_{-\infty}^{\infty} W(x'-x)[f_{\mathrm{O}}(x') - f_{\mathrm{B}}(x')]\, dx' \tag{3.3.1}$$

where x is the analysis point and is analogous to x_i in (3.1.7), and x' is the observation location and is analogous to x_k in (3.1.7). L is a constant with dimension of length, which will be specified subsequently. $W(x'-x)$ is the a posteriori weight and consistent with (3.1.7) can be written as

$$W(x'-x) = \frac{w(x'-x)}{\dfrac{1}{2\pi L}\int_{-\infty}^{\infty} w(x'-x)\, dx' + \varepsilon_{\mathrm{O}}^2}$$

where $w(x'-x)$ is the specified weight function. The inclusion of L in (3.3.1) ensures that W and w are both nondimensional, as in the discrete case (3.1.4). Defining

$\tilde{x} = x' - x$ gives an alternate form of (3.3.1):

$$f_A(x) = f_B(x) + \frac{1}{2\pi L}\int_{-\infty}^{\infty} W(\tilde{x})[f_O(\tilde{x} + x) - f_B(\tilde{x} + x)]\, d\tilde{x} \tag{3.3.2}$$

with

$$W(\tilde{x}) = \frac{w(\tilde{x})}{\frac{1}{2\pi L}\int_{-\infty}^{\infty} w(\tilde{x})\, d\tilde{x} + \varepsilon_O^2}$$

A simple way to examine the spectral response of (3.3.2) is through Fourier transforms (Sneddon 1951). Define the following Fourier transform pair:

$$f(x) = \int_{-\infty}^{\infty} \hat{f}(m)e^{imx}\, dm, \qquad \hat{f}(m) = \frac{1}{2\pi}\int_{-\infty}^{\infty} f(x)e^{-imx}\, dx \tag{3.3.3}$$

The two Fourier transforms (3.3.3) are the forms of (2.4.2) and (2.4.5), respectively, appropriate for an infinite domain. The spectral amplitude $\hat{f}(m)$ is now a continuous function of wavenumber.

Fourier transform pairs can be defined for the observed, background, and analyzed values of f: $f_O(x) \leftrightarrow \hat{f}_O(m)$, $f_B(x) \leftrightarrow \hat{f}_B(m)$, and $f_A(x) \leftrightarrow \hat{f}_A(m)$. Inserting (3.3.3) into (3.3.2) gives

$$L\int_{-\infty}^{\infty} [\hat{f}_A(m) - \hat{f}_B(m)]e^{imx}\, dm = \frac{1}{2\pi}\int_{-\infty}^{\infty} W(\tilde{x})\int_{-\infty}^{\infty} [\hat{f}_O(m) - \hat{f}_B(m)]e^{im(\tilde{x}+x)}\, dm\, d\tilde{x}$$

or

$$L\int_{-\infty}^{\infty} e^{imx}\{\hat{f}_A(m) - \hat{f}_B(m) - \alpha(m)[\hat{f}_O(m) - \hat{f}_B(m)]\}\, dm = 0 \tag{3.3.4}$$

where

$$\alpha(m) = \frac{1}{2\pi L}\int_{-\infty}^{\infty} W(\tilde{x})e^{im\tilde{x}}\, d\tilde{x} \tag{3.3.5}$$

Equation (3.3.4) is in the same form as the first equation of (3.3.3). Because (3.3.4) equals zero for all values of x, the term in curly brackets must equal zero for all values of m. Thus,

$$\hat{f}_A(m) - \hat{f}_B(m) = \alpha(m)[\hat{f}_O(m) - \hat{f}_B(m)] \tag{3.3.6}$$

or

$$\hat{f}_A(m) = [1 - \alpha(m)]\hat{f}_B(m) + \alpha(m)\hat{f}_O(m) \tag{3.3.7}$$

Equation (3.3.6) relates the spectrum of the analysis increment to the spectrum of the observation increment, and $\alpha(m)$ is defined as the spectral response of the algorithm. From (3.3.7), it can be seen that for wavenumber m, $\alpha(m) = 1$ implies that the algorithm draws exactly for the observations, whereas if $\alpha(m) = 0$, the analysis reverts to the background.

For the sake of simplicity, drop the ($\tilde{}$) notation in (3.3.5). Thus, x is now defined as the distance between observation point and analysis point. The specified weights (3.1.8–9) were functions only of the absolute distance between analysis point and observation point. Therefore, assume that $w(-x) = w(x)$ and consequently that $W(-x) = W(x)$. Then,

$$\alpha(m) = \frac{1}{2\pi L} \int_{-\infty}^{\infty} W(x) e^{imx}\, dx = \frac{1}{\pi L} \int_{0}^{\infty} W(x) \cos(mx)\, dx \qquad (3.3.8)$$

and $\alpha(-m) = \alpha(m)$.

In the same way, a Fourier transform pair can be defined for $w(x)$:

$$w(x) = 2L \int_{0}^{\infty} g(m) \cos(mx)\, dm, \qquad g(m) = \frac{1}{\pi L} \int_{0}^{\infty} w(x) \cos(mx)\, dx \qquad (3.3.9)$$

where $g(m)$ is the *transfer function*. From (3.3.2), (3.3.8), and (3.3.9), it is evident that

$$\alpha(m) = \frac{g(m)}{g(0) + \varepsilon_O^2} \qquad (3.3.10)$$

Thus, both $\alpha(m)$ and $g(m)$ are nondimensional.

We determine the transfer function for three specified weight functions $w(x)$, where x is now the distance between analysis point and observation station. First,

$$\begin{aligned} w(x) &= 1, \qquad x \le L \\ w(x) &= 0, \qquad x > L \end{aligned} \qquad (3.3.11)$$

where L is the length scale defined in (3.3.1). L can be considered as a characteristic scale of the function $w(x)$.

The weight function (3.3.11) corresponds to a simple unweighted mean of all the observations within a distance L of the analysis point. It is plotted as curve 1 of Figure 3.3(a) for $L = 1$ (in arbitrary length units). From (3.3.9), $g(0) = 1/\pi$ and

$$\frac{g(m)}{g(0)} = \frac{1}{L} \int_{0}^{L} \cos(mx)\, dx = \frac{\sin(mL)}{mL} \qquad (3.3.12)$$

The function (3.3.12) is often referred to as the diffraction function because of its role in optics (Blackman and Tukey 1958). It is also called the cardinal or composing function and is plotted as curve 1 of Figure 3.3(b). By definition, $g(m)/g(0) = 1$ at $m = 0$ and decreases in amplitude for increasing m, crossing zero every time mL equals some multiple of π.

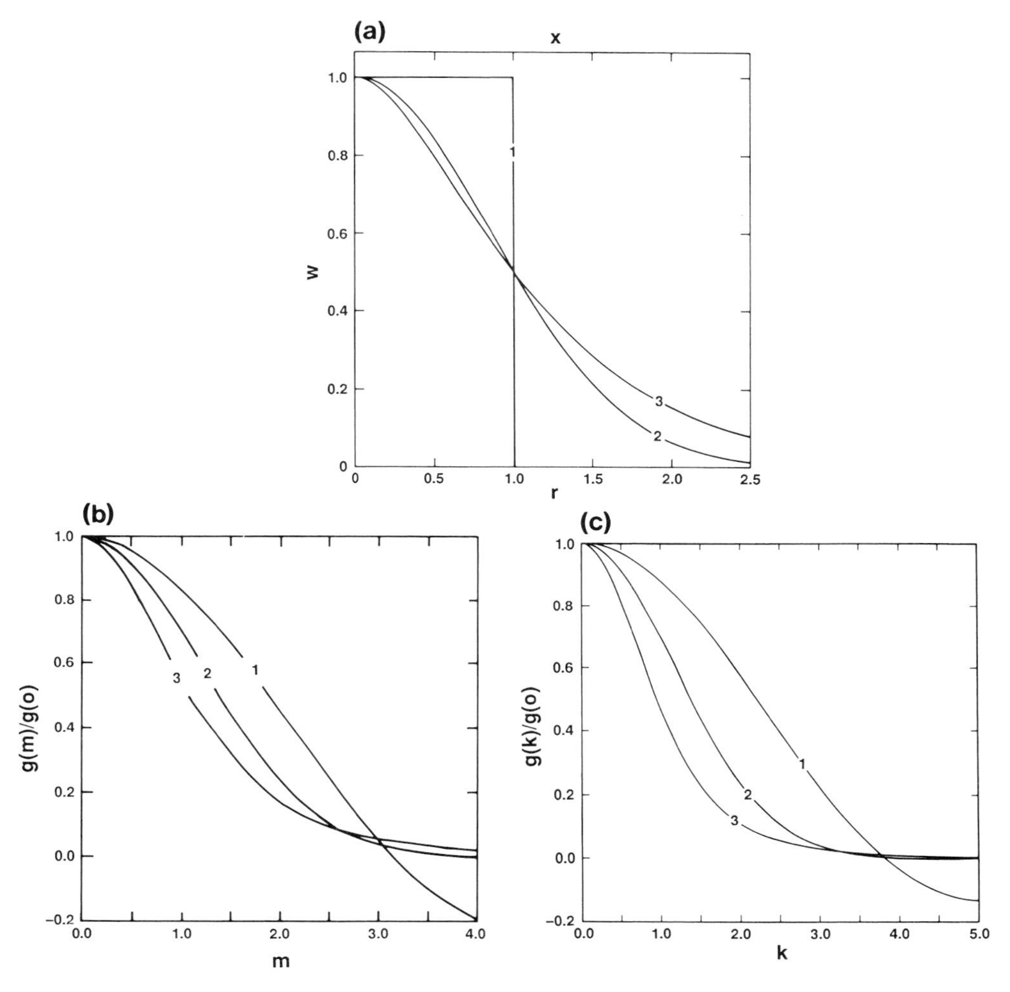

Figure 3.3 (a) Weight functons and (b,c) transfer functions. (b) is the one-dimensional case; (c) is the axisymmetric case.

Now consider the weight function given by (3.1.9),

$$w(x) = \exp\left(-\frac{x^2}{2L^2}\right) \tag{3.3.13}$$

which is plotted as curve 2 of Figure 3.3(a), with L chosen so that $w(1) = 0.5$. Equation (3.3.13) can be transformed using (3.3.9). The transformation involves well-known exponential integrals that can be found in integral tables (Abramovitz and Stegun 1965). In this case $g(0) = 1/\sqrt{2\pi}$ and

$$\frac{g(m)}{g(0)} = \exp\left(-\frac{m^2L^2}{2}\right) \tag{3.3.14}$$

Equation (3.3.14) is plotted (for the same value of L as Figure 3.3(a), curve 2) as curve 2 of Figure 3.3(b). Note that $g(m) > 0$ for all finite m.

A third weight function is given by

$$w(x) = \left(1 + \frac{|x|}{L}\right) \exp\left(-\frac{|x|}{L}\right) \tag{3.3.15}$$

which is plotted as curve 3 on Figure 3.3(a). L has been chosen to equal approximately 1/1.678 so that $w(1) = 0.5$. From (3.3.15) and (3.3.9), $g(0) = 2/\pi$ and

$$\frac{g(m)}{g(0)} = (1 + m^2L^2)^{-2} \tag{3.3.16}$$

which is plotted as curve 3 in Figure 3.3(b) for $L = 1/1.678$. Note that $g(m) > 0$ for all finite m in this case as well.

The effect of the spectral response function $\alpha(m)$ on the analyzed spectrum can be considered using (3.3.6–7) together with (3.3.12), (3.3.14), and (3.3.16). In all cases, $\alpha(m) \leq 1$ and equals 1 only if the observations are perfect ($\varepsilon_{\mathrm{O}}^2 = 0$) and $m = 0$. From (3.3.6), the observation increments are always filtered by the single correction algorithm for these three weight functions. In general, the spectral response $\alpha(m)$ decreases for increasing wavenumber, monotonically in the case of (3.3.14) and (3.3.16). Thus, from (3.3.6), the single correction algorithm filters in a scale-selective manner with the small-scale observation increments being the most heavily filtered. This is a low-pass filter. For the transfer functions (3.3.12), (3.3.14), and (3.3.16), Equation (3.3.7) suggests that the analyzed large scales will be taken from the observations whereas the analyzed small scales will be taken from the background field. As L is increased, the weight functions (3.3.11), (3.3.13), and (3.3.15) become broader, and the transfer functions (3.3.12), (3.3.14), and (3.3.16) become narrower. Note that the weight function (3.3.11) actually causes 180 degree phase shifts for some values of m, as indicated by negative values in (3.3.12).

The spectral responses, thus obtained, can be generalized to two dimensions. The two-dimensional analogue of (3.3.2) is

$$f_{\mathrm{A}}(x, y) = f_{\mathrm{B}}(x, y) + \frac{1}{2\pi L^2} \int_{-\infty}^{\infty} \int_{-\infty}^{\infty} W(\tilde{x}, \tilde{y})[f_{\mathrm{O}}(x + \tilde{x}, y + \tilde{y}) - f_{\mathrm{B}}(x + \tilde{x}, y + \tilde{y})] \, d\tilde{x} \, d\tilde{y} \tag{3.3.17}$$

where (x, y) defines the location of the analysis point, and $(\tilde{x}, \tilde{y})$ is the displacement between observation point and analysis point. L is a constant with dimensions of length and has been included for reasons of dimensional consistency, as in (3.3.1). The a posteriori weight $W(\tilde{x}, \tilde{y})$ is given by

$$W(\tilde{x}, \tilde{y}) = \frac{w(\tilde{x}, \tilde{y})}{\frac{1}{2\pi L^2} \int_{-\infty}^{\infty} \int_{-\infty}^{\infty} w(\tilde{x}, \tilde{y}) \, d\tilde{x} \, d\tilde{y} + \varepsilon_{\mathrm{O}}^2} \tag{3.3.18}$$

There is a two-dimensional analogue to the Fourier transform pair (3.3.3). Thus, define the double Fourier transform

$$f(x, y) = \int_{-\infty}^{\infty} \int_{-\infty}^{\infty} \hat{f}(m, n) e^{i(mx+ny)}\, dx\, dy \tag{3.3.19}$$

Inserting (3.3.19) into (3.3.18) and following the procedure (3.3.4) gives

$$\hat{f}_{\mathrm{A}}(m, n) = [1 - \alpha(m, n)]\hat{f}_{\mathrm{B}}(m, n) + \alpha(m, n)\hat{f}_{\mathrm{O}}(m, n) \tag{3.3.20}$$

where

$$\alpha(m, n) = \frac{1}{2\pi L} \int_{-\infty}^{\infty} \int_{-\infty}^{\infty} W(\tilde{x}, \tilde{y}) e^{i(m\tilde{x}+n\tilde{y})}\, d\tilde{x}\, d\tilde{y} \tag{3.3.21}$$

The specified weights (3.1.8–9) were functions only of the absolute distance between observation point and analysis point. Thus, it is advantageous to examine (3.3.21) in plane polar coordinates (r, ϕ). Define $\tilde{x} = r \cos \phi$ and $\tilde{y} = r \sin \phi$. As in (3.1.5), $r^2 = \tilde{x}^2 + \tilde{y}^2$, but the ($\tilde{}$) notation has been dropped for r and ϕ for the sake of simplicity. Following Equations (G17–G20) of Appendix G, we can show that

$$\alpha(m, n) = \sum_{p=-\infty}^{\infty} \alpha(k, p) e^{ip\lambda}$$

and

$$\alpha(k, p) = \frac{1}{2\pi L^2} \int_{-\pi}^{\pi} \int_{0}^{\infty} W(r) J_p(kr) e^{-ip\phi} r\, dr\, d\phi \tag{3.3.22}$$

Here $k^2 = m^2 + n^2$ and $\lambda = -\tan^{-1}(m/n)$. $J_p(kr)$ is a Bessel function of the first kind of integer order p, and $\alpha(k, p)$ is given by a Fourier–Bessel integral. The Bessel functions $J_0(r)$ and $J_1(r)$ are illustrated in Figure G1, and a number of relevant properties of the Bessel functions are given in Appendix G. It can be seen from (3.3.22) that the relationship between $\alpha(m, n)$ and $\alpha(k, p)$ is the same as that for the Fourier expansion (2.4.2).

Now, $W(r)$ is not a function of ϕ, so (3.3.22) simplifies to

$$\alpha(m, n) = \alpha(k, 0) = \alpha(k) \qquad \text{and} \qquad \alpha(k, p) = 0,\ p \neq 0$$

where

$$\alpha(k) = \frac{1}{L^2} \int_{0}^{\infty} W(r) J_0(kr) r\, dr \tag{3.3.23}$$

The expression for $\alpha(k)$ is called a Hankel transform and is discussed in Appendix G. Equation (3.3.23) is the two-dimensional equivalent of (3.3.8) for an axisymmetric weight function. Exactly analogous expressions to (3.3.9–10) exist for the two-dimensional case:

$$\alpha(k)=\frac{g(k)}{g(0)+\varepsilon_0^2},\qquad \text{where}\qquad g(k)=\frac{1}{L^2}\int_0^\infty w(r)J_0(kr)r\,dr \tag{3.3.24}$$

The Hankel transform pairs corresponding to (3.3.11–16) are

$$\begin{aligned} w(r)&=1, \quad r\le L,\\ w(r)&=0, \quad r>L, \end{aligned}\qquad \frac{g(k)}{g(0)}=\frac{2J_1(kL)}{kL} \tag{3.3.25}$$

$$w(r)=\exp\left(-\frac{r^2}{2L^2}\right),\qquad \frac{g(k)}{g(0)}=\exp\left(-\frac{k^2L^2}{2}\right) \tag{3.3.26}$$

$$w(r)=\left(1+\frac{r}{L}\right)\exp\left(-\frac{r}{L}\right),\qquad \frac{g(k)}{g(0)}=(1+k^2L^2)^{-5/2} \tag{3.3.27}$$

In deriving (3.3.25), we made use of the identity (G8), followed by an integration by parts. Result (3.3.26) can be derived by expanding $J_0(kr)$ in (3.3.24) in the series (G2) and integrating term by term. Equation (3.3.27) can be derived the same way or by using the identity (G5) and reversing the order of integration.

The $w(r)$ for (3.3.25–27) is plotted in Figure 3.3(a), and the transfer functions $g(k)/g(0)$ are plotted in Figure 3.3(c) in the same format as (b). Equation (3.3.25) is given by curve 1, (3.3.26) by curve 2, and (3.3.27) by curve 3. The values of L chosen for each curve were the same as for the corresponding curves in (b). The curves are similar but not identical to their counterparts in (b). Note that the transfer functions for (3.3.26–27) are always positive, but the $g(k)$ for (3.3.25) can be negative.

3.4 Spectral response: finite discrete networks

The infinite, continuous observation network discussed in the previous section bears little resemblance to the present, highly irregular time-dependent global network discussed in Section 1.3. It is conceivable that the spectral response for a realistic network might differ substantially from that of Section 3.3. Spectral responses can be obtained for realistic networks, though it may be necessary to resort to numerical procedures in some cases.

In a uniform observation network, the discrete observation station locations can be calculated using a simple algebraic formula. For this type of network, the spectral responses can usually be obtained analytically. As a simple example, consider the following uniform infinite, discrete one-dimensional observation network. Suppose x denotes the analysis point and the lth observation station is located at x_l. Define $\tilde{x}_l=x_l-x$ to be the displacement between observation station and analysis point. Then, define the uniform observation network as

$$\tilde{x}_l=\Delta x(l+\tfrac{1}{2}),\qquad l\ \text{integer},\qquad -\infty\le l\le\infty \tag{3.4.1}$$

where Δx is the (constant) distance between adjacent stations. Note that no observation station exists at the analysis point.

Assume there is a single type of observation instrument with spatially uncorrelated observation errors. Then, (3.1.7) can be written in the notation of (3.3.2) as

$$f_A(x) = f_B(x) + \sum_{l=-\infty}^{\infty} W(\tilde{x}_l)[f_O(x+\tilde{x}_l) - f_B(x+\tilde{x}_l)] \tag{3.4.2}$$

where

$$W(\tilde{x}_l) = \frac{w(\tilde{x}_l)}{\sum_{l=-\infty}^{\infty} w(\tilde{x}_l) + \varepsilon_O^2}$$

Define Fourier transform pairs for f_A, f_B, and f_O similar to (3.3.3). Then, a discrete analogue to (3.3.4) is

$$L\int_{-\infty}^{\infty} e^{imx}\{\hat{f}_A(m) - \hat{f}_B(m) - \alpha_d(m)[\hat{f}_O(m) - \hat{f}_B(m)]\}\, dm = 0 \tag{3.4.3}$$

where $\alpha_d(m) = \sum_{l=-\infty}^{\infty} W(\tilde{x}_l)e^{im\tilde{x}_l}$. The $\alpha_d(m)$ is the response of the single correction algorithm to the network (3.4.1) and can be compared with (3.3.5). The subscript d on α indicates that this is the spectral response for a discrete network. Assume $W(-x) = W(x)$ as before and then, because of the symmetry of the network about the analysis point,

$$\alpha_d(m) = 2\sum_{l=0}^{\infty} W[\Delta x(l+\tfrac{1}{2})] \cos[m(l+\tfrac{1}{2})\,\Delta x] \tag{3.4.4}$$

Discrete analogues of (3.3.9–10) are given by

$$g_d(m) = 2\sum_{l=0}^{\infty} w[\Delta x(l+\tfrac{1}{2})] \cos[m(l+\tfrac{1}{2})\,\Delta x] \tag{3.4.5}$$

and

$$\alpha_d(m) = \frac{g_d(m)}{g_d(0) + \varepsilon_O^2} \tag{3.4.6}$$

Now,

$$\cos\left[\left(m + \frac{2\pi j}{\Delta x}\right)\Delta x(l+\tfrac{1}{2})\right] = (-1)^j \cos[m\,\Delta x(l+\tfrac{1}{2})]$$

for any positive or negative integer j. Thus,

$$g_d\left(m + \frac{2\pi j}{\Delta x}\right) = (-1)^j g_d(m) \quad \text{and} \quad \frac{g_d(2\pi j/\Delta x)}{g_d(0)} = (-1)^j \tag{3.4.7}$$

Consider a specified weight function $w(x)$ with Fourier transform given by (3.3.9).

Take (3.3.15–16) as an example. Equation (3.4.7) indicates that the spectral response $g_\mathrm{d}(m)/g_\mathrm{d}(0)$ for the same weight function will be very different from (3.3.16):

$$\frac{g_\mathrm{d}(2\pi j/\Delta x)}{g_\mathrm{d}(0)} = 1$$

for all even values of j *regardless of the choice of* $w(x)$. Unlike (3.3.16), which tends to zero as $m \to \pm\infty$, $g_\mathrm{d}(m)/g_\mathrm{d}(0)$ has a perfect response for an infinite number of wavenumbers m. This is a manifestation of the aliasing phenomenon discussed in more detail in Appendix H. Basically, on the network (3.4.1), wavenumber m is indistinguishable from wavenumber $(m + 4\pi j/\Delta x)$. Equation (3.4.7) also demonstrates that $g_\mathrm{d}(m)/g_\mathrm{d}(0)$ is frequently negative, even though (3.3.16) is always positive. By defining a characteristic length scale L as in (3.3.1) and comparing (3.3.9) and (3.4.5), it is evident that

$$\lim_{\Delta x \to 0} \frac{\Delta x \, g_\mathrm{d}(m)}{2\pi L} = g(m) \qquad \text{and} \qquad \lim_{\Delta x \to 0} \frac{g_\mathrm{d}(m)}{g_\mathrm{d}(0)} = \frac{g(m)}{g(0)}$$

Now consider a two-dimensional example. Equation (3.1.7) can be written in the notation of (3.3.17) as

$$f_\mathrm{A}(x, y) = f_\mathrm{B}(x, y) + \sum_{l=1}^{K} W(\tilde{x}_l, \tilde{y}_l)[f_\mathrm{O}(x + \tilde{x}_l, y + \tilde{y}_l) - f_\mathrm{B}(x + \tilde{x}_l, y + \tilde{y}_l)]$$

where

$$W(\tilde{x}_l, \tilde{y}_l) = \frac{w(\tilde{x}_l, \tilde{y}_l)}{\sum_{l=1}^{K} w(\tilde{x}_l, \tilde{y}_l) + \varepsilon_\mathrm{O}^2} \tag{3.4.8}$$

Equation (3.4.8) is a general expression for the discrete two-dimensional single correction algorithm with a finite number of observation stations $(\tilde{x}_l, \tilde{y}_l)$, $1 \le l \le K$. Here $(\tilde{x}_l, \tilde{y}_l)$ indicates the displacement between analysis point (x, y) and the lth observation station. Using the double Fourier transform (3.3.19), we obtain for (3.4.8),

$$\hat{f}_\mathrm{A}(m, n) = [1 - \alpha_\mathrm{d}(m, n)]\hat{f}_\mathrm{B}(m, n) + \alpha_\mathrm{d}(m, n)\hat{f}_\mathrm{O}(m, n) \tag{3.4.9}$$

where

$$\alpha_\mathrm{d}(m, n) = \sum_{l=1}^{K} W(\tilde{x}_l, \tilde{y}_l) e^{i(m\tilde{x}_l + n\tilde{y}_l)}$$

Introduce plane polar coordinates (r_l, ϕ_l) where $\tilde{x}_l = r_l \cos \phi_l$ and $\tilde{y}_l = r_l \sin \phi_l$ and the ($\tilde{}$) notation has been dropped from r and ϕ. Following Section 3.3 and Equations (G15–G19) of Appendix G, we can show that

$$\alpha_\mathrm{d}(m, n) = \sum_{p=-\infty}^{\infty} \alpha_\mathrm{d}(k, p) e^{ip\lambda}$$

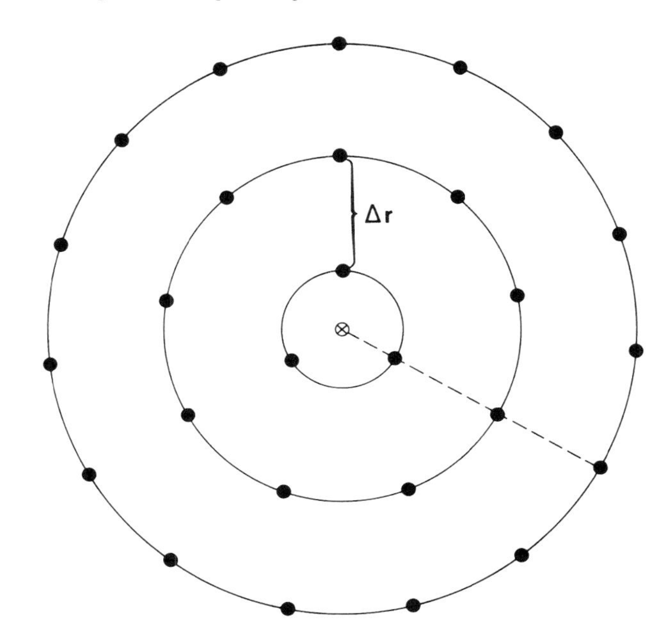

Figure 3.4 Regular observation network (solid dots) surrounding analysis gridpoint (circled cross).

where

$$\alpha_d(k, p) = \sum_{l=1}^{K} W(r_l, \phi_l) J_p(kr_l) e^{-ip\phi_l} \tag{3.4.10}$$

with $k^2 = m^2 + n^2$ and $\lambda = -\tan^{-1}(m/n)$. Equation (3.4.10) is the discrete analogue of (3.3.22). Now suppose W is a function only of

$$r_l = \sqrt{\tilde{x}_l^2 + \tilde{y}_l^2}$$

Then

$$\alpha_d(k, p) = \sum_{l=1}^{K} W(r_l) J_p(kr_l) e^{-ip\phi_l} \tag{3.4.11}$$

By analogy with (3.4.5–6),

$$\begin{aligned} g_d(k, p) &= \sum_{l=1}^{K} w(r_l) J_p(kr_l) e^{-ip\phi_l} \\ \alpha_d(k, p) &= \frac{g_d(k, p)}{g_d(0,0) + \varepsilon_0^2} \end{aligned} \tag{3.4.12}$$

Consider now the regular discrete network shown in Figure 3.4. The observation locations are shown by solid dots, and the analysis point by the circled cross. The dashed line indicates $\phi = 0$. Introduce the two-dimensional observation station location index, $l = (q, s)$, defined by

$$r_l = \frac{2q-1}{2}\Delta r, \qquad 1 \le q \le Q$$
$$\phi_l = \frac{2\pi s}{c(2q-1)}, \qquad 1 \le s \le c(2q-1) \tag{3.4.13}$$

In Figure 3.4, Δr is indicated and $c = 3$. The network in (3.4.13) is discrete, regular, and approximately equally spaced. On this network, (3.4.12) can be written as

$$g_{\mathrm{d}}(k, p) = \sum_{q=1}^{Q} \sum_{s=1}^{c(2q-1)} w\left[\frac{(2q-1)\,\Delta r}{2}\right] J_p\left[\frac{k(2q-1)\Delta r}{2}\right] \exp\left[-\frac{2\pi ips}{c(2q-1)}\right] \tag{3.4.14}$$

But

$$\sum_{s=1}^{c(2q-1)} \exp\left[-\frac{2\pi ips}{c(2q-1)}\right] = c(2q-1) \quad \text{if } p = jc(2q-1)$$
$$= 0 \quad \text{otherwise} \tag{3.4.15}$$

Here j is zero or any positive or negative integer. In Figure 3.4, where $c = 3$, $g_{\mathrm{d}}(k, p) = 0$ for any value of p that is not a multiple of 3.

$$g_{\mathrm{d}}(k, 0) = 3 \sum_{q=1}^{Q} (2q-1) w\left[\left(\frac{2q-1}{2}\right)\Delta r\right] J_0\left[k\left(\frac{2q-1}{2}\right)\Delta r\right] \tag{3.4.16}$$

and

$$g_{\mathrm{d}}(k, 3) = 3w\left(\frac{\Delta r}{2}\right) J_3\left(\frac{k\,\Delta r}{2}\right) \tag{3.4.17}$$

The $g_{\mathrm{d}}(0, 0)$ can be determined from (3.4.16); $g_{\mathrm{d}}(k, 6)$ involves contributions only from the innermost ring ($q = 1$) of Figure 3.4, and $g_{\mathrm{d}}(k, 9)$ involves contributions from the two innermost rings. Thus, unlike (3.3.23), the $\alpha_{\mathrm{d}}(k, p)$ are not all zero for $p \ne 0$. This is another example of aliasing (Appendix H). Define a response:

$$R_p(k, \Delta r, Q) = \frac{g_{\mathrm{d}}(k, p)}{g_{\mathrm{d}}(0, 0)} \tag{3.4.18}$$

The response (3.4.18) is plotted as a function of wavenumber k for the weight function (3.3.27) for a number of values of p, Δr, and Q. The value of L has been chosen, as before, to be 1/1.678.

Consider first the case $Q = \infty$, corresponding to an infinite discrete observation network. In Figure 3.5, $R_0(k, \Delta r, \infty)$ is plotted for several values of Δr. Curve a shows the asymptotic limit, $\Delta r \to 0$ and is identical to curve c of Figure 3.3(c). In curve b, $\Delta r = 1$; and in curve c, $\Delta r = 2$ in the same arbitrary length units as L. As might have been expected from the discrete one-dimensional network discussed earlier in this section, as Δr is increased, R_0 changes dramatically. Even though $g(k)/g(0)$ is never

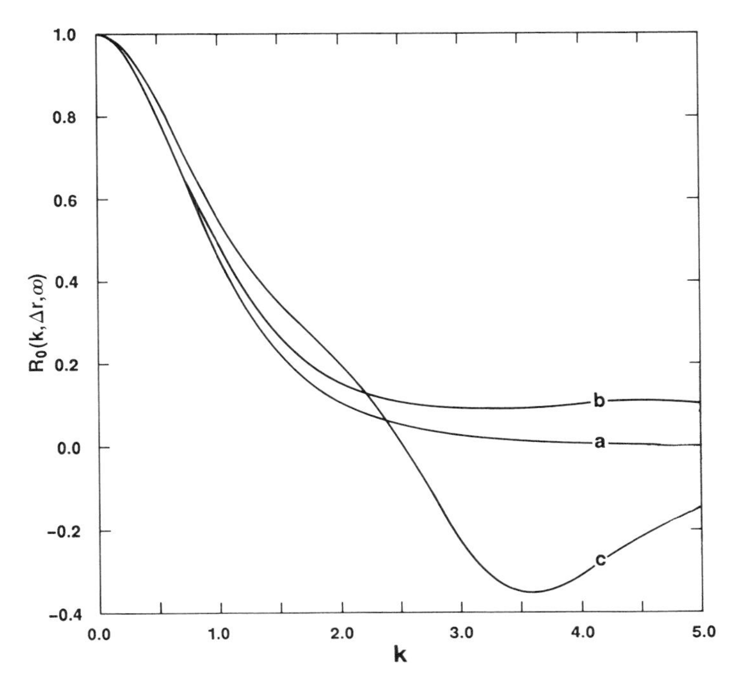

Figure 3.5 The axisymmetric spectral response for the infinite ($Q = \infty$) discrete network of Figure 3.4 as a function of wavenumber k.

negative, R_0 becomes negative for some k. The effect of discretization is not large for smaller values of k until $\Delta r/2$ approaches L, which it does for curves b and c.

Figure 3.6 shows $R_p(k, \Delta r, \infty)$ as function of wavenumber k for $p = 3$, 6, and 9. Δr has been chosen to be equal to 2 in this case, corresponding to curve c of Figure 3.5. As one would expect from the asymptotic limit of Bessel functions [(G9) of Appendix G], as $k \to 0$, then $R_p(k, \Delta r, \infty)$ becomes increasingly small for increasing p. Thus, the terms $R_p(k, \Delta r, \infty)$, for large p, become important only for larger values of k.

Figure 3.7 plots the response $R_0(k, \Delta r, Q)$ for a finite discrete network, that is, Q finite. For this network, Δr has been chosen to be equal to 1. Curve a shows the Hankel transform (3.3.27) for reference. Curve b shows the case $Q = 2$; that is, the network consists only of the innermost and second rings of Figure 3.5. For curve c, $Q = 1$, corresponding to the innermost ring alone. Both curves b and c are negative for some values of k. From (3.4.16), it can be seen that when $Q = 1$ (curve c),

$$R_0(k, \Delta r, 1) = J_0\left(\frac{\mathrm{k}\ \Delta \mathrm{r}}{2}\right) \tag{3.4.19}$$

which is independent of the choice of weight $w(r)$. Equation (3.4.19) indicates that if all the observations within the radius of influence are approximately the same distance from the analysis point, then the spectral response (and indeed the analysis itself) is

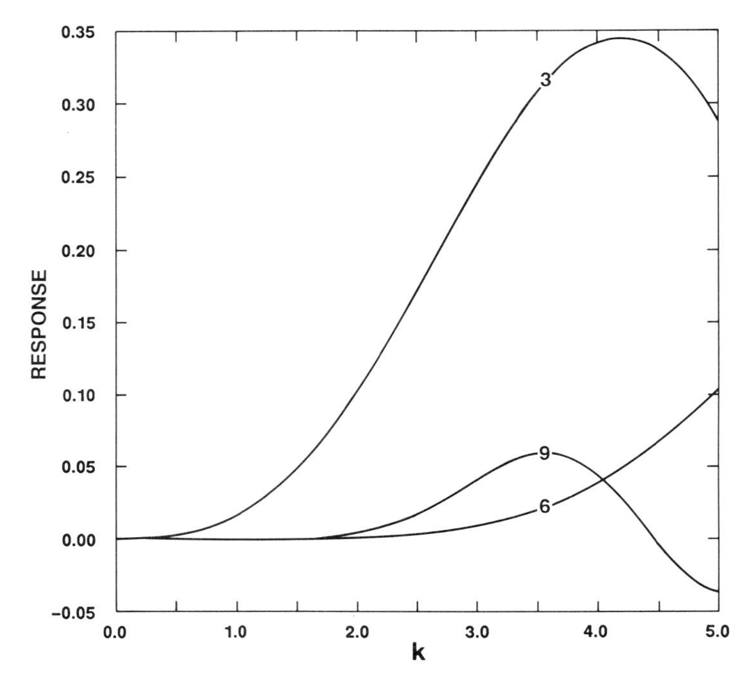

Figure 3.6 Selected nonaxisymmetric spectral responses for the infinite ($Q = \infty$) discrete network of Figure 3.4.

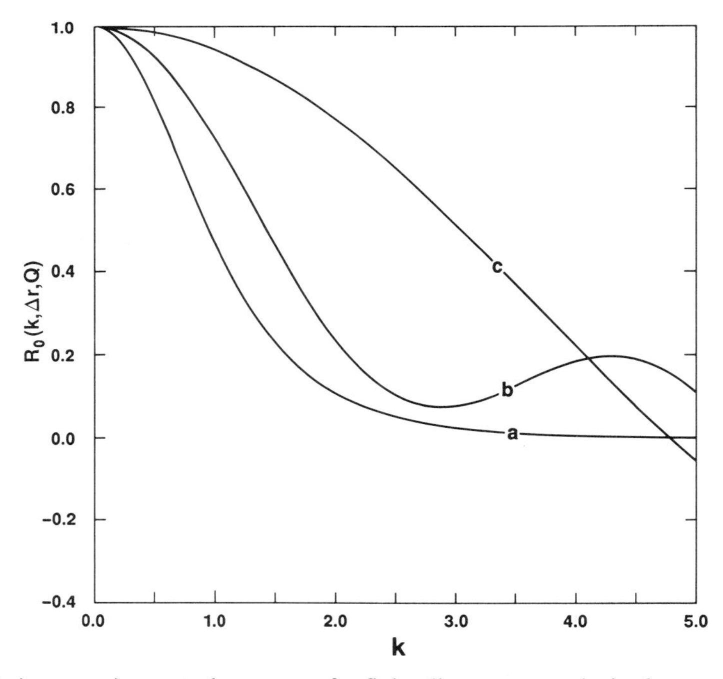

Figure 3.7 Axisymmetric spectral responses for finite discrete networks in the same format as Figure 3.5.

virtually independent of the choice of weight function. An important example of this phenomenon occurs when all the observations within the radius of influence are very close to the analysis point. Conversely, if the observations are located at varying distances from the analysis point, then the spectral response is more dependent on the choice of weight function. From (3.4.19), it can be seen that as Δr is reduced, the response increases for the smaller spatial scales (larger values of k), as might be expected.

The results plotted in Figures 3.5–3.7 are for a finite or infinite discrete network on an infinite domain. When the domain is finite, then the spectral responses would be discrete rather than continuous.

Networks such as that shown in Figure 3.4 are more realistic than the continuous, infinite networks of the previous section. Yet, such regular, discrete networks are themselves far removed from the global observation network of Section 1.3.

Spectral responses for somewhat more realistic networks can be obtained by statistical means. For example, the North American radiosonde network can be modeled by dividing the domain into equal-area boxes and placing one radiosonde station randomly within each box. Such a network is said to be homogeneous because, even though the observation stations are irregularly distributed, the observation density is approximately uniform. Stephens and Stitt (1970) and Stephens and Polan (1971) have studied the spectral responses for such networks. They assumed that within the radius of influence surrounding each analysis gridpoint are N observing stations distributed randomly. The use of probability theory leads to estimates for the spectral response of the objective analysis algorithm as a function of station density.

However, the global system of Section 1.3 does not have a uniform data density; and furthermore, the local data density is time dependent. We can determine the spectral response for the actual global network only by numerical means. This has been done by Leary and Thompson (1973), and a more recent experiment is described in Daley, Wergen, and Cats (1986).

Perhaps the most important lesson to be learned from this section is that the spectral response of the single correction algorithm depends primarily on the data distribution and only secondarily on the specified weight functions. In some cases, (3.4.19), the spectral response does not depend on the specified weight function at all. Stephens' (1967) remark sums up the situation perfectly: "with arbitrary data spacing, almost any response may be generated."

At first sight, it might seem that the spectral responses of the continuous, infinite networks of Section 3.3 have no more than a curiosity value. This may be true if the actual spectral response of a given discrete observation network is required. However, if the goal sought is knowledge of the general properties of an objective analysis algorithm, then the Fourier and Hankel transforms of Section 3.3 play a much more important role. This will become increasingly evident in Chapters 4 and 5; but a more immediate application of the theory of Section 3.3 is in deriving convergence properties of the SCM algorithm itself.

3.5 Convergence properties

Three important questions can be asked about the iteration procedure (3.2.4). Under what conditions does it converge? What does it converge to? How fast does it converge? These questions will be examined by two mathematical techniques. First, consider the convergence of the continuous, infinite one-dimensional version of (3.2.1–2). The single correction algorithm in this case is given by (3.3.2) and its Fourier transform by (3.3.6). Consider (3.3.6) as the Fourier transform of the first iteration of the SCM algorithm. Then, the Fourier transform of the jth iteration can be written as

$$\begin{aligned}\hat{f}_{\mathrm{A}}^{j+1}(m) - \hat{f}_{\mathrm{O}}(m) &= [1 - \alpha(m)][\hat{f}_{\mathrm{A}}^{j}(m) - \hat{f}_{\mathrm{O}}(m)] \\ &= [1 - \alpha(m)]^{j+1}[\hat{f}_{\mathrm{B}}(m) - \hat{f}_{\mathrm{O}}(m)]\end{aligned} \quad (3.5.1)$$

where $\hat{f}_{\mathrm{B}}(m) = \hat{f}_{\mathrm{A}}^{0}(m)$.

The iteration diverges unless $|1 - \alpha(m)| < 1$ or $0 < \alpha(m) < 2$. The $\alpha(m)$ is defined in (3.3.10). From Exercise 3.3, for any weight function $w(x)$ that is nonnegative, the corresponding transfer function $|g(m)| \leq g(0)$, and consequently $\alpha(m) < 1$. Thus convergence occurs when

$$g(m) > 0 \qquad \text{for all finite } m \quad (3.5.2)$$

The weight functions (3.3.13) and (3.3.15) satisfy condition (3.5.2), but (3.3.11) does not. Similar arguments hold in the two-dimensional case; convergence occurs with weight functions (3.3.26–27) but not necessarily with (3.3.25). It is clearly important that the Fourier or Hankel transform of the specified weight function be positive for all finite wavenumbers.

We now explore convergence of the SCM algorithm on realistic discrete networks, using eigensystem analysis, a technique that will also be useful in Chapter 4. This technique is based on examination of the objective analysis produced by the objective analysis algorithm at the *observation stations*.

Consider (3.2.6), and write the matrix $\underline{\underline{W}}^{\mathrm{T}}$ in the form

$$\underline{\underline{W}}^{\mathrm{T}} = \underline{\underline{M}}^{-1}\underline{\underline{C}} \quad (3.5.3)$$

where $\underline{\underline{C}}$ is a real symmetric matrix with elements $c_{kl} = w(r_{kl})$ and $\underline{\underline{M}}$ a real diagonal matrix with elements $M_k = \sum_{l=1}^{K} |c_{kl}| + \varepsilon_{\mathrm{O}}^2 = \sum_{l=1}^{K} |w(r_{kl})| + \varepsilon_{\mathrm{O}}^2$, and r_{kl} the magnitude of the distance between observation stations k and l, analogous to r_{ik} of (3.2.2). Note that M_k is defined using $|c_{kl}|$ rather than c_{kl}. This is a slightly more general expression than (3.2.2), in that M_k will never become vanishingly small even if some of the specified weights $w(r_{kl})$ are negative. $\underline{\underline{W}}^{\mathrm{T}}$ is not a symmetric matrix, which makes investigation of the convergence properties of (3.2.6) more difficult. However, a symmetric form is easily obtained from (3.2.7):

$$\underline{\mathrm{d}}_{j+1} = (\underline{\underline{I}} - \underline{\underline{S}})\underline{\mathrm{d}}_j = (\underline{\underline{I}} - \underline{\underline{S}})^{j+1}\underline{\mathrm{d}}_0 \quad (3.5.4)$$

where $\underline{\mathrm{d}}_0 = \underline{\underline{M}}^{1/2}(\underline{\mathrm{f}}_{\mathrm{B}} - \underline{\mathrm{f}}_{\mathrm{O}})$, $\underline{\underline{S}}$ is the symmetric real matrix $\underline{\underline{M}}^{-1/2}\underline{\underline{C}}\underline{\underline{M}}^{-1/2}$, and $\underline{\mathrm{d}}_j = \underline{\underline{M}}^{1/2}(\underline{f}_{\mathrm{A}}^{j} - \underline{\mathrm{f}}_{\mathrm{O}})$.

Convergence occurs if the observation increments are reduced after each iteration. This can be shown if the quadratic forms $\underline{d}^T\underline{d}$ satisfy

$$\underline{d}_{j+1}^T\underline{d}_{j+1} < \underline{d}_j^T\underline{d}_j \tag{3.5.5}$$

Left multiplication of (3.5.4) by $\underline{d}_{j+1}^T$ gives

$$\underline{d}_{j+1}^T\underline{d}_{j+1} = \underline{d}_j^T(\underline{\underline{I}} - \underline{\underline{S}})^2\underline{d}_j \tag{3.5.6}$$

$\underline{\underline{S}}$ is a real symmetric $K \times K$ matrix, and consequently its eigenvalues are real and its eigenvectors are real and orthogonal. Define $\underline{e}_p$ and $\underline{e}_q$ to be two eigenvectors of $\underline{\underline{S}}$ with corresponding eigenvalues μ_p and μ_q. Then $\underline{e}_p^T\underline{e}_q = \delta_{pq}$ (assuming the eigenvectors have been normalized). Expand $\underline{d}_{j+1}$ and $\underline{d}_j$ in (3.5.6) as a linear combination of the eigenvectors $\underline{e}_p$, $1 \le p \le K$ of $\underline{\underline{S}}$:

$$d_j = \sum_{p=1}^{K} d_j^p \underline{e}_p \quad \text{and} \quad d_{j+1} = \sum_{p=1}^{K} d_{j+1}^p \underline{e}_p \tag{3.5.7}$$

where $d_j^p = \underline{e}_p^T d_j$ and $d_{j+1}^p = \underline{e}_p^T d_{j+1}$ are the expansion coefficients. Inserting (3.5.7) into (3.5.6) gives

$$\sum_{p=1}^{K} (d_{j+1}^p)^2 = \sum_{p=1}^{K} (d_j^p)^2 (1 - \mu_p)^2 \tag{3.5.8}$$

making use of the orthonormality of the eigenvectors. Thus,

$$\underline{d}_{j+1}^T\underline{d}_{j+1} = \sum_{p=1}^{K} (d_{j+1}^p)^2 < \sum_{p=1}^{K} (d_j^p)^2 = \underline{d}_j^T\underline{d}_j \quad \text{if} \quad 0 < \mu_p < 2 \quad \text{for} \quad 1 \le p \le K \tag{3.5.9}$$

It is easy to verify that though matrices $\underline{\underline{W}}^T$ and $\underline{\underline{S}}$ do not have common eigenvectors (nonorthogonal in the case of $\underline{\underline{W}}^T$), they share eigenvalues μ_p, $1 \le p \le K$. Thus, any property of the eigenvalues μ_p applies equally to $\underline{\underline{S}}$ and $\underline{\underline{W}}^T$. Convergence of (3.2.6) depends on the functional form of $w(r_{kl})$.

The iteration procedure (3.2.6) can be guaranteed to converge ($0 < \mu_p < 2$, for $1 \le p \le K$) when

$$\underline{\underline{C}} \text{ is positive semidefinite} \tag{3.5.10}$$

We first show that the eigenvalues of $\underline{\underline{S}}$ and $\underline{\underline{W}}^T$ are nonnegative. Suppose $\underline{e}$ is an eigenvector of $\underline{\underline{W}}^T$ with elements e_l, $1 \le l \le K$, with corresponding eigenvalue μ. Then,

$$\underline{\underline{W}}^T\underline{e} = \underline{\underline{M}}^{-1}\underline{\underline{C}}\underline{e} = \mu\underline{e} \quad \text{or} \quad \underline{\underline{C}}\underline{e} = \mu\underline{\underline{M}}\underline{e} \tag{3.5.11}$$

Left multiplication of (3.5.11) by $\underline{e}$ gives

$$\mu = \frac{\underline{e}^T\underline{\underline{C}}\underline{e}}{\underline{e}^T\underline{\underline{M}}\underline{e}} \ge 0 \tag{3.5.12}$$

because $\underline{\underline{C}}$ is positive semidefinite and $\underline{\underline{M}}$ is positive definite, since it is diagonal with all positive elements. Equation (3.5.12) shows that $\mu \geq 0$; the special case $\mu = 0$ is discussed later.

To prove $\mu < 2$ requires application of a theorem due to Gershgorin (Wilkinson 1965; Bratseth 1986). The kth row of (3.5.11) can be written as

$$\sum_{l=1}^{K} c_{kl} e_l = \mu M_k e_k \tag{3.5.13}$$

Choose k such that $|e_l| \leq |e_k|$, $1 \leq l \leq K$. Because $c_{kk} = w(r_{kk}) = w(0) = 1$, (3.5.13) can be written as

$$(\mu M_k - 1) e_k = \sum_{l \neq k} c_{kl} e_l$$

or

$$|\mu M_k - 1| \, |e_k| = \left| \sum_{l \neq k} c_{kl} e_l \right| \leq \sum_{l \neq k} |c_{kl}| \, |e_l|$$

Thus,

$$|\mu M_k - 1| \leq \sum_{l \neq k} \frac{|c_{kl}| \, |e_l|}{|e_k|} \leq \sum_{l \neq k} |c_{kl}|$$

But

$$\mu \sum_{l=1}^{K} |w(r_{kl})| + \mu \varepsilon_{\mathrm{O}}^2 - 1 = \mu M_k - 1 \leq \sum_{l=1}^{K} |w(r_{kl})| - 1$$

$$\mu \leq \frac{\sum_{l=1}^{K} |w(r_{kl})|}{\sum_{l=1}^{K} |w(r_{kl})| + \varepsilon_{\mathrm{O}}^2} < 1 \tag{3.5.14}$$

Condition (3.5.14) is obviously satisfied when $w(r_{kl}) > 0$. However, when $\underline{\underline{M}}$ is defined using the absolute values of the specified weights as in (3.5.3), then the specified weights do not have to be positive everywhere to satisfy (3.5.14).

If the eigenvalues of $\underline{\underline{S}}$ satisfy $0 < \mu_p < 2$, $1 \leq p \leq K$, then, from (3.5.5) and (3.5.9),

$$\lim_{j \to \infty} \underline{d}_j^{\mathrm{T}} \underline{d}_j = \lim_{j \to \infty} \sum_{p=1}^{K} (d_j^p)^2 = 0 \tag{3.5.15}$$

Because $\underline{d}_j^{\mathrm{T}} \underline{d}_j$ is a quadratic form, all the elements of $\underline{d}_\infty$ and all the expansion coefficients d_∞^p, $1 \leq p \leq K$, are equal to zero. Therefore, $\underline{f}_{\mathrm{A}}^\infty = \underline{f}_{\mathrm{O}}$. In other words, in the limit the analyzed values at the observation stations are equal to the observations themselves. This is the correct solution if the observations are perfect (no observation error) and/or there are no background estimates. However, (2.2.7) suggests that if both an imperfect observation and a background estimate exist at a given observation station (and the error statistics are normally distributed), then the most probable value is a linear combination of the two, weighted according to their expected error variances. Clearly, the iteration cycle (3.2.4) does not converge to the most probable value and essentially gives no weight to the background estimate in the limit.

The special case in which one (or more) of the eigenvalues of $\underline{\underline{S}}$ is equal to zero also converges. This situation would occur if $\underline{\underline{W}}^{\mathrm{T}}$, $\underline{\underline{S}}$, and $\underline{\underline{C}}$ were singular because two (or more) rows of $\underline{\underline{C}}$ were identical. This event is unlikely but could occur if two (or more) observations were exactly co-located. For the case in which two observations (with equal expected observation error variance) were co-located, the iterative procedure (3.2.6) would converge to the observation as before, except for the two co-located observations. For these two observations (3.2.6) would converge to the simple average of the two observations (see Exercise 3.6).

It can be shown that the conditions of the continuous infinite case (3.5.2) and the discrete case (3.5.10) are related. There is a powerful theorem due to Bochner (see Yaglom 1962; Gel'fand and Vilenkin 1964) from which it can be shown that if the matrix $\underline{\underline{C}}$ has elements $c_{kl} = w(r_{kl})$ and the function $w(r)$ has a positive spectrum, then $\underline{\underline{C}}$ is positive semidefinite. A rigorous proof is beyond the scope of the book, but a simple demonstration can be made for the one-dimensional case. Define an arbitrary column vector $\underline{z}$ with elements z_k, $1 \le k \le K$. Then,

$$\underline{z}^{\mathrm{T}}\underline{\underline{C}}\underline{z} = \sum_{k=1}^{K}\sum_{l=1}^{K} c_{kl} z_k z_l = \sum_{k=1}^{K}\sum_{l=1}^{K} w(x_l - x_k) z_k z_l$$

$$= \sum_{k=1}^{K}\sum_{l=1}^{K} L \int_{-\infty}^{\infty} g(m) e^{im(x_l - x_k)}\, dm\, z_k z_l$$

$$= 2L \int_{0}^{\infty} \left[\left(\sum_{k=1}^{K} z_k \cos m x_k \right)^2 + \left(\sum_{k=1}^{K} z_k \sin m x_k \right)^2 \right] g(m)\, dm \ge 0 \quad (3.5.16)$$

if $g(m)$ is real, symmetric, and positive. From the definition (2.3.12), $\underline{\underline{C}}$ is positive semidefinite.

Thus, the convergence conditions (3.5.2) and (3.5.10) are equivalent, and the iteration procedure (3.2.6) is guaranteed to converge if the Fourier or Hankel transform of the specified weight function is positive for all finite wavenumbers. Clearly, convergence of the SCM algorithm can be guaranteed for specified weight functions (3.3.13), (3.3.15), (3.3.26), and (3.3.27) but not for (3.3.11) and (3.3.25). The fact that weight functions do not satisfy (3.5.10) does not necessarily imply that the SCM algorithm will always diverge, merely that it cannot be guaranteed to converge.

Specified weight functions such as (3.3.13), (3.3.15), (3.3.26), and (3.3.27) have Fourier or Hankel transforms that are positive, monotonically decreasing functions of increasing wavenumber. For such functions, (3.5.1) suggests that, for the continuous infinite case, the fastest convergence occurs for $m = 0$, and the slowest convergence occurs as $m \to \infty$. Thus, for weight functions with a positive, monotonically decreasing spectrum, convergence would be fastest/slowest for the large/small scales. In the same way, (3.5.9) demonstrates that for the discrete case, convergence would be fastest/slowest for the largest/smallest eigenvalues of $\underline{\underline{W}}^{\mathrm{T}}$. In Section 4.5, we demonstrate that for matrices like $\underline{\underline{W}}^{\mathrm{T}}$, which are constructed from specified weight functions with positive, monotonically decreasing spectra, the largest/smallest eigenvalues of $\underline{\underline{W}}^{\mathrm{T}}$ can be associated with the largest/smallest scale eigenvectors or spatial structures (see also Exercise 3.5). Thus, the largest scales usually converge most rapidly for the SCM algorithm and the smallest scales most slowly. Consequently, the analysis increments for an incompletely iterated SCM algorithm would be smoother than the completely iterated analysis increments (see Appendix F).

The convergence conditions derived for the SCM algorithm in this section will be useful in discussing Barnes' method in the next section.

3.6 Barnes' algorithm

The discussion of the iteration cycle in the preceding section was limited to the case for which the specified weight function was fixed during the iteration. In other words, $w(r_{ik})$ in (3.2.1–4) was independent of iteration number j. One of the advantages of SCM is that the specified weight function can be changed after each iteration; and it is often advantageous to do so.

Specified weight functions with positive, monotonically decreasing spectra imply rapid/slow convergence for small/large wavenumbers in the continuous analogue (3.5.1). Consider, as an example, the weight function (3.3.15). Then,

$$\alpha(m) = \frac{g(m)}{g(0) + \varepsilon_O^2} = \frac{2}{2 + \pi\varepsilon_O^2(1 + m^2L^2)^2}$$

and an increase/decrease of L causes a decrease/increase of $\alpha(m)$, for a given value of m. Consequently, convergence of (3.5.1) can be accelerated in this case by decreasing the value of L.

In the SCM algorithm described by Cressman (1959), the value of L was successively reduced after each iteration. On the first iteration, the large value of L led to rapid convergence of the very largest scales and suppression of the smallest scales. On the next iteration, the value of L was reduced. This had virtually no effect on the largest scales, which had already converged, but it helped to speed up convergence of the smaller scales.

Figure 3.8 shows an example of an SCM algorithm applied to the analysis of mean sea level pressure over the Atlantic Ocean. The (unlabeled) isopleths are contours of constant mean sea level pressure and highs and lows are indicated. There was no background field. Figure 3.8(a) shows the objective analysis at the end of the first iteration, where L was chosen to be 1800 km. It is apparent that only the largest scale features have been resolved. The value of L was then successively reduced after each iteration. The result is shown in (b) after the fifth iteration, when L was set equal to 300 km. The smaller scale features have now been resolved, as might be expected.

The concept of an iteration-dependent weight function has been extended and refined in Barnes' algorithm. This variant of SCM has been employed in mesoscale analysis of radar and satellite data for many years. First developed by Barnes (1964, 1978), it has undoubtedly been one of the most successful SCM algorithms and is still applied widely today.

Barnes' algorithm is usually used in situations in which no reasonable background field is available. The iteration is as follows:

$$f_A^0(\mathbf{r}_i) = \sum_{k=1}^{K_i} W_O(r_{ik}) f_O(\mathbf{r}_k) \tag{3.6.1}$$

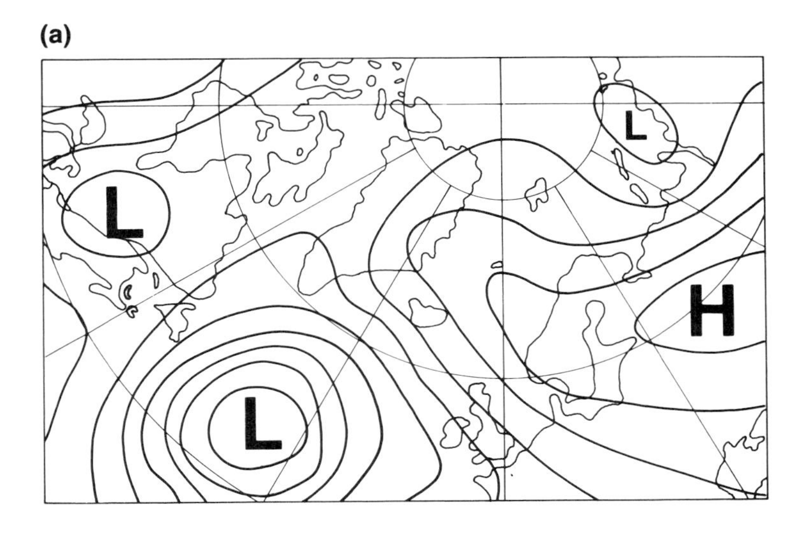

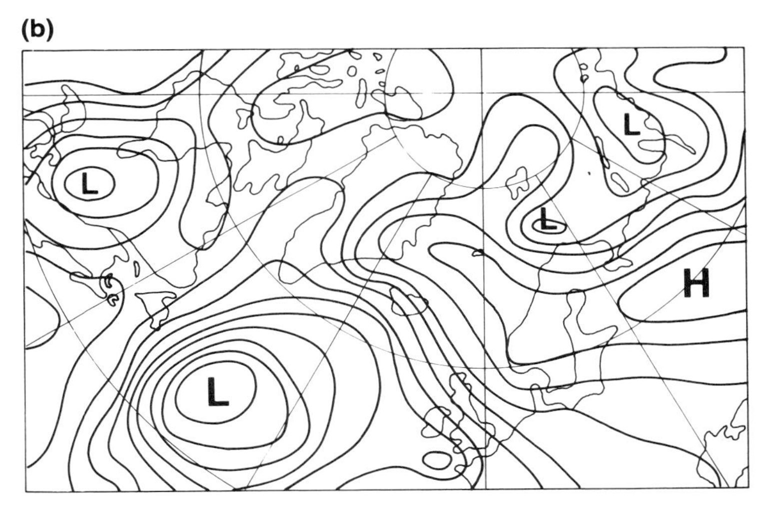

Figure 3.8 Mean sea level pressure analyses produced by an SCM algorithm with an iteration-dependent weight function (a) after one iteration and (b) after five iterations.

$$f_A^{j+1}(\mathbf{r}_i) = f_A^j(\mathbf{r}_i) + \sum_{k=1}^{K_i} W_{j+1}(r_{ik})[f_O(\mathbf{r}_k) - f_A^j(\mathbf{r}_k)] \tag{3.6.2}$$

$$W_j(r_{ik}) = \frac{w_j(r_{ik})}{\sum_{k=1}^{K_i} w_j(r_{ik})} \tag{3.6.3}$$

Here $\mathbf{r}_i$ is the location of the ith analysis point, and $\mathbf{r}_k$, $1 \le k \le K_i$, are the observation station locations within its region of influence. f_A^j is the jth iteration of the analyzed value of f; and $f_O(\mathbf{r}_k)$ is the observation at $\mathbf{r}_k$. $W_j(r_{ik})$ is the specified weight function as a function of the absolute distance between observation and analysis gridpoint (3.1.5) during the jth iteration, and $f^j_A(\mathbf{r}_k)$ is obtained by forward interpolation of the analyzed value $f^j_A(\mathbf{r}_i)$ to the observation location. Note that ε^2_O has been set equal to zero in (3.6.3) because there is no background estimate. The specified weight function is given by (3.3.26)

$$w_j(r_{ik}) = \exp\left(-\frac{r^2_{ik}}{2L^2_j}\right) \tag{3.6.4}$$

and is a function of iteration number j.

Note some of the advantages of (3.6.1–4). The spectrum of the weight function (3.6.4), given by (3.3.26), is positive and monotonically decreasing. Thus, from Section 3.5, the Barnes' algorithm would be guaranteed to converge if $L_j = L_0$ for all j. Moreover, no background estimate is required; the value of the zeroth iterate (3.6.1) is obtained as a weighted sum of the observations within the radius of influence. Barnes' algorithm converges to the observations at the observation stations, which is appropriate because there is no background estimate.

L_j can be chosen as follows. Define $\gamma = L^2_{j+1}/L^2_j$ for all j, where $0 \leq \gamma \leq 1$. Then,

$$L^2_{j+1} = \gamma^{j+1} L^2_0 \tag{3.6.5}$$

Koch, Desjardins, and Kocin (1983) have examined the convergence of (3.6.1–4) with L_j defined by (3.6.5). The case $\gamma = 1$ was discussed in Section 3.5 and is guaranteed to converge. If $\gamma < 1$, then convergence is also guaranteed, and convergence is more rapid. This is in accord with the opening remarks of this section.

In one widely used form of Barnes' algorithm (Koch et al. 1983), there are only two iterations ($j = 0, 1$). L_0 and L_1 are chosen in the following way. Consistent with the arguments of Appendix H, define L_N to be a measure of the average distance between adjacent stations in the observation network. The shortest wave that can be effectively resolved by the network has a wavelength $2L_N$; shorter waves can be interpreted as observational noise. The algorithm requires that the application of (3.6.1), followed by one application of (3.6.2), reduces the amplitude of wavenumber $k_N = 2\pi/2L_N$ to $1/e$ of its value in the original observations. Imposition of this condition effectively filters out the unresolvable scales from the network. This constraint, together with a suitable choice of γ, uniquely defines L_0 and L_1. If the spectrum of the signal is perceived to fall off rapidly with increasing wavenumber and/or the observation error is thought to be large, then a large value of γ is chosen to filter out the observational noise. On the other hand, if the signal spectrum is perceived to fall off slowly with increasing wavenumber and the observation error is thought to be small, then a small value of γ can be chosen to draw for maximum detail. This scheme is most effective when used interactively.

Four analyses of windspeed over the southeastern United States from radiosonde observations are shown in Figure 3.9. The contours are isotachs in meters per second (ms^{-1}), and the minimum resolvable scale $2L_N$ is also indicated. Analyses for four

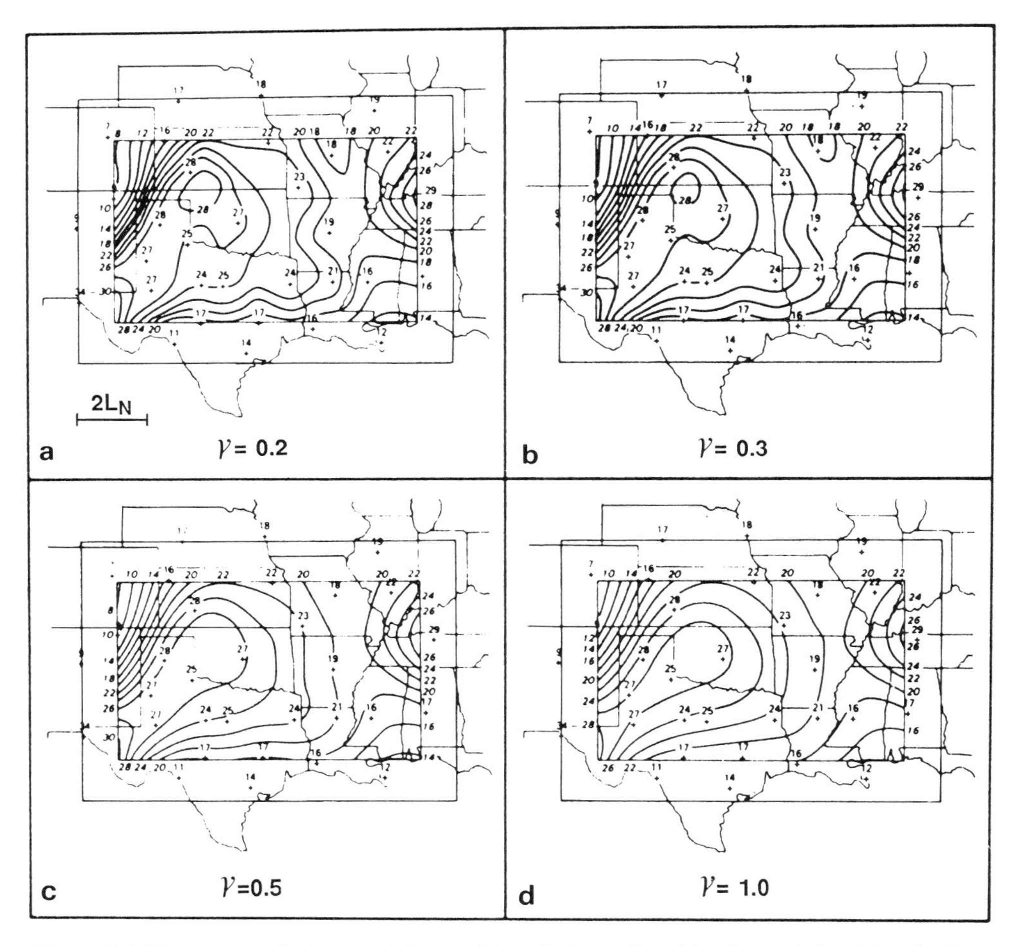

Figure 3.9 Illustration of windspeed (isotsach) analysis produced by Barnes' algorithm for four different values of γ. (From Koch, Desjardins, and Kocin, *J. Clim. Appl. Meteor.* **22**: 1487, 1983. The American Meteorological Society.)

values of γ are shown, $\gamma = 0.2$, 0.3, 0.5, and 1.0. The increasing detail in the objective analyses for lower values of γ is apparent.

3.7 Iteration to the optimal solution

The SCM algorithm (3.2.4) converges to the observed values at the observation stations, as shown in Section 3.5. This asymptotic limit is satisfactory only if the observations are perfect or there is no background estimate. When imperfect observations and imperfect background estimates both exist, the mathematical analysis of Section 2.2 shows that the analyzed values at the observation stations should be a linear combination of both (2.2.15).

A very simple modification of (3.2.4) converges to (2.2.15) at the observation stations. Assume that the background error is homogeneous with expected variance E_B^2 and that all observation errors are spatially uncorrelated with expected variance E_O^2. Suppose there are K observation stations and that each analyzed value is a function of all K observations. Then (3.2.4) can be written as

$$f_A^{j+1}(\mathbf{r}_i) - f_A^j(\mathbf{r}_i) = (1 + q_i)^{-1} \sum_{k=1}^{K} E_B^2 w(r_{ik}) E_O^{-2} [f_O(\mathbf{r}_k) - f_A^j(\mathbf{r}_k)]$$

and

$$q_i = \sum_{k=1}^{K} E_B^2 |w(r_{ik})| E_O^{-2} \tag{3.7.1}$$

Here $w(r_{ik})$ is the specified weight function, $f_O(\mathbf{r}_k)$, $1 \le k \le K$ are the observed values, $f_A^j(\mathbf{r}_i)$ is the analyzed value at the gridpoint $\mathbf{r}_i$ after the jth correction, and $f_A^j(\mathbf{r}_k)$ is the corresponding forward interpolated value at observation station $\mathbf{r}_k$. Also $f_A^0(\mathbf{r}_i) = f_B(\mathbf{r}_i)$, $f_A^0(\mathbf{r}_k) = f_B(\mathbf{r}_k)$, and $\varepsilon_O^2 = E_O^2/E_B^2$. Note that q_i has been defined with the absolute values of the specified weights, following (3.5.3).

Now consider a revised scheme:

$$f_A^{j+1}(\mathbf{r}_i) - f_A^j(\mathbf{r}_i) = (1 + q_i)^{-1} \sum_{k=1}^{K} E_B^2 w(r_{ik}) E_O^{-2} [f_O(\mathbf{r}_k) - f_A^j(\mathbf{r}_k)] + (1 + q_i)^{-1} [f_B(\mathbf{r}_i) - f_A^j(\mathbf{r}_i)] \tag{3.7.2}$$

Note that, unlike (3.7.1), the sum of the a posteriori weights in (3.7.2) is equal to one. The second term in (3.7.2) is equal to zero when $j = 0$ and thus would not appear in the single correction algorithm of Section 3.1.

Figure 3.10 is a simple illustration of an analysis using the SCM algorithm (3.7.2). The domain is the same periodic one-dimensional domain as in Figure 3.2. There are 20 equally spaced observations, indicated by solid dots. The background $f_B = f_A^0$ is indicated by the curve marked 0, and the analysis is performed at the observation stations. The specified weights and radii of influence are the same as in Figure 3.2, but $\varepsilon_O^2 = 1$. The SCM algorithm (3.7.2) was applied to these observations and iterated to convergence. The asymptotic analysis f_A^∞ is shown by the curve marked ∞. It can be seen that the resulting analysis is a weighted combination of observed and background values.

The observation error $\varepsilon_O(\mathbf{r}_k)$ satisfies $\langle \varepsilon_O(\mathbf{r}_k) \varepsilon_O(\mathbf{r}_l) \rangle = E_O^2 \delta_{kl}$. Now assume that the specified weights are given by

$$w(r_{ik}) = E_B^{-2} \langle \varepsilon_B(\mathbf{r}_k) \varepsilon_B(\mathbf{r}_i) \rangle \tag{3.7.3}$$

where $\varepsilon_B(\mathbf{r}_i)$ and $\varepsilon_B(\mathbf{r}_k)$ are the background errors at $\mathbf{r}_i$ and $\mathbf{r}_k$. That is, the specified weights with respect to the analysis gridpoint $\mathbf{r}_i$ and the observation station $\mathbf{r}_k$ are given by the background error covariance normalized by the expected background error variance. Now define the diagonal observation error covariance matrix $\underline{\underline{O}}$ (2.2.9) with nonzero elements $\langle \varepsilon_O(\mathbf{r}_k) \varepsilon_O(\mathbf{r}_k) \rangle$. Also define $\underline{B}_i$ as the column vector of length

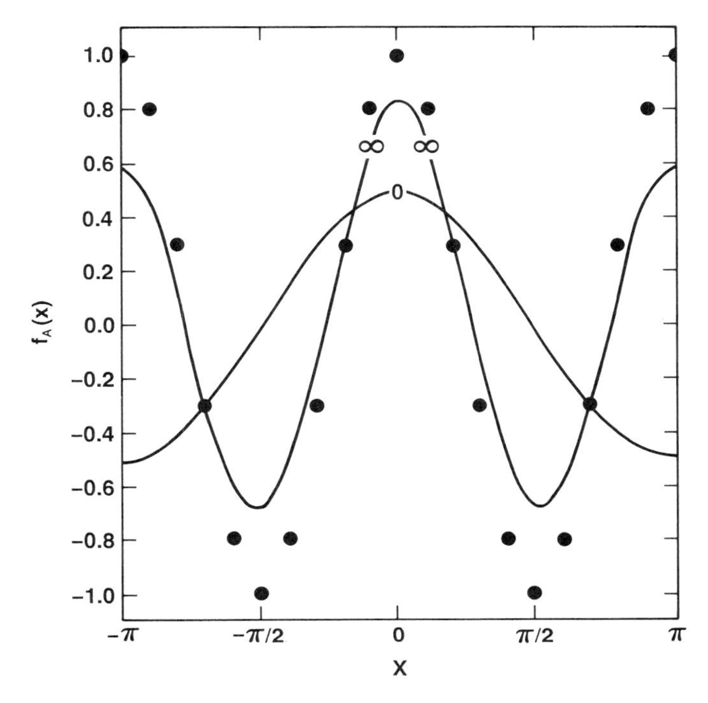

Figure 3.10 Illustration of the application of an SCM algorithm that converges to the optimal analysis. The format is the same as for Figure 3.2(a), with the background estimate denoted 0 and the asymptotic analysis ∞.

K with elements $\langle \varepsilon_B(\mathbf{r}_i)\varepsilon_B(\mathbf{r}_k)\rangle$. Then (3.7.2) can be written as

$$f_A^{j+1}(\mathbf{r}_i) - f_A^j(\mathbf{r}_i) = (1 + q_i)^{-1}\{\underline{\mathrm{B}}_i^{\mathrm{T}}\underline{\underline{\mathrm{O}}}^{-1}[\underline{\mathrm{f}}_{\mathrm{O}} - \underline{\mathrm{f}}_{\mathrm{A}}^j] + [f_B(\mathbf{r}_i) - f_A^j(\mathbf{r}_i)]\} \tag{3.7.4}$$

Following (3.2.6), when the analysis is performed at the set of observation stations $\mathbf{r}_k$, then (3.7.4) becomes

$$\underline{\mathrm{f}}_{\mathrm{A}}^{j+1} - \underline{\mathrm{f}}_{\mathrm{A}}^j = [\underline{\underline{\mathrm{I}}} + \underline{\underline{\mathrm{Q}}}]^{-1}\{\underline{\underline{\mathrm{B}}}\underline{\underline{\mathrm{O}}}^{-1}[\underline{\mathrm{f}}_{\mathrm{O}} - \underline{\mathrm{f}}_{\mathrm{A}}^j] + [\underline{\mathrm{f}}_{\mathrm{B}} - \underline{\mathrm{f}}_{\mathrm{A}}^j]\} \tag{3.7.5}$$

where $\underline{\underline{\mathrm{B}}}$ is the background error covariance matrix (2.2.11) with elements $b_{kl} = \langle \varepsilon_B(\mathbf{r}_k)\varepsilon_B(\mathbf{r}_l)\rangle$ and $\mathbf{r}_k$ and $\mathbf{r}_l$ are observation station locations, $\underline{\mathrm{f}}_{\mathrm{A}}^j$ is the column vector of analyzed values at the observation stations after the jth iteration, and $\underline{\underline{\mathrm{Q}}}$ is the diagonal matrix with elements

$$q_k = \sum_{l=1}^{K} |b_{kl}| E_O^{-2}$$

In the limit, if (3.7.5) converges, then $\underline{\mathrm{f}}_{\mathrm{A}}^{j+1} = \underline{\mathrm{f}}_{\mathrm{A}}^j = \underline{\mathrm{f}}_{\mathrm{A}}^{\infty}$, and

$$\underline{\mathrm{f}}_{\mathrm{A}}^{\infty} = \underline{\mathrm{f}}_{\mathrm{B}} + \underline{\underline{\mathrm{B}}}[\underline{\underline{\mathrm{B}}} + \underline{\underline{\mathrm{O}}}]^{-1}[\underline{\mathrm{f}}_{\mathrm{O}} - \underline{\mathrm{f}}_{\mathrm{B}}] \tag{3.7.6}$$

which is the same as (2.2.15). By subtracting $\underline{\mathrm{f}}_{\mathrm{A}}^{\infty}$ from both sides of (3.7.5) and following the procedure of Section 3.5, it can be easily shown that (3.7.5) converges when (3.2.6) converges. In this case, $\underline{f}_A^\infty$ depends on $\underline{\underline{B}}$ and $\underline{\underline{O}}$ but not on $\underline{\underline{Q}}$. The matrix $\underline{\underline{Q}}$ is chosen to ensure that (3.7.5) converges rapidly, but the asymptotic solution does not depend on the choice.

Now consider (3.7.4) in the limit as $j \to \infty$. Then, $f_A^{j+1}(\mathbf{r}_i) = f_A^j(\mathbf{r}_i) = f_A^\infty(\mathbf{r}_i)$, and we can write

$$f_A^\infty(\mathbf{r}_i) = f_B(\mathbf{r}_i) + \underline{B}_i^T \underline{\underline{O}}^{-1}[\underline{f}_O - \underline{f}_A^\infty] \tag{3.7.7}$$

Suppose there is no forward interpolation as such; instead, the analyzed values $\underline{f}_A^j$ at the observation stations are obtained by applying the algorithm directly at the observation stations at each iteration. Then, $\underline{f}_A^\infty$ is given by (3.7.6), and

$$f_A^\infty(\mathbf{r}_i) = f_B(\mathbf{r}_i) + \underline{B}_i^T [\underline{\underline{B}} + \underline{\underline{O}}]^{-1}[\underline{f}_O - \underline{f}_B] \tag{3.7.8}$$

This analysis is called the optimal analysis. A different SCM algorithm that converges to the same limit is described in Appendix F. We shall obtain the optimal analysis noniteratively in the next chapter using minimum variance estimation.

Exercises

3.1 Suppose the observation errors at observation stations $\mathbf{r}_k$, $1 \le k \le K$, are random, unbiased, and normally distributed. Show that if each observation station $\mathbf{r}_k$ is co-located with the analysis gridpoint $\mathbf{r}_i$, then the estimate $f_A(\mathbf{r}_i)$ produced by the single correction algorithm (3.1.6) is the maximum likelihood estimate.

3.2 Suppose there are two observation stations at $\mathbf{r}_1$ and $\mathbf{r}_3$ that are equally distant from an analysis gridpoint $\mathbf{r}_2$. Assume that the weight functions are of the form (3.1.9) and that $w(\mathbf{r}_1, \mathbf{r}_2) = w(\mathbf{r}_3, \mathbf{r}_2) = \mu$ whereas the weight function $w(\mathbf{r}_3, \mathbf{r}_1) = \sigma$. Denote the normalized observation error variance as ε_O^2, and assume the background field is equal to zero everywhere. There is no forward interpolation; rather the analyzed values $f_A^j(\mathbf{r}_1)$ and $f_A^j(\mathbf{r}_3)$ are obtained by applying the single correction algorithm (3.1.7) directly at the observation stations at each iteration. Find the analyzed values $f_A^1(\mathbf{r}_2)$ and $f_A^2(\mathbf{r}_2)$ as a function of the observed values $f_O(\mathbf{r}_1)$ and $f_O(\mathbf{r}_3)$. Show that after the second iteration, the a posteriori weights for co-located observations are smaller than when the observation stations are far apart.

3.3 For the one-dimensional weight function $w(x)$, show that if $w(x) \ge 0$, then for the corresponding transfer function, $|g(m)| \le g(0)$. For the two-dimensional problem, show that $w(r) \ge 0$ implies $|g(k)| \le g(0)$. Derive the Hankel transform (3.3.27). Use integral tables if necessary.

3.4 In Section 3.4, it was shown that the response of a discrete network was substantially different from that of a continuous infinite network. The finiteness of the network, in itself, is sufficient to alter substantially the continuous infinite response. Consider the weight function (3.3.15) and a *finite* continuous network, $-R \le x \le R$. Find the transfer function $g(m)$ for this case. Show that $g(m)$ can be negative for some values of m. Show that when $R \to \infty$, then $g(m)/g(0)$ approaches (3.3.16), and that when $L \to \infty$, it approaches $\sin(mR)/(mR)$.

3.5 Suppose there are two observation stations at $\mathbf{r}_1$ and $\mathbf{r}_2$ with uncorrelated observation errors and normalized expected observation error variances $\varepsilon_O^2 = 0.25$. Calculate the eigenvectors and eigenvalues of the a posteriori weight matrix $\underline{\underline{W}}^T$ defined in (3.5.3).

Using the specified weight functions (3.3.26), plot the values of the eigenvalues as a function of r/L where r is the magnitude of the relative distance between $\mathbf{r}_1$ and $\mathbf{r}_2$.

3.6 Suppose two stations $\mathbf{r}_1$ and $\mathbf{r}_2$ of a K station network are co-located, that is, $f_{\mathrm{O}}(\mathbf{r}_1) \neq f_{\mathrm{O}}(\mathbf{r}_2)$ but $f_{\mathrm{B}}(\mathbf{r}_1) = f_{\mathrm{B}}(\mathbf{r}_2)$. Assume that the observation errors are uncorrelated and that the expected observation error variance at every station is equal. The matrix $\underline{\underline{C}}$ of (3.5.3) has a zero eigenvalue in this case. Show that $f_{\mathrm{A}}(\mathbf{r}_k) = f_{\mathrm{O}}(\mathbf{r}_k)$, $3 \leq k \leq K$, and $f_{\mathrm{A}}(\mathbf{r}_1) = f_{\mathrm{A}}(\mathbf{r}_2) = 0.5[f_{\mathrm{O}}(\mathbf{r}_1) + f_{\mathrm{O}}(\mathbf{r}_2)]$ at convergence of the iteration process (3.2.6).

4

Statistical interpolation: univariate

Statistical interpolation is a powerful and widely used technique for the objective analysis of atmospheric data. At the time of this writing, most numerical weather prediction centers use variants of this method in their data assimilation cycles. Statistical interpolation has been widely studied, and many of its properties are now well understood. For these reasons, we devote two chapters to this subject. This chapter considers the univariate scalar problem, and Chapter 5 will examine multivariate analysis with implicit physical constraints.

The technique of statistical interpolation can be traced back to Kolmogorov (1941) and Wiener (1949), who applied it to problems in various branches of science and engineering. It has often been referred to in the literature as *optimal interpolation*, a term apparently coined by Wiener. In practice, the technique is rarely optimal so the more appropriate term *statistical interpolation* is used in this book. In the oceanographic literature, the method is called the Gauss–Markov method (Section 13.8).

In the atmospheric sciences, attempts to use statistical interpolation for objective analysis date back to the 1950s. Variants of this technique were studied in the West by Eliassen (1954), Kruger (1964), Eddy (1964, 1967), and Petersen and Middleton (1964). In the Soviet Union, the development of statistical interpolation techniques for the objective analysis of atmospheric data was pursued more vigorously. In 1963, L. S. Gandin published a textbook that appeared in English translation in 1965 as *Objective Analysis of Meteorological Fields*. This book had an enormous influence on the subsequent development of objective analysis techniques in both the Soviet Union and the West. The practical use of this technique awaited the development of adequate computer power, but by the mid-1970s it was being used operationally in Canada, and most of the major western meteorological services followed suit shortly thereafter.

Statistical interpolation is a minimum variance method that is closely related to the Kriging technique widely used in seismology. Before we derive the statistical interpolation algorithm, a brief discussion of minimum variance estimation is in order.

4.1 Minimum variance estimation

Consider the problem of estimating a parameter s, which was discussed at the beginning of Section 2.2. Define s to be the true value and consider measurements s_n, $1 \leq n \leq N$, each with a measurement error ε_n. Denote the expected measurement error variances $\sigma_n^2 = \langle \varepsilon_n^2 \rangle$ and suppose the errors are unbiased and uncorrelated:

$$\langle \varepsilon_n \rangle = 0 \qquad \text{and} \qquad \langle \varepsilon_m \varepsilon_n \rangle = 0 \qquad \text{for } m \neq n \tag{4.1.1}$$

Consider the following linear estimate:

$$s_{\mathrm{e}} = \sum_n c_n s_n, \qquad \text{with } c_n \geq 0, \qquad \text{for } 1 \leq n \leq N \tag{4.1.2}$$

The c_n are (as yet) unspecified weights. Define the error of the estimate as ε_{e}. The bias of the estimate is given by

$$\langle \varepsilon_{\mathrm{e}} \rangle = \langle s_{\mathrm{e}} \rangle - \langle s \rangle = \sum_n c_n \langle \varepsilon_n \rangle - \langle s \rangle \left[1 - \sum_n c_n \right]$$

Because $\langle \varepsilon_n \rangle = 0$, $1 \leq n \leq N$,

$$\langle \varepsilon_{\mathrm{e}} \rangle = 0 \qquad \text{if} \qquad \sum_n c_n = 1 \tag{4.1.3}$$

The estimate s_{e} of (4.1.2) is said to be an *unbiased linear estimate* of s if the sum of the weights is equal to 1.

Now define the expected error variance of the unbiased linear estimate s_{e},

$$\begin{aligned} \langle \varepsilon_{\mathrm{e}}^2 \rangle = \langle (s_{\mathrm{e}} - s)^2 \rangle &= \left\langle \left(\sum_n c_n s_n - \sum_n c_n s \right)^2 \right\rangle \\ &= \left\langle \left(\sum_n c_n \varepsilon_n \right)^2 \right\rangle = \sum_n c_n^2 \sigma_n^2 \end{aligned} \tag{4.1.4}$$

using (4.1.1) and (4.1.3).

Define $\sigma_{\max}^2$ as the largest of the expected measurement errors σ_n^2, $1 \leq n \leq N$. Then,

$$\langle \varepsilon_{\mathrm{e}}^2 \rangle = \sum_n c_n^2 \sigma_n^2 \leq \sigma_{\max}^2 \sum_n c_n^2$$

But all $c_n \geq 0$, and from (4.1.3), it is clear that

$$c_n \leq 1, \qquad c_n^2 \leq c_n, \qquad \text{and} \qquad \sum_n c_n^2 \leq 1$$

Thus,

$$\langle \varepsilon_e^2 \rangle \leq \sigma_{max}^2 \tag{4.1.5}$$

The expected error of the estimate is smaller than the largest expected measurement error. This is a gratifying but rather weak result.

Now consider all unbiased linear estimates of s of the form (4.1.2). The *minimum variance estimate* is defined to be the estimate s_a that minimizes (4.1.4). The value of s_a can be found by minimizing

$$\langle \varepsilon_e^2 \rangle = \sum_n c_n^2 \sigma_n^2 \qquad \text{subject to the constraint} \qquad \sum_n c_n = 1 \tag{4.1.6}$$

The most straightforward way to solve the problem (4.1.6) is by the method of Lagrange's undetermined multipliers (see Chapter 8). Introduce a Lagrange multiplier λ and define a functional:

$$J = \langle \varepsilon_e^2 \rangle + \lambda\left(1 - \sum_n c_n\right) = \sum_n c_n^2 \sigma_n^2 + \lambda\left(1 - \sum_n c_n\right) \tag{4.1.7}$$

Note that $J = \langle \varepsilon_e^2 \rangle$, so that minimizing J is equivalent to minimizing $\langle \varepsilon_e^2 \rangle$. The goal is to find the set of weights c_n, $1 \leq n \leq N$, that minimizes (4.1.7). Differentiating (4.1.7) with respect to each of the weights in turn and setting the results to zero gives

$$\frac{\partial J}{\partial c_n} = 2c_n \sigma_n^2 - \lambda = 0, \qquad 1 \leq n \leq N$$

or

$$c_n = \frac{\lambda}{2\sigma_n^2} \tag{4.1.8}$$

Now, sum (4.1.8) over n and use (4.1.3) to determine the multiplier λ:

$$\lambda = \frac{2}{\sum_n \sigma_n^{-2}}$$

or

$$c_n = \frac{\sigma_n^{-2}}{\sum_n \sigma_n^{-2}}, \qquad 1 \leq n \leq N \tag{4.1.9}$$

The minimum variance estimate is

$$s_a = \frac{\sum_n \sigma_n^{-2} s_n}{\sum_n \sigma_n^{-2}} \tag{4.1.10}$$

The weights (4.1.9) are called the *optimum weights*, and they minimize the expected estimation error variance (4.1.4). Note that the minimum variance estimate (4.1.10) is the same as the estimate (2.2.3); thus, the minimum variance estimate also minimizes the weighted sum of the residuals (2.2.2). If the measurement errors are *normally distributed* in addition to being unbiased and uncorrelated, then the minimum variance estimate (4.1.10) is also the maximum likelihood estimate.

If the optimal weights (4.1.9) are inserted into (4.1.4), then the minimum error variance $\langle \varepsilon_a^2 \rangle$ can be defined as

$$\langle \varepsilon_a^2 \rangle = \sum_n c_n^2 \sigma_n^2 = \frac{1}{\sum_n \sigma_n^{-2}} \tag{4.1.11}$$

which is the same as (2.2.4).

Now define $\sigma_{\min}^2$ to be the smallest of the expected measurement errors σ_n^2, $1 \leq n \leq N$. From (4.1.11),

$$\frac{1}{\langle \varepsilon_a^2 \rangle} = \sum_n \sigma_n^{-2}$$

or

$$\frac{1}{\langle \varepsilon_a^2 \rangle} \geq \sigma_{\min}^{-2} \qquad \text{and} \qquad \langle \varepsilon_a^2 \rangle \leq \sigma_{\min}^2 \tag{4.1.12}$$

Thus, although the expected error variance of any linear unbiased estimate is less than the *greatest* expected measurement error $\sigma_{\max}^2$, the expected error variance of the minimum variance estimate is less than the *least* expected error variance $\sigma_{\min}^2$. Clearly, (4.4.12) is a much stronger and more desirable result than (4.1.5). Moreover, each additional observation, no matter how inaccurate, *always reduces* the expected error variance $\langle \varepsilon_a^2 \rangle$ for the minimum variance estimate (4.1.10). This is *not* necessarily so for all unbiased linear estimates of the form (4.1.2–3). When $N = 2$, (4.1.10–11) are given by (2.2.7).

4.2 The statistical interpolation algorithm

The statistical interpolation algorithm can be formulated in several ways. The form derived here is the *unnormalized form*, in which the dimensionality of the dependent variables is retained throughout. This is consistent with the convention adopted earlier in Chapters 2 and 3 (and, apart from portions of Chapter 7, is the usual rule in this text). The relationship with the *normalized* form of the algorithm will be indicated at the end of this section. The statistical interpolation algorithm can be derived easily in just a few short lines of linear algebra (see Section 5.1). However, to make the derivation as clear as possible, "longhand" methods are used here. The relationship with the matrix form of the algorithm will be indicated later.

The basic form of the analysis equation is the same as that used in the method of successive corrections. Consider the univariate analysis of a variable $f(\mathbf{r})$, where $\mathbf{r}$ indicates the three-dimensional spatial coordinates. Define $f_A(\mathbf{r}_i)$ as the analyzed value of f at the analysis gridpoint $\mathbf{r}_i$, $f_B(\mathbf{r}_i)$ as the background (first-guess or trial) value of f at $\mathbf{r}_i$, and $f_O(\mathbf{r}_k)$ and $f_B(\mathbf{r}_k)$ as the observed and background values, respectively, at the observation station $\mathbf{r}_k$. Then,

$$f_A(\mathbf{r}_i) = f_B(\mathbf{r}_i) + \sum_{k=1}^{K} W_{ik}[f_O(\mathbf{r}_k) - f_B(\mathbf{r}_k)] \tag{4.2.1}$$

where K is the number of observation points and W_{ik} the as yet undetermined weight function indicating the weight given to each observation increment $[f_O(\mathbf{r}_k) - f_B(\mathbf{r}_k)]$ in the analysis at gridpoint $\mathbf{r}_i$. W_{ik} is an a posteriori weight and is therefore capitalized according to the convention established in Chapter 1.

It is convenient to introduce a simpler notation. Thus, defining $A_i = f_A(\mathbf{r}_i)$, $O_k = f_O(\mathbf{r}_k)$, $B_k = f_B(\mathbf{r}_k)$, . . ., (4.2.1) can be written as

$$A_i = B_i + \sum_{k=1}^{K} W_{ik}[O_k - B_k] \tag{4.2.2}$$

Here, $A_i - B_i$ is the analysis increment (or correction), and $O_k - B_k$ is the observation increment (or innovation). We assume here that the points $\mathbf{r}_i$ belong to a regular grid in three-dimensional space and the points $\mathbf{r}_k$ to the locations of observations that are irregularly distributed in space. Both observation and background values are assumed to contain error. We define the true value of the function at the points i and k, that is, $f_T(\mathbf{r}_i)$ or T_i and $f_T(\mathbf{r}_k)$ or T_k.

It is appropriate here to add a cautionary note concerning the correct interpretation of the values T_i and T_k. The observation error $O_k - T_k$ was assumed in Section 1.3 to contain both instrument error and error of representativeness. This implies that all the variance in the true spectrum that cannot be resolved by the observation network is represented as part of the observation error. Consequently, T_i and T_k must consist only of those components of the true spectrum that are of sufficiently large scale to be adequately resolved by the observation network (see Appendix H). Thus, T_i and T_k should be interpreted as spectrally truncated or smoothed versions of the truth.

The concept of background field goes back to the time of Gauss (see quotation at the beginning of Section 2.2). The background field $f_B(\mathbf{r})$ used by Gandin (1963) was the climatological field: $f_B(\mathbf{r}) = \langle f_T(\mathbf{r}) \rangle$, where $\langle \ \rangle$ indicates expectation value. In most applications now, $f_B(\mathbf{r})$ is obtained by integrating a numerical forecast model from an earlier objective analysis. Because of the increasing accuracy of numerical prediction models, forecast values of $f_B(\mathbf{r})$ are likely to be much closer to $f_T(\mathbf{r})$ than are the climatological values (see Section 4.3). Numerical forecasts are generally available only on a regular grid (usually, but not necessarily, the analysis gridpoints $\mathbf{r}_i$). Thus, to obtain $f_B(\mathbf{r}_k)$, it is necessary to interpolate $f_B(\mathbf{r}_i)$ to the observation location using procedures such as those in Appendix E. This interpolation is referred to in the literature as the *forward problem* or *forward interpolation*. In conventional statistical interpolation theory, the forward problem is not considered, and $f_B(\mathbf{r}_k)$ is taken as given. In this section, we follow the conventional formulation and assume that the forward interpolation has already been performed. We shall return to the forward problem in the more general treatment given in Section 5.6.

Now, subtract the true values at the analysis gridpoint $\mathbf{r}_i$ from each side of (4.2.2):

$$A_i - T_i = B_i - T_i + \sum_{k=1}^{K} W_{ik}[O_k - B_k] \tag{4.2.3}$$

It is convenient, though not necessary, to introduce a second assumption:

$$\langle f_\mathrm{B}(\mathbf{r}) - f_\mathrm{T}(\mathbf{r})\rangle = 0, \qquad \langle f_\mathrm{O}(\mathbf{r}) - f_\mathrm{T}(\mathbf{r})\rangle = 0$$

or

$$\langle B_i - T_i\rangle = \langle B_k - T_k\rangle = \langle O_k - T_k\rangle = 0 \tag{4.2.4}$$

In other words, both the background field and the observations are assumed to be unbiased. Because $\langle O_k - B_k\rangle = \langle (O_k - T_k) - (B_k - T_k)\rangle$, the observation increment must be unbiased. From (4.2.3), this implies that the analyzed value A_i must also be unbiased.

Now $\langle T\rangle$ is the true climate; and if the background field is climatology, then $\langle B\rangle = \langle T\rangle$ and is thus unbiased. However, if the background field is given by a model forecast, $\langle B\rangle \neq \langle T\rangle$ because the model climate usually differs from the true climate. Most instruments contain some bias, so $\langle O\rangle \neq \langle T\rangle$. It is possible to proceed with the statistical interpolation algorithm when $\langle B - T\rangle$ and/or $\langle O - T\rangle$ are not equal to zero. The final algorithm in this case differs slightly from the one we derive subsequently (see Section 2.2). However, in most meteorological applications, the biases $\langle B - T\rangle$ and $\langle O - T\rangle$ are known a priori; so it is simpler to modify B_i, B_k, and O_k in advance by removing the known biases and then using the modified values in the statistical interpolation algorithm. It will be assumed henceforth that B_i, B_k, and O_k are unbiased or that their biases have been removed in advance.

With these assumptions, the statistical interpolation algorithm is derived as follows. Square both sides of (4.2.3) and apply the expectation operator:

$$\langle (A_i - T_i)^2\rangle = \langle (B_i - T_i)^2\rangle + 2\sum_{k=1}^{K} W_{ik}\langle (O_k - B_k)(B_i - T_i)\rangle + \sum_{k=1}^{K}\sum_{l=1}^{K} W_{ik}W_{il}\langle (O_k - B_k)(O_l - B_l)\rangle \tag{4.2.5}$$

Here, $E_\mathrm{A}^2 = \langle (A_i - T_i)^2\rangle$ is the expected analysis error at gridpoint i, and $E_\mathrm{B}^2 = \langle (B_i - T_i)^2\rangle$ is the expected background error at gridpoint i.

Because

$$\langle O_k - B_k\rangle = \langle O_l - B_l\rangle = 0$$

$\langle (O_k - B_k)(O_l - B_l)\rangle$ is the covariance between observation increments at observation locations $\mathbf{r}_k$ and $\mathbf{r}_l$. Similarly, $\langle (O_k - B_k)(B_i - T_i)\rangle$ is the covariance between background error at $\mathbf{r}_i$ and the observation increment at $\mathbf{r}_k$. l is a dummy index running over the observation locations.

As noted earlier, statistical interpolation is a minimum variance estimation procedure and attempts to minimize the expected analysis error variance. The problem is to find the weights W_{ik} that minimize (4.2.5). Thus, differentiating (4.2.5) with respect to each of the weights W_{ik}, $1 \leq k \leq K$, gives

$$0 = \frac{\partial E_A^2}{\partial W_{ik}} = 2\langle(O_k - B_k)(B_i - T_i)\rangle + 2\sum_{l=1}^{K} W_{il}\langle(O_k - B_k)(O_l - B_l)\rangle \quad (4.2.6)$$

or

$$\sum_{l=1}^{K} W_{il}\langle(O_k - B_k)(O_l - B_l)\rangle = -\langle(O_k - B_k)(B_i - T_i)\rangle \quad (4.2.7)$$

Consider the right-hand term:

$$\langle(O_k - B_k)(B_i - T_i)\rangle = \langle(O_k - T_k)(B_i - T_i)\rangle - \langle(B_k - T_k)(B_i - T_i)\rangle$$

Terms of the form $\langle(O_m - T_m)(B_n - T_n)\rangle$, which occur on both sides of (4.2.7), are covariances between the background error and the observation error. Suppose the background field was climatology and the observations were obtained from radiosondes. Then, one would expect that the background and observation errors would be uncorrelated and that covariances of the form $\langle(O_m - T_m)(B_n - T_n)\rangle$ would vanish. The vanishing of these covariances would be reasonable for most observation systems and background fields. There are exceptions, however. Procedures for deriving temperatures from satellite-borne radiometers can use background information obtained from numerical forecasts. In this case, it cannot be assumed that covariances between observation and background error would vanish in (4.2.7). Covariances between background and observation error are not easy to determine a priori. If they are known, however, then it is sometimes possible to remove them from the observations by preprocessing (see Exercise 4.2).

Assuming no correlation between background and observation error, we can write (4.2.7) as

$$\sum_{l=1}^{K} W_{il}[\langle(B_k - T_k)(B_l - T_l)\rangle + \langle(O_k - T_k)(O_l - T_l)\rangle] = \langle(B_k - T_k)(B_i - T_i)\rangle \quad (4.2.8)$$

The statistical interpolation algorithm consists of (4.2.2) with the weights determined by (4.2.8). Following the notation of Chapters 2 and 3, these equations will now be written in matrix form. Define $\underline{f}_O$ and $\underline{f}_B$ to be the column vectors of length K of observations $f_O(\mathbf{r}_k)$ and background values $f_B(\mathbf{r}_k)$, respectively. Define $\underline{W}_i$ to be the column vector of length K of a posteriori weights. Then, (4.2.2) and (4.2.8) can be written

$$f_A(\mathbf{r}_i) = f_B(\mathbf{r}_i) + \underline{W}_i^T[\underline{f}_O - \underline{f}_B] \quad (4.2.9)$$

and

$$[\underline{\underline{B}} + \underline{\underline{O}}]\underline{W}_i = \underline{B}_i \quad (4.2.10)$$

where $\underline{B}_i$ is the column vector of length K whose elements are $\langle(B_k - T_k)(B_i - T_i)\rangle$, that is, the background error covariance between the observation station $\mathbf{r}_k$ and the analysis gridpoint $\mathbf{r}_i$. $\underline{\underline{\mathrm{B}}}$ and $\underline{\underline{\mathrm{O}}}$ are the background and observation error covariance matrices whose elements are $\langle (B_k - T_k)(B_l - T_l) \rangle$ and $\langle (O_k - T_k)(O_l - T_l) \rangle$. $\underline{\underline{\mathrm{O}}}$ and $\underline{\underline{\mathrm{B}}}$ are the symmetric $K \times K$ matrices defined in (2.2.9) and (2.2.11), respectively, and only involve the observation locations. Substitution of (4.2.10) into (4.2.9) yields (3.7.8).

Equation (4.2.9) is an unbiased linear estimate of $f_{\mathrm{T}}(\mathbf{r}_i)$. If *arbitrary* weights $\tilde{\underline{\mathrm{W}}}_i$ are specified, then an expression for the expected analysis error at gridpoint $\mathbf{r}_i$ is given by (4.2.5). Assuming no correlation between observation and background error, we can write (4.2.5) in matrix form as

$$E_{\mathrm{A}}^2 = E_{\mathrm{B}}^2 - 2\tilde{\underline{\mathrm{W}}}_i^{\mathrm{T}} \underline{\mathrm{B}}_i + \tilde{\underline{\mathrm{W}}}_i^{\mathrm{T}} [\underline{\underline{\mathrm{B}}} + \underline{\underline{\mathrm{O}}}] \tilde{\underline{\mathrm{W}}}_i \tag{4.2.11}$$

The sets of weights $\underline{\mathrm{W}}_i$ of (4.2.10) are those weights that minimize (4.2.11); they are called the *optimum weights*, analogously to (4.1.9). Assuming for the moment that the matrix $\underline{\underline{\mathrm{B}}} + \underline{\underline{\mathrm{O}}}$ is not singular, insert (4.2.10) into (4.2.11). This gives an expression for the *minimum expected analysis error* (denoted E_{A}^2), in the same way as (4.1.11). This expression is

$$E_{\mathrm{A}}^2 = E_{\mathrm{B}}^2 - \underline{\mathrm{W}}_i^{\mathrm{T}} \underline{\mathrm{B}}_i = E_{\mathrm{B}}^2 - \underline{\mathrm{B}}_i^{\mathrm{T}} [\underline{\underline{\mathrm{B}}} + \underline{\underline{\mathrm{O}}}]^{-1} \underline{\mathrm{B}}_i \tag{4.2.12}$$

Thus, when the optimum weights (4.2.10) are used in (4.2.9), $f_{\mathrm{A}}(\mathbf{r}_i)$ is a minimum variance estimate of $f_{\mathrm{T}}(\mathbf{r}_i)$, and E_{A}^2 is the least of all expected analysis error variances of the form (4.2.11). Interpolation of the form (4.2.9) using weights (4.2.10) is called *optimum interpolation*. However, these weights are optimal only if the observation and background error variances $\underline{\underline{\mathrm{B}}}$, $\underline{\underline{\mathrm{O}}}$, and $\underline{\mathrm{B}}_i$ are *correct*.

If (4.2.10) is used to derive the weights, but the assumed observation and background error variances and correlations are *not* correct, then (4.2.11) is not strictly minimized. In this case, interpolation of the form (4.2.9) using these weights is not optimal and is called statistical interpolation. This distinction between optimal and statistical interpolation is important. The correct values of $\underline{\underline{\mathrm{B}}}$, $\underline{\underline{\mathrm{O}}}$, and $\underline{\mathrm{B}}_i$ cannot be known precisely because they involve differences between observations and truth and background and truth; and since the truth is not known, $\underline{\underline{\mathrm{B}}}$, $\underline{\underline{\mathrm{O}}}$, and $\underline{\mathrm{B}}_i$ cannot be known precisely and must be estimated. How this is done will be discussed in Section 4.3. Section 4.9 will consider the sensitivity of the statistical interpolation procedure to errors in the estimates of the statistical quantities $\underline{\underline{\mathrm{B}}}$, $\underline{\underline{\mathrm{O}}}$, and $\underline{\mathrm{B}}_i$.

Although the weights $\underline{\mathrm{W}}_i$ in (4.2.9–10) are nondimensional f_{A}, f_{B}, $\underline{\mathrm{f}}_{\mathrm{O}}$, $\underline{\mathrm{f}}_{\mathrm{B}}$, $\underline{\mathrm{B}}_i$, $\underline{\underline{\mathrm{B}}}$, and $\underline{\underline{\mathrm{O}}}$ are dimensional. A widely-used nondimensional or normalized version of these equations was introduced by Lorenc (1981). The normalized form can be derived following the procedures of this section, or it can be derived directly from (4.2.9–10). Following Appendix D, define the diagonal matrix $\underline{\underline{\sigma}}_{\mathrm{B}}$ of background error standard deviations at the observation stations, $\langle (B_k - T_k)^2 \rangle^{1/2}$. Then, define a new set of weights ω_{ik} by $\underline{\mathrm{W}}_i^{\mathrm{T}} = E_{\mathrm{B}} \underline{\omega}_i^{\mathrm{T}} \underline{\underline{\sigma}}_{\mathrm{B}}^{-1}$. Equation (4.2.9) can then be written as

$$f_{\mathrm{A}}(\mathbf{r}_i) - f_{\mathrm{B}}(\mathbf{r}_i) = E_{\mathrm{B}} \underline{\omega}_i^{\mathrm{T}} \underline{\underline{\sigma}}_{\mathrm{B}}^{-1} [\underline{\mathrm{f}}_{\mathrm{O}} - \underline{\mathrm{f}}_{\mathrm{B}}]$$

or

$$\frac{A_i - B_i}{\sqrt{\langle (B_i - T_i)^2 \rangle}} = \sum_{k=1}^{K} \frac{\omega_{ik}[O_k - B_k]}{\sqrt{\langle (B_k - T_k)^2 \rangle}} \tag{4.2.13}$$

It can be seen that in this form, the analysis increment $f_A(\mathbf{r}_i) - f_B(\mathbf{r}_i)$ is normalized by the background error standard deviation at $\mathbf{r}_i$, whereas the observation increment $f_O(\mathbf{r}_k) - f_B(\mathbf{r}_k)$ is normalized by the background error standard deviation at $\mathbf{r}_k$. This form is completely nondimensional. Inserting the previous relation between $\underline{W}_i$ and $\underline{\omega}_i$ into (4.2.10) and left multiplying both sides by $\underline{\underline{\sigma}}_B^{-1}$ gives

$$[\underline{\underline{\sigma}}_B^{-1}\underline{\underline{B}}\underline{\underline{\sigma}}_B^{-1} + \underline{\underline{\sigma}}_B^{-1}\underline{\underline{O}}\underline{\underline{\sigma}}_B^{-1}]\underline{\omega}_i = E_B^{-1}\underline{\underline{\sigma}}_B^{-1}\underline{B}_i$$

or

$$[\underline{\underline{\rho}}_B + \underline{\underline{\sigma}}_B^{-1}\underline{\underline{O}}\underline{\underline{\sigma}}_B^{-1}]\underline{\omega}_i = \underline{\rho}_B^i \tag{4.2.14}$$

where $\underline{\underline{\rho}}_B$ is the background error correlation matrix (Appendix D) with elements

$$\frac{\langle (B_k - T_k)(B_l - T_l) \rangle}{[\langle (B_k - T_k)^2 \rangle \langle (B_l - T_l)^2 \rangle]^{1/2}}$$

and $\underline{\rho}_B^i$ is a background error correlation vector with elements

$$\frac{\langle (B_k - T_k)(B_i - T_i) \rangle}{[\langle (B_k - T_k)^2 \rangle \langle (B_i - T_i)^2 \rangle]^{1/2}}$$

The normalized expected analysis error variance (for optimum weights) can be obtained from (4.2.12):

$$\varepsilon_A^2 = 1 - \underline{\omega}_i^T \underline{\rho}_B^i \tag{4.2.15}$$

where $\varepsilon_A^2 = E_A^2 / E_B^2$.

The unnormalized form (4.2.9–10) is usually used in this treatment. However, under some circumstances the normalized and unnormalized forms become identical, as will be seen in the next section.

In practice, the a posteriori weights $\underline{W}_i$ of (4.2.9–10) or $\underline{\omega}_i$ of (4.2.13–14) are not usually calculated explicitly. Instead, the algorithm (4.2.9–10) is written in the form

$$f_A(\mathbf{r}_i) - f_B(\mathbf{r}_i) = \underline{B}_i^T \underline{q} \tag{4.2.16}$$

where $\underline{q} = [\underline{\underline{B}} + \underline{\underline{O}}]^{-1}[\underline{f}_O - \underline{f}_B]$, with a similar form corresponding to the normalized version (4.2.13–14). The form (4.2.16) has the advantage that $\underline{q}$ is a function of the covariances of the background and observation error between observation stations and the observed and background values at the observation stations, but does *not* depend on the position of the analysis gridpoint. Consequently, after $\underline{q}$ has been determined, an analysis increment $f_A(\mathbf{r}_i) - f_B(\mathbf{r}_i)$ can be produced at any point $\mathbf{r}_i$ simply by an appropriate choice of $\underline{B}_i^T$.

4.3 Background error covariances and correlations

The most important element in the statistical interpolation algorithm (4.2.9–10) is the background error covariance $\underline{\underline{B}}$ (and $\underline{B}_i$). To a large extent, the form of this matrix governs the resulting objective analysis.

At the observation stations, there is usually a discrepancy between the background values and the observations (the observation increment). The purpose of the objective analysis procedure is to interpolate or spread out the increment to the analysis gridpoints. In the statistical interpolation algorithm, the observation increment is spread out using the spatial structure of the background error covariance (4.2.16). Thus, it is appropriate to examine the properties of $\underline{\underline{B}}$ and $\underline{B}_i$.

The background error covariances can be simplified if the horizontal and vertical structures are assumed to be separable. Define the three-dimensional coordinates $\mathbf{r} = (x, y, P)$ where x and y indicate the horizontal coordinate and P (pressure) is the vertical coordinate. In hydrostatic flow, pressure rather than geometric altitude is generally used as the vertical coordinate. Define the background error covariance at the locations $\mathbf{r}_k$ and $\mathbf{r}_i$ as $C_B(\mathbf{r}_k, \mathbf{r}_l) = \langle (B_k - T_k)(B_l - T_l) \rangle$. Assuming separability, we have

$$C_B(\mathbf{r}_k, \mathbf{r}_l) = C_B(x_k, y_k, P_k, x_l, y_l, P_l) = C_B^H(x_k, y_k, x_l, y_l) C_B^V(P_k, P_l) \tag{4.3.1}$$

where C_B^V and C_B^H are the vertical and horizontal covariances, respectively. The separability between horizontal and vertical structures implied by (4.3.1) will be encountered again in Chapter 9. We postpone discussion of the vertical background error covariance until Section 4.7 and concentrate on the horizontal covariance. We drop the superscript H and assume for the remainder of this section that all covariances are horizontal. We continue to assume that the background errors are unbiased.

The horizontal (but not the vertical) background error covariances are generally assumed to be homogeneous. The homogeneity assumption is discussed in Appendix D, but additional consequences of this assumption will be encountered throughout Chapters 4 and 5.

To make the discussion more concrete, assume the dependent variable to be analyzed is the geopotential Φ. To simplify the notation, we suppress the subscript B for background, and all covariances (and correlations) will be understood to be background error covariances and correlations. Thus, $\Phi(x, y) = \Phi_B(x, y) - \Phi_T(x, y)$. Then, under homogeneous conditions,

$$C_{\Phi\Phi}(x_k, y_k, x_l, y_l) = C_{\Phi\Phi}(\mathbf{r}_k, \mathbf{r}_l) = C_{\Phi\Phi}(\mathbf{r}_l - \mathbf{r}_k) = C_{\Phi\Phi}(x_l - x_k, y_l - y_k) \tag{4.3.2}$$

Now introduce plane polar coordinates r and ϕ,

$$r = \sqrt{(x_l - x_k)^2 + (y_l - y_k)^2}, \qquad \phi = \tan^{-1}\left[\frac{y_l - y_k}{x_l - x_k}\right] \tag{4.3.3}$$

Thus, under homogeneous conditions, the covariance $C_{\Phi\Phi}$ can be written $C_{\Phi\Phi}(r, \phi)$; it depends only on the relative displacement vector between $\mathbf{r}_l$ and $\mathbf{r}_k$ and not on the absolute locations.

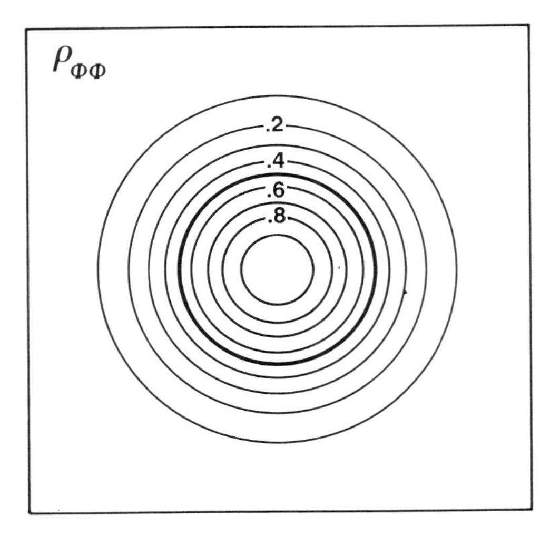

Figure 4.1 Background error correlation function (4.3.21) for geopotential.

From Equation (D20) of Appendix D, $C_{\Phi\Phi}(0) = \langle \Phi^2 \rangle$ is independent of location. Following (D13) and (D20) of Appendix D, introduce a correlation (actually an autocorrelation) $\rho_{\Phi\Phi}(r, \phi)$. Thus,

$$C_{\Phi\Phi}(r, \phi) = C_{\Phi\Phi}(0)\rho_{\Phi\Phi}(r, \phi) = \langle \Phi^2 \rangle \rho_{\Phi\Phi}(r, \phi) \tag{4.3.4}$$

From (D19) and (D20) of Appendix D,

$$\rho_{\Phi\Phi}(r, \phi) = \rho_{\Phi\Phi}(r, \phi \pm \pi)$$

and

$$|\rho_{\Phi\Phi}(r, \phi)| \leq \rho_{\Phi\Phi}(0) \tag{4.3.5}$$

If $\rho_{\Phi\Phi}$ is independent of ϕ, it is said to be *isotropic*; otherwise, it is *anisotropic*. Figure 4.1 shows an example of a continuous two-dimensional isotropic error correlation on the (r, ϕ) plane. The correlation is a maximum (1.0) at $r = 0$.

Insight can be obtained from investigating the homogeneous one-dimensional case, as in Chapters 2 and 3. Thus, define $x = x_l - x_k$ to be the displacement between two locations x_l and x_k. Then, $C_{\Phi\Phi}(x) = C_{\Phi\Phi}(0)\rho_{\Phi\Phi}(x)$, and the one-dimensional counterparts of (4.3.5) are

$$\rho_{\Phi\Phi}(x) = \rho_{\Phi\Phi}(-x) \qquad \text{and} \qquad |\rho_{\Phi\Phi}(x)| \leq \rho_{\Phi\Phi}(0) \tag{4.3.6}$$

A number of important properties of correlation and covariances under homogeneous conditions can be obtained from the theory of random variables (Yaglom 1962; Panchev 1971). The general theory is not covered here, but a few results relevant to statistical interpolation will be illustrated.

Assume that $\rho_{\Phi\Phi}$ is a continuous function of r and ϕ in two dimensions and a continuous function of x in one dimension. Thus, in the homogeneous two-dimensional case, the elements of the covariance matrices can be obtained by evaluating $\langle\Phi^2\rangle\rho_{\Phi\Phi}$ at values of r and ϕ given by (4.3.3).

For the one-dimensional homogeneous case, the spectrum $g(m)$ of the autocorrelation function $\rho_{\Phi\Phi}(x)$ is given by the Fourier transform:

$$g(m) = \frac{1}{\pi L}\int_0^\infty \rho_{\Phi\Phi}(x)\cos(mx)\,dx, \qquad \text{with} \qquad \rho_{\Phi\Phi}(x) = 2L\int_0^\infty g(m)\cos(mx)\,dm \tag{4.3.7}$$

where m is the wavenumber. L is a characteristic length to be defined subsequently, and the normalization is consistent with (3.3.9). $g(m)$ is known as the *spectral density function*. Because $\rho_{\Phi\Phi}(0) = 1$ and $C_{\Phi\Phi}(0) = \langle\Phi^2\rangle$, $2L\langle\Phi^2\rangle g(m)\,dm$ is the variance in the spectral interval between m and $m + dm$. Because $\rho_{\Phi\Phi}(x)$ is symmetric about $x = 0$, $g(m)$ is symmetric about $m = 0$.

For the two-dimensional, homogeneous isotropic case, the spectrum $g(k)$ of $\rho_{\Phi\Phi}(r)$ is given by the Hankel transform (Appendix G):

$$g(k) = \frac{1}{L^2}\int_0^\infty \rho_{\Phi\Phi}(r)J_0(kr)\,rdr, \qquad \text{with} \qquad \rho_{\Phi\Phi}(r) = L^2\int_0^\infty g(k)J_0(kr)k\,dk \tag{4.3.8}$$

where k is the two-dimensional wavenumber defined in Section 3.3, L is a characteristic scale and the normalization is consistent with (3.3.24), $g(k)$ the spectral density, and $L^2\langle\Phi^2\rangle g(k)\,dk$ the variance of the field in the spectral interval between k and $k + dk$. To be physically realizable, the variance $L^2\langle\Phi^2\rangle g(k)\,dk$ for any finite scale k must be positive. Thus, for correlation functions like $\rho_{\Phi\Phi}(x)$ and $\rho_{\Phi\Phi}(r)$, $g(m)$ and $g(k)$ are positive.

In the one-dimensional case, it can easily be demonstrated that the strict positiveness of the spectrum $g(m) > 0$ is consistent with $|\rho_{\Phi\Phi}(x)| \le \rho_{\Phi\Phi}(0)$. Thus,

$$\begin{aligned}|\rho_{\Phi\Phi}(x)| &= \left|2L\int_0^\infty g(m)\cos(mx)\,dm\right| \le 2L\int_0^\infty |g(m)|\,|\cos(mx)|\,dm \\ &\le 2L\int_0^\infty g(m)\,dm = \rho_{\Phi\Phi}(0)\end{aligned} \tag{4.3.9}$$

Very powerful theorems by Khinchine and Bochner (see Yaglom 1962) state that if $\rho_{\Phi\Phi}(r)$ is the autocorrelation of the function Φ, then its spectral density $g(k)$ is continuous, real, and positive. Moreover, the covariance and correlation matrices obtained by evaluating $\rho_{\Phi\Phi}$ at various locations are strictly positive definite (2.3.12), provided the locations are distinct. Thus, there is a three-way correspondence among the autocorrelation function, the positiveness of its spectral density,and the positive definiteness of covariance matrices constructed from it. The proof of the general theorem is beyond the scope of this book. However, an idea of the validity of the theorem can be obtained from examination of (3.5.16) and the discussion following Equation (D14) of Appendix D.

The importance of strictly positive definite error covariance matrices for statistical interpolation cannot be overstressed. Matrices that are merely positive semidefinite may contain zero (or near zero) eigenvalues and may, therefore, be effectively singular. Obtaining the weights $\underline{W}_i$ from (4.2.10) involves the inversion of the matrix $\underline{\underline{B}} + \underline{\underline{O}}$. Provided the observation locations are distinct, the background and observation error matrices are each strictly positive definite, and the matrix $\underline{\underline{B}} + \underline{\underline{O}}$ has *no* zero eigenvalues and is always nonsingular. The case of co-located observations will be dealt with shortly.

The characteristic length scale L of (4.3.7–8) can be defined in several ways. The length scale used in this treatment is often called the turbulent microscale. In the homogeneous one-dimensional case, it is defined as

$$L^2 = -\left.\frac{\rho_{\Phi\Phi}}{d^2\rho_{\Phi\Phi}/dx^2}\right|_{x=0} = \frac{\int_0^\infty g(m)\,dm}{\int_0^\infty m^2 g(m)\,dm} \tag{4.3.10}$$

Because $g(m)$ and $m^2 g(m)$ are positive for all finite m, both integrals in (4.3.10) are strictly positive and so is L^2. This implies that the second derivative of $\rho_{\Phi\Phi}$ must be strictly negative at $x = 0$, if continuous there. Because $\rho_{\Phi\Phi}$ is symmetric about $x = 0$, all odd derivatives are antisymmetric and must vanish at $x = 0$, if continuous there. Expanding in a Taylor series about $x = 0$ leads to

$$\rho_{\Phi\Phi}(x) = 1 + \frac{x^2}{2}\left.\frac{d^2\rho_{\Phi\Phi}}{dx^2}\right|_{x=0} + \cdots \quad \text{or} \quad \rho_{\Phi\Phi}(x) = 1 - \frac{x^2}{2L^2} + \cdots \tag{4.3.11}$$

Thus, for small x, $\rho_{\Phi\Phi}(x)$ must be parabolic. In the two-dimensional homogeneous isotropic case, L is defined by

$$L^2 = -\left.\frac{2\rho_{\Phi\Phi}}{\nabla^2\rho_{\Phi\Phi}}\right|_{r=0} \tag{4.3.12}$$

where the Laplacian operator ∇^2 in plane polar coordinates is given in Appendix G. The factor 2 that appears in (4.3.12) crops up widely in the two-dimensional case (see Sections 5.2 and 5.3). For the general homogeneous two-dimensional case, define L_x and L_y as component length scales or characteristic scales in the x and y directions. The scale L in (4.3.12) can be defined by $L^2 = L_x^2 + L_y^2$, which is a two-dimensional length scale. Under isotropic conditions, $L_x = L_y$ and $L^2 = 2L_x^2$. Thus, in (4.3.12), L is to be interpreted as a two-dimensional length scale rather than a component length scale.

In (4.3.2–12), the geopotential was used as an example in discussing the properties of background error covariances and correlations. We now revert to the more general notation used in Section 4.2. Thus, $\underline{\underline{B}}$ and $\underline{\underline{\rho}}_B$ are the background error covariance and correlation matrices. In the two-dimensional homogeneous case, E_B^2 is the background error variance, which is independent of location, and $\rho_B(r, \phi)$ is the background error correlation function, which is a continuous function of r and ϕ. $C_B(r, \phi) = E_B^2\rho_B(r, \phi)$ is the background error covariance. $\underline{\underline{\rho}}_B$ is constructed by evaluating $\rho_B(r, \phi)$ at the observation stations and $\underline{\underline{B}}$ by evaluating $C_B(r, \phi)$.

We noted earlier in this section that the error covariance matrices $\underline{\underline{O}}$ and $\underline{\underline{B}}$ are strictly positive definite when the observation stations are distinct. When two observation stations are very close together, there is the danger that the matrix $\underline{\underline{B}} + \underline{\underline{O}}$ will have a very small eigenvalue and be poorly conditioned. One way of avoiding this problem is by a pre-pass through the observations to combine separate observations that are very close together into a single *superobservation*. This procedure, introduced by Lorenc (1981), actually uses the statistical interpolation philosophy to combine the observations in an optimum fashion (see Section 4.6 and Exercise 8.4). Conditioning problems with the matrix $\underline{\underline{B}} + \underline{\underline{O}}$ are also ameliorated if the observation error is spatially uncorrelated or, at least, contains a substantial uncorrelated component. If the observation error is uncorrelated, then $\underline{\underline{O}}$ is a diagonal matrix; $\underline{\underline{B}} + \underline{\underline{O}}$ is therefore much more diagonally dominant, and its conditioning improves substantially.

Under homogeneous conditions, (4.3.4) suggests that the background error covariance matrix is equal to a simple constant (the variance) times the background error correlation matrix. Then, all elements of the diagonal background error variance matrix $\underline{\underline{\sigma}}_{\mathrm{B}}$ of (4.2.13) become equal to E_{B} or $\underline{\underline{\sigma}}_{\mathrm{B}} = E_{\mathrm{B}}\underline{\underline{I}}$, where $\underline{\underline{I}}$ is the identity matrix. Thus, the weights $\underline{\omega}_i$ of the normalized form (4.2.12–13) are equal to the weights $\underline{W}_i$ of the unnormalized formulation (4.2.9–10). In practice, the normalized form is often preferred because actual background error correlations are more nearly homogeneous than actual background error covariances (see Section 4.9).

We noted earlier that the accuracy of the statistical interpolation algorithm largely depends on a faithful characterization of the background error covariance. In practice, the horizontal background error covariances $C_{\mathrm{B}}(r, \phi)$ or correlations $\rho_{\mathrm{B}}(r, \phi)$ are determined from prespecified correlation functions. Improving the specification of $\rho_{\mathrm{B}}(r, \phi)$ has been approached both observationally and theoretically, and an extensive literature now exists on the subject.

Observations can be used to estimate both E_{B}^2 (the background error variance) and $\rho_{\mathrm{B}}(r, \phi)$. The most satisfactory way to estimate these quantities is to use data from a dense homogeneous observation network (Section 3.4) with uncorrelated observation errors. For example, in estimating horizontal correlations, the relatively dense radiosonde networks over North America, Europe, or Australia are very useful. E_{B}^2 and ρ_{BB}^2 have been estimated with respect to two types of background field: climatology and numerical forecasts. Both types are important and will be discussed later in this section.

The observational procedure (Drozdov and Shepelevskii 1946; Rutherford 1972) works by accumulating a large number of observations over a long period of time from this dense radiosonde network. The observations are stratified by vertical level, variable (geopotential, temperature, water vapor, etc.), season of the year, and local region. The values of the background field (climatology or forecast) are obtained at the observation stations, and biases are removed from the observations and fields. Now consider a network of K stations, with many observations at each station. Calculate

$$R_{lk} = \frac{\overline{(O_k - B_k)(O_l - B_l)}}{\sqrt{\overline{(O_k - B_k)^2}\ \overline{(O_l - B_l)^2}}} \qquad (4.3.13)$$

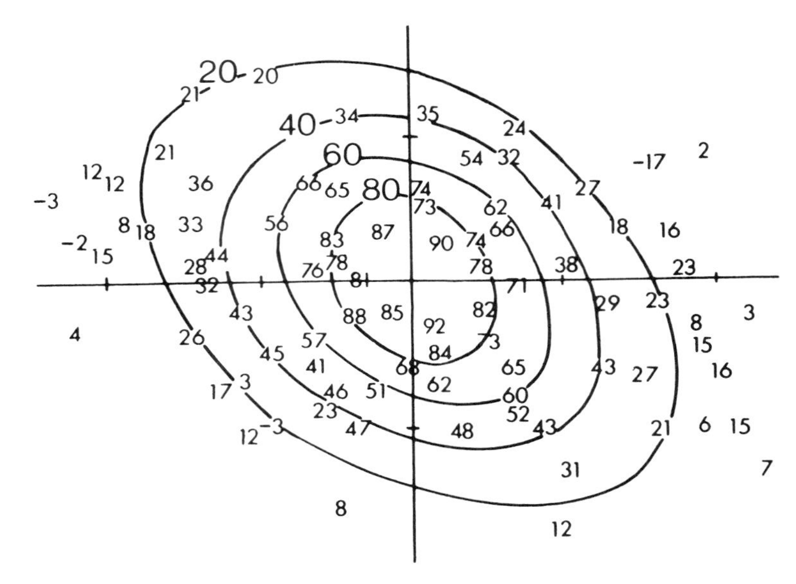

Figure 4.2 Observed-minus-background (climatology) correlation for the 500 mb geopotential field over Australia. All correlations are with respect to the observation station at the origin. (After Seaman, *Aus. Met. Mag.* **30**, 133, 1982. AGPS Canberra, reproduced by permission of Commonwealth of Australia copyright.)

where the overbar indicates the average over a long sequence of observations taken at the stations l and k. R_{lk} is the correlation between the observation increments at stations l and k.

Figure 4.2 shows a typical correlation R_{lk} for the 500 mb geopotential field using a climatological background. The numbers indicate the correlation R_{lk} (with decimal point omitted) between a station at location l and a station k located at the intersection of the two axes. In other words, k is fixed, and R_{lk} is calculated for all other stations (l) in the network. Isopleths of constant correlation are drawn at values of 0.2 (20), 0.4 (40), 0.6 (60), and 0.8 (80). The axes are oriented north–south and east–west and are graduated at 1000 km intervals. This example is for Australia during the winter season.

The isotropic component of the correlation can be determined by first plotting the correlation R_{lk} of each station pair (l, k) on a scatter diagram as a function of the absolute distance r between stations l and k. Figure 4.3 is an example of such a scatter diagram. This example is again for the 500 mb geopotential with a climatological background; but this time it is for the North American radiosonde network. Each point in Figure 4.3 represents the correlation of a single station pair. The next step is to fit a curve through all the points using an appropriate fitting procedure to produce a single curve $R(r)$. Two examples of such curves are shown in Figure 4.4. Both are for the 500 mb geopotential over the North American network. Curve c (Schlatter 1975) is for a climatological background and is the curve fitted to the

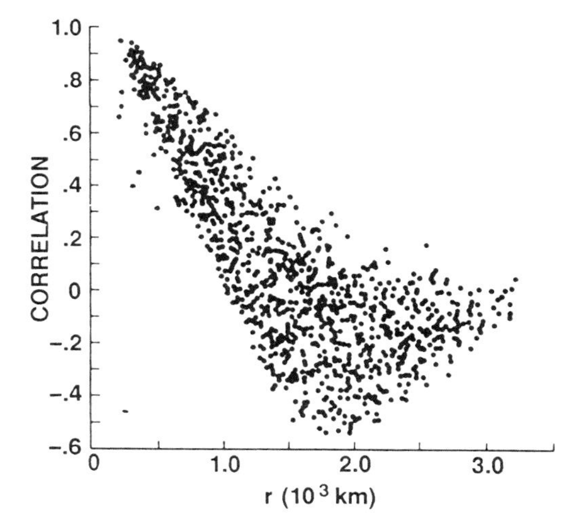

Figure 4.3 Observed-minus-background (climatology) correlation for the 500 mb height as a function of distance between station pairs. (After Schlatter, *Mon. Wea. Rev.* **103**: 246, 1975. The American Meteorological Society.)

points in Figure 4.3. Curve f (Lönnberg and Hollingsworth 1986) is for a forecast background field.

Under homogeneous conditions, the variance $\overline{(O_k - B_k)^2}$ should be independent of observation station k. Moreover, if the background and observation errors are not mutually correlated (which is the case for radiosondes) then

$$\frac{1}{K}\sum_{k=1}^{K}\overline{(O_k - B_k)^2} = \frac{1}{K}\sum_{k=1}^{K}\overline{(O_k - T_k)^2} + \frac{1}{K}\sum_{k=1}^{K}\overline{(B_k - T_k)^2} = E_{\mathrm{O}}^2 + E_{\mathrm{B}}^2 \quad (4.3.14)$$

where E_{O}^2 is the observation error variance. $R_{kk} = 1$, by definition. However, in Figure 4.3, it can be seen that there are no pairs of stations separated by an absolute distance of less than about 200 km, and the observation network is incapable of resolving variances on smaller scales. Consequently, when we fit the curve to the points in Figure 4.3 to produce curve c of Figure 4.4, it must be extrapolated to the origin ($r = 0$). For neither curve c nor f of Figure 4.4 is this zero intercept at $R = 1$. Define the zero intercept of these curves as

$$R_z = \lim_{r \to 0} R(r) \quad (4.3.15)$$

The observation error is horizontally uncorrelated, so its only contribution to $R(r)$ is at $r = 0$. On the other hand, the background error is horizontally correlated. R_z is a measure of the horizontally correlated part of the total error:

$$R_z = \frac{E_{\mathrm{B}}^2}{E_{\mathrm{B}}^2 + E_{\mathrm{O}}^2} \quad (4.3.16)$$

The horizontally uncorrelated part of the error (E_{O}^2) is made up of the observation

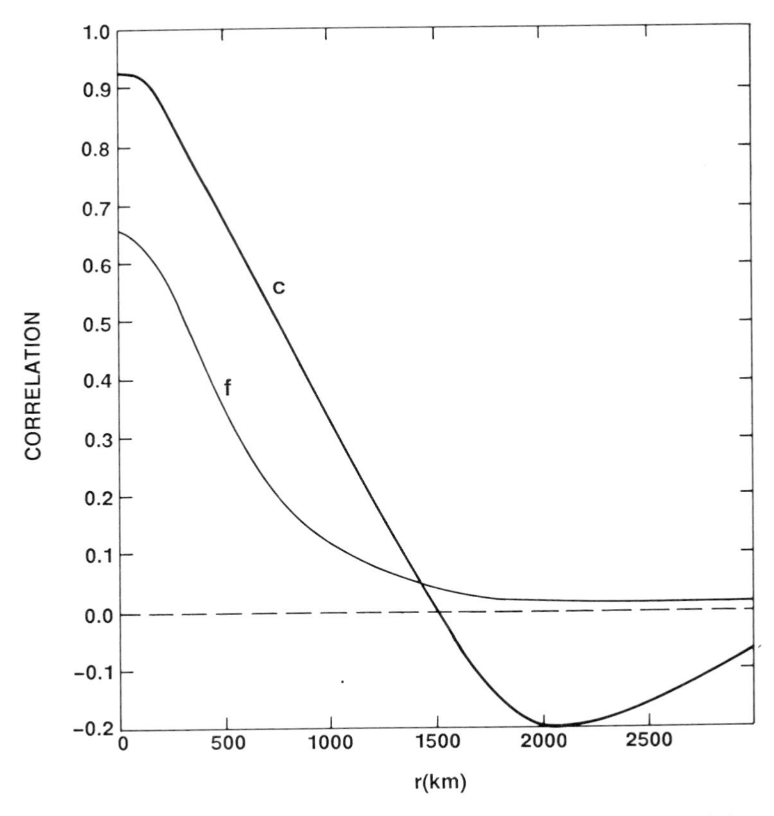

Figure 4.4 Observed-minus-background correlation for the 500 mb geopotential as a function of distance between stations. Curve c is for a climatological background and curve f is for a forecast background. (Adapted from Schlatter, *Mon. Wea. Rev.* **103**: 246, 1975. The American Meteorological Society.)

error plus the error in horizontal scales which are too small to be resolved by the observation network (i.e., the errors of representativeness discussed in Section 1.3). E_O^2 and E_B^2 can be obtained indirectly by solving (4.3.16) and (4.3.14), knowing R_z and $(1/K)\sum_{k=1}^{K}\overline{(O_k - B_k)^2}$. The isotropic component of the background error correlation is then defined as

$$\rho_B(r) = \frac{R(r)}{R_z} \qquad \text{with} \qquad \rho_B(0) = \lim_{r\to 0}\frac{R(r)}{R_z} = 1 \tag{4.3.17}$$

Figure 4.4 shows that for the 500 mb geopotential, a climatological background has a much larger zero intercept R_z than the forecast background, and consequently, the background error variance E_B^2 is much smaller for a forecast background than a climatological background. Because forecast backgrounds are closer to the "truth" than their climatological counterparts, they are generally preferred for statistical interpolation (despite possible correlations with the observations as discussed in

Section 4.2). In the forecast background case, E_B^2 and ρ_B are known reasonably well in data-rich regions and are usually assumed to be time invariant (at least over a season). However, in data voids, there may be little or no data for many cycles of the data assimilation process. The background error variance in these regions will likely increase with time until it reaches the level of the climatological error variance (see Ghil et al. 1981). It is also known that for short-range numerical forecasts, most of the error is in smaller scales; longer-range forecasts have increasing errors in the larger scales. Consequently, in the absence of new observations, the characteristic scale of forecast background error correlations increases slowly with time (Bengtsson and Gustavsson 1971). In present operational practice, the background error variance E_B^2 is allowed to increase with time in the absence of new observations, though the background error correlation is usually specified to be time-invariant (McPherson et al. 1979).

Correlation functions of atmospheric variables have been calculated many times (see Schlatter 1975 for a table of references). Most of these calculations have been done for a climatological background. This is not surprising because in this case only the properties of the atmosphere are involved. When the background is a numerical forecast, then the correlations and covariances depend on the properties of the numerical model as well as the atmospheric properties. Thus, spatial correlations with respect to a forecast background are less fundamental and ultimately less valuable because they relate to forecast models that will eventually be modified or superseded.

Background error correlations using a climatological background differ substantially from those with a forecast background. This is clearly evident in Figure 4.4. It has already been noted that E_B^2 is relatively much larger for the climatological background than for the forecast background. It can also be seen that the background error correlation is everywhere positive for a forecast background, but it goes negative at large distances for the climatological background. The reason for this is discussed in Appendix H.

We have concentrated so far on the isotropic component of the correlation. The contours of constant correlation in plane polar coordinates are concentric circles in the isotropic case (see Figure 4.1). However, the example of Figure 4.2 (which had a climatological background) indicates a substantial anisotropy with elliptical contours whose major axis tilts from upper left (northwest) to lower right (southeast). Seaman (1982) and Buell and Seaman (1983) have shown that for a climatological background, the autocorrelation of the 500 mb geopotential field is elliptic with the major axis tilted southwest–northeast in the northern hemisphere and northwest–southeast in the southern hemisphere. This is consistent with the characteristic tilts of transient troughs and ridges in the atmospheric circulation that transport heat and momentum poleward. Anisotropic correlations for a climatological background have also been examined by Thiebaux (1976, 1977, 1985). The anisotropic component of correlations with a forecast background is discussed by Hollingsworth (1987).

Correlations derived with both forecast and climatological backgrounds vary in different parts of the globe. Figure 4.5 shows estimates of the isotropic component

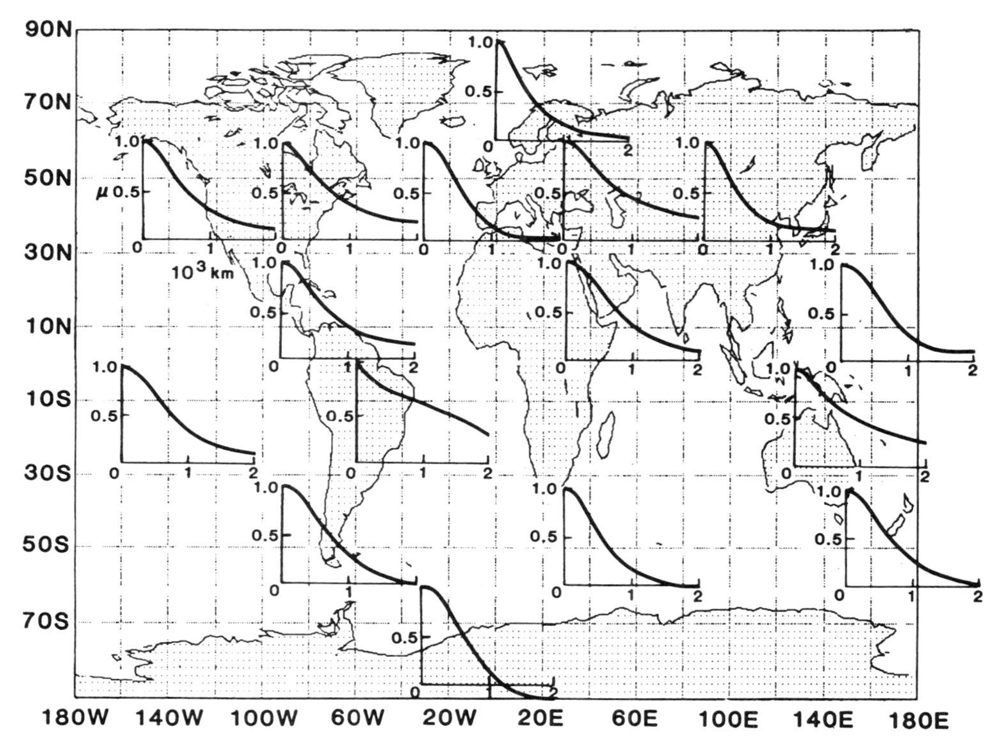

Figure 4.5 Isotropic component of 500 mb geopotential background (forecast) error correlation in different parts of the globe. (After Baker et al. *Mon. Wea. Rev.* **115**: 272, 1987. The American Meteorological Society.)

of the 500 mb geopotential correlation $\rho_B(r)$ with a forecast background. The reliability of these estimates is highest over land in the extratropics and is lowest over the oceans and in the tropics. It can be seen that $\rho_B(r)$ is generally positive and is somewhat broader (has a larger characteristic scale) in the tropics. If the atmosphere were strictly homogeneous, then $\rho_B(r)$ would be constant over the globe. The validity of the homogeneity assumption is examined in Section 4.9.

Calculation of background error correlations can be made considerably more efficient if the correlations are specified as analytic functions. A number of functional forms have been proposed for the continuous two-dimensional homogeneous isotropic correlation function $\rho_B(r)$. These functional forms are sometimes called *correlation models* or candidate correlation models; a number of them are reviewed and examined in Buell (1972a) and Julian and Thiebaux (1975). The spectral densities of these correlation models must be strictly positive. In addition, certain other conditions (discussed in Sections 5.2 and 5.3) must be satisfied by spatial derivatives of the correlation model at the origin ($r = 0$). Two correlation models are introduced here.

The first correlation model is

$$\rho_B(r) = \left[\cos(cr) + \frac{\sin(cr)}{Lc}\right]e^{-r/L} \tag{4.3.18}$$

where c and L are specified constants. This correlation model was derived by Thiebaux (1976) and is for a climatological background. The fit of this model to atmospheric data is discussed in Thiebaux (1985). Note that $\rho_B(r)$ can be negative for some values of r, as one would expect from Figure 4.4 (curve c). Correlation models, such as (4.3.18), that are not strictly positive functions of r can have spectral densities that are *not* strictly monotonically decreasing functions of wavenumber k (see Exercise 3.3). In fact, the spectral densities of climatological background error correlations may well have maximum variance at nonzero wavenumbers (see Appendix H).

The limit of $\rho_B(r)$ as the constant c approaches zero is

$$\rho_B(r) = \left(1 + \frac{r}{L}\right)e^{-r/L} \tag{4.3.19}$$

which is the same form as (3.3.27) plotted in Figure 3.3. This correlation model is positive everywhere and is more appropriate for a forecast background (curve f of Figure 4.4). The spectral density corresponding to (4.3.19) can be obtained from (3.3.27) and is plotted in Figure 3.3. In the one-dimensional case, the corresponding correlation model is given by (3.3.15) and the spectral density by (3.3.16):

$$\rho_B(x) = \left(1 + \frac{|x|}{L}\right)\exp\left(-\frac{|x|}{L}\right) \quad \text{and} \quad g(m) = \frac{2}{\pi(1 + m^2L^2)^2} \tag{4.3.20}$$

In both the one- and two-dimensional cases, the spectral densities are always a positive monotonically decreasing function of wavenumber.

The second correlation model that has been used frequently (for a forecast background) is

$$\rho_B(r) = \exp\left(\frac{-r^2}{2L^2}\right) \tag{4.3.21}$$

The spectrum of (4.3.21) can be obtained from (3.3.26) in two dimensions and from (3.3.14) in one dimension and is always positive. Figure 4.1 is derived from (4.3.21). The characteristic horizontal scale defined by (4.3.10) and (4.3.12) is equal to L for (4.3.19–21) in one and two dimensions (see Exercise 4.4).

Several attempts have been made to develop correlation models from first principles. For example, the correlation function (4.3.18) was derived by Thiebaux (1976) from a stochastically forced differential equation that occurs in the study of Brownian motion. The correlation function (4.3.20) was derived by Balgovind et al. (1983), who used a stochastic model based on a barotropic version of the equation of conservation of potential vorticity (7.8.8). This approach has the advantage, over the purely stochastic model, that the dynamics of the atmosphere are at least partially accounted for.

Perhaps the most ambitious and successful attempt to determine correlation functions from first principles was that of Phillips (1986). Phillips considered midlatitude flow with a forecast background. It was assumed that background errors were small, random, and spatially correlated. In the terminology to be introduced in Chapter 9, the background error projected completely on the Rossby or geostrophic modes; that is, it was geostrophically balanced. The system was closed by assuming that the expected amplitude of each Rossby model was equal (equipartition assumption). The model of Phillips was three dimensional and coupled the wind and mass fields (through the geostrophic assumption).

Earlier in this section, it was noted that under homogeneous conditions, the statistical interpolation algorithm (4.2.9–10) or (4.2.13–14) simplified considerably. In (4.2.13–14), $\underline{\underline{\sigma}}_B = E_B\underline{\underline{I}}$, $\underline{\omega}_i = \underline{W}_i$, and $\underline{\underline{B}} = E_B^2\underline{\underline{\rho}}_B$. Now assume that all observations are spatially correlated, but have the same expected observation error variance E_O^2. This would occur if all observations were made with the same type of instrument. Then, $\underline{\underline{O}} = E_O^2\underline{\underline{\rho}}_O$, where $\underline{\underline{\rho}}_O$ is an observation error correlation matrix. Define $\varepsilon_O^2 = E_O^2/E_B^2$. Then (4.2.10) and (4.2.14) are identical and can be written as

$$[\underline{\underline{\rho}}_B + \varepsilon_O^2\underline{\underline{\rho}}_O]\underline{W}_i = \underline{\rho}_B^i$$

or

$$\sum_{l=1}^{K} W_{il}[\rho_B(\mathbf{r}_l - \mathbf{r}_k) + \varepsilon_O^2\rho_O(\mathbf{r}_l - \mathbf{r}_k)] = \rho_B(\mathbf{r}_i - \mathbf{r}_k) \quad (4.3.22)$$

If it is further assumed that the observation errors are uncorrelated and are all made with the same type of instrument, $\underline{\underline{\rho}}_O = \underline{\underline{I}}$. Then Equation (4.3.22) becomes

$$[\underline{\underline{\rho}}_B + \varepsilon_O^2\underline{\underline{I}}]\underline{W}_i = \underline{\rho}_B^i$$

or

$$\sum_{l=1}^{K} W_{il}\rho_B(\mathbf{r}_l - \mathbf{r}_k) + \varepsilon_O^2 W_{ik} = \rho_B(\mathbf{r}_i - \mathbf{r}_k) \quad (4.3.23)$$

These very simple forms of the statistical interpolation algorithm will prove useful in the coming section.

4.4 A continuous analogue

The background error characteristics discussed in Sections 4.2 and 4.3 can now be used to investigate the properties of the statistical interpolation algorithm. As in earlier chapters, these properties are best illustrated by simple examples or limiting cases. In Section 4.6, statistical interpolation for the limiting case of 1 and 2 observations are considered. The present section considers a quite different limit, an infinitely dense observation network: in other words, a continuous analogue of statistical interpolation. This limit was encountered previously in Section 3.3 in connection with the spectral response for the method of successive corrections.

The continuous form of statistical interpolation was first examined by Ikawa (1984b). We consider the one-dimensional case and define the analysis location to be at $x = 0$. This simplification does not result in a loss of generality but does require the introduction of one less dummy variable. At $x = 0$, the continuous analogue of (4.2.2) is given by a simplified form of (3.3.1),

$$f_A(0) = f_B(0) + \frac{1}{2\pi L_B} \int_{-\infty}^{\infty} W(x)[f_O(x) - f_B(x)]\, dx \tag{4.4.1}$$

where $f_A(0)$ and $f_B(0)$ are the analyzed and background values at the analysis point $x = 0$, x the distance between analysis gridpoint and observation station, and $W(x)$ an a posteriori weight; L_B is a characteristic length scale to be defined later and is included to ensure that $W(x)$ is nondimensional.

Assume that the background and observation errors are homogeneous and spatially correlated with expected error variances E_B^2 and E_O^2, respectively. Denote $\varepsilon_O^2 = E_O^2/E_B^2$ and define covariances and correlations for the background and observation errors following (4.3.4):

$$C_B(x) = E_B^2 \rho_B(x) \qquad \text{and} \qquad C_O(x) = \varepsilon_O^2 E_B^2 \rho_O(x) \tag{4.4.2}$$

where $\rho_B(0) = \rho_O(0) = 1$. Now assume that the weight function is given by a one-dimensional continuous form of (4.3.22):

$$\frac{1}{2\pi L_B} \int_{-\infty}^{\infty} W(x')[\rho_B(x' - x) + \varepsilon_O^2 \rho_O(x' - x)]\, dx' = \rho_B(-x) \tag{4.4.3}$$

In (4.4.3), it has been assumed that the covariances C_B and C_O are correct, so $W(x')$ defines the optimum weight function. In (4.4.3), x, x', and 0 correspond to x_k, x_l, and x_i, respectively, in (4.3.22).

Now introduce Fourier transform pairs for the background and observation error spectral densities:

$$\rho_B(x) = L_B \int_{-\infty}^{\infty} g_B(m) e^{imx}\, dm \qquad \text{with} \qquad g_B(m) = \frac{1}{2\pi L_B} \int_{-\infty}^{\infty} \rho_B(x) e^{-imx}\, dx \tag{4.4.4}$$

$$\rho_O(x) = L_O \int_{-\infty}^{\infty} g_O(m) e^{imx}\, dm \qquad \text{with} \qquad g_O(m) = \frac{1}{2\pi L_O} \int_{-\infty}^{\infty} \rho_O(x) e^{-imx}\, dx \tag{4.4.5}$$

where $g_B(m)$ and $g_O(m)$ are the spectral densities for the background and observation errors, and L_B and L_O are characteristic scales defined by (4.3.10) for the background and observation error correlations. Now insert (4.4.4–5) into (4.4.3), giving

$$\frac{1}{2\pi L_{\mathrm{B}}}\int_{-\infty}^{\infty} W(x')\left\{L_{\mathrm{B}}\int_{-\infty}^{\infty} g_{\mathrm{B}}(m)e^{im(x'-x)}\,dm+\varepsilon_{\mathrm{O}}^{2}L_{\mathrm{O}}\int_{-\infty}^{\infty} g_{\mathrm{O}}(m)e^{im(x'-x)}\,dm\right\}dx'$$

$$=L_{\mathrm{B}}\int_{-\infty}^{\infty} g_{\mathrm{B}}(m)e^{-imx}\,dm$$

Interchanging the order of integration gives

$$L_{\mathrm{B}}\int_{-\infty}^{\infty} e^{-imx}\left\{\left[g_{\mathrm{B}}(m)+\frac{\varepsilon_{\mathrm{O}}^{2}L_{\mathrm{O}}g_{\mathrm{O}}(m)}{L_{\mathrm{B}}}\right]\alpha(m)-g_{\mathrm{B}}(m)\right\}dm=0 \qquad (4.4.6)$$

where

$$\alpha(m)=\frac{1}{2\pi L_{\mathrm{B}}}\int_{-\infty}^{\infty} W(x')e^{imx'}\,dx'$$

$\alpha(m)$ is the Fourier transform of the a posteriori weight function and corresponds to (3.3.5). Because the first equation of (4.4.6) must equal zero for all values of x, the expression in curly brackets must equal zero for all values of m,

$$\alpha(m)=\frac{L_{\mathrm{B}}g_{\mathrm{B}}(m)}{L_{\mathrm{B}}g_{\mathrm{B}}(m)+\varepsilon_{\mathrm{O}}^{2}L_{\mathrm{O}}g_{\mathrm{O}}(m)} \qquad (4.4.7)$$

Following (4.3.7), define

$$E_{\mathrm{B}}^{2}(m)=\eta_{\mathrm{B}}g_{\mathrm{B}}(m), \qquad E_{\mathrm{O}}^{2}(m)=\eta_{\mathrm{B}}\varepsilon_{\mathrm{O}}^{2}g_{\mathrm{O}}(m)\frac{L_{\mathrm{O}}}{L_{\mathrm{B}}}, \qquad \eta_{\mathrm{B}}=L_{\mathrm{B}}E_{\mathrm{B}}^{2} \qquad (4.4.8)$$

Thus, $E_{\mathrm{B}}^{2}(m)\,dm$ and $E_{\mathrm{O}}^{2}(m)\,dm$ are the expected background and observation errors in the spectral intervals between m and $m+dm$. Consequently,

$$C_{\mathrm{B}}(0)=E_{\mathrm{B}}^{2}=\int_{-\infty}^{\infty} E_{\mathrm{B}}^{2}(m)\,dm, \qquad C_{\mathrm{O}}(0)=\int_{-\infty}^{\infty} E_{\mathrm{O}}^{2}(m)\,dm \qquad (4.4.9)$$

Equation (4.4.7) then becomes

$$\alpha(m)=\frac{1}{1+E_{\mathrm{O}}^{2}(m)/E_{\mathrm{B}}^{2}(m)} \qquad (4.4.10)$$

Because $E_{\mathrm{B}}^{2}(m)$ and $E_{\mathrm{O}}^{2}(m)$ are always positive, $0\le\alpha(m)\le 1$, for all m. Moreover, $g_{\mathrm{B}}(m)\to 0$ as $m\to\infty$, so $\alpha(m)\to 0$ as $m\to\infty$. $\alpha(m)$ can be interpreted as a spectral response function as in Section 3.3. Because (4.4.1) is of the same form as (3.3.1), Equations (3.3.6–7) relating spectral components of the analyzed field to spectral components of the background and observed fields also apply to the continuous form of the statistical interpolation algorithm. Thus, for a given wavelength m, if the expected observation error variance is much less than the expected background error variance, $\alpha(m)\to 1$, and the algorithm draws closely for the observations. Conversely, when $E_{\mathrm{B}}^{2}(m)\ll E_{\mathrm{O}}^{2}(m)$, $\alpha(m)\to 0$, and the analysis reverts to the background field.

Define now an expected analysis error

$$E_{\mathrm{A}}^2 = \int_{-\infty}^{\infty} E_{\mathrm{A}}^2(m)\, dm \tag{4.4.11}$$

where $E_{\mathrm{A}}^2(m)\, dm$ is the expected analysis error in the spectral interval between m and $m + dm$. For the continuous case, the analysis error is given by a continuous form of (4.2.12):

$$E_{\mathrm{A}}^2 = E_{\mathrm{B}}^2 - \frac{1}{2\pi L_{\mathrm{B}}} \int_{-\infty}^{\infty} W(x) C_{\mathrm{B}}(x)\, dx \tag{4.4.12}$$

Introducing (4.4.2) and (4.4.4) and reversing the order of integration gives

$$E_{\mathrm{A}}^2 = E_{\mathrm{B}}^2 - \frac{E_{\mathrm{B}}^2}{2\pi} \int_{-\infty}^{\infty} g_{\mathrm{B}}(m) \int_{-\infty}^{\infty} W(x) e^{imx}\, dx\, dm$$

Applying (4.4.6) and (4.4.8) leads to

$$E_{\mathrm{A}}^2 = E_{\mathrm{B}}^2 - \int_{-\infty}^{\infty} E_{\mathrm{B}}^2(m) \alpha(m)\, dm \tag{4.4.13}$$

With the help of (4.4.9–11), Equation (4.4.13) becomes

$$E_{\mathrm{A}}^2(m) = \frac{E_{\mathrm{B}}^2(m) E_{\mathrm{O}}^2(m)}{E_{\mathrm{B}}^2(m) + E_{\mathrm{O}}^2(m)} \tag{4.4.14}$$

Note the resemblance between (4.4.14) and (2.2.7). Thus, for any wavenumber m, the expected analysis error variance $E_{\mathrm{A}}^2(m)$ always satisfies

$$E_{\mathrm{A}}^2(m) \le E_{\mathrm{B}}^2(m) \qquad \text{and} \qquad E_{\mathrm{A}}^2(m) \le E_{\mathrm{O}}^2(m) \tag{4.4.15}$$

and is consequently smaller than the smallest of the expected observation or background error variances. When $E_{\mathrm{O}}^2(m) \ll E_{\mathrm{B}}^2(m)$, the expected analysis error variance becomes similar to the expected observation error variance; and when $E_{\mathrm{B}}^2(m) \ll E_{\mathrm{O}}^2(m)$, it becomes similar to the expected background error variance. In other words, when the spectral response is large, the expected analysis error variance is similar to the expected observation error variance; but when the spectral response is small, the expected analysis error variance becomes similar to the expected background error variance. From (4.4.9) and (4.4.11), it is also clear that $E_{\mathrm{A}}^2 \le E_{\mathrm{B}}^2$ and $E_{\mathrm{A}}^2 \le C_{\mathrm{O}}(0) = \varepsilon_{\mathrm{O}}^2 E_{\mathrm{B}}^2$.

In (4.4.2–15), we have considered the continuous analogue of the discrete equation (4.3.22) in which the observation error is spatially correlated. It is also possible to derive a continuous analogue for (4.3.23), in which the observation error is uncorrelated. This can be done by deriving equations corresponding to (4.4.2–15) for the uncorrelated case. However, the uncorrelated case can also be derived directly in the spectral domain.

Suppose the spectral density of the expected observation error is independent of wavenumber:

$$g_{\rm O}(m) = c \tag{4.4.16}$$

where c is an arbitrary constant. This type of spectrum is called a *continuous white noise spectrum* in analogy with the visible part of the electromagnetic spectrum. From (4.4.5), the observation error correlation function can be written as

$$\rho_{\rm O}(x) = cL_{\rm O} \int_{-\infty}^{\infty} e^{imx}\, dm \tag{4.4.17}$$

However, $\int_{-\infty}^{\infty} e^{imx}\, dm = 2\pi\delta(x)$, where $\delta(x)$ is the Dirac delta function, which is infinite at $x = 0$ and equal to zero elsewhere (see Appendix G). Thus, in this case,

$$\rho_{\rm O}(x) = 2\pi c L_{\rm O}\delta(x) \tag{4.4.18}$$

From (4.4.2), the expected observation error variance $C_{\rm O}(0) = 2\pi c L_{\rm O}\varepsilon_{\rm O}^2 E_{\rm B}^2\delta(0)$ is infinite. Even though the spectral density (4.4.16) is *constant and finite*, the expected observation error variance requires an integration over m between the limits of $\pm\infty$ and is consequently *infinite*. Thus, the continuous white noise spectrum (4.4.16) describes observation error that is infinite and yet spatially uncorrelated. Clearly, this situation is not physically realizable. Remember, however, that the continuous observation network is only a limiting case and not a situation that could actually occur in practice. Consider two observation stations, separated by Δx, that have spatially uncorrelated observation errors. In the limit, as $\Delta x \to 0$, the two stations will be simultaneously co-located and yet have uncorrelated observation errors – an obvious physical impossibility.

Clearly, no real spectrum can be a continuous white noise spectrum. In general, it is the smaller values of m that are of interest, and real spectra are often nearly flat over the range of interest. In signal processing (Blackman and Tukey 1958; Yaglom 1962), it is often convenient to assume that the real spectrum is a continuous white noise spectrum in these circumstances. It is a nonphysical idealization, not unlike the point-mass assumption (which implies an infinite density) used in classical mechanics. As will be seen shortly, though $\rho_{\rm O}(x)$ is a delta function, the statistical interpolation algorithm is well behaved in this case.

For the spectral density (4.4.18), $L_{\rm O}$ clearly has no physical meaning. Choose $c = L_{\rm B}/L_{\rm O}$. Then, from (4.4.16–17),

$$E_{\rm O}^2(m) = \eta_{\rm B}\varepsilon_{\rm O}^2 \qquad \text{and} \qquad \rho_{\rm O}(x) = 2\pi L_{\rm B}\delta(x) \tag{4.4.19}$$

Inserting the second equation of (4.4.19) into (4.4.3) gives

$$\frac{1}{2\pi L_{\rm B}} \int_{-\infty}^{\infty} W(x')\rho_{\rm B}(x'-x)\, dx' + \varepsilon_{\rm O}^2 W(x) = \rho_{\rm B}(-x) \tag{4.4.20}$$

or

$$\frac{1}{2\pi L_{\rm B}} \int_{-\infty}^{\infty} W(x')C_{\rm B}(x'-x)\, dx' + \varepsilon_{\rm O}^2 E_{\rm B}^2 W(x) = C_{\rm B}(-x)$$

because

$$\int_{-\infty}^{\infty} W(x')\delta(x'-x)\,dx' = W(x)$$

Equation (4.4.20) is the continuous analogue of (4.3.23), which describes the weights for the statistical interpolation algorithm for a discrete observation network with spatially uncorrelated observation errors. In this case, the observation error correlation function is a Dirac delta function.

The spectral response in this case can be obtained by substituting the first equation of (4.4.19) into (4.4.10):

$$\alpha(m) = \frac{g(m)}{g(m) + \varepsilon_{\mathrm{O}}^2} \tag{4.4.21}$$

Here $g(m) = g_{\mathrm{B}}(m)$, and the subscript B can be safely dropped because the observation error spectral density $g_{\mathrm{O}}(m)$ no longer appears explicitly in the problem. Equation (4.4.21) is the spectral filter function for the statistical interpolation algorithm over a continuous network with uncorrelated observation errors. It can be derived directly from (4.4.20) by following the steps of (4.4.2–7) *without* invoking the Dirac delta function. Equation (4.4.21) can be compared with its counterparts (3.3.10) for the single correction algorithm. For a monotonically decreasing function $g(m)$, the single correction and statistical interpolation algorithms have similar spectral responses for small m. However, for large m, the single correction algorithm filters much more heavily than the statistical interpolation algorithm.

The derivative of (4.4.21) with respect to m is given by

$$\frac{d\alpha(m)}{dm} = \frac{\varepsilon_{\mathrm{O}}^2}{[g(m) + \varepsilon_{\mathrm{O}}^2]^2} \frac{dg(m)}{dm} \tag{4.4.22}$$

Thus, the slope of $\alpha(m)$ has the same sign as the slope of the background error spectral density $g(m)$, and the relative maxima and minima of $\alpha(m)$ and $g(m)$ occur at the same values of m. If $g(m)$ is a monotonically decreasing function of m, then $\alpha(m)$ is also. $\alpha(m)$ tends to filter the smaller scales most strongly and is, therefore, a low-pass filter. Consequently, for uncorrelated observation errors, one would expect the statistical interpolation algorithm to draw for the observations at large scales and for the background field at small scales.

It is simple to confirm that (4.4.14) holds when the observation error is uncorrelated; thus, for any wavenumber m, $E_{\mathrm{A}}^2(m) \leq E_{\mathrm{B}}^2(m)$ and $E_{\mathrm{A}}^2(m) \leq E_{\mathrm{O}}^2(m)$. This is possible because the spectral density of the expected observation error is always finite. The expected background error variance E_{B}^2 is, of course, finite whereas the expected observation error variance – as given by (4.4.19) or the second equation of (4.4.9) – is $C_{\mathrm{O}}(0) = 2\pi L_{\mathrm{B}}\varepsilon_{\mathrm{O}}^2 E_{\mathrm{B}}^2 \delta(0)$, which is infinite. However, the expected analysis error variance $E_{\mathrm{A}}^2 \leq E_{\mathrm{B}}^2$ and is thus finite. Consequently, the expected analysis error variance is less than either of the expected background or observation error variances, as was the case for correlated observation error. It is interesting to note that, even though the expected observation error variance is infinite, the expected analysis error variance is not only finite but smaller than the expected background error variance. The reason for this can be seen as follows. $\alpha(m) \to 0$ as $m \to \infty$. Define m_α as the value of m at which $\alpha(m)$ effectively vanishes. Then, for $|m| > |m_\alpha|$, the statistical interpolation algorithm filters out all observation error variance. It is only the larger scales of the observation field that are used to modify the background field to produce the analysis. Because of the filtering properties of the algorithm, the observation error variance at large m causes no difficulty.

The weight function $W(x)$ can be obtained from the inverse Fourier transform of $\alpha(m)$ given by (4.4.6):

$$W(x) = L_B \int_{-\infty}^{\infty} \alpha(m) e^{-imx}\, dm = 2L_B \int_0^{\infty} \alpha(m) \cos(mx)\, dm \tag{4.4.23}$$

Consider the uncorrelated case with $\alpha(m)$ given by (4.4.21) and background error spectral density $g(m)$ given by (4.3.20). Then,

$$W(x) = 4L_B \int_0^{\infty} \frac{\cos(mx)}{2 + \pi\varepsilon_O^2(1 + L_B^2 m^2)}\, dm \tag{4.4.24}$$

where L_B is the characteristic scale for the background error correlation function. From integral tables (Gradshteyn and Ryzhik 1965),

$$W(x) = \frac{\sqrt{2\pi L_B}}{\varepsilon_O q} e^{-Ax}[B \cos(Bx) + A \sin(Bx)] \tag{4.4.25}$$

where

$$A = \frac{1}{L_B}\sqrt{\frac{q+1}{2}}, \qquad B = \frac{1}{L_B}\sqrt{\frac{q-1}{2}}, \qquad q = \sqrt{\frac{2 + \varepsilon_O^2\pi}{\varepsilon_O^2\pi}}$$

In Figure 4.6, $W(x)$ is plotted as a function of x for $\varepsilon_O^2 = 0.1$ and $\varepsilon_O^2 = 2.0$. $L_B = 0.5$, and the correlation function $\rho_B(x)$ of (4.3.20) has been plotted for comparison.

The parameter ε_O^2 increases/decreases as the expected background error variance E_B^2 decreases/increases. From Figure 4.6, it is evident that as ε_O^2 increases, $W(x)$ decreases and less and less weight is given to the observation increments. On the other hand, as ε_O^2 decreases, $W(x)$ increases very sharply for observations close to the analysis gridpoint. In the limit, using (4.4.1) and (4.4.20),

$$\begin{aligned} &\varepsilon_O^2 \to \infty, \qquad W(x) \to \frac{\rho_B(x)}{\varepsilon_O^2} \qquad \text{and} \qquad f_A(0) \to f_B(0) \\ &\varepsilon_O^2 \to 0, \qquad W(x) \to 2\pi L_B \delta(x) \qquad \text{and} \qquad f_A(0) \to f_O(0) \end{aligned} \tag{4.4.26}$$

The analogous limit for a discrete observation network is examined in Exercise 4.5. Note that even though $\rho_B(x)$ is always positive, $W(x)$ can be negative. This phenomenon, called *screening*, also occurs in Figure 3.2(b).

The continuous analogue just introduced is most instructive when viewed in

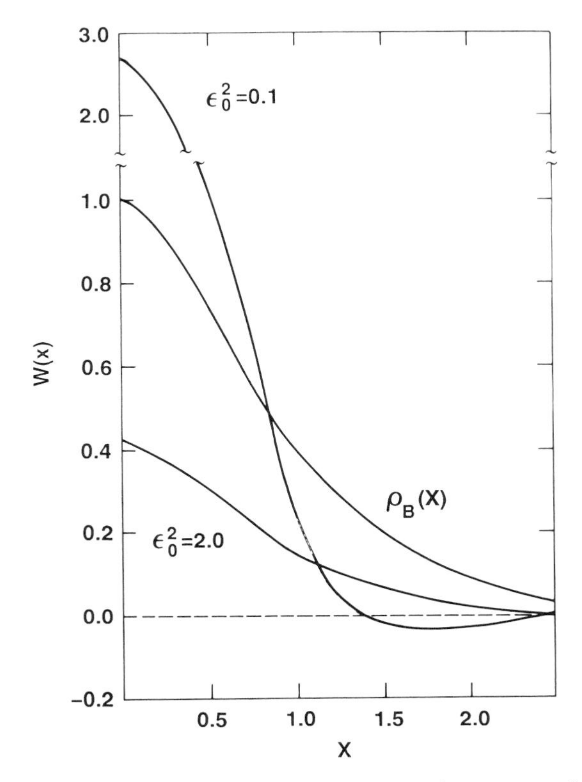

Figure 4.6 A posteriori weight function $W(x)$ of (4.4.25) for $\varepsilon_O^2 = 2.0$ and $\varepsilon_O^2 = 0.1$. Also plotted is the background error correlation $\rho_B(x)$ corresponding to (4.3.20). Note break in the curve at $W(x) = 1.1$.

spectral space and will serve as the basis for an examination of the spectral properties of the multivariate algorithm in Chapter 5. For the univariate algorithm, we next examine the filtering properties on a discrete network, using the results of this section as a guide.

4.5 Filtering and interpolation properties

The infinitely dense observation network discussed in the previous section bears little resemblance to the present highly irregular global observation network discussed in Chapter 1. An obvious corollary is that the spectral response of a statistical interpolation algorithm to observations on a realistic network differs substantially from the response to an infinitely dense network. As shown in Chapter 3, discrete observations and a finite area of influence alter the response considerably even when the network is uniform.

Hollingsworth (1987) introduced a methodology for examining the response on any observation network. Consider (4.2.9–10), which relate the analysis increment

$f_A(\mathbf{r}_i) - f_B(\mathbf{r}_i)$ to the observation increments $f_O(\mathbf{r}_k) - f_B(\mathbf{r}_k)$ at the observation locations $1 \le k \le K$:

$$f_A(\mathbf{r}_i) - f_B(\mathbf{r}_i) = \underline{\underline{B}}_i^T[\underline{\underline{B}} + \underline{\underline{O}}]^{-1}[\underline{f}_O - \underline{f}_B] \tag{4.5.1}$$

where $\underline{f}_O$ and $\underline{f}_B$ are the column vectors of observed and background values at the observation stations. The covariance matrices $\underline{\underline{B}}$ and $\underline{\underline{O}}$ are considered to be correct as in the previous section.

Equation (4.5.1) can be written as

$$f_A(\mathbf{r}_i) - f_B(\mathbf{r}_i) = \underline{\underline{B}}_i^T\underline{\underline{B}}^{-1}\underline{d} \tag{4.5.2}$$

where $\underline{d} = \underline{\underline{B}}[\underline{\underline{B}} + \underline{\underline{O}}]^{-1}[\underline{f}_O - \underline{f}_B]$ is a column vector of length k. $\underline{d}$ is a vector of *filtered* observation increments at the K observation locations. The vector $\underline{\underline{B}}_i^T\underline{\underline{B}}^{-1}$ can be regarded as the operator that *interpolates* the filtered observation increments $\underline{d}$ to the analysis gridpoint to produce the analysis increment $f_A(\mathbf{r}_i) - f_B(\mathbf{r}_i)$. The decomposition (4.5.2) permits the separate examination of the interpolation and filtering properties of statistical interpolation. The remainder of this section is concerned with the filtering properties; interpolation properties will be discussed in the next section.

The mathematical procedures of Section 3.5 provide an alternative perspective on (4.5.1–2). If the set of analysis gridpoints and observation stations coincide, then, following (3.7.6), we have

$$\underline{f}_A - \underline{f}_B = \underline{d} = \underline{\underline{B}}[\underline{\underline{B}} + \underline{\underline{O}}]^{-1}[\underline{f}_O - \underline{f}_B] \tag{4.5.3}$$

where $\underline{f}_A$ is the column vector of analyzed values at the observation stations. First note that (4.5.3) is the same as (2.2.15) derived by minimizing (2.2.11). Thus, the analyzed values at the observation stations are the same whether derived by minimizing the expected analysis error variance as in Section 4.2 or by least squares minimization of the residuals as in Section 2.2.

As in Section 3.5, the filtering properties of (4.5.3) can be illustrated by examining the eigenstructures of the covariance matrices $\underline{\underline{B}}$ and $\underline{\underline{O}}$. $\underline{\underline{B}}[\underline{\underline{B}} + \underline{\underline{O}}]^{-1}$ is the a posteriori weight matrix in this case. However, to facilitate comparison with the previous section, denote

$$\underline{\underline{\alpha}} = \underline{\underline{B}}[\underline{\underline{B}} + \underline{\underline{O}}]^{-1} = [\underline{\underline{I}} + \underline{\underline{O}}\underline{\underline{B}}^{-1}]^{-1} \tag{4.5.4}$$

Suppose $\underline{e}$ is an eigenvector of the matrix $\underline{\underline{O}}\underline{\underline{B}}^{-1}$ with associated eigenvalue v. Then,

$$\underline{\underline{O}}\underline{\underline{B}}^{-1}\underline{e} = v\underline{e} \tag{4.5.5}$$

Defining $\underline{y} = \underline{\underline{B}}^{-1}\underline{e}$, (4.5.5) becomes

$$\underline{\underline{O}}\underline{y} = v\underline{\underline{B}}\underline{y} \tag{4.5.6}$$

Multiplying both sides of (4.5.6) by $\underline{y}^T$ gives

$$v = \frac{\underline{y}^T\underline{\underline{O}}\underline{y}}{\underline{y}^T\underline{\underline{B}}\underline{y}} \tag{4.5.7}$$

Because $\underline{\underline{O}}$ and $\underline{\underline{B}}$ are both symmetric positive definite matrices, ν must be real and positive. Now $\underline{e}$ is also an eigenvector of the identity matrix $\underline{\underline{I}}$, with eigenvalue equal to 1. Moreover, the eigenvalues of the inverse of a matrix are equal to the reciprocals of the eigenvalues of the original matrix (if they are nonzero); consequently,

$$\underline{\underline{\alpha}}\underline{e} = [\underline{\underline{I}} + \underline{\underline{O}}\underline{\underline{B}}^{-1}]^{-1}\underline{e} = [1 + \nu]^{-1}\underline{e} \tag{4.5.8}$$

The matrix $\underline{\underline{\alpha}}$ can be regarded as a response matrix, the discrete counterpart to the function $\alpha(m)$ defined by (4.4.6) for a continuous observation network. It can be seen that the eigenvalues of $\underline{\underline{\alpha}}$ have the same form as (4.4.10). Because ν is positive, $0 < (1 + \nu)^{-1} < 1$ and all the eigenvalues of $\underline{\underline{\alpha}} = \underline{\underline{B}}[\underline{\underline{B}} + \underline{\underline{O}}]^{-1}$ lie between zero and one. Following (3.5.4–9), we expand $\underline{f}_A - \underline{f}_B$ and $\underline{f}_O - \underline{f}_B$ in the eigenvectors of $\underline{\underline{\alpha}}$. In Section 3.5, the analyzed-minus-observed increments decreased after each iteration if the eigenvalues of the matrix $\underline{\underline{I}} - \underline{\underline{S}}$ were real, positive, and less than one. In the same way, because all the eigenvalues of $\underline{\underline{\alpha}}$ lie between 0 and 1,

$$\underline{d}^T\underline{d} = [\underline{f}_A - \underline{f}_B]^T[\underline{f}_A - \underline{f}_B] < [\underline{f}_O - \underline{f}_B]^T[\underline{f}_O - \underline{f}_B] \tag{4.5.9}$$

In other words, after statistical interpolation (with optimum weights), the analysis increments at the observation stations are smaller than the observation increments. The analyzed values at the observation stations are closer to the background values than are the observations.

Now consider the special case in which the background and observation error covariance matrices commute. That is, $\underline{\underline{B}}\underline{\underline{O}} = \underline{\underline{O}}\underline{\underline{B}}$. Here the eigenvectors (though not necessarily the eigenvalues) of $\underline{\underline{B}}$ and $\underline{\underline{O}}$ are the same.

Suppose $\underline{e}$ is an eigenvector of $\underline{\underline{B}}$ and $\underline{\underline{O}}$ with associated eigenvalues μ_B^2 for $\underline{\underline{B}}$ and μ_O^2 for $\underline{\underline{O}}$. From Appendix D, μ_B^2 and μ_O^2 are the expected background and observation error variance for the eigenvector $\underline{e}$. The application of (4.5.5) indicates that

$$\underline{\underline{O}}\underline{\underline{B}}^{-1}\underline{e} = \underline{\underline{B}}^{-1}\underline{\underline{O}}\underline{e} = \frac{\mu_O^2}{\mu_B^2}\underline{e} = \nu\underline{e} \tag{4.5.10}$$

Equation (4.5.8) can then be written as

$$\underline{\underline{\alpha}}\underline{e} = \left[1 + \frac{\mu_O^2}{\mu_B^2}\right]^{-1}\underline{e} \tag{4.5.11}$$

which is the discrete equivalent of (4.4.10). Consequently, the eigenvalues μ_B^2 and μ_O^2 play the same role as $E_B^2(m)$ and $E_O^2(\mathrm{m})$ in the continuous case.

When the background error is homogeneous (with expected variance E_B^2) and the observation error is spatially uncorrelated (with expected variance E_O^2, independent of observation station), then $\underline{\underline{B}} = E_B^2\underline{\underline{\rho}}_B$ and $\underline{\underline{O}} = E_O^2\underline{\underline{I}}$. $\underline{\underline{B}}$ and $\underline{\underline{O}}$ commute in this case; so if $\underline{e}$ is an eigenvector of $\underline{\underline{\rho}}_B$ with corresponding eigenvalue λ, then

$$\mu_O^2 = E_O^2, \qquad \mu_B^2 = E_B^2\lambda, \qquad \underline{\underline{\alpha}}\underline{e} = \lambda[\lambda + \varepsilon_O^2]^{-1}\underline{e} \tag{4.5.12}$$

where $\varepsilon_O^2 = E_O^2/E_B^2$. The resemblance between the spectral response for the continuous network with uncorrelated observation error (4.4.21) and Equation (4.5.12) is obvious.

Eigenvalues λ of the background error correlation matrix $\underline{\rho}_B$ play the same role as the background error spectral density $g_B(m) = g(m)$ in the continuous case.

Because $[1 + \varepsilon_O^2/\lambda]^{-1}$ is always less than or equal to one, the statistical interpolation algorithm always acts as a filter of the observation increments. The eigenvectors of $\underline{\rho}_B$ with the largest eigenvalues $(\lambda \gg \varepsilon_O^2)$ are filtered the least, and those with the smallest eigenvalues $(\lambda \ll \varepsilon_O^2)$ are filtered the most.

In the continuous case (4.4.21), when the background error spectral density $g_B(m)$ is a monotonically decreasing function of wavenumber m, the greatest/least filtering occurs for small/large scales. The concept of scale is less well defined for the eigenvectors of a matrix, but it is nonetheless possible to make the same statement for a discrete network.

Consider a very simple example, a two-station network, $\mathbf{r}_1$ and $\mathbf{r}_2$. Denote the background error correlation between the two stations as $\rho_B(\mathbf{r}_1, \mathbf{r}_2) = \rho$ and write the background error correlation matrix as

$$\underline{\rho}_B = \begin{pmatrix} 1 & \rho \\ \rho & 1 \end{pmatrix}$$

The eigensystem of $\underline{\rho}_B$ is given by

$$\begin{aligned} \underline{e}_1 &= \frac{1}{\sqrt{2}} \begin{vmatrix} 1 \\ 1 \end{vmatrix}, & \underline{e}_2 &= \frac{1}{\sqrt{2}} \begin{vmatrix} 1 \\ -1 \end{vmatrix} \\ \lambda_1 &= 1 + \rho, & \lambda_2 &= 1 - \rho \end{aligned} \tag{4.5.13}$$

The eigenvalues $\underline{e}_1$ and $\underline{e}_2$ have been normalized so that $\underline{e}_l^T \underline{e}_k = \delta_{lk}$. The eigenvector with the larger eigenvalue $\underline{e}_1$ corresponds to the mean value of the observations, and $\underline{e}_2$ corresponds to the gradient information in the observations. When the two observations are co-located, $\rho = 1$, $\lambda_1 = 2$, and $\lambda_2 = 0$. On the other hand, when the observations are far apart, $\rho \to 0$ and $\lambda_1 = \lambda_2 = 1$. For correlation models such as (4.3.19–21), $\rho \geq 0$ and λ_1 is always greater than or equal to λ_2.

An effective scale for each of the eigenvectors $\underline{e}_1$ and $\underline{e}_2$ can be thought of as being inversely proportional to the number of zero crossings of the eigenvector. $\underline{e}_1$ has no zero crossings, and $\underline{e}_2$ has one. Thus, $\underline{e}_1$ could be thought of as the largest scale eigenvector of $\underline{\rho}_B$ and $\underline{e}_2$ as the smallest scale eigenvector. Now, the background error correlation matrix $\underline{\rho}_B$ is obtained by evaluating the correlation function at the observation stations. The spectral densities of correlation functions such as (4.3.19–21) are monotonically decreasing functions of increasing wavenumber. In the same way, the eigenvectors of $\underline{\rho}_B$ have decreasing eigenvalues as their effective scale decreases. When the observation error is spatially uncorrelated, (4.5.12) implies that the smallest scale eigenvector $\underline{e}_2$ is filtered the most and the largest scale eigenvector $\underline{e}_1$ is filtered the least. From this simple example, it is reasonable to suppose that, provided the observation error is uncorrelated and the spectral density of the background error is monotonically decreasing, the statistical interpolation algorithm minimally/-maximally filters the largest/smallest scale eigenstructures of the background error correlation matrix. The analysis of the eigenstructure of a slightly more complicated observation network is left as Exercise 4.6.

The parameter ε_O^2 controls how closely the observations are fitted and plays a similar role to the parameter γ in the Barnes' algorithm (Section 3.6). Thus, when $\varepsilon_O^2 \ll \lambda$ for all the eigenvectors of the correlation matrix $\underline{\underline{\rho}}_B$, the observations are well fitted, and there is little filtering. This is similar to Figure 3.9(a). On the other hand, when $\lambda \ll \varepsilon_O^2$ for all the eigenvectors of $\underline{\underline{\rho}}_B$, the statistical interpolation algorithm ignores the observations and the analysis reverts to the background field. This situation is more akin to Figure 3.9(d).

Let us now return to the more general case of (4.5.1–9) and consider the analysis error covariance matrix defined in (2.2.16–19). Consider the form given by (2.2.19),

$$\underline{\underline{A}}^{-1} = \underline{\underline{B}}^{-1} + \underline{\underline{O}}^{-1} \tag{4.5.14}$$

Now $\underline{\underline{A}}$, $\underline{\underline{B}}$, and $\underline{\underline{O}}$ are covariance matrices and are real, symmetric, positive definite $K \times K$ matrices. Except in special circumstances, the set of K eigenvectors of each of these matrices differs. However, define $\mu_A^2(l)$, $\mu_B^2(l)$, and $\mu_O^2(l)$, $1 \le l \le K$, as the eigenvalues of $\underline{\underline{A}}$, $\underline{\underline{B}}$, and $\underline{\underline{O}}$, respectively, ordered in nondecreasing order. From Appendix D, $\mu_A^2(l)$, $\mu_B^2(l)$, and $\mu_O^2(l)$ can be regarded as the expected variances of the lth eigenvectors of the analysis, background, and observation error covariance matrices, respectively.

The eigenvalues $\mu_A^2(l)$, $\mu_B^2(l)$, and $\mu_O^2(l)$ are real and positive, and so are $\mu_A^{-2}(l)$, $\mu_B^{-2}(l)$, and $\mu_O^{-2}(l)$. Consequently, $\underline{\underline{A}}^{-1}$, $\underline{\underline{B}}^{-1}$, and $\underline{\underline{O}}^{-1}$ are real, symmetric, positive definite matrices. From (4.5.14),

$$\underline{z}^T \underline{\underline{A}}^{-1} \underline{z} = \underline{z}^T \underline{\underline{B}}^{-1} \underline{z} + \underline{z}^T \underline{\underline{O}}^{-1} \underline{z}$$

where $\underline{z}$ is an arbitrary column vector. Thus,

$$\underline{z}^T \underline{\underline{A}}^{-1} \underline{z} > \underline{z}^T \underline{\underline{B}}^{-1} \underline{z} \qquad \text{and} \qquad \underline{z}^T \underline{\underline{A}}^{-1} \underline{z} > \underline{z}^T \underline{\underline{O}}^{-1} \underline{z} \tag{4.5.15}$$

A well-known result from the minimax characterization of algebraic eigenvalues (Wilkinson 1965) is that if $\underline{\underline{G}}$ and $\underline{\underline{H}}$ are real, symmetric, positive definite $K \times K$ matrices, with eigenvalues g_l and h_l, $1 \le l \le K$, arranged in nondecreasing order, then

$$g_l < h_l, \qquad 1 \le l \le K, \qquad \text{provided} \qquad \underline{z}^T \underline{\underline{G}} \underline{z} < \underline{z}^T \underline{\underline{H}} \underline{z} \tag{4.5.16}$$

where $\underline{z}$ is an arbitrary column vector.

From (4.5.15–16),

$$\mu_A^{-2}(l) > \mu_B^{-2}(l), \qquad \mu_A^{-2}(l) > \mu_O^{-2}(l)$$

or

$$\mu_A^2(l) < \mu_B^2(l), \qquad \mu_A^2(l) < \mu_O^2(l) \tag{4.5.17}$$

Thus, if the eigenvectors of $\underline{\underline{A}}$, $\underline{\underline{B}}$, and $\underline{\underline{O}}$ are ordered by their relative variance, the expected variance (eigenvalue) of the lth eigenvector of $\underline{\underline{A}}$ is always less than the expected variance of the corresponding eigenvectors of $\underline{\underline{B}}$ and $\underline{\underline{O}}$, even though the eigenvectors themselves are different. Equation (4.5.17) is the discrete counterpart of (4.4.15).

Application of (4.5.17) and (4.5.16) in reverse shows that

$$\underline{z}^{T}\underline{\underline{A}}\underline{z} < \underline{z}^{T}\underline{\underline{B}}\underline{z} \qquad \text{and} \qquad \underline{z}^{T}\underline{\underline{A}}\underline{z} < \underline{z}^{T}\underline{\underline{O}}\underline{z}$$

or

$$\underline{z}^{T}[\underline{\underline{B}} - \underline{\underline{A}}]\underline{z} > 0 \qquad \text{and} \qquad \underline{z}^{T}[\underline{\underline{O}} - \underline{\underline{A}}]\underline{z} > 0 \tag{4.5.18}$$

Thus $\underline{\underline{B}} - \underline{\underline{A}}$ and $\underline{\underline{O}} - \underline{\underline{A}}$ are real, symmetric, positive definite matrices. It can be shown that the diagonal elements of positive definite matrices are positive. Suppose that the real diagonal element c_{kk} of the positive definite matrix $\underline{\underline{C}}$ were negative. Define the column vector $\underline{z}$ with all zero elements except the kth element, which is equal to 1. Then, $\underline{z}^{T}\underline{\underline{C}}\underline{z} = c_{kk} < 0$, which violates the definition of a positive definite matrix (2.3.12). Thus, the diagonal elements of $\underline{\underline{B}} - \underline{\underline{A}}$ and $\underline{\underline{O}} - \underline{\underline{A}}$ are positive. But the diagonal elements of $\underline{\underline{A}}$, $\underline{\underline{B}}$, and $\underline{\underline{O}}$ are simply the $E_{A}^{2}(\mathbf{r}_k)$, $E_{B}^{2}(\mathbf{r}_k)$, and $E_{O}^{2}(\mathbf{r}_k)$, the expected analysis, background, and observation error variances, respectively, at the kth observation station, $1 \leq k \leq K$. So,

$$E_{A}^{2}(\mathbf{r}_k) < E_{B}^{2}(\mathbf{r}_k) \qquad \text{and} \qquad E_{A}^{2}(\mathbf{r}_k) < E_{O}^{2}(\mathbf{r}_k) \tag{4.5.19}$$

and the expected analysis error at the observation stations is always less than either the expected background or observation errors, consistent with the continuous case discussed in the previous section. Note that (4.5.17) and (4.5.19) are vector generalizations (for $N = 2$) of the scalar result (4.1.12).

Now consider the case when $\underline{\underline{B}}$ and $\underline{\underline{O}}$ commute as in (4.5.10). Suppose $\underline{e}$ is an eigenvector of $\underline{\underline{B}}$ and $\underline{\underline{O}}$ with corresponding eigenvalues μ_{B}^{2} and μ_{O}^{2}. Then, from (4.5.14), $\underline{e}$ is also an eigenvector of $\underline{\underline{A}}$. Define μ_{A}^{2} as the corresponding eigenvalue of $\underline{\underline{A}}$ given by

$$\frac{1}{\mu_{A}^{2}} = \frac{1}{\mu_{O}^{2}} + \frac{1}{\mu_{B}^{2}} \qquad \text{or} \qquad \mu_{A}^{2} = \frac{\mu_{B}^{2}\mu_{O}^{2}}{\mu_{B}^{2} + \mu_{O}^{2}} \tag{4.5.20}$$

which is the discrete counterpart of (4.4.14). Define $v = \mu_{O}^{2}/\mu_{B}^{2}$ as in (4.5.10) and rewrite (4.5.20), giving

$$\frac{\mu_{A}^{2}}{\mu_{B}^{2}} = v[1 + v]^{-1} \tag{4.5.21}$$

Obviously, $0 < \mu_{A}^{2}/\mu_{B}^{2} < 1$ as demonstrated earlier. μ_{A}^{2}/μ_{B}^{2} is a measure of the reduction of the expected background error variance for the eigenvector $\underline{e}$ after the statistical interpolation has been completed. From (4.5.21), it can be seen that when $\mu_{B}^{2} \gg \mu_{O}^{2}$, then $\mu_{A}^{2} \ll \mu_{B}^{2}$; and when $\mu_{B}^{2} \ll \mu_{O}^{2}$, then $\mu_{A}^{2} \lessapprox \mu_{B}^{2}$. Thus, the maximum/minimum expected error reduction by the algorithm occurs for eigenmodes in which the expected background error is much larger/smaller than the expected observation error. In the example (4.5.13), the maximum error reduction by the algorithm would occur for eigenmode $\underline{e}_1$ and the minimum for $\underline{e}_2$.

4.6 One, two, and multiple observation problems

In Section 4.4, the limiting case of an infinitely dense observation network was instructive in examining the spectral response of the statistical interpolation algorithm. This section examines a number of limiting cases in which there are a small number of discrete observations. All cases consider univariate analysis with uncorrelated observation errors and homogeneous background error. We introduce a simplified notation.

Denote $\mathbf{r}_0$ as the location of the analysis gridpoint and $\mathbf{r}_k$, $1 \leq l \leq K$ as the position of each observation station. Denote the expected background error variance as E_B^2 and the expected observation error variance as $E_O^2(\mathbf{r}_k)$. Then for each observation, define normalized observation errors $\varepsilon_k^2 = E_O^2(\mathbf{r}_k)/E_B^2$ and a posteriori weight W_k. The background error correlation between any two locations m and n is denoted ρ_{mn}. Because the analysis is univariate, $\rho_{mn} = \rho_{nm}$ and $\rho_{nn} = 1$.

The statistical interpolation algorithm (4.2.9–10) can then be written in the form

$$f_A(\mathbf{r}_0) - f_B(\mathbf{r}_0) = \sum_{k=1}^{K} W_k[f_O(\mathbf{r}_k) - f_B(\mathbf{r}_k)] \tag{4.6.1}$$

$$\sum_{l=1}^{K} W_l[\rho_{kl} + \varepsilon_k^2] = \rho_{k0}, \qquad 1 \leq k \leq K \tag{4.6.2}$$

The normalized expected analysis error variance (assuming the error covariances are correct) is given by (4.2.12),

$$\varepsilon_A^2 = \frac{E_A^2}{E_B^2} = 1 - \sum_{k=1}^{K} \rho_{k0} W_k \tag{4.6.3}$$

Case 1: A single observation

Consider first the analysis at gridpoint 0 with a single observation at point 1. The statistical interpolation algorithm given by (4.6.1–3) becomes

$$f_A(\mathbf{r}_0) = f_B(\mathbf{r}_0) + W_1[f_O(\mathbf{r}_1) - f_B(\mathbf{r}_1)] \tag{4.6.4}$$

$$W_1 = \rho_{10}/(1 + \varepsilon_1^2) \tag{4.6.5}$$

$$\varepsilon_A^2 = 1 - \frac{\rho_{10}^2}{1 + \varepsilon_1^2} \tag{4.6.6}$$

The weight equation (4.6.5) can be compared with its counterpart for the single correction algorithm (3.1.7). For a single observation and $w(\tilde{r}) = \rho_{01}$, the a posteriori weight is $\rho_{10}/(\rho_{10} + \varepsilon_1^2)$. Thus, the single correction algorithm tends to give more weight to distant observations than (4.6.5).

Now consider the special case in which the observation station and the analysis gridpoint coincide ($\mathbf{r}_1 = \mathbf{r}_0$). Then, $\rho_{10} = 1$ and

$$W_1 = \frac{1}{1 + \varepsilon_1^2}, \qquad \varepsilon_A^2 = \frac{\varepsilon_1^2}{1 + \varepsilon_1^2} \tag{4.6.7}$$

It can be seen that (4.6.4) with $\mathbf{r}_1 = \mathbf{r}_0$ and (4.6.7) are equivalent to (2.2.7). The value of W_1 in (4.6.7) is the maximum value of (4.6.5). If the observation $f_O(\mathbf{r}_0)$ is perfect, then $\varepsilon_1^2 = \varepsilon_A^2 = 0$, $W_1 = 1$, and from (4.6.4), $f_A(\mathbf{r}_0) = f_O(\mathbf{r}_0)$, as would be expected.

Now consider the case in which the observation station and the analysis gridpoint are widely separated. Then, $\rho_{10} = 0$, and

$$W_1 = 0, \qquad \varepsilon_A^2 = 1, \qquad \text{or} \qquad E_A^2 = E_B^2 \tag{4.6.8}$$

In this case, $f_A(\mathbf{r}_0) = f_B(\mathbf{r}_0)$, and the expected analysis error variance is no smaller than the expected background error variance.

Case 2: A network with two observing stations

In this case, the analysis weights given by (4.6.2) are

$$\begin{aligned} W_1(1 + \varepsilon_1^2) + W_2\rho_{12} &= \rho_{10} \\ W_1\rho_{12} + W_2(1 + \varepsilon_2^2) &= \rho_{20} \end{aligned} \tag{4.6.9}$$

Solving (4.6.9) for W_1 and W_2 yields

$$W_1 = \frac{\rho_{10}(1 + \varepsilon_2^2) - \rho_{20}\rho_{12}}{(1 + \varepsilon_1^2)(1 + \varepsilon_2^2) - \rho_{12}^2} \tag{4.6.10}$$

$$W_2 = \frac{\rho_{20}(1 + \varepsilon_1^2) - \rho_{10}\rho_{12}}{(1 + \varepsilon_1^2)(1 + \varepsilon_2^2) - \rho_{12}^2} \tag{4.6.11}$$

with expected analysis error (from 4.6.3) of

$$\varepsilon_A^2 = 1 - \frac{\rho_{10}^2(1 + \varepsilon_2^2) + \rho_{20}^2(1 + \varepsilon_1^2) - 2\rho_{10}\rho_{20}\rho_{12}}{(1 + \varepsilon_1^2)(1 + \varepsilon_2^2) - \rho_{12}^2} \tag{4.6.12}$$

Before considering the general two-station case (4.6.9–12), we discuss two special cases. In both, the observation error variances are assumed to be equal, $\varepsilon_1^2 = \varepsilon_2^2 = \varepsilon^2$, and both observations are assumed to be an equal distance from the observation gridpoint, $\rho_{10} = \rho_{20} = \tilde{\rho}$.

Case 3: Two isolated observations

If the two observations are on opposite sides of the analysis gridpoint, then they are much less strongly correlated with each other than they are with the analysis gridpoint. Suppose $\mathbf{r}_1$ and $\mathbf{r}_2$ are sufficiently separated that ρ_{12} is vanishing small and $\tilde{\rho}$ is still finite. Then,

$$W_1 = W_2 \approx \frac{\tilde{\rho}}{1 + \varepsilon^2} \qquad \text{and} \qquad \varepsilon_A^2 \approx 1 - \frac{2\tilde{\rho}^2}{1 + \varepsilon^2} \tag{4.6.13}$$

Case 4: Two adjacent observations

If the two observations are very close to each other, then they will be strongly correlated. In the limit when points 1 and 2 are co-located, $\rho_{12} = 1$, and

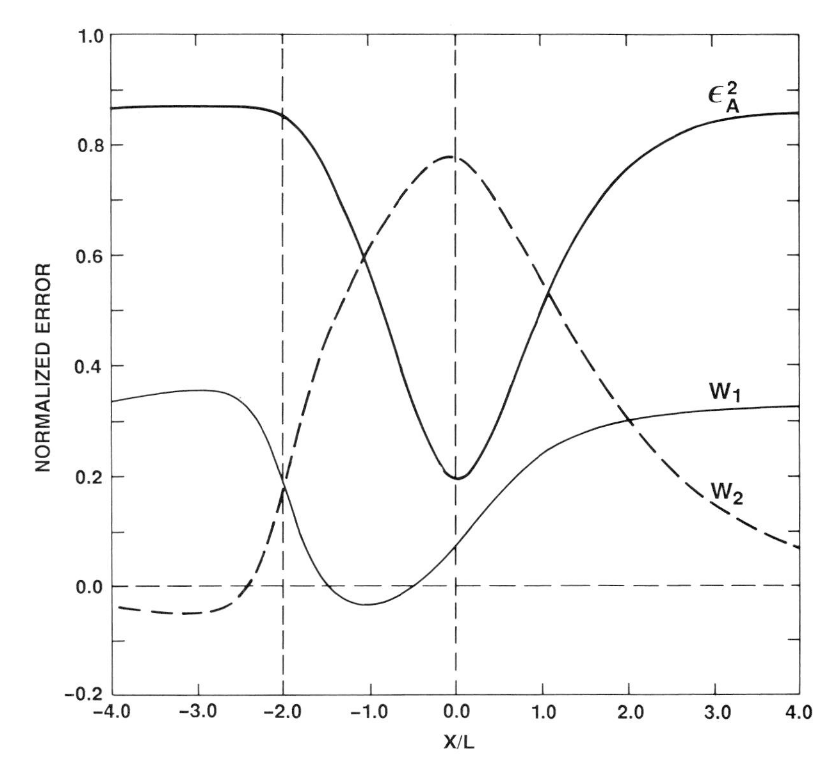

Figure 4.7 A posteriori weights W_1 and W_2 and normalized expected analysis error variance ε_A^2 for the analysis gridpoint at $x = 0$, observation 1 at $x = -2.0$ and the position of observation 2 varying between $x = \pm\infty$.

$$W_1 = W_2 = \frac{\tilde{\rho}}{2 + \varepsilon^2} \qquad \text{and} \qquad \varepsilon_A^2 = 1 - \frac{2\tilde{\rho}^2}{2 + \varepsilon^2} \tag{4.6.14}$$

When two observations are very close to each other (case 4), the weights are smaller and the analysis error is larger than when the two observations are far apart (case 3). For the single correction algorithm discussed in Section 3.1, these two cases would be indistinguishable because the method does not account for correlations among the observations.

We now return to the more general two-observation case (4.6.9–12). Figure 4.7 is a plot of W_1, W_2, and ε_A^2 for a two-observation network. The domain is one dimensional, and the normalized observation error variances at the two stations are equal: $\varepsilon_1^2 = \varepsilon_2^2 = 0.25$. The background error correlation model is given by (4.3.20), and the observing station and analysis gridpoint locations are given in terms of x/L. The analysis gridpoint is at $x_0/L = 0$, and observation station 1 is at $x_1/L = -2.0$. The location of observation station 2 is allowed to vary with x; $-\infty \leq x_2/L \leq \infty$. The ordinate indicates the values of the normalized analysis error ε_A^2 and weights W_1 and W_2 as a function of the position of observation station 2. From (4.3.20), $\rho_{10} = 0.406$.

As $x_2/L \to \pm\infty$, $W_2 = 0$, and W_1 and ε_A^2 tend to their single observation values, given by (4.6.5) and (4.6.6), respectively. At $x_2/L = +2.0$, $W_1 = W_2$, and the weights and normalized analysis error are given approximately by (4.6.13). At $x_2/L = -2.0$, $W_1 = W_2$, and the weights and normalized analysis errors are given exactly by (4.6.14) and, as indicated, are less than when $x_2/L = +2.0$. When point 1 lies between point 2 and the analysis gridpoint, W_2 can be negative. Similarly, when point 2 lies between point 1 and the analysis gridpoint, W_1 can be negative. These are examples of screening, in which the presence of a closer observation causes a more distant observation to have a negative analysis weight. Compare this with Figure 4.6 or Figure 3.2(b).

Case 5: Two perfect observations equidistant from the analysis gridpoint
This case illustrates the interpolation properties of statistical interpolation. For perfect observations, $\varepsilon_1^2 = \varepsilon_2^2 = 0$. Defining $\rho_{12} = \rho$ and $\rho_{10} = \rho_{20} = \tilde{\rho}$ yields

$$W_1 = W_2 = \frac{\tilde{\rho}}{1+\rho}, \qquad \varepsilon_A^2 = 1 - \frac{2\tilde{\rho}^2}{1+\rho} \tag{4.6.15}$$

First, consider interpolation when, in the one-dimensional case, the two observations are on opposite sides of the analysis gridpoint. Assume that the background error correlation model is given by

$$\tilde{\rho} = \exp\left(-\frac{\Delta x}{2L}\right), \qquad \rho = \exp\left(-\frac{\Delta x}{L}\right) \tag{4.6.16}$$

where Δx is the (positive) distance between observations, and L is the characteristic scale of the background error correlation function. The functional form of the correlation model (4.6.16) is a monotonically decreasing function of Δx similar to those given in Section 4.3. However, it is not used operationally because of the behavior of its spatial derivatives at the origin (see Section 5.2). This drawback of the correlation model (4.6.16) does not affect the present example, which is univariate. Substitution of (4.6.16) into (4.6.15) gives

$$W_1 = W_2 = \frac{1}{2\cosh(\Delta x/2L)}, \qquad \varepsilon_A^2 = \frac{\sinh(\Delta x/2L)}{\cosh(\Delta x/2L)} \tag{4.6.17}$$

When $\Delta x = 0$, $W_1 = W_2 = 0.5$, $\varepsilon_A^2 = 0$, and the analyzed value is equal to the average of the two observed values. On the other hand, as $\Delta x \to \infty$, $W_1 = W_2 \to 0$, $\varepsilon_A^2 = E_A^2/E_B^2 \to 1$, and the analyzed values is equal to the background value. Compare these results with those given by the Lagrange interpolation formula, (E1) of Appendix E, and the single correction algorithm (3.1.7). For both these algorithms $W_1 = W_2 = \frac{1}{2}$ regardless of the value of Δx. Both algorithms give a finite weight to the observations even if they are an infinite distance away from the analysis gridpoint.

Now, consider extrapolation when, in the one-dimensional case, the two observations are close together, and the analysis gridpoint is a large distance from either of them ($\tilde{\rho} \approx 0$). Equation (4.6.15) implies that W_1 and W_2 would be very small and that the analysis would revert to the background. In the single correction algorithm, the weights would again equal $\frac{1}{2}$; but in the Lagrange interpolation formula, the weights would, in general, be unbounded (see Figure E1).

For atmospheric objective analysis, the extrapolation properties of Lagrange interpolation are totally unacceptable, and those of the single correction algorithm are seriously flawed.

Case 6: A network with K clustered observations

Under certain conditions, simple solutions can be obtained for a multiple station network. Assume the K observation stations are co-located and thus $\rho_{lk} = 1$ and $\rho_{k\mathrm{O}} = \tilde{\rho}$. Define $\bar{\mathbf{r}}$ as the position of the K co-located observations and $f_{\mathrm{O}}^{k}(\bar{\mathbf{r}})$ as the kth observation at $\bar{\mathbf{r}}$. Then, (4.6.1–2) becomes

$$f_{\mathrm{A}}(\mathbf{r}_{\mathrm{O}}) - f_{\mathrm{B}}(\mathbf{r}_{\mathrm{O}}) = \sum_{k=1}^{K} W_k[f_{\mathrm{O}}^{k}(\bar{\mathbf{r}}) - f_{\mathrm{B}}(\bar{\mathbf{r}})] \tag{4.6.18}$$

and

$$\sum_{l=1}^{K} W_l + \varepsilon_k^2 W_k = \tilde{\rho}, \qquad 1 \le k \le K \tag{4.6.19}$$

First, note that the elements of the background error correlation matrix $\underline{\underline{\rho}}_{\mathrm{B}}$ of (4.6.19) are all equal to 1. The matrix is singular and has one eigenvector whose elements are all equal to 1 with associated eigenvalue equal to K. Because $\mathrm{Trace}(\underline{\underline{\rho}}_{\mathrm{B}}) = K$ and all eigenvalues are nonnegative, the remaining eigenvalues must all be identically equal to zero. Thus, the only eigenvector of $\underline{\underline{\rho}}_{\mathrm{B}}$ with nonzero eigenvalue is the mean or average eigenvector.

To find the a posteriori weights, divide each equation of (4.6.19) by ε_k^2 and sum over K, which gives

$$\sum_{k=1}^{K} W_k = \frac{\tilde{\rho} \sum_{k=1}^{K} \varepsilon_k^{-2}}{1 + \sum_{k=1}^{K} \varepsilon_k^{-2}} \tag{4.6.20}$$

Back substitution of (4.6.20) into (4.6.19) yields

$$W_k = \frac{\tilde{\rho}\varepsilon_k^{-2}}{1 + \sum_{k=1}^{K} \varepsilon_k^{-2}}, \qquad 1 \le k \le K \tag{4.6.21}$$

Define

$$\bar{\varepsilon}^2 = \left[\sum_{k=1}^{K} \varepsilon_k^{-2}\right]^{-1} \qquad \text{and} \qquad \bar{f}_{\mathrm{O}}(\bar{\mathbf{r}}) = \bar{\varepsilon}^2 \sum_{k=1}^{K} \varepsilon_k^{-2} f_{\mathrm{O}}^{k}(\bar{\mathbf{r}})$$

Then, (4.6.18) becomes

$$f_{\mathrm{A}}(\mathbf{r}_{\mathrm{O}}) - f_{\mathrm{B}}(\mathbf{r}_{\mathrm{O}}) = \frac{\tilde{\rho}}{1 + \bar{\varepsilon}^2} [\bar{f}_{\mathrm{O}}(\bar{\mathbf{r}}) - f_{\mathrm{B}}(\bar{\mathbf{r}})] \tag{4.6.22}$$

Note the resemblance between (4.6.22) and (4.6.4–5). In effect, the K co-located observations have been replaced by a weighted linear combination or superobservation $\bar{f}_{\mathrm{O}}(\bar{\mathbf{r}})$, in which the weights are given by (4.1.9). This superobservation is then treated as a single observation with expected observation error $\bar{\varepsilon}^2$ given by (4.1.11). This is the simplest version of the super-obbing algorithm. The more general case, when the observations are not exactly co-located, is left as Exercise 8.4.

If all expected observation errors are the same and are equal to ε^2, then $\bar{\varepsilon}^2 = \varepsilon^2/K$ and $\bar{f}_{\mathrm{O}}(\bar{\mathbf{r}}) = (1/K)\sum_{k=1}^{K} f_{\mathrm{O}}^{k}(\bar{\mathbf{r}})$. The expected analysis error variance (4.6.3) in this case is

$$\varepsilon_{\mathrm{A}}^2 = 1 - \frac{\tilde{\rho}^2 K}{K + \varepsilon^2} \tag{4.6.23}$$

In the special case when the observation location and the analysis gridpoint coincide, $\mathbf{r}_{\mathrm{O}} = \bar{\mathbf{r}}$, $\tilde{\rho} = 1$, and (4.6.23) becomes

$$\varepsilon_{\mathrm{A}}^2 = \frac{\varepsilon^2}{K + \varepsilon^2} \tag{4.6.24}$$

The expected analysis error (4.6.24) is the lower limit for K uncorrelated observations and continues to decrease as K increases.

4.7 The vertical question

In Section 4.3, it was assumed that background error covariances were horizontally and vertically separable, that is,

$$C_{\mathrm{B}}(\mathbf{r}_k, \mathbf{r}_l, P_k, P_l) = C_{\mathrm{B}}^{\mathrm{H}}(\mathbf{r}_k, \mathbf{r}_l) C_{\mathrm{B}}^{\mathrm{V}}(P_k, P_l) \tag{4.7.1}$$

where $\mathbf{r}$ indicates the horizontal coordinate, P the vertical coordinate, k and l indicate two stations, and superscripts H and V indicate horizontal and vertical, respectively. The validity of assumption (4.7.1) has been questioned (see Section 4.9); but it is widely invoked in the application of statistical interpolation to the analysis of global and synoptic scales. The horizontal background error covariance $C_{\mathrm{B}}^{\mathrm{H}}$ has received considerable theoretical and practical attention and was discussed in some detail in Section 4.3. The theory and practice of the vertical background error covariance is in a much less satisfactory state and will be dealt with here in a perfunctory manner.

The vertical aspects of objective analysis (like the vertical aspects of numerical modeling) have never been very satisfactory, for several reasons, all related to the vertical structure of the atmosphere. As noted in Section 1.1, vertical scales are much shorter than global and synoptic horizontal scales, and therefore the vertical variation of most variables is rapid. The most rapid vertical variation occurs in the planetary boundary layer, but large vertical gradients also exist near the tropopause. The atmosphere is bounded from below by the earth's surface and extends indefinitely above. However, in practice there are few regular observations above 30 km, and most numerical forecast models have explicit or implicit lids at about this same height. Consequently, the vertical domain can be regarded as bounded from below by a real physical boundary and from above by a nonphysical but nonetheless effective barrier. The rapid vertical variation and the upper and lower boundaries imply statistical inhomogeneity in the vertical; thus, the homogeneity assumption is rarely invoked.

Examples of vertical background error correlations are shown in Figure 4.8. These are the geopotential background error correlations determined from observations and a forecast background using the methods of Section 4.3. The correlations are plotted for eight vertical pressure levels (100, 150, 200, 300, 400, 500, 700, and 850 mb). The shape of the curves is different for each pressure level as would be expected in the inhomogeneous case. In addition, the variances $C_B^V(P_k, P_k)$ also differ for each value of k.

Lönnberg and Hollingsworth (1986) determined the vertical correlation of the geopotential background error as a function of horizontal scale. They found that the largest horizontal (global) scales had a broader (i.e., more barotropic) vertical scale than the synoptic scales. The vertical correlations plotted in Figure 4.8 are a combination of both the large-scale and the synoptic-scale vertical correlations.

One of the most serious difficulties in vertical analysis is caused by single level data (see Hollingsworth 1987). Cloud track and aircraft wind observations (Section 1.3) are often available in remote oceanic regions where no other sources of data exist. Unfortunately, often only a single observation is available in a given vertical column. In this case, the analysis problem is underdetermined; that is, the number of observations is less than the number of degrees of freedom in the analysis. The same problem arises in the horizontal with isolated island observations. Several approaches to this problem are possible, none particularly satisfactory.

One extreme approach is to draw for the observation at its own vertical level and allow the objective analysis to revert to the background field at all other vertical levels. This is tantamount to assuming that vertical error correlations for both observations and background field are given by the Kronecker delta function. At the other extreme, one can allow the single level observation to affect all vertical levels equally, $C_B^V(P_k, P_l)$ independent of P_k and P_l. This is equivalent to projecting the single observation increment only onto the vertically averaged or barotropic component of the atmospheric flow. Barwell and Lorenc (1985) showed that a broader vertical correlation gave better results than the delta function approach when used in operational analysis. Separate vertical correlations for the largest horizontal scales and the synoptic scales have also found their way into operational practice.

4.8 Correlated observation error

In this chapter, the observation error has generally been assumed to be spatially uncorrelated (4.3.23). This assumption can always be made when the observations are taken with separate immovable instruments such as those in the surface observation networks. Radiosonde measurements are taken in a vertical column (apart from some horizontal drift) using the same instrument, and the vertical

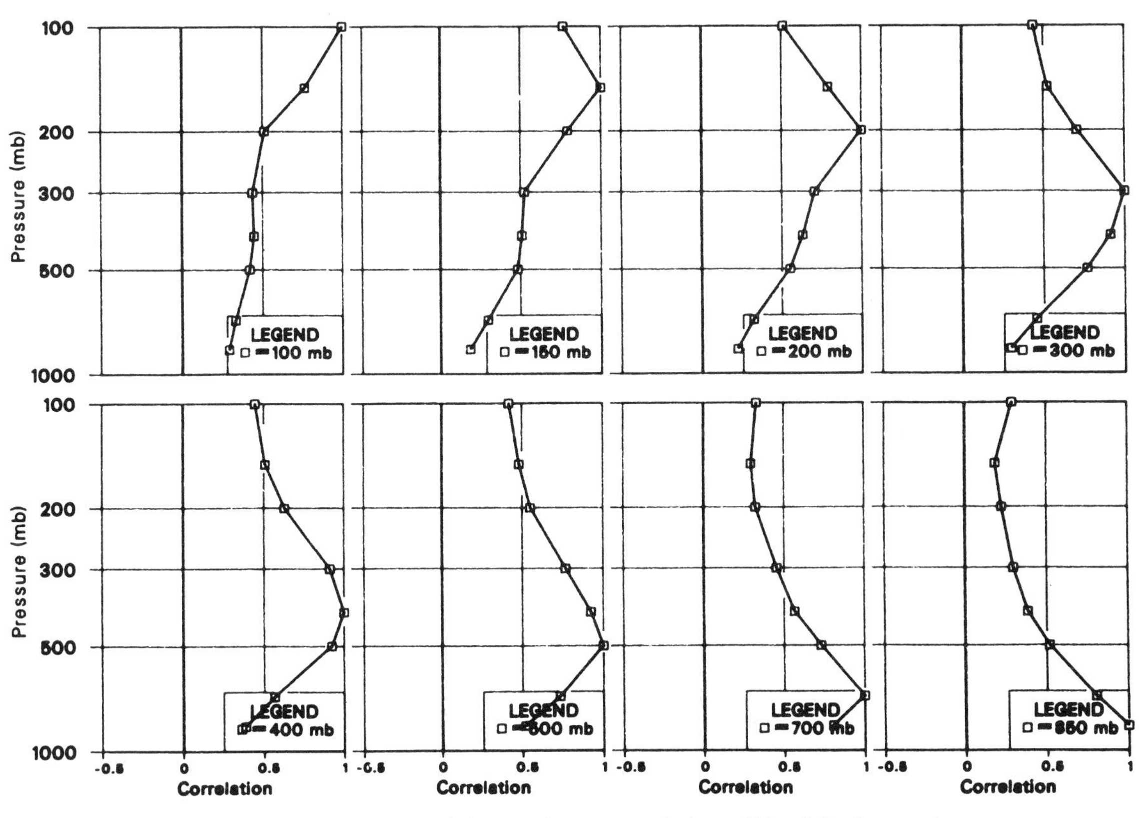

Figure 4.8 Vertical geopotential background (forecast) error correlations. (After Lönnberg and Hollingsworth 1986)

correlation of the observation errors cannot be ignored. The newer satellite-borne instruments often have both horizontally and vertically correlated observation error.

The vertically correlated radiosonde observation errors for the geopotential are illustrated in Figure 4.9, which shows the observation error covariance normalized by the observation error variances in the same format as in Figure 4.8. The covariances are plotted for 100, 150, 200, 300, 400, 500, 700, and 850 mb. Figure 4.10 gives an example of the horizontally correlated observation error for the 500 mb geopotential field for a satellite-borne radiometer; these data are for western Europe, but other horizontal observation error correlations are similar. The vertical and horizontal observation error correlations, Figures 4.9 and 4.10, can be compared with their background error counterparts, Figures 4.8 and 4.4, respectively.

Correlated observation errors change the nature of the univariate and multivariate algorithms. The accuracy and spatial responses of the algorithm can be affected substantially, as has been shown by Bergman and Bonner (1976), Seaman (1977), Schlatter and Branstator (1979), and Hollingsworth (1987).

We first examine the effect of correlated observation error on the expected analysis error of the univariate algorithm. Assume that the background error is homogeneous, with expected error variance E_{B}^2, and the observation error is spatially correlated, but with equal expected observation error variance E_{O}^2 at each observation station. In this case, the weight equation (4.6.2) can be written as

$$\sum_{l=1}^{K} W_l[\rho_{kl}^{\mathrm{B}} + \varepsilon^2 \rho_{kl}^{\mathrm{O}}] = \rho_{k\mathrm{O}}^{\mathrm{B}} \tag{4.8.1}$$

where ρ_{kl}^{B}, $\rho_{k\mathrm{O}}^{\mathrm{B}}$, and ρ_{kl}^{0} are the background and observation error correlations and $\varepsilon^2 = E_{\mathrm{O}}^2/E_{\mathrm{B}}^2$ are the normalized observation error variances. As in Section 4.6, subscripts k and l indicate observation stations and subscript zero indicates analysis gridpoint. Equations (4.6.1) and (4.6.3) remain the same.

Assume that all observation stations and the analysis gridpoint coincide. Then, ρ_{kl}^{B}, $\rho_{k\mathrm{O}}^{\mathrm{B}}$, and $\rho_{kl}^{\mathrm{O}} = 1$, and

$$(1 + \varepsilon^2) \sum_{k=1}^{K} W_k = 1$$

Thus, all weights are the same, and from (4.6.3) we have

$$\varepsilon_{\mathrm{A}}^2 = 1 - \sum_{k=1}^{K} W_k = \frac{\varepsilon^2}{1 + \varepsilon^2} \tag{4.8.2}$$

Equation (4.8.2) can be compared with (4.6.24), which is its counterpart for uncorrelated observation error. Unlike (4.6.24), (4.8.2) does not converge to zero as $K \rightarrow \infty$. Thus, the lower limit of the expected analysis error (4.8.2) for correlated observation error is larger than it is for uncorrelated observation error.

The spectral response with correlated observation error can be examined using the continuous analogue of Section 4.4. The continuous form of the weight equation

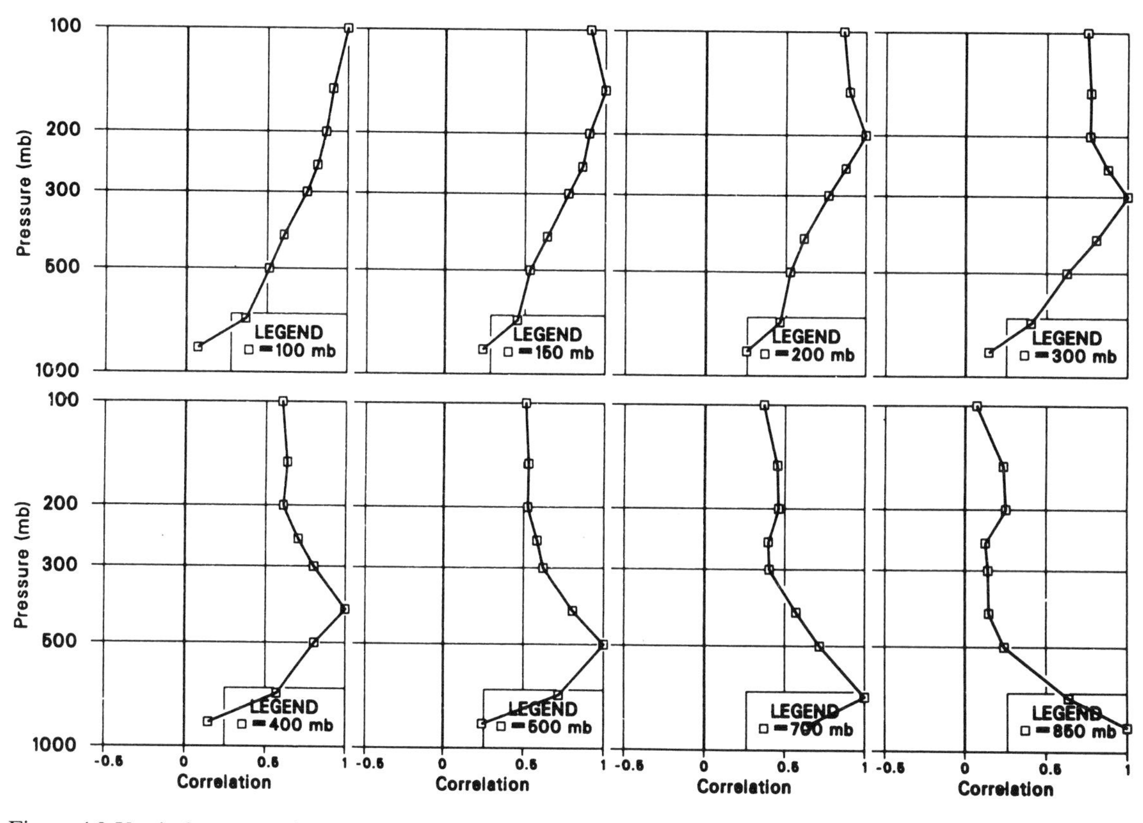

Figure 4.9 Vertical geopotential observation error correlations for radiosondes. (After Lönnberg and Hollingsworth 1986)

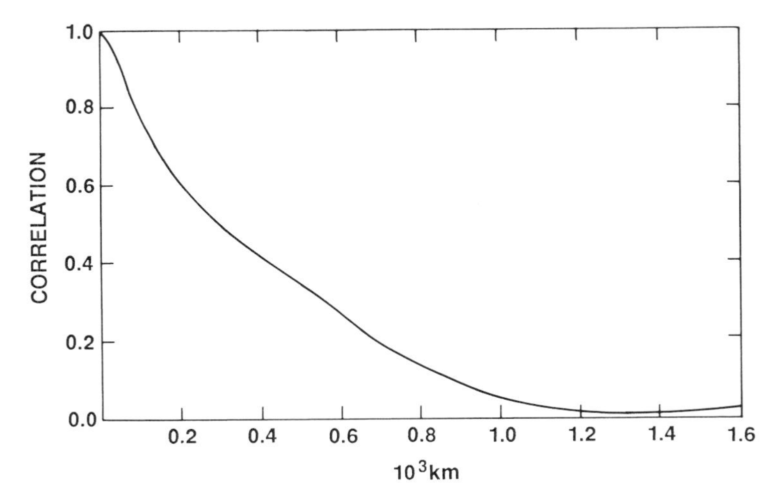

Figure 4.10 Horizontally correlated observation error for the 500 mb geopotential over Western Europe. (After Baker et al. *Mon. Wea. Rev.* **115**: 272, 1987. The American Meteorological Society.)

is given by (4.4.3) and the spectral response $\alpha(m)$ by (4.4.7):

$$\alpha(m) = \frac{L_B g_B(m)}{L_B g_B(m) + \varepsilon^2 L_O g_O(m)} \tag{4.8.3}$$

where L_B, L_O, g_B, and g_O have the same meaning as in Section 4.4, and $\varepsilon^2 = \varepsilon_O^2$. In Section 4.4, for uncorrelated observation errors, (4.4.22) indicated that if the background error spectral density $g_B(m)$ was a monotonically decreasing function of wavenumber m, then so was the response function $\alpha(m)$. This is not necessarily true when the observation error is spatially correlated. The derivative of (4.8.3) with respect to m is

$$\alpha'(m) = \frac{\varepsilon^2 L_B L_O [g_O(m) g_B'(m) - g_B(m) g_O'(m)]}{[L_B g_B(m) + \varepsilon^2 L_O g_O(m)]^2} \tag{4.8.4}$$

where (′) indicates differentiation with respect to m. For monotonically decreasing functions $g_B(m)$ and $g_O(m)$, $g_B'(m)$ and $g_O'(m)$ are negative. Thus, $\alpha'(m)$ is a monotonically decreasing function of m when

$$\left|\frac{g_B'(m)}{g_B(m)}\right| \geq \left|\frac{g_O'(m)}{g_O(m)}\right|, \qquad \text{for all } m \tag{4.8.5}$$

Consider the correlation function (4.3.20). For this correlation model,

$$\left|\frac{g'(m)}{g(m)}\right| = \frac{4m}{m^2 + L^{-2}} \tag{4.8.6}$$

where L is the characteristic scale. Thus, for the correlation model (4.3.20), $|g'(m)/g(m)|$ increases when L increases and decreases when L decreases.

Suppose the observation and background error correlations are given by (4.3.20) with characteristic scales L_O and L_B, respectively. Then, from (4.8.3–6),

$$\begin{aligned} \alpha'(m) &< 0, \qquad L_O < L_B \\ \alpha'(m) &= 0, \qquad L_O = L_B \\ \alpha'(m) &> 0, \qquad L_O > L_B \end{aligned} \tag{4.8.7}$$

If the characteristic scale of the observation error correlation is less than that of the background error, the univariate algorithm acts like a low-pass filter on the observation increments, as in (4.4.21). When $L_O = L_B$, $\alpha'(m) = 0$, and

$$\alpha(m) = \frac{1}{1 + \varepsilon^2} \tag{4.8.8}$$

which is a filter with no scale selectivity at all. When $L_O > L_B$, $\alpha'(m) > 0$, and the algorithm behaves like a high-pass filter of the observation increments. In this last case, the algorithm draws for the observations at small scales and for the background at large scales, which is the reverse of (4.4.21). This can be easily explained because when $L_O > L_B$, the observation error is mostly in the large scales and only the smaller scales of the observations are reliable.

The effect of correlated observation error on the spectral characteristics of the statistical interpolation algorithm can be illustrated further by plotting the errors $E_B^2(m)$, $E_O^2(m)$, and $E_A^2(m)$ of (4.4.8) and (4.4.14). Define

$$E_B^2(m) = \eta_B g_B(m), \qquad E_O^2(m) = \frac{\eta_B \varepsilon^2 g_O(m) L_O}{L_B} \tag{4.8.9}$$

and

$$E_A^2(m) = \frac{E_B^2(m) E_O^2(m)}{E_B^2(m) + E_O^2(m)}$$

where $\eta_B = L_B E_B^2$.

$E_B^2(m)$, $E_O^2(m)$, and $E_A^2(m)$, scaled by η_B, are plotted as a function of m for two cases in Figure 4.11. The correlation model is (4.3.20) with $L_B = 1.0$ and $\varepsilon^2 = 0.50$. In (a) the characteristic scale of the observation error is shorter than the characteristic scale of the background error, $L_O = 0.5 L_B$. In (b), the algorithm acts as a high-pass filter, $L_O = 3 L_B$. The background error spectrum $E_B^2(m)$ is the same for both panels, but the peak in the observation error spectrum $E_O^2(m)$ shifts increasingly to larger scales as L_O increases relative to L_B. The response of the statistical interpolation algorithm is to reduce the expected analysis error variance in the shorter scales while increasing it in the larger scales.

In the discrete case, the filtering properties with correlated observation error can be examined using (4.5.4). Suppose the observation and background error covariance matrices are equal, apart from a constant of proportionality, $\underline{\underline{O}} = \varepsilon^2 \underline{\underline{B}}$. Then, (4.5.8)

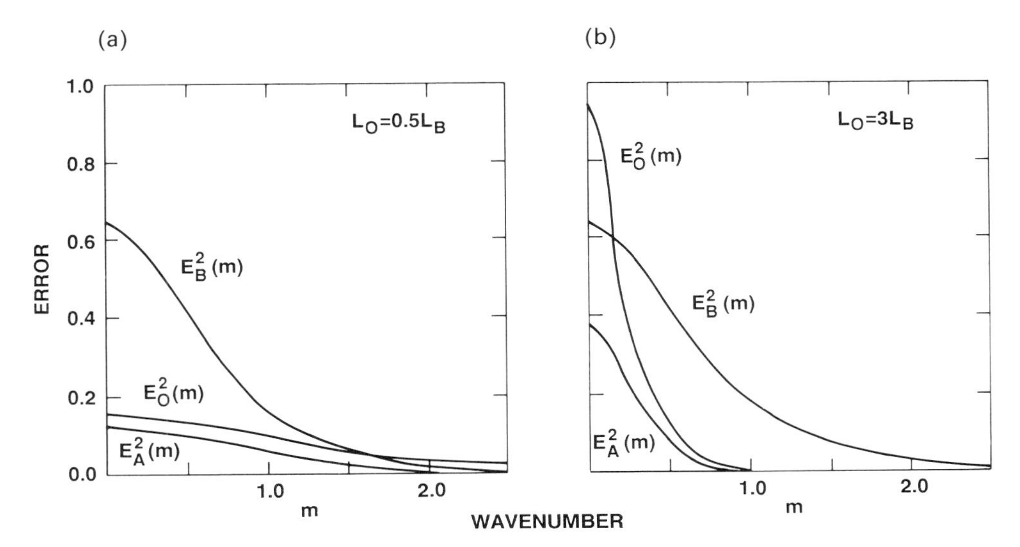

Figure 4.11 Plot of error versus wavenumber for the one-dimensional continuous network with correlated observation error. The background, observation, and analysis errors are denoted $E_B^2(m)$, $E_O^2(m)$, and $E_A^2(m)$, respectively. (a) shows $L_O = 0.5L_B$; (b) shows $L_O = 3L_B$.

becomes

$$\underset{\approx}{\alpha}\underset{\sim}{e} = [1 + \varepsilon^2]^{-1}\underset{\sim}{e} \tag{4.8.10}$$

Equation (4.8.10) implies that all eigenvalues of $\underset{\approx}{\alpha}$ are the same and thus there is no scale selectivity in the filter, consistent with (4.8.8).

The effect of correlated observation error in the discrete case was investigated in more detail by Seaman (1977). The expected geopotential analysis error for a simple two-dimensional observation network was examined as a function of the ratio L_O/L_B. Consistent with (4.8.2), it was found that geopotential analysis error increased as L_O/L_B increased. Seaman (1977) also showed that correlated observation error increased the accuracy of the analysis of the gradient of the geopotential field. This result is consistent with Figure 4.11. An accurate estimate of a gradient requires an accurate estimate of the smaller scales. From Figure 4.11, it is clear that the analysis error is reduced in the shorter scales as L_O/L_B increases, thus improving the accuracy of the analysis of the gradient.

4.9 Misspecification of the background and observation error covariances

Equations (4.2.12) or (4.2.15) provide internally consistent estimates of the expected analysis error produced by the statistical interpolation algorithm. They are obtained by inserting the optimum a posteriori weights (4.2.10) into the general expression for the expected analysis error of an unbiased linear estimate (4.2.11). Equation (4.2.12) gives the smallest expected analysis error of all linear unbiased estimates of the form (4.2.9).

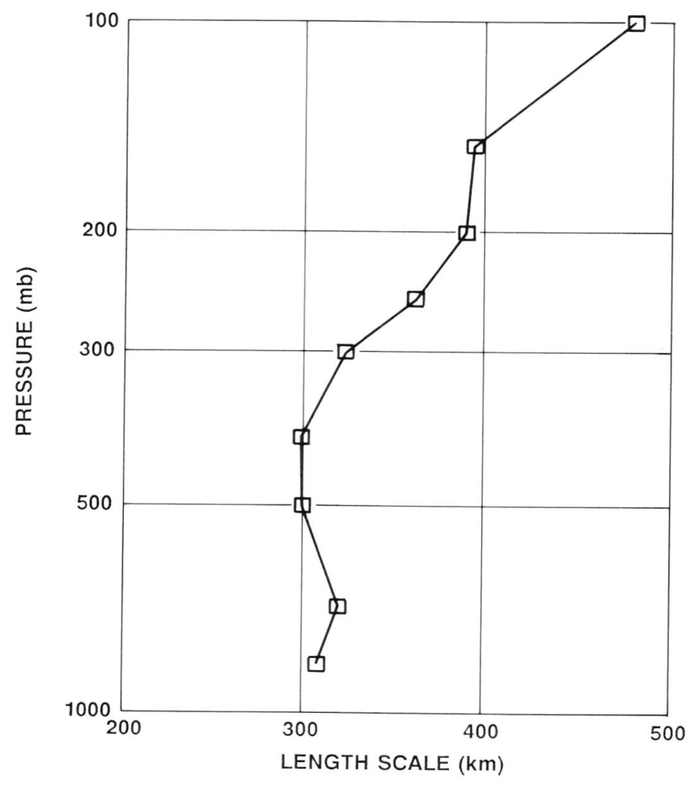

Figure 4.12 The characteristic scale of the geopotential background (forecast) error correlation for the North American radiosonde network as a function of pressure. (After Lönnberg and Hollingsworth 1986)

The accuracy of the estimates of analysis error (4.2.12) cannot really be determined in data-sparse regions. In data-rich areas, however, the internally generated analysis error estimates can be checked by withholding observations. The results of such experiments invariably indicate that the estimated analysis error is somewhat lower than the actual analysis error.

Why should this be? The problem is that the analysis error estimate (4.2.12) implicitly assumes that the specified error covariance matrices $\underline{\underline{B}}$ and $\underline{\underline{O}}$ are *correct*. However, we already know from Sections 4.2 and 4.3 that a considerable number of assumptions have to be made to derive usable forms of the background error covariance matrix $\underline{\underline{B}}$. For example, in Section 4.2 it was assumed that the background and observation error are not mutually correlated, and in Section 4.3 it was assumed that the background error correlations are separable and horizontally homogeneous or even isotropic. Moreover, the observation error includes both a relatively well-known instrument error and a somewhat more imprecise error of representativeness. Consequently, both the observation and background errors are likely to be in error. This means that the a posteriori weights given by (4.2.10) cannot be the true

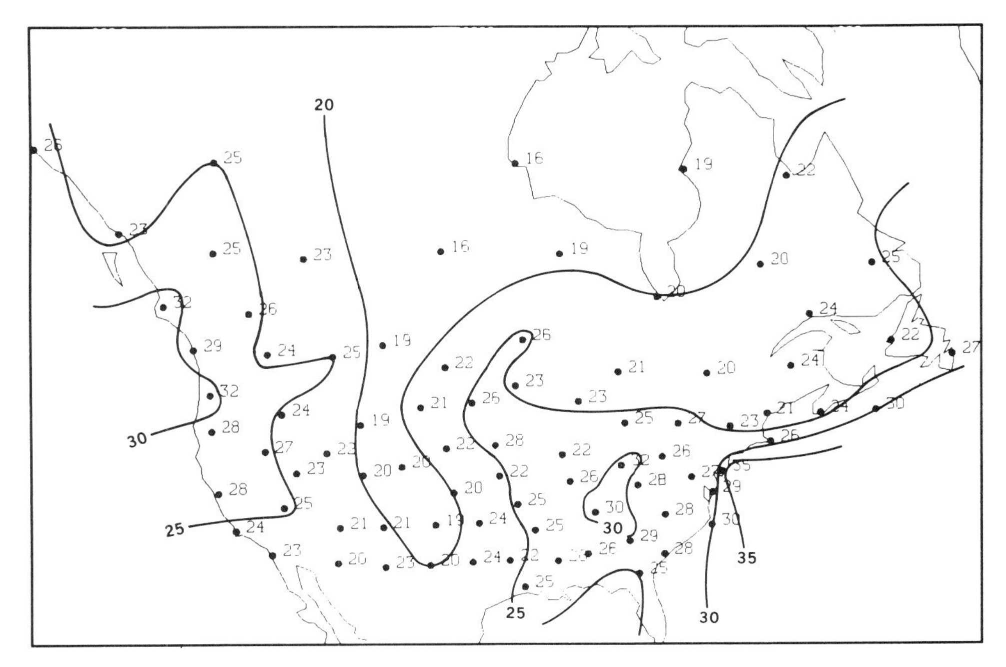

Figure 4.13 The standard deviation of the 250 mb geopotential background (forecast) error over North America (m). (After Lönnberg and Hollingsworth 1986)

optimum weights, and the estimated analysis error given by (4.2.12) is likely to be an underestimate of the true analysis error.

Because the most serious problem seems to be the incorrect specification of the background error covariances (or correlations), most of this section is devoted to this problem. First, we examine the validity of the assumptions of separability and homogeneity of the background error.

Figure 4.12 is a plot of the characteristic horizontal scale (4.3.12) of the background error correlation function of the geopotential calculated over the North American radiosonde network using a forecast background. The ordinate is pressure (mb) and the abscissa is $L/\sqrt{2}$, where L is the characteristic scale defined in (4.3.12). The separability assumption (4.3.1) would imply that the characteristic horizontal scale is independent of the vertical coordinate P. The figure, however, indicates that this is not the case; in fact, L increases with altitude. The stochastic background error correlation models of Balgovind et al. (1983) and Phillips (1986), discussed in Section 4.3, also suggest that the separability assumption is dubious.

Figure 4.13 shows the horizontal variation of the background error standard deviation (E_B) for the geopotential field at 250 mb over North America. Under strictly horizontally homogeneous conditions, E_B and E_B^2 would be horizontally invariant. It is apparent, in Figure 4.13 that the background field is not strictly homogeneous. However, the normalized background error (produced by dividing by the square root of the background error variance) tends to be more homogeneous than the background error itself. That is why the normalized form (4.2.13–14) of the statistical interpolation algorithm is preferred to the unnormalized form (4.2.9–10). This does not mean, of course, that the background error correlations are completely homogeneous. It is quite clear from Figure 4.5 that this is not the case. However, the inhomogeneities in the background error correlation are on a very large scale (several thousand kilometers), whereas the characteristic scales of the background error correlation function themselves – as seen in Figure 4.4 or given by Equation (4.3.12) – are much smaller (several hundred kilometers). When the scale of the inhomogeneity is much larger than the scale of the correlation, it is not a major source of error.

Next, we examine the effect of misspecification of the background error covariances (or correlations) on the statistical interpolation algorithm. We begin with misspecified expected background error variances, using the simple example of (2.2.7). Suppose there are two estimates s_o and s_b of the true value s. Denote s_o as the observed value with correct estimated error variance σ_o^2. s_b is the background value with correct expected error variance σ_b^2 and *incorrectly specified* background error variance $\tilde{\sigma}_b^2$. Then a supposedly optimal estimate of s is obtained from (2.2.7):

$$s_s = \frac{1}{\tilde{\sigma}_b^2 + \sigma_o^2}[\sigma_o^2 s_b + \tilde{\sigma}_b^2 s_o] \tag{4.9.1}$$

with perceived expected analysis error variance

$$\langle \tilde{\varepsilon}_a^2 \rangle = \frac{\tilde{\sigma}_b^2 \sigma_o^2}{\tilde{\sigma}_b^2 + \sigma_o^2} \tag{4.9.2}$$

and actual expected analysis error variance, given by (4.1.4),

$$\langle \varepsilon_a^2 \rangle = \frac{\tilde{\sigma}_b^4 \sigma_o^2 + \sigma_o^4 \sigma_b^2}{(\sigma_o^2 + \tilde{\sigma}_b^2)^2} \tag{4.9.3}$$

If $\tilde{\sigma}_b = \sigma_b$, then Equations (4.9.2–3) are the same and $\langle \varepsilon_a^2 \rangle$ is less than either σ_o^2 or σ_b^2. However, because $\tilde{\sigma}_b$ is not correct, from (4.1.5), the expected analysis error variance is only guaranteed to be smaller than the larger of σ_o^2 and σ_b^2.

Now consider two extreme cases. First, $\sigma_b^2 \ll \sigma_o^2 \ll \tilde{\sigma}_b^2$. In this case, the algorithm (4.9.1) would draw for the observations. The actual and perceived analysis error variances $\langle \varepsilon_a^2 \rangle \approx \langle \tilde{\varepsilon}_a^2 \rangle \lessapprox \sigma_o^2$. The only trouble with this specification of the background error variance is that a perfectly good background value would have been discarded and the analysis error variance not reduced to as small a value as it might have been. Second, $\tilde{\sigma}_b^2 \ll \sigma_o^2 \ll \sigma_b^2$. In this case, (4.9.1) would draw to a very inaccurate background value, the perceived analysis error $\langle \tilde{\varepsilon}_a^2 \rangle \lessapprox \tilde{\sigma}_b^2$ would be very impressive, but the actual analysis error variance $\langle \varepsilon_a^2 \rangle \lessapprox \sigma_b^2$ would be unacceptably large. This type of error is very serious because the objective analysis would have been more accurate if the background had been completely ignored. Thus, if a background is to be used in the objective analysis, it is wiser to overestimate the expected background error variance than to underestimate it.

In statistical interpolation, the spatial correlation structure as well as the expected

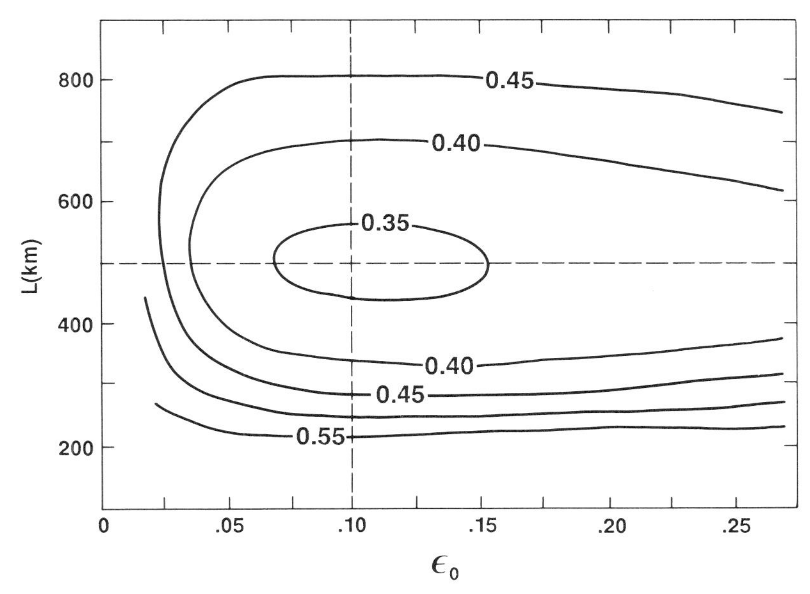

Figure 4.14 Standard deviation of the normalized expected analysis error ε_A as the background error characteristic scale and normalized observation error ε_O are allowed to vary around their correct values, $L = 500$ km and $\varepsilon_O = 0.10$. (After Seaman, *Aus. Met. Mag.* **31**: 225, 1983. AGPS Canberra, reproduced by permission of Commonwealth of Australia copyright.)

variances of the background and observation errors can be misspecified. Seaman (1977, 1983) and Franke (1985) have examined the effect on the analysis error of incorrectly specified background and observation error covariances. We examine here an example taken from Seaman (1983). The general estimate of analysis error variance given by (4.2.11) is

$$E_A^2 = E_B^2 - 2\underline{W}_i^T\underline{B}_i + \underline{W}_i^T[\underline{\underline{B}} + \underline{\underline{O}}]\underline{W}_i \tag{4.9.4}$$

Interpret $\underline{B}_i$, $\underline{\underline{B}}$, and $\underline{\underline{O}}$ as the *true* background and observation error covariances. Now assume $\underline{W}_i$ are the a posteriori weights given by *any* objective analysis scheme of the form (4.2.2). Interpreted in this light, (4.9.4) is the estimated analysis error for many of the analysis schemes of Chapters 2 and 3, as well as for statistical interpolation.

Now specify incorrect values $\tilde{\underline{B}}_i$, $\tilde{\underline{\underline{B}}}$, and $\tilde{\underline{\underline{O}}}$ of the background and observation error covariances, and then calculate a posteriori weights using (4.2.10):

$$\underline{W}_i = [\tilde{\underline{\underline{B}}} + \tilde{\underline{\underline{O}}}]^{-1}\tilde{\underline{B}}_i \tag{4.9.5}$$

If $\tilde{\underline{\underline{B}}} = \underline{\underline{B}}$, $\tilde{\underline{B}}_i = \underline{B}_i$, and $\tilde{\underline{\underline{O}}} = \underline{\underline{O}}$, then the algorithm (4.9.5) yields optimal weights and (4.9.4) simplifies to (4.2.12). When $\tilde{\underline{\underline{B}}}$, $\tilde{\underline{B}}_i$, and $\tilde{\underline{\underline{O}}}$ are incorrectly specified, the analysis is no longer optimal, and the estimated analysis error (4.9.4) is larger than that given by (4.2.12).

Figure 4.14 illustrates the effect on the expected analysis error when the background or observation error variances and/or spatial correlation structure are misspecified. In this example, the background error is homogeneous with expected variance E_B^2, and the background error correlation model is given by (4.3.21). The observation error is spatially uncorrelated and has the same expected variance E_O^2 at each observation station of a simple network. The *correct* background error has a characteristic scale $L = 500$ km, and the *correct* normalized root mean square observation error $\varepsilon_O = E_O/E_B = 0.1$. These values of ε_O and L produce the optimum weights and the minimum value of E_A. Incorrectly specified background and observation covariances $\tilde{\underline{B}}_i$, $\tilde{\underline{\underline{B}}}$, and $\tilde{\underline{\underline{O}}}$ are simulated by varying L and ε_O around their correct values. The expected analysis error is then calculated using (4.9.4) and plotted as a function of L and ε_O. Isopleths of constant normalized root mean square analysis error $\varepsilon_A = E_A/E_B$ are drawn at intervals of 0.05. Note that both ε_O and ε_A are root mean square values rather than the variances ε_O^2 and ε_A^2 that have generally been used in this chapter. The minimum value of ε_A is 0.343, and the increase of analysis error with incorrectly specified background or observation error covariances is apparent. Other sources of error such as specifying an isotropic background error correlation when the true correlation is anisotropic are also examined in Seaman (1983).

A more general method for examination of the accuracy of objective analyses is described by Hollingsworth (1989). Consider an observation network with K observation stations $\mathbf{r}_k$, $1 \leq k \leq K$. For an objective analysis algorithm of the form (4.2.1), the analyzed values at the observation stations are given by

$$\underline{f}_A = \underline{f}_B + \underline{\underline{W}}^T[\underline{f}_O - \underline{f}_B] \tag{4.9.6}$$

where $\underline{f}_A$, $\underline{f}_B$, and $\underline{f}_O$ are column vectors of length K of analyzed, background, and observed values at the observing stations; $\underline{\underline{W}}^T$ are the a posteriori weights. If the weights are optimal, then (4.9.6) becomes equal to (4.5.3). An analysis error covariance matrix $\underline{\underline{A}}$ can be obtained from (4.9.6) simply by following derivation (2.2.10–19). The result is

$$\underline{\underline{A}} = [\underline{\underline{I}} - \underline{\underline{W}}]^T\underline{\underline{B}}[\underline{\underline{I}} - \underline{\underline{W}}] + \underline{\underline{W}}^T\underline{\underline{O}}\underline{\underline{W}} \tag{4.9.7}$$

where $\underline{\underline{B}}$ and $\underline{\underline{O}}$ are the background and observation error covariance matrices. It can be seen that (4.9.7) is a vector generalization of the scalar result (4.1.4) for the case $N = 2$. Substitution of the optimum weights $\underline{\underline{W}}^T = \underline{\underline{B}}[\underline{\underline{B}} + \underline{\underline{O}}]^{-1}$ into (4.9.7) yields (2.2.19). In general, (2.2.19) is not the correct analysis error covariance matrix because $\underline{\underline{B}}$ and $\underline{\underline{O}}$ are imprecisely specified.

The analysis error covariance matrix $\underline{\underline{A}}$ of (4.9.7) is the true analysis error covariance matrix (at the observation stations) for any objective analysis algorithm of the form (4.2.1). Knowedge of this analysis error covariance would be extremely useful in determining the properties of the analysis algorithm. Unfortunately, the analysis errors are not known because the true values are not known. Moreover, the determination of $\underline{\underline{A}}$ from (4.9.7) requires precise knowledge of $\underline{\underline{B}}$ and $\underline{\underline{O}}$. Hollingsworth (1989) estimates the elements of $\underline{\underline{A}}$ by a procedure that is very similar to that used in Section 4.3 for estimating the background error correlations and covariances.

Observed-minus-analysis increments are collected from an observation network with spatially uncorrelated errors (a radiosonde network, for example). These observed-minus-analysis increments can be processed in much the same way as the observed-minus-background increments in (4.3.13–17), to produce estimates of expected analysis error variances and spatial correlations.

Examination of these analysis error variances and correlations provides both qualitative and quantitative information on analysis accuracy and indicates how effectively the observations have been used in the objective analysis. Observed-minus-analysis increments can be routinely calculated for operational data assimilation systems, thus providing running diagnostics of analysis accuracy.

Exercises

4.1 Using Equation (4.2.7), show that there is no correlation between the analysis error $A_i - T_i$ and the observation increment $O_k - B_k$.

4.2 Suppose the background and observation errors at station l are mutually correlated. Assuming that $\langle (O_l - T_l)(B_l - T_l) \rangle$ and $\langle (B_l - T_l)^2 \rangle$ are known, find a procedure that produces a modified observation O_l', whose error is *not* correlated with the background error.

4.3 For the homogeneous isotropic two-dimensional case, show that as the spectral density $g(k) > 0$ for finite k, then for the corresponding autocorrelation, $|\rho_{\Phi\Phi}(r)| \leq \rho_{\Phi\Phi}(0)$.

4.4 Show that the scale L in the one- and two-dimensional forms of the correlation functions (4.3.19) and (4.3.21) is equal to the characteristic scale defined by (4.3.10) and (4.3.12).

4.5 Consider an observation network $\mathbf{r}_k$, $1 \leq k \leq K$, with constant normalized observation error $\varepsilon_O^2 = E_O^2/E_B^2$. Suppose there is, in addition, a *perfect* observation at the analysis gridpoint $\mathbf{r}_i$. Using the statistical interpolation algorithm (4.2.1) with optimum weights (4.3.22), find the analyzed value at $\mathbf{r}_i$. Compare this value with that obtained from the single correction algorithm (3.1.6). What happens to the analyzed values produced by the two algorithms when all the observations are perfect.

4.6 Consider a collinear observation network of three equally spaced stations with x_2 in the middle, $x_1 = x_2 - \Delta x$, and $x_3 = x_2 + \Delta x$. Define background error correlations $q = \rho_B(x_1, x_2) = \rho_B(x_2, x_3)$ and $p = \rho_B(x_1, x_3)$. Find the three eigenvalues λ_1, λ_2, and λ_3 of the 3×3 background error correlation matrix $\underline{\rho}_B$ and show they are $\lambda_1 = 1 + p/2 + r$, $\lambda_2 = 1 - p$, and $\lambda_3 = 1 + p/2 - r$, where $r = [(p/2)^2 + 2q^2]^{1/2}$. Under what conditions are all three eigenvalues positive? Plot λ_1, λ_2, and λ_3 as a function of $\Delta x/L$ for the one-dimensional form of (4.3.21) and show that this condition is always satisfied. Find the corresponding eigenvectors and show they are orthogonal.

4.7 Following the simple networks of Section 4.6, consider the three-dimensional case and four observations that are equidistant from each other ($\rho_{lk} = \rho$, $l \neq k$) and equidistant from the analysis gridpoint ($\rho_{k0} = \tilde{\rho}$). Assume the background error is homogeneous, the observation error is uncorrelated, and the expected observation error variance is the same at each observation station. Find an expression for the a posteriori weights and expected analysis error variance for this network.

4.8 Consider the two-dimensional case with homogeneous background error and three perfect observations arrayed as follows. In plane polar coordinates (r, ϕ), the analysis gridpoint is at $r = 0$, and the three observation stations are at $(\tilde{r}, \pi)$, $(\tilde{r}, \tilde{\phi})$, and $(\tilde{r}, -\tilde{\phi})$. Assume the background error correlation function is given by (4.3.21) with $L = \tilde{r}$. Plot the normalized expected analysis error variance $\varepsilon_A^2 = E_A^2/E_B^2$ for this network as $\tilde{\phi}$ varies between 0 and π.

5

Statistical interpolation: multivariate

The objective analysis techniques discussed in Chapters 2–4 are largely mathematical and statistical concepts and could be applied to analysis problems in a variety of scientific disciplines. What distinguishes applications in the atmospheric sciences from other applications is the exact or approximate *physical* relationships or constraints satisfied by the dependent or state variables of the problem. Apart from cursory discussions in Sections 2.6 and 4.3, atmospheric physics has been hardly mentioned thus far. The remaining chapters of the book will be increasingly focused on the physical aspects of the objective analysis and data assimilation problem.

The main reason for the popularity of statistical interpolation in the atmospheric sciences is the ease and elegance with which simple physical constraints can be incorporated into the methodology. This process, originally called field matching by Gandin (1963) and now known as *multivariate analysis*, permits internally consistent analyses of several state variables simultaneously. The multivariate algorithm follows logically from the univariate theory, with the constraints built right into the algorithm. In fact, the analysis increments are supposed to satisfy, exactly or approximately, the imposed constraints (Lorenc, 1981).

The multivariate statistical interpolation algorithm is used by most of the major operational short- and medium-range forecasting centers. Recent applications of the algorithm are discussed by Baker et al. (1987), Dey and Morone (1985), and Shaw et al. (1987).

5.1 The multivariate algorithm

The form of the multivariate algorithm depends on the state variables to be analyzed and the particular constraints that interrelate them. As in Section 4.1, a simple scalar form of the problem can serve as a guide to the development of the general vector algorithm. Suppose h and v are parameters that must be optimally estimated using observed values h_O and v_O and background values h_B and v_B. Thus, there are two variables but only $N=2$ independent estimates of each variable and there is no spatial dependence.

The observation and background estimates are assumed to have errors $\varepsilon_h^{\mathrm{O}}$, $\varepsilon_v^{\mathrm{O}}$, $\varepsilon_h^{\mathrm{B}}$, and $\varepsilon_v^{\mathrm{B}}$. The errors are assumed to be unbiased, and there is no correlation between observed and background values:

$$\langle \varepsilon_h^{\mathrm{O}} \rangle = \langle \varepsilon_v^{\mathrm{O}} \rangle = \langle \varepsilon_h^{\mathrm{B}} \rangle = \langle \varepsilon_v^{\mathrm{B}} \rangle = 0$$

and

$$\langle \varepsilon_h^{\mathrm{O}} \varepsilon_h^{\mathrm{B}} \rangle = \langle \varepsilon_h^{\mathrm{O}} \varepsilon_v^{\mathrm{B}} \rangle = \langle \varepsilon_v^{\mathrm{O}} \varepsilon_v^{\mathrm{B}} \rangle = \langle \varepsilon_v^{\mathrm{O}} \varepsilon_h^{\mathrm{B}} \rangle = 0 \tag{5.1.1}$$

Estimates of h_{A} and v_{A} (the analyzed values) can be obtained from the following linear combination of observed and background values:

$$\begin{aligned} h_{\mathrm{A}} - h_{\mathrm{B}} &= W_{hh}[h_{\mathrm{O}} - h_{\mathrm{B}}] + W_{hv}[v_{\mathrm{O}} - v_{\mathrm{B}}] \\ v_{\mathrm{A}} - v_{\mathrm{B}} &= W_{vh}[h_{\mathrm{O}} - h_{\mathrm{B}}] + W_{vv}[v_{\mathrm{O}} - v_{\mathrm{B}}] \end{aligned} = \begin{pmatrix} W_{hh} & W_{hv} \\ W_{vh} & W_{vv} \end{pmatrix} \begin{vmatrix} h_{\mathrm{O}} - h_{\mathrm{B}} \\ v_{\mathrm{O}} - v_{\mathrm{B}} \end{vmatrix} \tag{5.1.2}$$

Define $\varepsilon_h^{\mathrm{A}}$ and $\varepsilon_v^{\mathrm{A}}$ as the errors of the analyzed values of h and v, respectively, and subtract the true value of h from each side of the first equation of (5.1.2):

$$\varepsilon_h^{\mathrm{A}} = \varepsilon_h^{\mathrm{B}} + W_{hh}[\varepsilon_h^{\mathrm{O}} - \varepsilon_h^{\mathrm{B}}] + W_{hv}[\varepsilon_v^{\mathrm{O}} - \varepsilon_v^{\mathrm{B}}] \tag{5.1.3}$$

Clearly $\langle \varepsilon_h^{\mathrm{A}} \rangle = 0$. Now square (5.1.3) and apply the expectation operator:

$$\begin{aligned} \langle (\varepsilon_h^{\mathrm{A}})^2 \rangle = {} & \langle (\varepsilon_h^{\mathrm{B}})^2 \rangle - 2W_{hh}\langle \varepsilon_h^{\mathrm{B}} \varepsilon_h^{\mathrm{B}} \rangle - 2W_{hv}\langle \varepsilon_v^{\mathrm{B}} \varepsilon_h^{\mathrm{B}} \rangle \\ & + W_{hh}[\langle \varepsilon_h^{\mathrm{B}} \varepsilon_h^{\mathrm{B}} \rangle + \langle \varepsilon_h^{\mathrm{O}} \varepsilon_h^{\mathrm{O}} \rangle] W_{hh} + W_{hv}[\langle \varepsilon_v^{\mathrm{B}} \varepsilon_v^{\mathrm{B}} \rangle + \langle \varepsilon_v^{\mathrm{O}} \varepsilon_v^{\mathrm{O}} \rangle] W_{hv} \\ & + 2W_{hh}[\langle \varepsilon_h^{\mathrm{B}} \varepsilon_v^{\mathrm{B}} \rangle + \langle \varepsilon_h^{\mathrm{O}} \varepsilon_v^{\mathrm{O}} \rangle] W_{hv} \end{aligned} \tag{5.1.4}$$

making use of (5.1.1).

$\langle (\varepsilon_h^{\mathrm{A}})^2 \rangle$ is the expected analysis error variance of h. A similar expression can be obtained for $\langle (\varepsilon_v^{\mathrm{A}})^2 \rangle$. Minimize (5.1.4) by differentiating with respect to the a posteriori weights W_{hh} and W_{hv} and setting the results to zero. Performing this operation with W_{hh} yields

$$[\langle \varepsilon_h^{\mathrm{B}} \varepsilon_h^{\mathrm{B}} \rangle + \langle \varepsilon_h^{\mathrm{O}} \varepsilon_h^{\mathrm{O}} \rangle] W_{hh} + [\langle \varepsilon_h^{\mathrm{B}} \varepsilon_v^{\mathrm{B}} \rangle + \langle \varepsilon_h^{\mathrm{O}} \varepsilon_v^{\mathrm{O}} \rangle] W_{hv} = \langle \varepsilon_h^{\mathrm{B}} \varepsilon_h^{\mathrm{B}} \rangle \tag{5.1.5}$$

Define the covariances $C_{hv}^{\mathrm{B}} = \langle \varepsilon_h^{\mathrm{B}} \varepsilon_v^{\mathrm{B}} \rangle$, $C_{vh}^{\mathrm{O}} = \langle \varepsilon_v^{\mathrm{O}} \varepsilon_h^{\mathrm{O}} \rangle$, and so on, and rewrite (5.1.5) as

$$[C_{hh}^{\mathrm{B}} + C_{hh}^{\mathrm{O}}] W_{hh} + [C_{hv}^{\mathrm{B}} + C_{hv}^{\mathrm{O}}] W_{hv} = C_{hh}^{\mathrm{B}} \tag{5.1.6}$$

In the same way, differentiation of (5.1.4) with respect to W_{hv} leads to

$$[C_{vh}^{\mathrm{B}} + C_{vh}^{\mathrm{O}}] W_{hh} + [C_{vv}^{\mathrm{B}} + C_{vv}^{\mathrm{O}}] W_{hv} = C_{vh}^{\mathrm{B}} \tag{5.1.7}$$

Minimizing $\langle (\varepsilon_v^{\mathrm{A}})^2 \rangle$ with respect to W_{vh} and W_{vv} gives

$$[C_{hh}^{\mathrm{B}} + C_{hh}^{\mathrm{O}}] W_{vh} + [C_{hv}^{\mathrm{B}} + C_{hv}^{\mathrm{O}}] W_{vv} = C_{hv}^{\mathrm{B}} \tag{5.1.8}$$

and

$$[C_{vh}^{\mathrm{B}} + C_{vh}^{\mathrm{O}}] W_{vh} + [C_{vv}^{\mathrm{B}} + C_{vv}^{\mathrm{O}}] W_{vv} = C_{vv}^{\mathrm{B}} \tag{5.1.9}$$

Equations (5.1.6–9) are four linear equations in four unknowns, which can be solved to obtain the a posterior weights W_{hh}, W_{hv}, W_{vh}, and W_{vv} if the covariances C^{B}_{hv}, C^{O}_{vh}, and so on, are known. When these optimal weights are inserted into (5.1.2), the estimates h_A and v_A are minimum variance estimates because $\langle(\varepsilon^{A}_{h})^2\rangle$ and $\langle(\varepsilon^{A}_{v})^2\rangle$ are minimized simultaneously. If there are no observations of v, then W_{hv}, W_{vh}, and W_{vv} are zero, and (5.1.2) and (5.1.6) become equivalent to the univariate algorithm (2.2.7).

Now consider the vector problem. Suppose $\mathbf{r}_i$ is the analysis gridpoint and $\mathbf{r}_k$, $1 \le k \le K$, are the observation stations. Define $h_A(\mathbf{r}_i)$ and $v_A(\mathbf{r}_i)$ as the analyzed values and $h_B(\mathbf{r}_i)$ and $v_A(\mathbf{r}_i)$ as the background estimates at the analysis gridpoint. The background and analysis errors at $\mathbf{r}_i$ are denoted $\varepsilon^{B}_{h}(\mathbf{r}_i)$, $\varepsilon^{B}_{v}(\mathbf{r}_i)$, $\varepsilon^{A}_{h}(\mathbf{r}_i)$, and $\varepsilon^{A}_{v}(\mathbf{r}_i)$, respectively.

Define $\underline{\mathrm{h}}_O$ and $\underline{\mathrm{v}}_O$ as the column vectors of length K whose elements are the observed values $h_O(\mathbf{r}_k)$ and $v_O(\mathbf{r}_k)$ at the observation stations. In the same way denote $\underline{\mathrm{h}}_B$ and $\underline{\mathrm{v}}_B$ as the column vector of background estimates at the observation stations. The observed and background errors at the observation stations are denoted $\underline{\varepsilon}^{O}_{h}$, $\underline{\varepsilon}^{O}_{v}$, $\underline{\varepsilon}^{B}_{h}$, and $\underline{\varepsilon}^{B}_{v}$, with elements $\varepsilon^{O}_{h}(\mathbf{r}_k)$, $\varepsilon^{O}_{v}(\mathbf{r}_k)$, $\varepsilon^{B}_{h}(\mathbf{r}_k)$, and $\varepsilon^{B}_{v}(\mathbf{r}_k)$, respectively. The observed and background errors are unbiased and uncorrelated with each other:

$$\begin{aligned}\langle\underline{\varepsilon}^{O}_{h}\rangle = \langle\underline{\varepsilon}^{O}_{v}\rangle = \langle\underline{\varepsilon}^{B}_{h}\rangle = \langle\underline{\varepsilon}^{B}_{v}\rangle \\ \langle\underline{\varepsilon}^{O}_{h}(\underline{\varepsilon}^{B}_{h})^{\mathrm{T}}\rangle = \langle\underline{\varepsilon}^{O}_{h}(\underline{\varepsilon}^{B}_{v})^{\mathrm{T}}\rangle = \langle\underline{\varepsilon}^{O}_{v}(\underline{\varepsilon}^{B}_{v})^{\mathrm{T}}\rangle = \langle\underline{\varepsilon}^{O}_{v}(\underline{\varepsilon}^{B}_{h})^{\mathrm{T}}\rangle = 0\end{aligned} \tag{5.1.10}$$

and

$$\langle\varepsilon^{B}_{h}(\mathbf{r}_i)\rangle = 0, \qquad \langle\underline{\varepsilon}^{O}_{v}\varepsilon^{B}_{h}(\mathbf{r}_i))\rangle = 0, \qquad \ldots$$

Then, the analysis increments $h_A(\mathbf{r}_i) - h_B(\mathbf{r}_i)$ and $v_A(\mathbf{r}_i) - v_B(\mathbf{r}_i)$ are linear combinations of the observation increments:

$$\begin{aligned} h_A(\mathbf{r}_i) - h_B(\mathbf{r}_i) &= \underline{\mathrm{W}}^{\mathrm{T}}_{hh}[\underline{\mathrm{h}}_O - \underline{\mathrm{h}}_B] + \underline{\mathrm{W}}^{\mathrm{T}}_{hv}[\underline{\mathrm{v}}_O - \underline{\mathrm{v}}_B] \\ v_A(\mathbf{r}_i) - v_B(\mathbf{r}_i) &= \underline{\mathrm{W}}^{\mathrm{T}}_{vh}[\underline{\mathrm{h}}_O - \underline{\mathrm{h}}_B] + \underline{\mathrm{W}}^{\mathrm{T}}_{vv}[\underline{\mathrm{v}}_O - \underline{\mathrm{v}}_B] \end{aligned} \tag{5.1.11}$$

where $\underline{\mathrm{W}}_{hh}$, $\underline{\mathrm{W}}_{hv}$, $\underline{\mathrm{W}}_{vh}$, and $\underline{\mathrm{W}}_{vv}$ are the a posteriori weight vectors with elements $W_{hh}(\mathbf{r}_i, \mathbf{r}_k)$, $W_{hv}(\mathbf{r}_i, \mathbf{r}_k)$, $W_{vh}(\mathbf{r}_i, \mathbf{r}_k)$, and $W_{vv}(\mathbf{r}_i, \mathbf{r}_k)$, respectively.

Derivation of vector analogues (5.1.6–9) proceeds in the same way as the scalar case. In the derivation we can use the longhand methods of Sections 2.2 or 4.2 or the following vector identities. If $\underline{\mathrm{x}}$, $\underline{\mathrm{c}}$, $\underline{\mathrm{y}}$, and $\underline{\mathrm{d}}$ are arbitrary column vectors of length K, then

$$(\underline{\mathrm{c}}^{\mathrm{T}}\underline{\mathrm{x}})(\underline{\mathrm{d}}^{\mathrm{T}}\underline{\mathrm{y}}) = \underline{\mathrm{c}}^{\mathrm{T}}(\underline{\mathrm{x}}\underline{\mathrm{y}}^{\mathrm{T}})\underline{\mathrm{d}} = \underline{\mathrm{d}}^{\mathrm{T}}(\underline{\mathrm{y}}^{\mathrm{T}}\underline{\mathrm{x}})\underline{\mathrm{c}}$$

and

$$\frac{d}{d\underline{\mathrm{x}}}[\underline{\mathrm{x}}^{\mathrm{T}}\underline{\mathrm{y}}] = \underline{\mathrm{y}}, \qquad \frac{d}{d\underline{\mathrm{y}}}[\underline{\mathrm{x}}^{\mathrm{T}}\underline{\mathrm{y}}] = \underline{\mathrm{x}} \tag{5.1.12}$$

The derivation of the vector counterparts of (5.1.6–9) using (5.1.12) is left as Exercise

5.1. The result is

$$[\underline{\underline{C}}^{B}_{hh} + \underline{\underline{C}}^{O}_{hh}]\underline{W}_{hh} + [\underline{\underline{C}}^{B}_{hv} + \underline{\underline{C}}^{O}_{hv}]\underline{W}_{hv} = \underline{C}^{B}_{hh}(\mathbf{r}_i) \tag{5.1.13}$$

$$[\underline{\underline{C}}^{B}_{vh} + \underline{\underline{C}}^{O}_{vh}]\underline{W}_{hh} + [\underline{\underline{C}}^{B}_{vv} + \underline{\underline{C}}^{O}_{vv}]\underline{W}_{hv} = \underline{C}^{B}_{vh}(\mathbf{r}_i) \tag{5.1.14}$$

$$[\underline{\underline{C}}^{B}_{hh} + \underline{\underline{C}}^{O}_{hh}]\underline{W}_{vh} + [\underline{\underline{C}}^{B}_{hv} + \underline{\underline{C}}^{O}_{hv}]\underline{W}_{vv} = \underline{C}^{B}_{hv}(\mathbf{r}_i) \tag{5.1.15}$$

$$[\underline{\underline{C}}^{B}_{vh} + \underline{\underline{C}}^{O}_{vh}]\underline{W}_{vh} + [\underline{\underline{C}}^{B}_{vv} + \underline{\underline{C}}^{O}_{vv}]\underline{W}_{vv} = \underline{C}^{B}_{vv}(\mathbf{r}_i) \tag{5.1.16}$$

where $\underline{\underline{C}}^{B}_{hv} = \langle \underline{\varepsilon}^{B}_{h}(\underline{\varepsilon}^{B}_{v})^{T} \rangle$ is the $K \times K$ background error covariance matrix between the observation stations with elements $C^{B}_{hv}(\mathbf{r}_k, \mathbf{r}_l) = \langle \varepsilon^{B}_{h}(\mathbf{r}_k)\varepsilon^{B}_{v}(\mathbf{r}_l) \rangle$, $1 \le k, l \le K$, and similarly for $\underline{\underline{C}}^{B}_{hh}$, $\underline{\underline{C}}^{O}_{vv}$, and so on. $\underline{C}^{B}_{vh}(\mathbf{r}_i) = \langle \underline{\varepsilon}^{B}_{v}\varepsilon^{B}_{h}(\mathbf{r}_i) \rangle$ is the column vector of background error covariances between the observation stations and analysis gridpoint with elements $C^{B}_{vh}(\mathbf{r}_k, \mathbf{r}_i) = \langle \varepsilon^{B}_{v}(\mathbf{r}_k)\varepsilon^{B}_{h}(\mathbf{r}_i) \rangle$, $1 \le k \le K$, and similarly for $\underline{C}^{B}_{vv}(\mathbf{r}_i)$, and so on.

Equations (5.1.13–16) can be solved for the weight vectors $\underline{W}_{hh}$, $\underline{W}_{hv}$, $\underline{W}_{hv}$, and $\underline{W}_{vv}$, and these optimum weights can be inserted into (5.1.11) to yield the analyzed values $h_A(\mathbf{r}_i)$ and $v_A(\mathbf{r}_i)$. The weight equations (5.1.13–16) can be written in a more familiar fashion as follows. Each matrix equation of (5.1.13–16) is comprised of K scalar equations. Take the kth equation from each of (5.1.13–16) and rewrite as

$$\sum_{l=1}^{K} [\underline{\underline{B}}_{kl} + \underline{\underline{O}}_{kl}]\underline{\underline{W}}_{il} = \underline{\underline{B}}_{ki} \tag{5.1.17}$$

where

$$\underline{\underline{B}}_{kl} = \begin{pmatrix} C^{B}_{hh}(\mathbf{r}_k, \mathbf{r}_l) & C^{B}_{hv}(\mathbf{r}_k, \mathbf{r}_l) \\ C^{B}_{vh}(\mathbf{r}_k, \mathbf{r}_l) & C^{B}_{vv}(\mathbf{r}_k, \mathbf{r}_l) \end{pmatrix}, \qquad \underline{\underline{O}}_{kl} = \begin{pmatrix} C^{O}_{hh}(\mathbf{r}_k, \mathbf{r}_l) & C^{O}_{hv}(\mathbf{r}_k, \mathbf{r}_l) \\ C^{O}_{vh}(\mathbf{r}_k, \mathbf{r}_l) & C^{O}_{vv}(\mathbf{r}_k, \mathbf{r}_l) \end{pmatrix}$$

$$\underline{\underline{W}}_{il} = \begin{pmatrix} W_{hh}(\mathbf{r}_i, \mathbf{r}_l) & W_{vh}(\mathbf{r}_i, \mathbf{r}_l) \\ W_{hv}(\mathbf{r}_i, \mathbf{r}_l) & W_{vv}(\mathbf{r}_i, \mathbf{r}_l) \end{pmatrix}, \qquad \underline{\underline{B}}_{ki} = \begin{pmatrix} C^{B}_{hh}(\mathbf{r}_k, \mathbf{r}_i) & C^{B}_{hv}(\mathbf{r}_k, \mathbf{r}_i) \\ C^{B}_{vh}(\mathbf{r}_k, \mathbf{r}_i) & C^{B}_{vv}(\mathbf{r}_k, \mathbf{r}_i) \end{pmatrix}$$

The algorithm (5.1.17) is in the same form as the univariate algorithm (4.2.8), except that the individual elements have been replaced by 2×2 submatrices.

Now, define $\underline{s}^{T}(\mathbf{r}_k) = [h(\mathbf{r}_k)\mathrm{v}(\mathbf{r}_k)]$. Then, a vector of length $2K$ with interleaved values of h and v at the observation stations can be defined as follows:

$$\begin{aligned} \underline{s}^{T} &= [\underline{s}^{T}(\mathbf{r}_1) \cdots \underline{s}^{T}(\mathbf{r}_K)] \\ &= [h(\mathbf{r}_1)\mathrm{v}(\mathbf{r}_1) \cdots \mathrm{h}(\mathbf{r}_K)\mathrm{v}(\mathbf{r}_K)] \end{aligned}$$

Equation (5.1.11) can now be written in the same form as (4.2.9),

$$\begin{vmatrix} h_A(\mathbf{r}_i) \\ v_A(\mathbf{r}_i) \end{vmatrix} = \underline{s}_A(\mathbf{r}_i) = \underline{s}_B(\mathbf{r}_i) + \underline{W}^{T}_{i}[\underline{s}_O - \underline{s}_B] \tag{5.1.18}$$

where $\underline{s}_O - \underline{s}_B$ is the column vector of observation increments with elements $\underline{s}_O(\mathbf{r}_k) - \underline{s}_B(\mathbf{r}_k)$, and the subscripts O, B, and A on $\underline{s}$ stand for observed, background, and analyzed values. $\underline{W}^{T}_{i}$ is the column vector of a posteriori weights, whose elements are the submatrices $\underline{\underline{W}}^{T}_{il}$ of (5.1.17). Equation (5.1.17) can then be written in the same form as (4.2.10),

$$[\underline{\underline{B}} + \underline{\underline{O}}]\underline{W}_i = \underline{B}_i \tag{5.1.19}$$

where $\underline{\underline{B}}$ and $\underline{\underline{O}}$ are the square covariance matrices whose elements are given by the submatrices $\underline{\underline{B}}_{kl}$ and $\underline{\underline{O}}_{kl}$, respectively. $\underline{B}_i$ is the column vector whose elements are the submatrices $\underline{\underline{B}}_{ki}$.

Normalized forms of (5.1.18–19) consistent with (4.2.13–14) can also be constructed. If no h or v observation exists at a particular observation station, then the order of $\underline{\underline{B}}$ and $\underline{\underline{O}}$ is reduced by one. If the observation error is spatially uncorrelated and there is no observation error correlation between variables, then all elements of $\underline{\underline{O}}_{kl}$ equal zero for $k \neq l$ and all elements of $\underline{\underline{O}}_{kk}$ equal zero except $C^{\mathrm{O}}_{hh}(\mathbf{r}_k, \mathbf{r}_k)$ and $C^{\mathrm{O}}_{vv}(\mathbf{r}_k, \mathbf{r}_k)$.

Extension of the algorithm (5.1.18–19) to three variables h, u, and v is straightforward. The 2×2 submatrices in (5.1.17) are replaced by 3×3 submatrices. For example, $\underline{\underline{B}}_{kl}$ becomes

$$\underline{\underline{B}}_{kl} = \begin{pmatrix} C^{\mathrm{B}}_{hh}(\mathbf{r}_k, \mathbf{r}_l) & C^{\mathrm{B}}_{hu}(\mathbf{r}_k, \mathbf{r}_l) & C^{\mathrm{B}}_{hv}(\mathbf{r}_k, \mathbf{r}_l) \\ C^{\mathrm{B}}_{uh}(\mathbf{r}_k, \mathbf{r}_l) & C^{\mathrm{B}}_{uu}(\mathbf{r}_k, \mathbf{r}_l) & C^{\mathrm{B}}_{uv}(\mathbf{r}_k, \mathbf{r}_l) \\ C^{\mathrm{B}}_{vh}(\mathbf{r}_k, \mathbf{r}_l) & C^{\mathrm{B}}_{vu}(\mathbf{r}_k, \mathbf{r}_l) & C^{\mathrm{B}}_{vv}(\mathbf{r}_k, \mathbf{r}_l) \end{pmatrix} \tag{5.1.20}$$

and similarly for $\underline{\underline{O}}_{kl}$, $\underline{\underline{W}}_{il}$, and $\underline{\underline{B}}_{ki}$.

The background error covariance matrix $\underline{\underline{B}}$ in (5.1.19) can be written as

$$\underline{\underline{B}} = \langle \underline{\varepsilon}^{\mathrm{B}}_s (\underline{\varepsilon}^{\mathrm{B}}_s)^{\mathrm{T}} \rangle, \qquad \text{where} \qquad \underline{\varepsilon}^{\mathrm{B}}_s = [\varepsilon^{\mathrm{B}}_h(\mathbf{r}_1)\varepsilon^{\mathrm{B}}_v(\mathbf{r}_1) \cdots \varepsilon^{\mathrm{B}}_h(\mathbf{r}_K)\varepsilon^{\mathrm{B}}_v(\mathbf{r}_K)]^{\mathrm{T}} \tag{5.1.21}$$

$\underline{\underline{B}}$ is clearly symmetric. It is also positive definite because if $\underline{z}$ is an arbitrary column vector of length $2K$, then

$$\underline{z}^{\mathrm{T}}\underline{\underline{B}}\underline{z} = \underline{z}^{\mathrm{T}}\langle \underline{\varepsilon}^{\mathrm{B}}_s(\underline{\varepsilon}^{\mathrm{B}}_s)^{\mathrm{T}} \rangle \underline{z} = \langle \underline{z}^{\mathrm{T}}\underline{\varepsilon}^{\mathrm{B}}_s(\underline{\varepsilon}^{\mathrm{B}}_s)^{\mathrm{T}}\underline{z} \rangle = \langle (\underline{z}^{\mathrm{T}}\underline{\varepsilon}^{\mathrm{B}}_s)^2 \rangle > 0 \tag{5.1.22}$$

if the observation stations are distinct. Thus, the eigenvalues of $\underline{\underline{B}}$ are real and positive. A similar result holds for the observation error covariance matrix $\underline{\underline{O}}$.

The algorithm (5.1.18–19) can also be put in a form similar to (4.2.16), for an arbitrary analysis gridpoint $\mathbf{r}$,

$$\begin{aligned} h_{\mathrm{A}}(\mathbf{r}) - h_{\mathrm{B}}(\mathbf{r}) &= \sum_{k=1}^{K} [q^{\mathrm{h}}_k C^{\mathrm{B}}_{hh}(\mathbf{r}_k, \mathbf{r}) + q^{\mathrm{v}}_k C^{\mathrm{B}}_{vh}(\mathbf{r}_k, \mathbf{r})] \\ v_{\mathrm{A}}(\mathbf{r}) - v_{\mathrm{B}}(\mathbf{r}) &= \sum_{k=1}^{K} [q^{\mathrm{h}}_k C^{\mathrm{B}}_{hv}(\mathbf{r}_k, \mathbf{r}) + q^{\mathrm{v}}_k C^{\mathrm{B}}_{vv}(\mathbf{r}_k, \mathbf{r})] \end{aligned} \tag{5.1.23}$$

where

$$\underline{q} = [\underline{\underline{B}} + \underline{\underline{O}}]^{-1}[\underline{s}_{\mathrm{O}} - \underline{s}_{\mathrm{B}}] = [q^{\mathrm{h}}_1 q^{\mathrm{v}}_1 \cdots q^{\mathrm{h}}_K q^{\mathrm{v}}_K]^{\mathrm{T}}$$

is a vector of coefficients that does *not* depend on the analysis gridpoint. $h_{\mathrm{A}}(\mathbf{r})$, $v_{\mathrm{B}}(\mathbf{r})$, $C^{\mathrm{B}}_{vh}(\mathbf{r}_k, \mathbf{r})$, and so on, can be thought of as continuous functions of $\mathbf{r}$. Suppose, for any observation station $\mathbf{r}_k$, the background error covariances satisfy linear functional relationships of the form

$$C^{\mathrm{B}}_{hv}(\mathbf{r}_k, \mathbf{r}) = F[C^{\mathrm{B}}_{hh}(\mathbf{r}_k, \mathbf{r})], \qquad C^{\mathrm{B}}_{vv}(\mathbf{r}_k, \mathbf{r}) = F[C^{\mathrm{B}}_{vh}(\mathbf{r}_k, \mathbf{r})] \tag{5.1.24}$$

Then, $v_{\mathrm{A}}(\mathbf{r}) - v_{\mathrm{B}}(\mathbf{r}) = F[h_{\mathrm{A}}(\mathbf{r}) - h_{\mathrm{B}}(\mathbf{r})]$. In other words, if the background error covariances satisfy a continuous, linear, multivariate relationship, then so do the analysis increments. Such multivariate relationships are the heart of the multivariate statistical interpolation algorithm and will be discussed in detail in the next two sections.

5.2 Statistics of a homogeneous two-dimensional windfield

The multivariate algorithm (5.1.19) for two variables contains four types of covariances for both the background and observation errors. Suppose Φ is the geopotential and u and v are the horizontal components of the windfield. Then, from (5.1.20), there are nine types of cross- or autocovariances. In Chapter 4, scalar background error covariances of the form $C^{\mathrm{B}}_{\Phi\Phi}(\mathbf{r}_k, \mathbf{r}_l)$ were discussed extensively. In this section, the properties of the background error cross- and autocovariances of the horizontal windfield (C^{B}_{uu}, C^{B}_{uv}, C^{B}_{vu}, and C^{B}_{vv}) are explored. The background error cross-covariances between wind and geopotential ($C^{\mathrm{B}}_{\Phi u}$, $C^{\mathrm{B}}_{\Phi v}$, $C^{\mathrm{B}}_{u\Phi}$, and $C^{\mathrm{B}}_{v\Phi}$) will be examined in Section 5.3.

We assume that wind covariances and wind/geopotential covariances are vertically and horizontally separable, as in Section 4.3. As in the univariate case discussed in Section 4.7, separability implies that the horizontal structures of each of the auto- and cross-covariances are independent of the vertical coordinate. In addition, however, the relations between the horizontal covariances derived later in this section require the vertical structures of all the covariances to be the same. The validity of the separability assumption for the background error in the multivariate case is discussed by Lönnberg and Hollingsworth (1986). The vertical wind covariances are not discussed explicitly here because they are treated operationally in much the same way as the vertical scalar covariances (geopotential, temperature, etc.) discussed in Section 4.7. Thus, both Sections 5.2 and 5.3 will concentrate on the horizontal or two-dimensional covariances.

The present formulation of the error statistics of a horizontal windfield is based on the theory of two-point velocity correlations, which is discussed in texts on turbulence theory such as Batchelor (1953), Panchev (1971), or Monin and Yaglom (1975). Turbulence theory, of course, is about the winds themselves rather than the wind errors. However, the theory of two-point velocity correlations can be applied equally well to winds and wind errors. Consequently, in the remainder of Sections 5.2 and 5.3, we do not refer specifically to errors, and the notations B for background and O for observations are not used. The treatment here is for two-dimensional two-point velocity correlations (which is a special case of the general three-dimensional theory) and is based on Daley (1983b, 1985) and particularly Hollingsworth and Lönnberg (1986).

Consider two spatial locations $\mathbf{r}_1 = (x_1, y_1)$ and $\mathbf{r}_2 = (x_2, y_2)$ on the (x, y) plane. Define $u_1 = u(x_1, y_1)$, $u_2 = u(x_2, y_2)$, $v_1 = v(x_1, y_1)$, and $v_2 = v(x_2, y_2)$ to be the eastward and northward components of the horizontal windfield at these two points. Then, there exist four two-point velocity covariances, $C_{uu}(\mathbf{r}_1, \mathbf{r}_2) = \langle u_1 u_2 \rangle$, $C_{uv}(\mathbf{r}_1, \mathbf{r}_2) = \langle u_1 v_2 \rangle$, $C_{vu}(\mathbf{r}_1, \mathbf{r}_2) = \langle v_1 u_2 \rangle$, and $C_{vv}(\mathbf{r}_1, \mathbf{r}_2) = \langle v_1 v_2 \rangle$. We assume that $\langle u_1 \rangle = \langle u_2 \rangle = \langle v_1 \rangle = \langle v_2 \rangle = 0$, which is appropriate for wind errors though not for the winds themselves. These four cross- and autocovariances can be related to the simple scalar autocovariances of Chapter 4, using Helmholtz's theorem,

$$u = -\frac{\partial \psi}{\partial y} + \frac{\partial \chi}{\partial x}, \qquad v = \frac{\partial \psi}{\partial x} + \frac{\partial \chi}{\partial y} \tag{5.2.1}$$

where ψ is the streamfunction and χ the velocity potential. The four covariances C_{uu}, C_{uv}, C_{vu}, and C_{vv} can be written in terms of ψ and χ. C_{uv} is used as an illustration (but the other covariances are very similar):

$$C_{uv}(x_1, y_1, x_2, y_2) = \langle u_1 v_2 \rangle$$

$$= -\left\langle \frac{\partial \psi_1}{\partial y_1} \frac{\partial \psi_2}{\partial x_2} \right\rangle + \left\langle \frac{\partial \chi_1}{\partial x_1} \frac{\partial \psi_2}{\partial x_2} \right\rangle - \left\langle \frac{\partial \psi_1}{\partial y_1} \frac{\partial \chi_2}{\partial y_2} \right\rangle + \left\langle \frac{\partial \chi_1}{\partial x_1} \frac{\partial \chi_2}{\partial y_2} \right\rangle \tag{5.2.2}$$

where $\psi_1 = \psi(x_1, y_1)$, $\chi_2 = \chi(x_2, y_2)$, and so on.

Now assume the flow is homogeneous,

$$C_{uv}(\mathbf{r}_1, \mathbf{r}_2) = C_{uv}(x_1, y_1, x_2, y_2) = C_{uv}(\tilde{x}, \tilde{y}) = C_{uv}(\tilde{\mathbf{r}})$$

where $\tilde{\mathbf{r}} = \mathbf{r}_2 - \mathbf{r}_1$, $\tilde{x} = x_2 - x_1$, and $\tilde{y} = y_2 - y_1$, following (4.3.2) and (D16). Consider the first term on the right-hand side of (5.2.2) under the assumption of homogeneity:

$$\frac{\partial \psi_1}{\partial y_1} \frac{\partial \psi_2}{\partial x_2} = \left\langle \lim_{\Delta y_1 \to 0} \left[\frac{\psi(x_1, y_1 + \Delta y_1) - \psi(x_1, y_1)}{\Delta y_1} \right] \frac{\partial \psi_2}{\partial x_2} \right\rangle$$

$$= \lim_{\Delta y_1 \to 0} \left[\frac{\left\langle \psi(x_1, y_1 + \Delta y_1) \frac{\partial \psi_2}{\partial x_2} \right\rangle - \left\langle \psi(x_1, y_1) \frac{\partial \psi_2}{\partial x_2} \right\rangle}{\Delta y_1} \right]$$

$$= \frac{\partial}{\partial y_1} \left\langle \psi_1 \frac{\partial \psi_2}{\partial x_2} \right\rangle$$

Now repeat the same operation for x_2 and

$$\left\langle \frac{\partial \psi_1}{\partial y_1} \frac{\partial \psi_2}{\partial x_2} \right\rangle = \frac{\partial^2}{\partial y_1 \, \partial x_2} \langle \psi_1 \psi_2 \rangle$$

But $y_1 = y_2 - \tilde{y}$ and $x_2 = \tilde{x} + x_1$, so

$$\left\langle \frac{\partial \psi_1}{\partial y_1} \frac{\partial \psi_2}{\partial x_2} \right\rangle = -\frac{\partial^2}{\partial \tilde{y} \, \partial \tilde{x}} \langle \psi_1 \psi_2 \rangle \tag{5.2.3}$$

Proceeding in this manner with each of the four terms constituting the four wind covariances, we can write

$$\langle u_1 u_2 \rangle = -\frac{\partial^2}{\partial \tilde{y}^2}\langle \psi_1 \psi_2 \rangle - \frac{\partial^2}{\partial \tilde{x}^2}\langle \chi_1 \chi_2 \rangle + \frac{\partial^2}{\partial \tilde{x}\,\partial \tilde{y}}[\langle \psi_1 \chi_2 \rangle + \langle \chi_1 \psi_2 \rangle] \tag{5.2.4}$$

$$\langle v_1 v_2 \rangle = -\frac{\partial^2}{\partial \tilde{x}^2}\langle \psi_1 \psi_2 \rangle - \frac{\partial^2}{\partial \tilde{y}^2}\langle \chi_1 \chi_2 \rangle - \frac{\partial^2}{\partial \tilde{x}\,\partial \tilde{y}}[\langle \psi_1 \chi_2 \rangle + \langle \chi_1 \psi_2 \rangle] \tag{5.2.5}$$

$$\langle u_1 v_2 \rangle = \frac{\partial^2}{\partial \tilde{x}\,\partial \tilde{y}}[\langle \psi_1 \psi_2 \rangle - \langle \chi_1 \chi_2 \rangle] + \frac{\partial^2}{\partial \tilde{y}^2}\langle \psi_1 \chi_2 \rangle - \frac{\partial^2}{\partial \tilde{x}^2}\langle \chi_1 \psi_2 \rangle \tag{5.2.6}$$

$$\langle v_1 u_2 \rangle = \frac{\partial^2}{\partial \tilde{x}\,\partial \tilde{y}}[\langle \psi_1 \psi_2 \rangle - \langle \chi_1 \chi_2 \rangle] + \frac{\partial^2}{\partial \tilde{y}^2}\langle \psi_1 \chi_2 \rangle - \frac{\partial^2}{\partial \tilde{x}^2}\langle \chi_1 \psi_2 \rangle \tag{5.2.7}$$

In deriving (5.2.4–7), we have implicitly assumed that $\langle \psi_1 \psi_2 \rangle$, $\langle \chi_1 \chi_2 \rangle$, $\langle \chi_1 \psi_2 \rangle$, and $\langle \psi_1 \chi_2 \rangle$ are expressed by continuous differentiable functions as in Section 4.3. The relatively simple forms of (5.2.4–7) are possible only because of the homogeneity assumption.

Now define $C_{uv}(\tilde{\mathbf{r}}) = \langle u_1 u_2 \rangle$, $C_{\chi\psi}(\tilde{\mathbf{r}}) = \langle \chi_1 \psi_2 \rangle$, and so on, and add and subtract (5.2.4–5) and (5.2.6–7). The result is

$$C_{uu} + C_{vv} = -\left(\frac{\partial^2}{\partial \tilde{y}^2} + \frac{\partial^2}{\partial \tilde{x}^2}\right)(C_{\psi\psi} + C_{\chi\chi}) \tag{5.2.8}$$

$$C_{uu} - C_{vv} = -\left(\frac{\partial^2}{\partial \tilde{y}^2} - \frac{\partial^2}{\partial \tilde{x}^2}\right)(C_{\psi\psi} - C_{\chi\chi}) + 2\frac{\partial^2}{\partial \tilde{x}\,\partial \tilde{y}}(C_{\psi\chi} + C_{\chi\psi}) \tag{5.2.9}$$

$$C_{uv} + C_{vu} = 2\frac{\partial^2}{\partial \tilde{y}\,\partial \tilde{x}}(C_{\psi\psi} - C_{\chi\chi}) + \left(\frac{\partial^2}{\partial \tilde{y}^2} - \frac{\partial^2}{\partial \tilde{x}^2}\right)(C_{\psi\chi} + C_{\chi\psi}) \tag{5.2.10}$$

$$C_{uv} - C_{vu} = \left(\frac{\partial^2}{\partial \tilde{y}^2} + \frac{\partial^2}{\partial \tilde{x}^2}\right)(C_{\psi\chi} - C_{\chi\psi}) \tag{5.2.11}$$

As in Section 4.2, we introduce planar polar coordinates. Define distance r and angle ϕ as in (4.3.3). Dropping the ($\tilde{\ }$) notation in r and ϕ for the sake of simplicity, we can write the Cartesian operators in (5.2.8–11) as

$$\begin{aligned}
\frac{\partial^2}{\partial \tilde{y}^2} + \frac{\partial^2}{\partial \tilde{x}^2} &= \nabla^2 = \frac{1}{r}\frac{\partial}{\partial r} r \frac{\partial}{\partial r} + \frac{1}{r^2}\frac{\partial^2}{\partial \phi^2} \\
\frac{\partial^2}{\partial \tilde{y}^2} - \frac{\partial^2}{\partial \tilde{x}^2} &= (\sin^2\phi - \cos^2\phi)R_1 - 4\sin\phi\cos\phi R_2 \\
\frac{\partial^2}{\partial \tilde{x}\,\partial \tilde{y}} &= \sin\phi\cos\phi\, R_1 + (\sin^2\phi - \cos^2\phi)R_2
\end{aligned} \tag{5.2.12}$$

where

$$R_1 = r\frac{\partial}{\partial r}\frac{1}{r}\frac{\partial}{\partial r} - \frac{1}{r^2}\frac{\partial^2}{\partial \phi^2}, \qquad R_2 = \frac{\partial}{\partial \phi}\left(\frac{1}{r^2} - \frac{1}{r}\frac{\partial}{\partial r}\right)$$

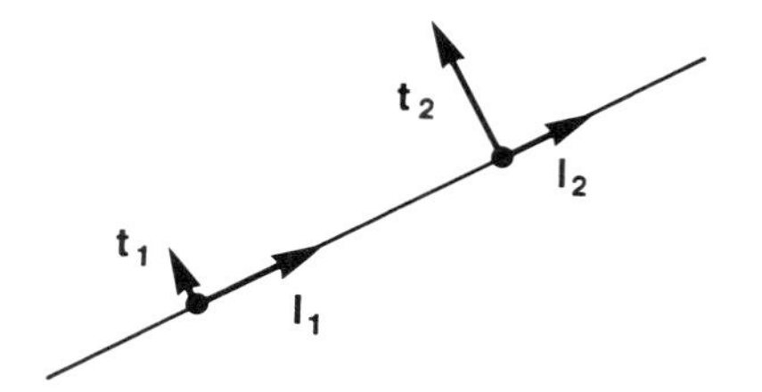

Figure 5.1 Longitudinal and transverse wind velocity components.

Equations (5.2.8–11) can then be written in plane polar coordinates as

$$C_{uu} + C_{vv} = -\nabla^2(C_{\psi\psi} + C_{\chi\chi}) \tag{5.2.13}$$

$$\begin{aligned} C_{uu} - C_{vv} = {} & 2 \sin\phi \cos\phi[2R_2(C_{\psi\psi} - C_{\chi\chi}) + R_1(C_{\psi\chi} + C_{\chi\psi})] \\ & + (\cos^2\phi - \sin^2\phi)[R_1(C_{\psi\psi} - C_{\chi\chi}) - 2R_2(C_{\psi\chi} + C_{\chi\psi})] \end{aligned} \tag{5.2.14}$$

$$\begin{aligned} C_{uv} + C_{vu} = {} & 2 \sin\phi \cos\phi[R_1(C_{\psi\psi} - C_{\chi\chi}) - 2R_2(C_{\psi\chi} + C_{\chi\psi})] \\ & - (\cos^2\phi - \sin^2\phi)[2R_2(C_{\psi\psi} - C_{\chi\chi}) + R_1(C_{\psi\chi} + C_{\chi\psi})] \end{aligned} \tag{5.2.15}$$

$$C_{uv} - C_{vu} = \nabla^2(C_{\psi\chi} - C_{\chi\psi}) \tag{5.2.16}$$

Equations (5.2.13–16) can be simplified by introducing the longitudinal and transverse velocity components shown in Figure 5.1. At the two spatial locations 1 and 2, define l to be the wind component along the line connecting the two points and t to be the wind component in the direction perpendicular to l, positive to the left. Thus, the eastward and northward wind components (u_1, v_1) and (u_2, v_2) are replaced by (l_1, t_1) and (l_2, t_2) as shown in Figure 5.1. l and t are related to u and v by

$$\begin{aligned} l &= u \cos\phi + v \sin\phi \\ t &= -u \sin\phi + v \cos\phi \end{aligned} \tag{5.2.17}$$

where ϕ is the angle between the x (east–west) axis and l, measured positive counterclockwise as in (5.2.12). If $\phi = 0$, then $u = l$ and $v = t$. Covariances $C_{ll}(r, \phi)$, $C_{lt}(r, \phi)$, $C_{tl}(r, \phi)$, and $C_{tt}(r, \phi)$ can be related to $C_{uu}(r, \phi)$, $C_{uv}(r, \phi)$, $C_{vu}(r, \phi)$, and $C_{vv}(r, \phi)$ by a simple application of (5.2.17):

$$\begin{pmatrix} C_{ll} & C_{lt} \\ C_{tl} & C_{tt} \end{pmatrix} = \begin{pmatrix} \cos\phi & \sin\phi \\ -\sin\phi & \cos\phi \end{pmatrix} \begin{pmatrix} C_{uu} & C_{uv} \\ C_{vu} & C_{vv} \end{pmatrix} \begin{pmatrix} \cos\phi & -\sin\phi \\ \sin\phi & \cos\phi \end{pmatrix} \tag{5.2.18}$$

It is left as Exercise (5.2) to show that

$$C_{ll} + C_{tt} = -\nabla^2(C_{\psi\psi} + C_{\chi\chi}) \tag{5.2.19}$$

$$C_{ll} - C_{tt} = R_1(C_{\psi\psi} - C_{\chi\chi}) - 2R_2(C_{\psi\chi} + C_{\chi\psi}) \tag{5.2.20}$$

$$C_{lt} + C_{tl} = -2R_2(C_{\psi\psi} - C_{\chi\chi}) - R_1(C_{\psi\chi} - C_{\chi\psi}) \tag{5.2.21}$$

$$C_{lt} - C_{tl} = \nabla^2(C_{\psi\chi} - C_{\chi\psi}) \tag{5.2.22}$$

The simple form (5.2.19–22) has been derived with only the assumption of horizontal homogeneity.

The background error covariances C_{uu}, C_{uv}, C_{vv}, and C_{vu} or C_{ll}, C_{tt}, C_{tl}, and C_{lt} have been measured observationally using the methods of Section 4.3 in a number of studies, most of which used a climatological background (Buell 1960, 1971, 1972a,b; Brown and Robinson 1979; Seaman and Gauntlett 1980; Buell and Seaman 1983). When the background is climatology, there is considerable anisotropy in the background error field (Seaman and Gauntlett 1980; Buell and Seaman 1983). Much of this anisotropy is related to the tilted-trough mechanism discussed in Section 4.3. The background error wind covariances in the case of a forecast background have been obtained observationally by Hollingsworth and Lönnberg (1986); the anisotropy in this case is discussed in Hollingsworth (1987).

We devote the remainder of this section to the isotropic case. Assume that the covariances $C_{\psi\psi}$, $C_{\chi\chi}$, $C_{\psi\chi}$, and $C_{\chi\psi}$ in (5.2.19–22) are independent of ϕ. Then, from (D19) of Appendix D, $C_{\psi\chi}(r) = C_{\chi\psi}(r)$, and from (5.2.22), $C_{lt}(r) = C_{tl}(r)$. To further simplify the discussion, we also assume that no correlation exists between the velocity potential and the streamfunction, $C_{\psi\chi}(r) = C_{\chi\psi}(r) = 0$. The justification for this last assumption is based on the observational work of Hollingsworth and Lönnberg (1986), which demonstrates that these terms are small for the background error with a forecast background. Panchev (1971) contains a more theoretical discussion of this point.

Under the assumption of isotropy and vanishing of the $C_{\psi\chi}$ covariances, (5.2.9–22) can be written as

$$C_{ll}(r) = -\frac{1}{r}\frac{d}{dr}C_{\psi\psi} - \frac{d^2}{dr^2}C_{\chi\chi} \tag{5.2.23}$$

$$C_{tt}(r) = -\frac{d^2}{dr^2}C_{\psi\psi} - \frac{1}{r}\frac{d}{dr}C_{\chi\chi} \tag{5.2.24}$$

$$C_{lt}(r) = C_{tl}(r) = 0 \tag{5.2.25}$$

If $C_{\psi\psi}$ and $C_{\chi\chi}$ are isotropic, then so are C_{ll} and C_{tt}; but from (5.2.18), C_{uu}, C_{uv}, and C_{vv} are functions of ϕ.

Correlation functions are now introduced. First, define $E_u^2 = C_{uu}(0)$, $E_v^2 = C_{vv}(0)$, $E_l^2 = C_{ll}(0)$, and $E_t^2 = C_{tt}(0)$, which are all equal and independent of location because of the assumptions of isotropy and homogeneity. In (5.2.13), $E_u^2 + E_v^2 = E_l^2 + E_t^2$ defines a kinetic energy (or error kinetic energy) of the horizontal flow.

Then, correlation functions are defined as $C_{uu}(r, \phi) = E_u^2 \rho_{uu}(r, \phi)$, $C_{vv}(r, \phi) = E_v^2 \rho_{vv}(r, \phi)$, $C_{ll}(r) = E_l^2 \rho_{ll}(r)$, and $C_{tt}(r) = E_t^2 \rho_{tt}(r)$. Define also $C_{\psi\psi}(r) = E_\psi^2 \rho_{\psi\psi}(r)$ and $C_{\chi\chi}(r) = E_\chi^2 \rho_{\chi\chi}(r)$ with characteristic horizontal length scales L_ψ and L_χ given by (4.3.12).

Adding (5.2.23) and (5.2.24), inserting the definitions of C_{ll}, C_{tt}, $C_{\psi\psi}$, and $C_{\chi\chi}$, and evaluating at $r = 0$ gives

$$E_u^2 = E_v^2 = \frac{E_\psi^2}{L_\psi^2} + \frac{E_\chi^2}{L_\chi^2} = E_l^2 = E_t^2 \tag{5.2.26}$$

Note that the factor 2, which would normally appear in this equation, cancels because of the definition of characteristic scale (4.3.12). See Exercise 4.4. We can then write

$$\frac{E_\psi^2}{L_\psi^2} = (1 - \nu^2)E_u^2, \qquad \frac{E_\chi^2}{L_\chi^2} = \nu^2 E_u^2, \qquad \text{where} \qquad \nu^2 = \frac{E_\chi^2/L_\chi^2}{E_u^2} \tag{5.2.27}$$

is the ratio of the divergent kinetic energy to the total horizontal kinetic energy. Insertion of (5.2.27) into (5.2.23)–24) yields

$$\rho_{ll}(r) = -\left[L_\psi^2(1 - \nu^2)\frac{1}{r}\frac{d}{dr}\rho_{\psi\psi} + L_\chi^2\nu^2\frac{d^2}{dr^2}\rho_{\chi\chi}\right] \tag{5.2.28}$$

$$\rho_{tt}(r) = -\left[L_\psi^2(1 - \nu^2)\frac{d^2}{dr^2}\rho_{\psi\psi} + L_\chi^2\nu^2\frac{1}{r}\frac{d}{dr}\rho_{\chi\chi}\right] \tag{5.2.29}$$

for the isotropic homogeneous case. The parameter ν^2 is a measure of the divergence in $\rho_{ll}(r)$ and $\rho_{tt}(r)$. If $\nu^2 = 0$, then the flow is completely nondivergent, whereas if $\nu^2 = 1$, it is completely irrotational. As will be seen in Chapter 7, the divergent component of the atmospheric flow is usually smaller than the rotational component for the global and synoptic scales, and thus ν^2 is quite small. The observational study of Hollingsworth and Lönnberg (1986) suggests a value of ν^2 between 0.1 and 0.2 for the background error wind covariances with a forecast background.

In Figure 5.2, (5.2.28–29) are used to plot the autocorrelations $\rho_{ll}(r)$ and $\rho_{tt}(\mathrm{r})$ for three values of ν^2 using the autocorrelation model (4.3.19) for $\rho_{\psi\psi}$ and $\rho_{\chi\chi}$. In this figure, $L_\chi = L_\psi = 1$, and the curves are plotted as a function of r. In the nondivergent case, ρ_{ll} is always positive, but ρ_{tt} becomes negative at $r = L_\psi$. In the irrotational case, ρ_{tt} is always positive, but ρ_{ll} becomes negative at $r = L_\psi$. For $\nu^2 = 0.5$, $\rho_{tt}(r) = \rho_{ll}(r)$ when $L_\psi = L_\chi$.

The behavior of $\rho_{tt}(r)$ and $\rho_{ll}(r)$ in the nondivergent and irrotational cases can be explained by Figure 5.3, in which (a) shows a nondivergent flow and (b) an irrotational flow. The isolines in (a)/(b) are lines of constant streamfunction/velocity potential with values increasing outward from the center. The flow in (a) is counterclockwise around the center ψ_0, and in (b) it is directed outward from the center χ_0. In each case, five values of the longitudinal component l and transverse component t are shown along the same straight line. It is clear that for the nondivergent flow, the effective correlation length is shorter for t than it is for l; the reverse is true for the irrotational flow.

The correlations $\rho_{uu}(r, \phi)$, $\rho_{uv}(r, \phi)$, and $\rho_{vv}(r, \phi)$ can be calculated from $\rho_{ll}(r)$ and $\rho_{tt}(r)$ using (5.2.18). Under isotropic conditions,

$$\rho_{uu}(r, \phi) = \rho_{ll}(r)\cos^2\phi + \rho_{tt}(r)\sin^2\phi \tag{5.2.30}$$

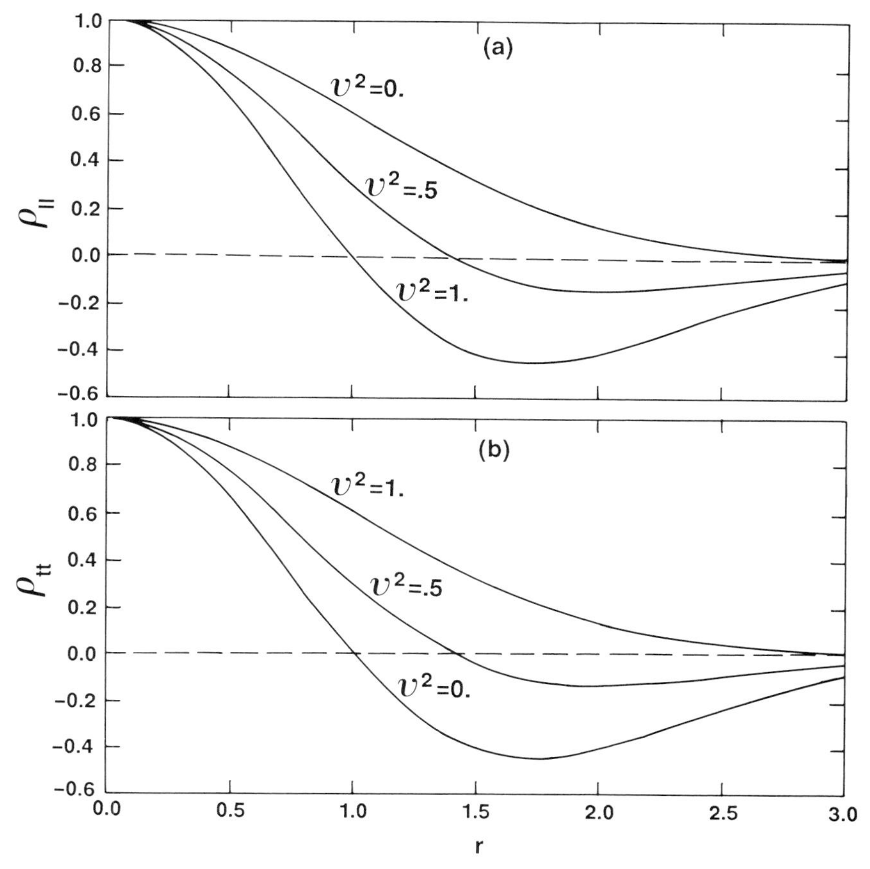

Figure 5.2 The (a) longitudinal and (b) transverse wind autocorrelations as a function of r, for the correlation model (4.3.19).

$$\rho_{uv}(r, \phi) = \rho_{vu}(r, \phi) = [\rho_{ll}(r) - \rho_{tt}(r)] \sin \phi \cos \phi \tag{5.2.31}$$

$$\rho_{vv}(r, \phi) = \rho_{ll}(r) \sin^2 \phi + \rho_{tt}(r) \cos^2 \phi \tag{5.2.32}$$

Note that $\rho_{vv}(r, \phi) = \rho_{uu}(r, \phi + \pi/2)$.

$\rho_{uu}(r, \phi)$, $\rho_{vv}(r, \phi)$, and $\rho_{uv}(r, \phi)$ are plotted for the case $v^2 = 0$ (nondivergent) as a function of r/L_ψ and ϕ in Figure 5.4. The correlation model (4.3.21) is used in calculating (5.2.30–32). Positive contours are solid, and negative contours are dashed. The horizontal scale is indicated under the ρ_{uv} correlation. The velocity correlations ρ_{uu}, ρ_{uv}, and ρ_{vv} for $L_\chi = L_\psi$ and $v^2 \neq 0$ can be inferred from (5.2.30–32). When $v^2 = 0.5$, $\rho_{uv} = \rho_{vu} = 0$, and $\rho_{uu} = \rho_{vv}$ are independent of ϕ, indicating no correlation between wind components.

Note that by subtracting (5.2.24) from (5.2.23) and evaluating the result at $r = 0$, we have

$$E_u^2[\rho_{ll}(0) - \rho_{tt}(0)] = -r \frac{d}{dr} \frac{1}{r} \frac{d^2}{dr^2} [E_\psi^2 \rho_{\psi\psi}(r) - E_\chi^2 \rho_{\chi\chi}(r)]\big|_{r=0} \tag{5.2.33}$$

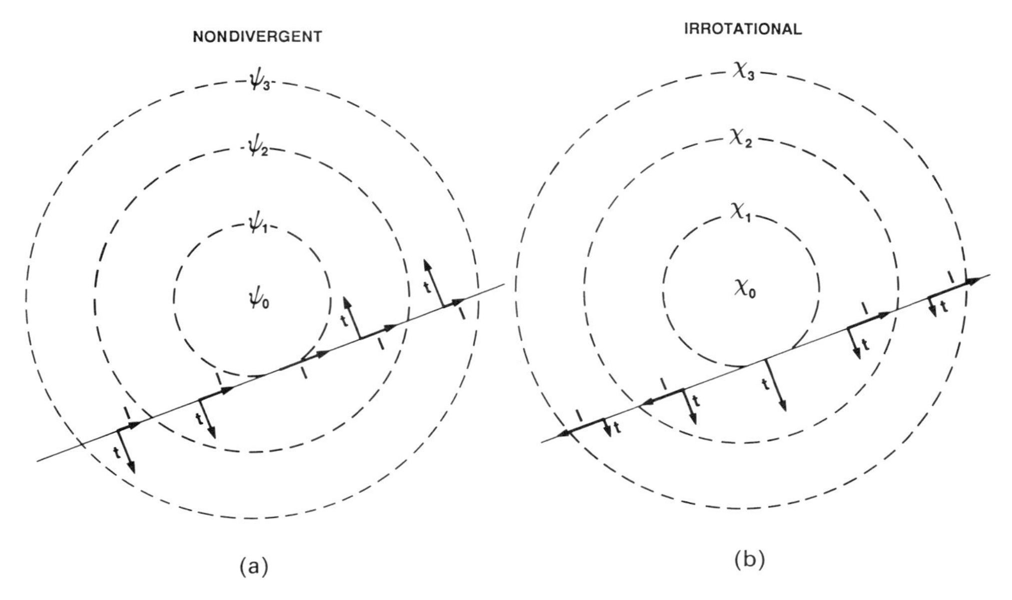

Figure 5.3 The behavior of ρ_{ll} and ρ_{tt} for (a) nondivergent flow and (b) irrotational flow.

The term in square brackets on the right-hand side is not usually equal to zero, and the left-hand side can only equal zero, as it should, if

$$r \frac{d}{dr} \frac{1}{r} \frac{d}{dr} \rho_{\psi\psi}\Big|_{r=0} = 0 \quad \text{and} \quad r \frac{d}{dr} \frac{1}{r} \frac{d}{dr} \rho_{\chi\chi}\Big|_{r=0} = 0 \tag{5.2.34}$$

Conditions (5.2.34), first derived by Buell (1972a) and Julian and Thiebaux (1975), are extra conditions that must be satisfed by autocorrelation functions $\rho_{\psi\psi}$ and $\rho_{\chi\chi}$ in addition to those discussed in Section 4.3. Autocorrelation models (4.3.19) and (4.3.21) satisfy (5.2.34), but the autocorrelation function (4.6.16) does not.

We complete this section by deriving an expression for the spectral density of the horizontal windfield kinetic energy under homogeneous isotropic conditions. Define Hankel transforms of $\rho_{\psi\psi}(r)$, $\rho_{\chi\chi}(r)$, and $\rho_{ll}(r) + \rho_{tt}(r)$, following (4.3.8):

$$\rho_{\psi\psi}(r) = L_\psi^2 \int_0^\infty g_\psi(k) J_0(kr) k \, dk, \qquad \rho_{\chi\chi}(r) = L_\chi^2 \int_0^\infty g_\chi(k) J_0(kr) k \, dk$$

and

$$\rho_{ll}(r) + \rho_{tt}(r) = \rho_{uu}(r) + \rho_{vv}(r) = L_w^2 \int_0^\infty g_w(k) J_0(kr) k \, dk$$

where L_w^2 is a characteristic scale defined in the usual way. Adding together (5.2.28–29) gives

$$\rho_{ll}(r) + \rho_{tt}(r) = -L_\psi^2(1 - v^2)\, \nabla^2 \rho_{\psi\psi} - L_\chi^2 v^2\, \nabla^2 \rho_{\chi\chi} \tag{5.2.35}$$

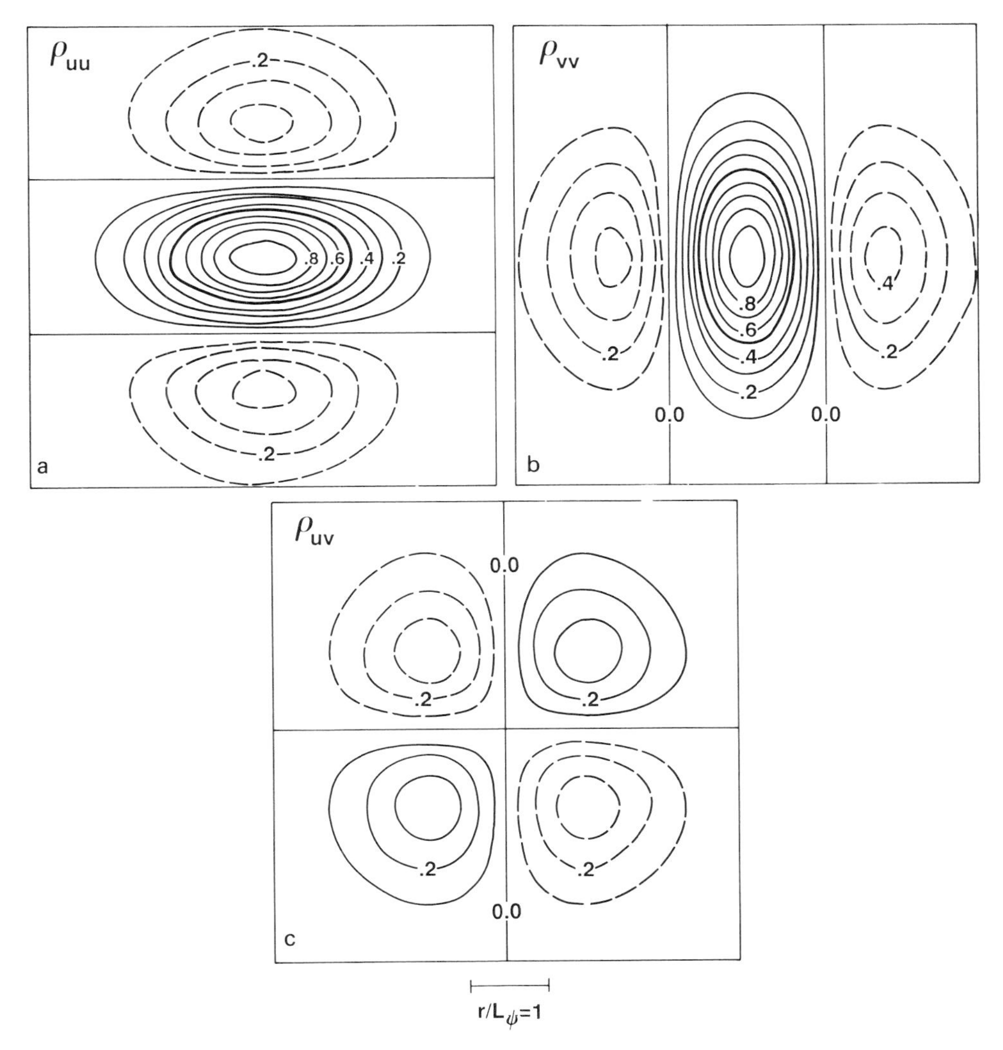

Figure 5.4 (a) $\rho_{uu}(r, \phi)$, (b) $\rho_{vv}(r, \phi)$, and (c) $\rho_{uv}(r, \phi)$ for the nondivergent isotropic case for the correlation model (4.3.21).

Inserting the Hankel transforms and applying (G16) yields

$$L_w^2 g_w(k) = k^2[L_\psi^4(1 - v^2)g_\psi(k) + L_\chi^4 v^2 g_\chi(k)] \tag{5.2.36}$$

If $\rho_{\psi\psi}(r)$ and $\rho_{\chi\chi}(r)$ are specified by correlation models such as (4.3.19) or (4.3.21), then their spectral densities $g_\psi(k)$ and $g_\chi(k)$ have a maximum at $k = 0$. From (5.2.6), however, it is clear that $g_w(k) \to 0$ as $k \to 0$. This is a direct consequence of the differential relation (5.2.35) and ultimately of Helmholtz's theorem (5.2.1). It might be noted from (5.2.35) that $-\nabla^2\rho_{\psi\psi}$ and $-\nabla^2\rho_{\chi\chi}$ for correlation models (4.3.19) and (4.3.21) are negative at large r, which from Exercise 3.3 is consistent with $g_w(0) < g_w(k)$.

5.3 The geopotential/wind covariances

In this section we derive expressions for the wind/geopotential background error covariances: $C^{\mathrm{B}}_{\Phi u}$, $C^{\mathrm{B}}_{\Phi v}$, $C^{\mathrm{B}}_{u\Phi}$, and $C^{\mathrm{B}}_{v\Phi}$. As in Section 5.2, the superscript B is not used, and we assume that the wind/geopotential covariances are separable. The cautionary note concerning the separability assumption in the second paragraph of Section 5.2 also applies to this section.

The discussion of the horizontal geopotential/wind covariances mirrors closely the treatment of Section 5.2. Define the horizontal cross-covariances $C_{\Phi u}(\mathbf{r}_1, \mathbf{r}_2) = \langle \Phi_1 u_2 \rangle$, $C_{\Phi v}(\mathbf{r}_1, \mathbf{r}_2) = \langle \Phi_1 v_2 \rangle$, $C_{u\Phi}(\mathbf{r}_1, \mathbf{r}_2) = \langle u_1 \Phi_2 \rangle$, and $C_{v\Phi}(\mathbf{r}_1, \mathbf{r}_2) = \langle v_1 \Phi_2 \rangle$ between the two spatial locations 1 and 2. As in Section (5.2), assume $\langle u_1 \rangle = \langle u_2 \rangle = \langle v_1 \rangle = \langle v_2 \rangle = \langle \Phi_1 \rangle = \langle \Phi_2 \rangle = 0$. Introducing Helmholtz's theorem (5.2.1) and the assumption of horizontal homogeneity gives

$$\langle \Phi_1 u_2 \rangle = -\frac{\partial}{\partial \tilde{y}} \langle \Phi_1 \psi_2 \rangle + \frac{\partial}{\partial \tilde{x}} \langle \Phi_1 \chi_2 \rangle \tag{5.3.1}$$

$$\langle \Phi_1 v_2 \rangle = \frac{\partial}{\partial \tilde{x}} \langle \Phi_1 \psi_2 \rangle + \frac{\partial}{\partial \tilde{y}} \langle \Phi_1 \chi_2 \rangle \tag{5.3.2}$$

$$\langle u_1 \Phi_2 \rangle = \frac{\partial}{\partial \tilde{y}} \langle \psi_1 \Phi_2 \rangle - \frac{\partial}{\partial \tilde{x}} \langle \chi_1 \Phi_2 \rangle \tag{5.3.3}$$

$$\langle v_1 \Phi_2 \rangle = -\frac{\partial}{\partial \tilde{x}} \langle \psi_1 \Phi_2 \rangle - \frac{\partial}{\partial \tilde{y}} \langle \chi_1 \Phi_2 \rangle \tag{5.3.4}$$

and $\tilde{\mathbf{r}} = \mathbf{r}_2 - \mathbf{r}_1$, $\tilde{x} = x_2 - x_1$, and $\tilde{y} = y_2 - y_1$. Equations (5.3.1–4) follow easily from the development in (5.2.2–7).

Define $C_{\phi\psi}(\tilde{\mathbf{r}}) = \langle \Phi_1 \psi_2 \rangle$, $C_{\chi\Phi}(\tilde{\mathbf{r}}) = \langle \chi_1 \Phi_2 \rangle$, and so on, and introduce plane polar coordinates r and ϕ, as in Section 5.2. Then, (5.3.1–2) can be written as

$$\begin{vmatrix} C_{\Phi u} \\ C_{\Phi v} \end{vmatrix} = \begin{pmatrix} -\sin\phi & \cos\phi \\ \cos\phi & \sin\phi \end{pmatrix} \begin{vmatrix} \dfrac{\partial}{\partial r} C_{\Phi\psi} + \dfrac{1}{r}\dfrac{\partial}{\partial \phi} C_{\Phi\chi} \\ \dfrac{\partial}{\partial r} C_{\Phi\chi} - \dfrac{1}{r}\dfrac{\partial}{\partial \phi} C_{\Phi\psi} \end{vmatrix} \tag{5.3.5}$$

with similar equations corresponding to (5.3.3–4). The 2×2 matrix in (5.3.5) can be inverted to produce

$$\begin{vmatrix} C_{\Phi t} \\ C_{\Phi l} \end{vmatrix} = \begin{pmatrix} -\sin\phi & \cos\phi \\ \cos\phi & \sin\phi \end{pmatrix} \begin{vmatrix} C_{\Phi u} \\ C_{\Phi v} \end{vmatrix} = \begin{vmatrix} \dfrac{\partial}{\partial r} C_{\Phi\psi} + \dfrac{1}{r}\dfrac{\partial}{\partial \phi} C_{\Phi\chi} \\ \dfrac{\partial}{\partial r} C_{\Phi\chi} - \dfrac{1}{r}\dfrac{\partial}{\partial \phi} C_{\Phi\psi} \end{vmatrix} \tag{5.3.6}$$

using the longitudinal l and transverse t wind components defined in (5.2.17).

Now make the isotropic assumption, and all scalar covariances $C_{\Phi\psi}$, $C_{\psi\Phi}$, $C_{\Phi\chi}$, and $C_{\chi\Phi}$ become independent of ϕ. From (5.3.5–6), it can be seen that $C_{\Phi u}$ and $C_{\Phi v}$

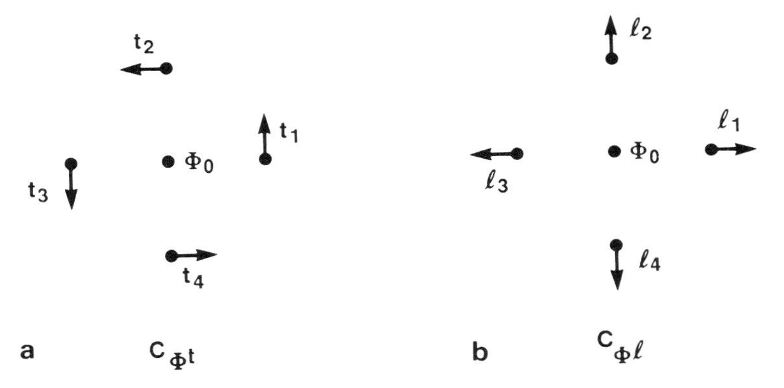

Figure 5.5 The covariances (a) $C_{\Phi t}$ and (b) $C_{\Phi l}$.

are functions of r and ϕ whereas $C_{\Phi t}$ and $C_{\Phi l}$ become functions of r only. Equation (D19) of Appendix D implies $C_{\Phi\psi}(r) = C_{\psi\Phi}(r)$ and $C_{\Phi\chi}(r) = C_{\chi\Phi}(r)$. Consequently, from (5.3.1–5), $C_{\Phi u}(r, \phi) = -C_{u\Phi}(r, \phi)$ and $C_{\Phi v}(r, \phi) = -C_{v\Phi}(r, \phi)$; and from (5.3.6), $C_{\Phi l}(r) = -C_{l\Phi}(r)$ and $C_{\Phi t}(r) = -C_{t\Phi}(r)$. Then, (5.3.6) can be written as

$$C_{\Phi t} = -C_{t\Phi} = \frac{d}{dr} C_{\Phi\psi}, \qquad C_{\Phi l} = -C_{l\Phi} = \frac{d}{dr} C_{\Phi\chi} \tag{5.3.7}$$

Thus, the isotropic component of the geopotential/streamfunction covariance only involves the swirl or transpose component of the wind about a geopotential observation point whereas the geopotential/velocity potential covariance involves only the radial component. This is illustrated in Figure 5.5, which shows a geopotential at point 0 and at points 1–4 the (a) transverse and (b) longitudinal wind components. The swirl component involved in the $C_{\Phi t}$ covariance and the radial wind component in the $C_{\Phi l}$ covariance are evident.

At this point, correlations $\rho_{\Phi t}$ and $\rho_{\Phi l}$ can be introduced. Variances E_u^2, E_Φ^2, E_ψ^2, and E_χ^2 can be defined as in the previous section. Define E_Φ^2 from $C_{\Phi\Phi}(r) = E_\Phi^2 \rho_{\Phi\Phi}(r)$, $C_{\Phi t}(r) = E_\Phi E_u \rho_{\Phi t}(r)$, $C_{\Phi l}(r) = E_\Phi E_u \rho_{\Phi l}(r)$, $C_{\Phi\psi}(r) = E_\Phi E_\psi \rho_{\Phi\psi}(r)$, and $C_{\Phi\chi}(r) = E_\Phi E_\chi \rho_{\Phi\chi}(r)$, noting that $E_u = E_v = E_l = E_t$ under isotropic homogeneous conditions.

The correlations $\rho_{\Phi t}$ and $\rho_{\Phi l}$ have been calculated for the background error using a forecast background by Lönnberg and Hollingsworth (1986). Following the observational procedure of Section 4.3, observed-minus-background correlations similar to (4.3.13) were calculated for each of the station pairs of the North American radiosonde network. The results are plotted in Figure 5.6 for the geopotential/transverse correlation and the geopotential/longitudinal correlation. The isotropic component at 850 mb as a function of distance between stations is in a format similar to Figure 4.4. The quantities

$$\frac{\overline{(\Phi_O - \Phi_B)(t_O - t_B)}}{\sqrt{\overline{(\Phi_O - \Phi_B)^2}\ \overline{(t_O - t_B)^2}}} \quad \text{and} \quad \frac{\overline{(\Phi_O - \Phi_B)(l_O - l_B)}}{\sqrt{\overline{(\Phi_O - \Phi_B)^2}\ \overline{(l_O - l_B)^2}}}$$

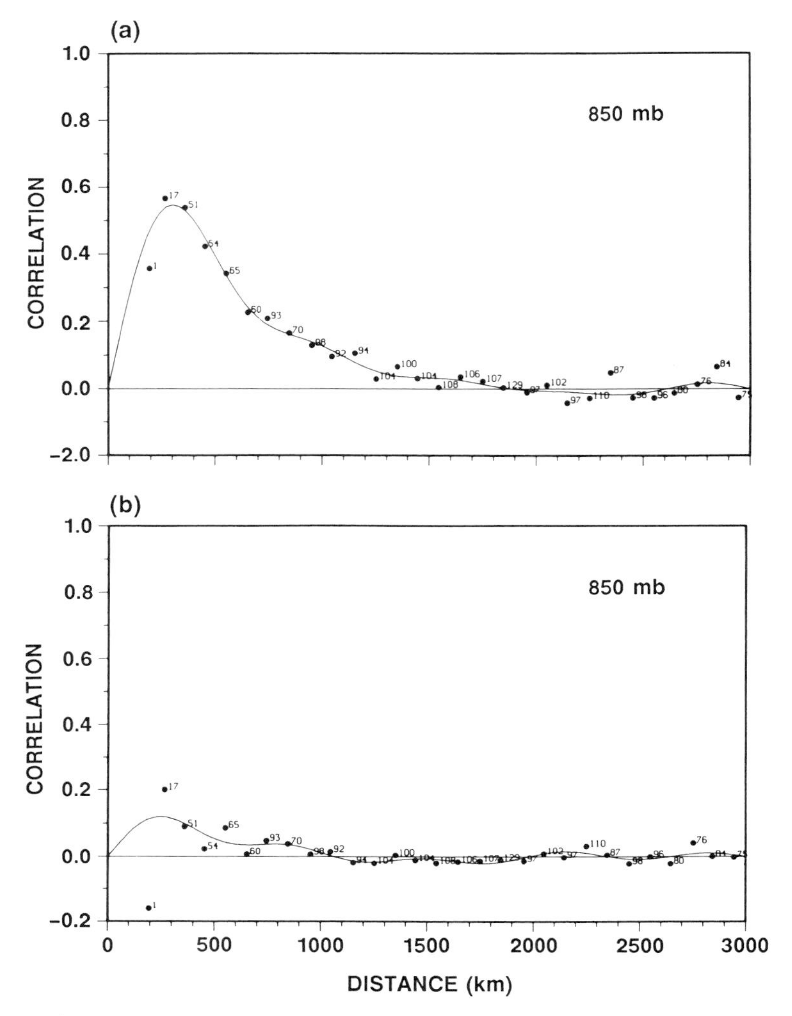

Figure 5.6 The isotropic component of the background error correlation (a) $\rho_{\Phi t}$ and (b) $\rho_{\Phi l}$ for a forecast background at 850 mb as a function of station separation. (After Lönnberg and Hollingsworth 1986)

are plotted in (a) and (b), respectively. Each dot represents a bin or group of station pairs with similar separations, and the associated figure is the number of stations pairs in that particular bin. The smooth curves represent a fit to the dots.

Introducing the definitions of $\rho_{\Phi t}$, $\rho_{\Phi l}$, $\rho_{\Phi\psi}$, and $\rho_{\Phi\chi}$ into (5.3.7) gives

$$\rho_{\Phi t} = L_\psi \sqrt{1 - v^2}\,\frac{d}{dr}\,\rho_{\Phi\psi}, \qquad \rho_{\Phi t} = L_\chi v\,\frac{d}{dr}\,\rho_{\Phi\chi} \tag{5.3.8}$$

where L_ψ and L_χ are defined in (5.2.26) and v in (5.2.27). Note that although $\rho_{\Phi\psi} = \rho_{\psi\Phi}$ and $\rho_{\Phi\chi} = \rho_{\chi\Phi}$, they are not autocorrelations and do not necessarily equal one at $r = 0$.

In mid and high latitudes, the geostrophic assumption (2.6.1) is often used in deriving expressions for the geopotential/wind background error covariances. In cartesian coordinates on an f-plane (constant Coriolis parameter), (2.6.1) can also be written as

$$\Phi = f_0 \psi, \qquad \chi = 0 \tag{5.3.9}$$

using Helmholtz's theorem (5.2.1). From (5.3.9), the correlations $C_{\Phi\Phi}(r)$, $C_{\psi\psi}(r)$, $C_{uu}(r, \phi)$, and $C_{vv}(r, \phi)$ can be related:

$$\begin{aligned} C_{\Phi\Phi} &= f_0^2 C_{\psi\psi} \\ C_{uu} + C_{vv} &= -\nabla^2 C_{\psi\psi} = -\frac{\nabla^2 C_{\Phi\Phi}}{f_0^2} \end{aligned} \tag{5.3.10}$$

using (5.2.13). Introduce a characteristic horizontal scale for the geopotential autocorrelation consistent with (4.3.12):

$$\frac{L_\Phi^2}{2} = -\frac{\rho_{\Phi\Phi}}{\nabla^2 \rho_{\Phi\Phi}}\bigg|_{r=0} \tag{5.3.11}$$

Then, evaluating (5.3.10) at $r = 0$ gives

$$E_u^2 = \frac{E_\psi^2}{L_\psi^2} = \frac{E_\Phi^2}{f_0^2 L_\Phi^2} \tag{5.3.12}$$

and $\rho_{\psi\psi}(r) = \rho_{\Phi\Phi}(r)$, $L_\psi = L_\Phi$.

Under the geostrophic assumption, (5.3.8) becomes

$$\rho_{\Phi t} = -\rho_{t\Phi} = L_\Phi \frac{d}{dr} \rho_{\Phi\psi}, \qquad \rho_{\Phi l} = -\rho_{l\Phi} = 0 \tag{5.3.13}$$

Now introduce a parameter μ such that

$$f_0 C_{\Phi\psi} = \mu C_{\Phi\Phi}, \qquad \text{which implies} \qquad \rho_{\Phi\psi}(r) = \rho_{\psi\Phi}(r) = \mu \rho_{\Phi\Phi}(r) \tag{5.3.14}$$

as (5.3.12) implies $|f_0| E_\Phi E_\psi = E_\Phi^2$.

The nondimensional parameter $\mu = 1$, if Φ and ψ are fully correlated, and $\mu = 0$, if ψ and Φ are completely uncorrelated. The parameter μ (introduced by Lorenc 1981) controls the coupling between wind and geopotential. The parameter ν^2 defined in (5.2.17) controls the divergence, and μ controls the geostrophy. Under assumption (5.3.14), $\rho_{\Phi\psi}(r)$ is similar to an autocorrelation except that it equals μ at $r = 0$. Lönnberg and Hollingsworth (1986) have shown observationally that for the background error with a forecast background, μ varies between about 0.75 and 0.95 in the northern hemisphere midlatitudes. In the southern hemisphere midlatitudes, exact geostrophic coupling is implied if $\mu = -1$. In the tropics, where geostrophy does not apply, $\mu = 0$. Lönnberg and Hollingsworth (1986) consider more general circumstances when the Coriolis parameter is latitudinally variable and μ is a function of horizontal scale.

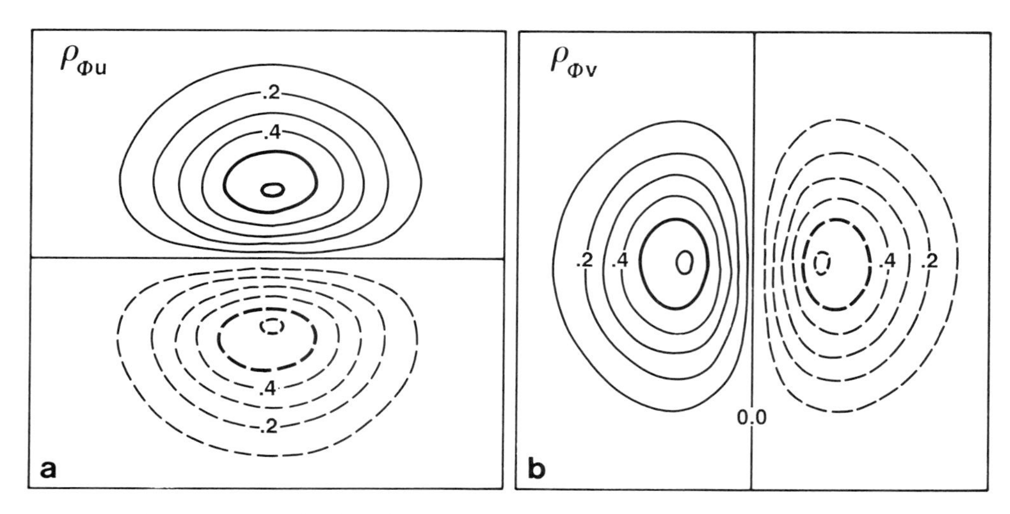

Figure 5.7 (a) $\rho_{\Phi u}(r, \phi)$ and (b) $\rho_{\Phi v}(r, \phi)$ for $\mu = 1$, using the correlation model (4.3.21).

Expressions for $\rho_{\Phi u}$ and $\rho_{\Phi v}$ under homogeneous, isotropic geostrophic conditions can be obtained from (5.3.5), (5.3.13), and (5.3.14):

$$\rho_{\Phi u} = -\rho_{u\Phi} = -\mu L_\Phi \sin\phi \frac{d}{dr} \rho_{\Phi\Phi} \tag{5.3.15}$$

$$\rho_{\Phi v} = -\rho_{v\Phi} = \mu L_\Phi \cos\phi \frac{d}{dr} \rho_{\Phi\Phi} \tag{5.3.16}$$

Figure 5.7 shows plots of $\rho_{\Phi u}$ and $\rho_{\Phi v}$ for the case $\mu = 1$, using correlation model (4.3.21) for $\rho_{\Phi\Phi}(r)$: (a) shows $\rho_{\Phi u}$ (5.3.15) and (b) shows $\rho_{\Phi v}$ (5.3.16), as a function of r/L_ψ and ϕ. Negative contours are dashed, and the horizontal scale is the same as in Figure 5.4. $\rho_{u\Phi}$ and $\rho_{v\Phi}$ are simply the negatives of (a) and (b), respectively.

From (5.3.15–16), $\rho_{\Phi u}(r, \phi)$ and $\rho_{\Phi v}(r, \phi)$ are continuous at $r = 0$ only if

$$\frac{d}{dr} \rho_{\psi\psi}(r)\Big|_{r=0} = \frac{d}{dr} \rho_{\Phi\Phi}(r)\Big|_{r=0} = 0 \tag{5.3.17}$$

Condition (5.3.17) must be satisfied by $\rho_{\psi\psi}$ in addition to those mentioned in Section 4.3. Autocorrelation models (4.3.19) and (4.3.21) satisfy (5.3.17), but (4.6.16) does not.

It is appropriate to summarize the conditions under which the background error covariances are *exactly* geostrophically coupled. For homogeneous isotropic fields, these conditions are:

$$\begin{aligned}
&(1) \quad \rho_{\psi\psi}(r) = \rho_{\Phi\Phi}(r), \qquad L_\psi = L_\Phi \\
&(2) \quad E_\Phi^2 = f_0^2 L_\Phi^2 E_u^2 = f_0^2 E_\psi^2 \\
&(3) \quad \rho_{\Phi\psi}(r) = \rho_{\psi\Phi}(r) = \mu\rho_{\Phi\Phi}(r) \qquad \text{with} \quad \mu = 1 \\
&(4) \quad v^2 = 0
\end{aligned} \tag{5.3.18}$$

Equations (5.3.18) imply a *continuous geostrophic* relationship among the geopotential, wind, and wind/geopotential covariances. Consequently, from (5.1.24), conditions (5.3.18) are sufficient to guarantee that the analysis increments for the continuous form of the analysis will also be exactly geostrophically coupled.

The background error covariances in this section were all derived under the f-plane assumption and have no validity in the tropics. Recently, Parrish (1988) has applied the ideas of Phillips (1986), discussed at the end of Section 4.3, to generate truly global background error covariances. He assumed that the background error projected entirely onto the Rossby–Hough modes described in Section 2.6 and Chapter 9. He further assumed that no correlation existed between the background errors of the individual Rossby–Hough modes, resulting in a diagonal matrix in Rossby–Hough mode space. Transforming these spectral covariances into real space produced correlations $\rho_{\Phi u}$, $\rho_{\Phi v}$, ρ_{uu}, and so on, which are similar to those shown in Figures 5.4 and 5.7 at high latitudes but have a completely different structure in the tropics.

5.4 Multivariate filtering properties

In Section 4.4, the spectral responses and filtering properties of univariate statistical interpolation were examined using a continuous analogue. The same technique can be employed to illustrate the spectral response of the multivariate algorithm. In this section, we extend the one-dimensional continuous analogue of Section 4.4 to the multivariate case. Section 5.5 examines the discrete case and, more particularly, the effect of the constraint parameters μ and v^2, defined in Sections 5.2 and 5.3.

Suppose $v(x)$ is a one-dimensional (north–south) wind component, and $h(x) = \Phi(x)/f_0$ is a scaled geopotential on the domain $-\infty \leq x \leq \infty$. Essentially, f_0 has been scaled out of the problem to reduce the number of constants in the equations. As in Section 4.4, consider the analysis gridpoint to be at $x = 0$. Then, following (5.1.11), the multivariate version of (4.4.1) is

$$
\begin{aligned}
h_A(0) &= h_B(0) + \frac{1}{2\pi L}\int_{-\infty}^{\infty} \{W_{hh}(x)[h_O(o) - h_B(x)] + W_{hv}(x)[v_O(x) - v_B(x)]\}\, dx \\
v_A(0) &= v_B(0) + \frac{1}{2\pi L}\int_{-\infty}^{\infty} \{W_{vh}(x)[h_O(o) - h_B(x)] + W_{vv}(x)[v_O(x) - v_B(x)]\}\, dx
\end{aligned}
\tag{5.4.1}
$$

where subscripts A, B, and O stand for analysis, background, and observation, respectively. W_{hh}, W_{hv}, W_{vh}, and W_{vv} are the a posteriori weights as in (5.1.11), and L is a characteristic length scale, analogous to L_B of (4.4.1), which will be defined later.

Assume that the background error is homogeneous and that the observation error is homogeneous and spatially uncorrelated. The univariate version of this case was discussed in (4.4.16–26). The observation error spectral densities are finite and independent of wavenumber, but the actual observation errors are infinite, as noted in Section 4.4. Under these circumstances, the continuous analogues of (5.1.13–16) have a form similar to the second equation of (4.4.20):

$$\frac{1}{2\pi L}\int_{-\infty}^{\infty}[W_{hh}(x')C_{hh}(\tilde{x})+W_{hv}(x')C_{hv}(\tilde{x})]\,dx'+\varepsilon_h^2E_h^2(\mathrm{B})W_{hh}(x)=C_{hh}(-x) \tag{5.4.2}$$

$$\frac{1}{2\pi L}\int_{-\infty}^{\infty}[W_{hh}(x')C_{vh}(\tilde{x})+W_{hv}(x')C_{vv}(\tilde{x})]\,dx'+\varepsilon_v^2E_v^2(\mathrm{B})W_{hv}(x)=C_{vh}(-x) \tag{5.4.3}$$

$$\frac{1}{2\pi L}\int_{-\infty}^{\infty}[W_{vh}(x')C_{hh}(\tilde{x})+W_{vv}(x')C_{hv}(\tilde{x})]\,dx'+\varepsilon_h^2E_h^2(\mathrm{B})W_{vh}(x)=C_{hv}(-x) \tag{5.4.4}$$

$$\frac{1}{2\pi L}\int_{-\infty}^{\infty}[W_{vh}(x')C_{vh}(\tilde{x})+W_{vv}(x')C_{vv}(\tilde{x})]\,dx'+\varepsilon_v^2E_v^2(\mathrm{B})W_{vv}(x)=C_{vv}(-x) \tag{5.4.5}$$

where $\tilde{x}=x'-x$. As in Section 4.4, x corresponds to x_k and x' is a dummy variable corresponding to x_l. $C_{hv}(\tilde{x})$, $C_{vh}(-x)$, and so on, are background error covariances whereas $E_h^2(\mathrm{B})$ and $E_v^2(\mathrm{B})$ are background error variances of height and wind, which are independent of position. ε_h^2 can be identified with ε_0^2 of the univariate equation (4.4.20), whereas ε_v^2 plays the same role for the windfield. Following (4.4.19), it can be seen that ε_h^2 is proportional to the (constant) spectral density of the height observation errors whereas ε_v^2 is proportional to the (constant) spectral density of the wind observation errors.

With $E_h^2(\mathrm{B})=C_{hh}(0)$ and $E_v^2(\mathrm{B})=C_{vv}(0)$, define a background error correlation $\rho(x)=\rho_{hh}(x)$ for the height by $C_{hh}(x)=E_h^2(\mathrm{B})\rho(x)$. Now, if the background error fields are nondivergent and geostrophically related as in Section 5.3, then,

$$\frac{d}{dx}[h_{\mathrm{B}}(x)-h_{\mathrm{T}}(x)]=\frac{d}{dx}[\psi_{\mathrm{B}}(x)-\psi_{\mathrm{T}}(x)]=[v_{\mathrm{B}}(x)-v_{\mathrm{T}}(x)] \tag{5.4.6}$$

where subscript T indicates true value. Equation (5.4.6) also implies $\rho(x)=\rho_{hh}(x)=\rho_{\psi\psi}(x)$. A characteristic length scale L can be defined for $\rho(x)$ following (4.3.10). Then, from the one-dimensional analogue of (5.3.10–12) and the definitions of E_v, E_h, L, and $\rho(x)$, we can write

$$E_v^2(\mathrm{B})=\frac{E_h^2(\mathrm{B})}{L^2}\quad\text{and}\quad C_{vv}(x)=-E_h^2(\mathrm{B})\frac{d^2}{dx^2}\rho(x) \tag{5.4.7}$$

Introducing the geostrophic coupling parameter μ, we obtain the one-dimensional analogue of (5.3.14–16):

$$C_{hv}(x)=-C_{vh}(x)=\mu E_h^2(\mathrm{B})\frac{d\rho(x)}{dx} \tag{5.4.8}$$

Because $\rho(x)$ is symmetric with respect to $x=0$,

$$\frac{d\rho(-x)}{dx}=-\frac{d\rho(x)}{dx}\quad\text{and}\quad\frac{d^2\rho(-x)}{dx^2}=\frac{d^2\rho(x)}{dx^2}$$

Equations (5.4.2–5) can then be rewritten in terms of ε_h^2, ε_u^2, L^2, and ρ_B. Rather than follow through the process for all four equations, we illustrate it for (5.4.3) only:

$$\frac{1}{2\pi L}\int_{-\infty}^{\infty}\left[-\mu L^2 W_{hh}(x')\frac{d\rho(\tilde{x})}{d\tilde{x}} - L^2 W_{hv}(x')\frac{d^2\rho(\tilde{x})}{d\tilde{x}^2}\right]dx' + \varepsilon_v^2 W_{hv}(x) = \mu L^2\frac{d\rho(x)}{dx} \tag{5.4.9}$$

Define

$$\alpha(m) = \frac{1}{2\pi L}\int_{-\infty}^{\infty} W_{hh}(x)e^{-imx}\,dx, \qquad \beta(m) = \frac{1}{2\pi L}\int_{-\infty}^{\infty} W_{hv}(x)e^{-imx}\,dx$$
$$\gamma(m) = \frac{1}{2\pi L}\int_{-\infty}^{\infty} W_{vh}(x)e^{-imx}\,dx, \qquad \delta(m) = \frac{1}{2\pi L}\int_{-\infty}^{\infty} W_{vv}(x)e^{-imx}\,dx \tag{5.4.10}$$

to be the Fourier transforms of W_{hh}, W_{hv}, W_{vh}, and W_{vv}, and note that the definitions of α, β, γ, and δ in (5.4.10) differ slightly from (4.4.6). Define the spectral density $g(m)$ of $\rho(x)$ as in (4.4.4),

$$\rho(x) = L\int_{-\infty}^{\infty} g(m)e^{imx}\,dm, \qquad g(m) = \frac{1}{2\pi L}\int_{-\infty}^{\infty}\rho(x)e^{-imx}\,dx \tag{5.4.11}$$

and note that

$$\frac{d\rho}{dx} = iL\int_{-\infty}^{\infty} mg(m)e^{imx}\,dm \quad \text{and} \quad \frac{d^2\rho}{dx^2} = -L\int_{-\infty}^{\infty} m^2 g(m)e^{imx}\,dm \tag{5.4.12}$$

Following (4.4.6), we obtain, for (5.4.9),

$$L\int_{-\infty}^{\infty} e^{imx}[i\mu L^2 mg(m)\alpha(m) + L^2m^2g(m)\beta(m) + \varepsilon_v^2\beta(m) - i\mu L^2 mg(m)]dm = 0 \tag{5.4.13}$$

which must equal zero for all values of x, and therefore the expression in square brackets must equal zero for all values of m.

Similar expressions can be derived for (5.4.2), (5.4.4), and (5.4.5). Consider the case $\mu = 1$, that is, exactly geostrophically coupled wind and height background error. Then, solving for $\alpha(m)$, $\beta(m)$, $\gamma(m)$, and $\delta(m)$ gives

$$\alpha(m) = \varepsilon_v^2 d(m)g(m), \qquad \beta(m) = i\varepsilon_h^2 L^2 md(m)g(m)$$
$$\gamma(m) = -i\varepsilon_v^2 md(m)g(m), \qquad \delta(m) = \varepsilon_h^2 L^2 m^2 d(m)g(m) \tag{5.4.14}$$

where $d(m) = [g(m)(\varepsilon_v^2 + \varepsilon_h^2 L^2 m^2) + \varepsilon_h^2\varepsilon_v^2]^{-1}$.

The spectral equivalents of (5.4.1) corresponding to (3.3.6–7) can be found as follows. Define

$$\Delta\hat{h}_A(m) = \frac{1}{2\pi L}\int_{-\infty}^{\infty}[h_A(x) - h_B(x)]e^{-imx}\,dx$$
$$\Delta\hat{h}_O(m) = \frac{1}{2\pi L}\int_{-\infty}^{\infty}[h_O(x) - h_B(x)]e^{-imx}\,dx \tag{5.4.15}$$

to be the Fourier transforms of the height analysis increments and observation increments respectively. In the same way define $\Delta v_A(m)$ and $\Delta v_O(m)$ to be the Fourier transforms of the wind analysis increments and observation increments. Substitution of (5.4.15) into (5.4.1) yields

$$\begin{aligned}\Delta\hat{h}_A(m) &= \alpha(m)\,\Delta\hat{h}_O(m) - \beta(m)\,\Delta\hat{v}_O(m)\\ \Delta\hat{v}_A(m) &= -\gamma(m)\,\Delta\hat{h}_O(m) + \delta(m)\,\Delta\hat{v}_O(m)\end{aligned} \tag{5.4.16}$$

which relates the spectral response of the analysis increments to the spectra of the observation increments $\Delta\hat{h}_O$ and $\Delta\hat{v}_O$ and the spectra of the weight functions W_{hh}, W_{hv}, W_{vh}, and W_{vv}. The background error was assumed to be geostrophically coupled (5.4.6). It is left for Exercise 5.3 to show that the analysis increments $v_A(x) - v_B(x)$ and $h_A(x) - h_B(x)$ are also geostrophically coupled as would be expected from (5.1.24).

Two important limiting cases for (5.4.14) are $\varepsilon_v^2 \to \infty$ and $\varepsilon_h^2 \to \infty$. The case $\varepsilon_v^2 \to \infty$ corresponds to nonexistent wind observations. In this limit,

$$\begin{aligned}\alpha(m) &= R_h(m) = \frac{g(m)}{g(m)+\varepsilon_h^2}, & \beta(m) &= 0\\ \gamma(m) &= \frac{-img(m)}{g(m)+\varepsilon_h^2}, & \delta(m) &= 0\end{aligned} \tag{5.4.17}$$

$R_h(m)$ is the response of the height field to the univariate algorithm (4.4.21). $\gamma(m)$ in (5.4.17) indicates the spectrum of the wind analysis increment derived purely from height observation increments through the assumption of geostrophy in the background error.

The other limiting case is $\varepsilon_h^2 \to \infty$, which corresponds to nonexistent height observations:

$$\begin{aligned}\alpha(m) &= 0, & \beta(m) &= \frac{iL^2mg(m)}{L^2m^2g(m)+\varepsilon_v^2}\\ \gamma(m) &= 0, & \delta(m) &= R_w(m) = \frac{L^2m^2g(m)}{L^2m^2g(m)+\varepsilon_v^2}\end{aligned} \tag{5.4.18}$$

The one-dimensional nondivergent analogue of (5.2.36) is $g_w(m) = L^2m^2g(m)$. Then, $R_w(m)$ can be written $R_w(m) = g_w(m)/(g_w(m) + \varepsilon_v^2)$, which has the same functional form as $R_h(m)$. Thus, we refer to $R_w(m)$ as the univariate wind response. $\beta(m)$ in (5.4.18) is the spectrum of the height analysis increments derived purely from wind observation increments through the assumption of geostrophy in the background error.

In Figure 5.8, $\alpha(m)$. $-i\beta(m)$, $i\gamma(m)$, and $\delta(m)$ are plotted as functions of wavenumber m for three choices of ε_h^2 and ε_v^2. The correlation model $\rho(x)$ with Fourier transform $g(m)$ is given by (4.3.20) with $L = 1$. For curves a, $\varepsilon_h^2 = 0.10$ and $\varepsilon_v^2 \to \infty$; for curves b, $\varepsilon_h^2 = \varepsilon_v^2 = 0.10$; and for curves c, $\varepsilon_h^2 \to \infty$ and $\varepsilon_v^2 = 0.10$. From (5.4.17), curves a are identically zero for $\beta(m)$ and $\delta(m)$; from (5.4.18), curves c are identically zero for $\alpha(m)$ and $\gamma(m)$.

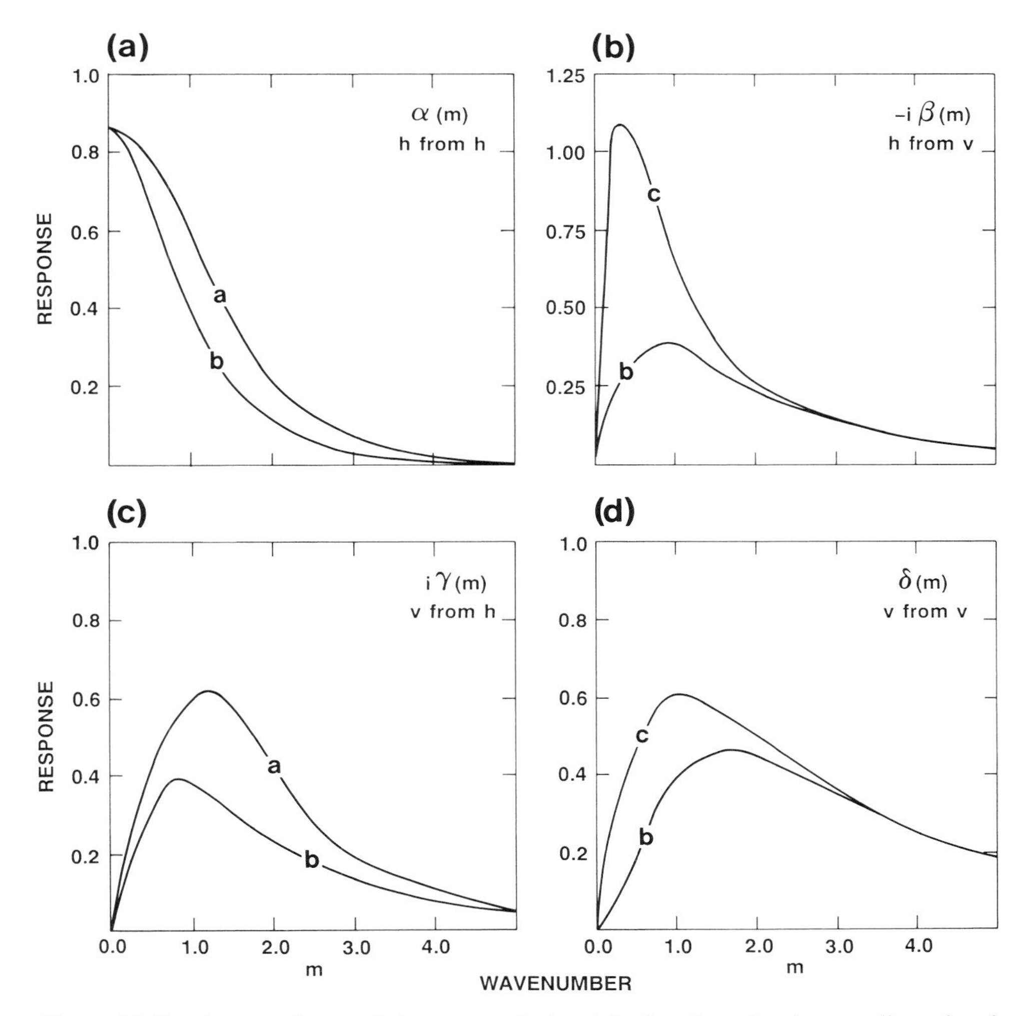

Figure 5.8 Fourier transforms of the a posteriori weight functions for the one-dimensional continuous observation network: (a) W_{hh}, (b) W_{hv}, (c) W_{vh}, and (d) W_{vv}.

Insight into the spectral responses of the multivariate statistical interpolation algorithm can be gained from a careful examination of Figure 5.8. Let us begin by comparing the two univariate cases: the univariate height case R_h plotted as curve a of Figure 5.8(a) and the univariate wind case R_w plotted as curve c of Figure 5.8(d). The univariate wind response is much better than the univariate height response for small scales and is much worse at large scales, which we expect from the form of the background error spectral density for the windfield $g_w(m) = L^2m^2g(m)$. As can be seen from (5.2.35–36) and (5.4.12), the quite different wavenumber dependence of $g_w(m)$ is a direct consequence of the application of Helmholtz's theorem (5.2.1).

The univariate wind response can be more clearly understood by examining the appropriate background, observation, and analysis error variances. Following (4.4.8),

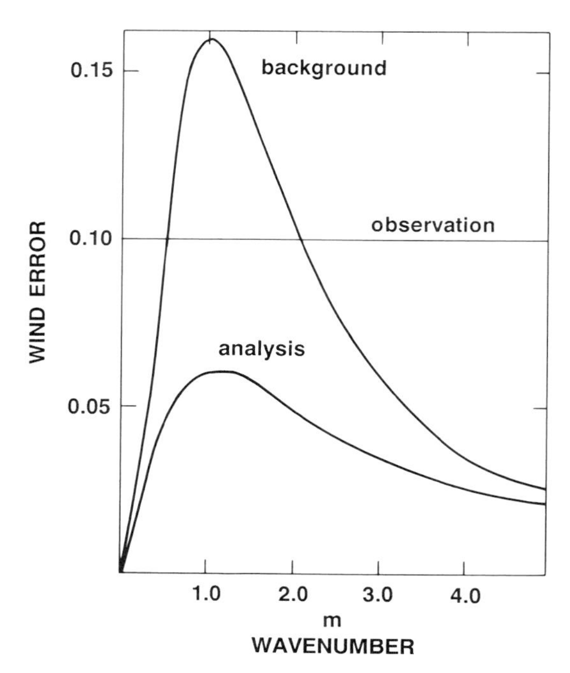

Figure 5.9 The background, observation, and analysis wind error spectra for the one-dimensional continuous observation network with uncorrelated observation error.

(4.4.14), and (4.4.19), define

$$E_{\rm B}^2(m) = \eta_{\rm B} g_w(m), \qquad E_{\rm O}^2(m) = \eta_{\rm B}\varepsilon_v^2, \qquad E_{\rm A}^2(m) = \frac{\eta_{\rm B}\varepsilon_v^2 g_w(m)}{g_w(m) + \varepsilon_v^2}$$

where $\eta_{\rm B} = LE_v^2({\rm B})$ and $g_w(m) = m^2L^2g(m)$. In Figure 5.9 are plotted $E_{\rm B}^2(m)$, $E_{\rm O}^2(m)$, and $E_{\rm A}^2(m)$, scaled by $\eta_{\rm B}$, as a function of m. $\varepsilon_v^2 = 0.10$, and the correlation model is (4.3.20) with $L = 1$. The analysis error variance is always less than either the observation or background error variance as expected from (4.4.14). The greatest reduction in the background error by the algorithm occurs at the wavenumber for which the background error variance is a maximum ($m = 1$, in this case), consistent with (4.5.21). At small m, the background error variance, consistent with (5.2.36), is relatively small to begin with; so at these scales the statistical interpolation algorithm does not draw as strongly for the observations as it does in the univariate height case.

Let us now return to Figure 5.8 for an examination of the multivariate case. Consider first the analysis of the height field from height observations $\alpha(m)$ and from wind observations $\beta(m)$. $\alpha(0) = g(0)/[g(0) + \varepsilon_h^2]$, which decreases as ε_h^2 increases whereas $\beta(0)$ is always equal to zero because of (5.4.12). For $m > 0$, it can be seen that as the wind observations become relatively more accurate than the height observations, $\alpha(m)$ decreases and the maximum response of $\beta(m)$ shifts to larger scales. This behavior can be explained as follows. The statistical interpolation algorithm (with uncorrelated observation errors) attempts to maximally reduce the background error variance in the scales for which it is a maximum. For the background height error, this maximum occurs at wavelength zero. Therefore, as ε_v^2 becomes relatively much smaller than ε_h^2, the burden of reducing the background error variance falls increasingly on the wind observations. Thus, the maximum response of $\beta(m)$ shifts toward smaller m when $\varepsilon_v^2 \ll \varepsilon_h^2$. However, because $\beta(0) = 0$, the mean (wavenumber zero) height field cannot be obtained from wind observations, consistent with Section 2.6. Figure 5.8(a) also implies that $W_{hh}\,(x)$ broadens as ε_h^2 increases.

The reverse situation occurs for the wind analysis, $\gamma(m)$ and $\delta(m)$. The maximum background wind error variance (Figure 5.9) occurs at $m = 1$ in this case. Consequently, when $\varepsilon_h^2 \gg \varepsilon_v^2$, the maximum of $\gamma(m)$ must shift to shorter wavelengths to maximally reduce the background wind error. $\delta(m) \to R_w(m)$ as $m \to \infty$, which is independent of height observations.

Equation (5.4.16) can be simplified substantially if the height and wind increments $\Delta\hat{h}_{\mathrm{O}}(m)$ and $\Delta\hat{v}_{\mathrm{O}}(m)$ are exactly geostrophically coupled. It is left for Exercise 5.3 to show that in this case the height response $\Delta\hat{h}_{\mathrm{A}}(m)/\Delta\hat{h}_{\mathrm{O}}(m)$ and the wind response $\Delta\hat{v}_{\mathrm{A}}(m)/\Delta\hat{v}_{\mathrm{O}}(m)$ are the same and are always greater than either of the univariate height and wind responses, $R_h(m)$ and $R_w(m)$.

It is appropriate to add a warning at this point. The developments in this section (and in Sections 3.3 and 4.4) assume an infinite domain. In practice, the domain is the earth's atmosphere, which is periodic and spherical. The infinite domain assumption has little effect on small scales (large m), but the results at large scales (small m) should be treated with more caution. The continuous case discussed in this section turns out to be a very useful guide to the discrete results in the next section.

5.5 Multivariate versus univariate analysis

The filtering properties of the multivariate algorithm discussed in the previous section strictly apply only to a continuous observation network on a one-dimensional infinite domain. The reader might well ask if these results have any applicability to more realistic networks. The answer to this question is yes, as might have been anticipated by the general agreement between the continuous and discrete filtering characteristics of Sections 4.4 and 4.5.

Before we discuss the results of discrete network filtering experiments it is necessary to introduce the subject of data selection. The results in the preceding sections have implicitly assumed that the analyzed value at each analysis gridpoint is a function of *all* of the observations. The $\underline{\underline{\mathrm{B}}} + \underline{\underline{\mathrm{O}}}$ matrix of (4.2.10) or (5.1.19), which must be inverted to produce the analysis weights, has an order equal to the number of observations. In practice, the size of a matrix that can be inverted is limited by computer capacity. In operational applications, which can have many thousands or even millions of observations, the matrix $\underline{\underline{\mathrm{B}}} + \underline{\underline{\mathrm{O}}}$ for the whole network is simply too large to invert. The usual solution to this difficulty has been to analyze each gridpoint separately, using only those observations that are likely to be highly correlated with that particular gridpoint. This process is called *data selection*; in practice, it involves only observations that are within a certain distance (the radius of influence) of the analysis gridpoint. Thus, the inversion of one huge $\underline{\underline{B}} + \underline{\underline{O}}$ matrix is replaced by the inversion of many smaller matrices.

A slightly different data selection methodology was introduced by Lorenc (1981). In this procedure, all the gridpoints in a local subdomain are analyzed simultaneously using the same set of observations, and then the subdomain analyses are combined to produce the whole domain analysis.

The spectral response of the multivariate algorithm to a discrete observation network has been investigated by Daley (1983b, 1983) using a numerical experimentation technique. Some of these experiments were conducted with midlatitude f-plane geometry using a modified operational algorithm and can be directly compared with the analytic results of the previous section. The observations were regularly spaced on a rectangular grid, and the observation spacing could be varied experimentally. The data selection procedure was based on Lorenc's (1981) method.

The background error was assumed to be homogeneous, isotropic, nondivergent ($v^2 = 0$), and geostrophically coupled ($\mu = 1$), with background error correlation given by (4.3.21). The observation error was assumed to be spatially uncorrelated, with normalized error $\varepsilon_h^2 = \varepsilon_v^2 = 0.25$ at each observation station.

The experimental procedure was simply to present the algorithm with observation increments constructed from a single two-dimensional Fourier component of the geopotential (h) and horizontal wind components (u, v). The analysis increments produced by the algorithms were then compared with the observation increments and a response calculated. By varying the wavenumber of the single Fourier component, the spectral response could be determined over a wide range of horizontal scales.

The results are summarized in Figure 5.10. Each part of the figure shows the spectral response (contours of 0.1, 0.3, 0.5, 0.7, 0.9) of the analysis increment as a function of wavenumber and observation spacing. The abscissa is linear in wavenumber, and the largest wavelength is 40,000 km, roughly corresponding to global wavenumber 1 (Section 1.1). Part (a) corresponds to univariate analysis of geopotential observation increments (h) and can be compared with the curve a of Figure 5.8(a). Part (b) corresponds to univariate, actually bivariate (u, v), analysis of nondivergent wind observation increments and can be compared with curve c of Figure 5.8(d). Part (c) is univariate, actually bivariate (u, v), analysis of irrotational wind observation increments and has no direct counterpart in the previous section. Part (d) corresponds to multivariate (h, u, v) analysis of geostrophically related wind and geopotential observation increments and corresponds to the case analyzed in Exercise 5.3.

Examination of the univariate geopotential response in Figure 5.10(a) indicates that response increases with increasing scale, consistent with Figure 5.8, and decreases with increasing observation spacing, consistent with (4.6.17). Figure 5.10(b) for the nondivergent wind observation increments shows the same tendency as curve c of Figure 5.8 for a maximum response at an intermediate scale. Because the largest-scale

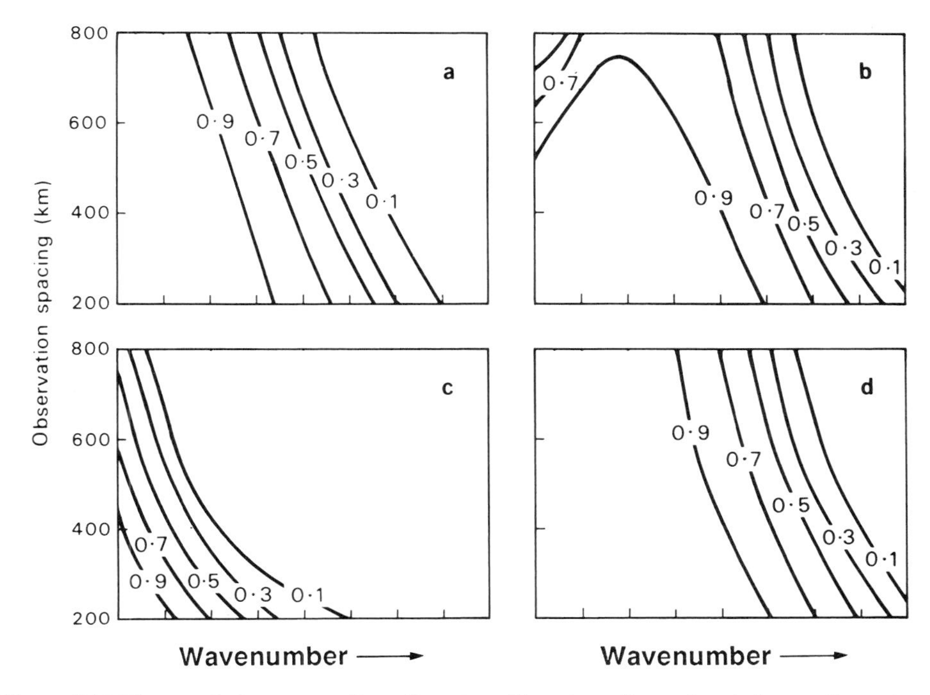

Figure 5.10 The spectral responses for a discrete uniform two-dimensional observation network as a function of wavenumber. (After Daley, *Mon. Wea. Rev.* **113**: 1066, 1985. The American Meteorological Society.)

waves of the experiment had a finite wavelength, the response does not go to zero as it did in the previous section. The response at smaller scales is better than in Figure 5.10(a), consistent with Figure 5.8.

The relatively poor response at large scales for univariate wind analysis is worth further comment. The analytic results of the previous section show that this phenomenon is a direct consequence of the application of Helmholtz's theorem in the derivation of background wind error statistics. The fact that this phenomenon also occurs on a discrete network is easily explained. In Section 4.5, it was noted that for background error correlation functions whose spectral densities are monotonically decreasing functions of wavenumber, the corresponding background error correlation matrices had eigenvectors whose eigenvalues increased with increasing scale. However, wind correlation functions are not positive everywhere (Figures 5.2 and 5.4) and have spectral densities that are *not* monotonically decreasing functions of wavenumber. Consequently, the eigenvectors of the background wind error correlation matrix do not necessarily have eigenvalues that increase with increasing scale. As an example, consider the two observation network of (4.5.13). Suppose ρ is the background *wind* error correlation. Then, from Figures 5.2 and 5.4, we know that for some observation spacing, ρ is *not* positive. From (4.5.13), the largest scale eigenvector $\underset{\sim}{e}_1$ could have a smaller eigenvalue than the smaller scale eigenvector $\underset{\sim}{e}_2$.

Equation (4.5.12) would then imply that the largest scale eigenvector $\underset{\sim}{e}_1$ was more strongly filtered than the smaller scale eigenvector $\underset{\sim}{e}_2$. (See also Exercise 5.4.)

The poor response at large scale of univariate wind analysis has been the subject of some concern. More realistic experiments (Daley et al. 1986) have shown that this phenomenon also occurs on the sphere with actual observation networks. Hollingsworth and Lonnberg (1986) have proposed and applied background wind error correlation models with a stronger response at larger scales.

Turning now to Figure 5.10(d), we can see that the multivariate algorithm produces a larger response at all scales than the separate application of the univariate algorithm to height and wind observations in Figure 5.10(a,b). This result is consistent with the analytic results of Exercise 5.3 and is due simply to the fact that the multivariate algorithm processes more observations than the univariate algorithm. This suggests that if there is a relationship among the observation increments of different variables, then a multivariate algorithm that reflects this relationship in its background error statistics will have a larger response at all scales than the separate univariate analyses of the same variables. One consequence is that analysis in the extratropics (where the geostrophic relationship can legitimately be applied to the background error statistics) is generally more successful than tropical analysis (where the analyses are essentially univariate).

The response shown in Figure 5.10(c) for irrotational wind observation increments is puzzling at first. After all, the background error statistics were assumed to be nondivergent ($v^2 = 0$), and Equations (5.1.23–24) would suggest that the analysis increments must also be nondivergent, implying a zero response at all scales. Clearly, the response is much smaller than for the nondivergent wind increments in Figure 5.10(b), but it is *not* zero, particularly in the largest scales. Obviously, setting $v^2 = 0$ in the background error statistics is not sufficient to guarantee that the analysis increments will be nondivergent.

In (5.1.23–24), it has been implicitly assumed that the analysis at every gridpoint has been made with the same set of observations. In practice, this is not usually the case, as noted earlier. Statistical interpolation is applied operationally in a local sense, with each gridpoint or group of neighboring gridpoints being analyzed separately using the local observation surrounding the gridpoint. This produces a minimum in the expected analysis error variance at each gridpoint, but not necessarily a minimum in the domain-averaged expected analysis error variance. The analysis at each gridpoint is not entirely consistent with those at other gridpoints; thus, the implicit assumption underlying (5.1.23–24) is not strictly valid when the data are selected in this way. From Figure 5.10(d), it is easy to see that the effect is most noticeable in the largest scales, which are most likely to be affected by inconsistencies between the analyses at different gridpoints. Phillips (1982) and Ikawa (1984a) have shown formally that multivariate relationships imposed on the background error statistics are identically satisfied by the analysis increments only if the same observation set is used for all analysis points.

Physical constraints can be used in either an exact or approximate manner in the specification of the background error statistics. If the relationship is used exactly, it is said to be a strong constraint. Thus, setting $\nu^2 = 0$ in (5.2.28–29) and $\mu = 1$ in (5.3.15–16) implies the strong (exact) constraints of nondivergence and geostrophy, respectively, in the background error statistics. Constraints can also be used in a weak fashion with ν^2 small but not exactly zero and μ close to but not exactly equal to 1 (-1) in the northern (southern) hemisphere midlatitudes. In the weak case, the constraints are only approximately satisfied in the background error statistics.

Weak constraints imposed on the background error statistics affect the analysis increments differently from strong constraints. Consider the variation of the spectral response for the one-dimensional continuous algorithm (5.4.1–5) as the geostrophic coupling parameter μ is varied. As an example, consider the spectral response $\gamma(m)$ defined in (5.4.10):

$$\gamma(m) = \frac{-i\mu e_v^2 m g(m)}{g(m)(\varepsilon_v^2 + \varepsilon_h^2 L^2 m^2) + \varepsilon_h^2 \varepsilon_v^2 + L^2(1-\mu)m^2 g^2(m)} \tag{5.5.1}$$

when $\mu \neq 1$. Equation (5.5.1) simplifies to (5.4.14) if $\mu = 1$. When there are no wind observations ($\varepsilon_v^2 \to \infty$), Equation (5.5.1) becomes

$$\gamma(m) = \frac{-i\mu m g(m)}{g(m) + \varepsilon_h^2} \tag{5.5.2}$$

The parameter μ enters linearly in (5.5.2); thus, when there are no wind observations, $\gamma(m)$ is directly proportional to μ. This implies that the magnitude of the wind component analysis increments analyzed from geopotential increments *only* is directly proportional to μ. This is also true for the reverse case of geopotential analyzed from wind only. The discrete case (5.1.17) or (5.1.19) is similar. When the geopotential is analyzed from the wind only, or vice versa, the parameter μ appears only in the matrices $\underline{\underline{B}}_{ki}$ on the right-hand side of (5.1.17) and *not* in the matrices $\underline{\underline{B}}_{kl}$ on the left-hand side.

When both wind and geopotential observations exist, (5.5.1) suggests that the situation is more complex. The extra term $L^2(1-\mu)m^2 g^2(m)$ appears in the denominator. This extra term also appears in $\alpha(m)$, $\beta(m)$, and $\delta(m)$. In the discrete case, μ appears on both the left- and right-hand sides of (5.1.17) when there are both wind and geopotential observations.

Daley (1983b, 1985) has shown that the spectral response for the multivariate algorithm has a highly nonlinear dependence on the coupling parameters μ and ν^2. Irrotational or ageostrophic flow components were filtered from the analysis increments only when ν^2 was very close to zero and μ was very close to 1. Thus, if it is considered desirable for the analysis increments to be nondivergent and geostrophic, then the appropriate coupling parameters must be set very nearly equal to their strong values.

5.6 A generalized algorithm

The multivariate statistical interpolation algorithm (Section 5.1) has largely superseded other procedures for the objective analysis of global- and synoptic-scale atmospheric flow. Statistical interpolation does have its limitations, however, and will eventually be replaced by the more advanced methods to be discussed in Chapter 13.

We have already alluded to many of the limitations of statistical interpolation. Some of these drawbacks are fundamental to the procedure, and others arise from the way the algorithm is applied in practice. We now discuss the major limitations of statistical interpolation, in more or less the same order they were encountered in Chapters 4 and 5, and then introduce a more general formulation of the algorithm, which addresses some of these limitations.

1. The forward problem. In Section 4.2, it was noted that the statistical interpolation algorithm assumed that the background B_k was known at the observation stations. A climatological background is primarily large scale and can be assumed to be known everywhere. A forecast background, however, is only known at the gridpoints of a regular grid, and B_k can only be obtained by interpolation. A forecast background has much more smaller-scale variance than a climatological background, and the forward interpolation can be a serious source of error (Franke 1985). The forward horizontal interpolation is not usually a problem because of the relative fineness of the horizontal analysis grid spacing. The real problem is in the forward vertical interpolation. As noted in Section 4.7, the vertical variation of most variables is very rapid. Moreover, most forecast models use a terrain-following coordinate system rather than the approximately horizontal pressure coordinate system. According to Hollingsworth (1987), an inferior forward vertical interpolation scheme can easily negate the advantages of statistical interpolation over competing methods.
2. Correlated background and observation error. The mutual correlation of background and observation error can cause difficulties with certain types of observing systems, as noted in Section 4.2.
3. Simplifications of $\underline{\underline{B}}$. The background error covariances must be correct if the expected analysis error is to be given by (4.2.12) rather than by the larger value (4.2.11). The assumptions of separability, homogeneity, and isotropy are all sources of error in generating the elements of $\underline{\underline{B}}$ (Section 4.9).
4. The specification of $\underline{\underline{B}}$ in data voids. The background error covariances can be determined reasonably well from historical data in data-rich regions such as North America or Europe. Because data are available from the many stations in these areas every few hours, $\underline{\underline{B}}$ can be assumed to be essentially time invariant (at least over a month or season). In data voids, the background error covariance $\underline{\underline{B}}$ in the forecast case depends critically on the accuracy of the numerical forecast model and on the age of the most recent observations in the area. The

background error variance E_B^2 can vary from a relatively low value when there have been recent observations to a climatological value when there have been no observations for many days.

5 Continuous correlation functions. Equations (5.1.23–24) showed that if the background error covariances satisfied a continuous multivariate relationship, then so did the analysis increments. However, analyzed values are produced at discrete regular gridpoints, and a continuous multivariate relationship would be applied on the analysis grid as a *discrete* approximation. For example, a possible discretization of

$$v = \frac{dh}{dx}$$

is

$$v(x_i) = \frac{h(x_{i+1}) - h(x_{i-1})}{2\,\Delta x}$$

where x_i are the gridpoints of the analysis grid and $\Delta x = x_i - x_{i-1}$. The error in this discrete approximation vanishes as $\Delta x \to 0$. Consequently, for finite Δx, there is an inconsistency between the use of a continuous form of a given multivariate relation for the background error statistics and the use of a discrete form for the analysis increments.

6 Local application of the algorithm. The statistical interpolation algorithm is usually applied locally rather than over the whole domain simultaneously. This means that the average analysis error over the whole analysis grid is not an absolute minimum and that contraints imposed on the background error statistics are not necessarily reflected in the analysis increments.

Problem 1 is addressed primarily by the use of more accurate forward interpolation routines. In addition, the generalized algorithm (to be discussed) explicitly accounts for the statistics of the forward interpolation error. Problem 2 can be avoided by using raw observations (radiometric measurements, say) rather than processed observations (temperature, say) which require background information. Problem 4 is addressed by the Kalman–Bucy filter (Section 13.3), which uses the forecast model itself to predict the background error statistics. Problems 3, 5, and 6 are to some extent addressed by the more general algorithm to be given. The univariate version of the algorithm is developed in this section, and the multivariate version is discussed in Section 13.1.

Suppose there is a three-dimensional domain with an observation network of K stations and a regular analysis grid of I gridpoints. Consider the analysis of a state variable $f_A(\mathbf{r}_i)$, $1 \le i \le I$, using observed values $f_O(\mathbf{r}_k)$, $1 \le k \le K$. We assume that there exist background estimates $f_B(\mathbf{r}_i)$, $1 \le i \le I$ of the variable f at the analysis gridpoints. Then, an estimate $f_B(\mathbf{r}_k)$ at the kth observation station can be represented as

$$f_B(\mathbf{r}_k) = \Omega(f_B(\mathbf{r}_1), f_B(\mathbf{r}_2), \ldots, f_B(\mathbf{r}_I)) \tag{5.6.1}$$

where Ω expresses some linear or nonlinear forward interpolation algorithm. Assume that Ω is a linear operator. Then, (5.6.1) becomes

$$f_B(\mathbf{r}_k) = \sum_{j=1}^{I} \Omega_{kj} f_B(\mathbf{r}_j) \tag{5.6.2}$$

where Ω_{kj} are the forward interpolation weights, and j is a dummy index over the analysis gridpoints. Then, following (4.2.1), the analysis increment at the ith gridpoint is

$$f_A(\mathbf{r}_i) = f_B(\mathbf{r}_i) + \sum_{k=1}^{K} W_{ik}\left[f_O(\mathbf{r}_k) - \sum_{j=1}^{I} \Omega_{kj} f_B(\mathbf{r}_j)\right]$$

or

$$f_A(\mathbf{r}_i) = f_B(\mathbf{r}_i) + \underline{\mathrm{W}}_i^{\mathrm{T}}[\underline{\mathrm{f}}_O - \underline{\underline{\Omega}}\underline{\mathrm{f}}_B] \tag{5.6.3}$$

where $\underline{\mathrm{W}}_i$ is the column vector of length K of a posteriori weights, $\underline{\underline{\Omega}}$ the $K \times I$ rectangular matrix with elements Ω_{kj}, $\underline{\mathrm{f}}_O$ the column vector of length K of observations, and $\underline{\mathrm{f}}_B$ the column vector of length I of background values at the analysis gridpoints. Subtract the true values of f from both sides of (5.6.3), giving

$$\varepsilon_A(\mathbf{r}_i) = \varepsilon_B(\mathbf{r}_i) + \underline{\mathrm{W}}_i^{\mathrm{T}}[\underline{\varepsilon}_O - \underline{\underline{\Omega}}\underline{\varepsilon}_B + \underline{\varepsilon}_F] \tag{5.6.4}$$

where $\varepsilon_A(\mathbf{r}_i)$ and $\varepsilon_B(\mathbf{r}_i)$ are the analysis and background errors at analysis gridpoints $\mathbf{r}_i$, $\underline{\varepsilon}_O$ is the column vector of length K of observation errors at the observation stations, and $\underline{\varepsilon}_B$ the column vector of length I of background errors at the analysis gridpoints. $\underline{\varepsilon}_F$ is the column vector of length K with elements $\varepsilon_F(\mathbf{r}_k)$ defined as

$$\varepsilon_F(\mathbf{r}_k) = f_T(\mathbf{r}_k) - \sum_{j=1}^{I} \Omega_{kj} f_T(\mathbf{r}_j) \tag{5.6.5}$$

Here $f_T(\mathbf{r}_j)$ and $f_T(\mathbf{r}_k)$ are the true values of f at the analysis gridpoint $\mathbf{r}_j$ and observation station $\mathbf{r}_k$. $\varepsilon_F(\mathbf{r}_k)$ is the error in the forward interpolation, which is only equal to zero if $\underline{\underline{\Omega}}$ interpolates without error.

Now assume that $\underline{\varepsilon}_O$, $\underline{\varepsilon}_B$, and $\underline{\varepsilon}_F$ are unbiased and not correlated with each other:

$$\langle \underline{\varepsilon}_O \rangle = \langle \underline{\varepsilon}_B \rangle = \langle \underline{\varepsilon}_F \rangle = 0$$

and

$$\langle \underline{\varepsilon}_O \underline{\varepsilon}_B^{\mathrm{T}} \rangle = \langle \underline{\varepsilon}_O \underline{\varepsilon}_F^{\mathrm{T}} \rangle = \langle \underline{\varepsilon}_B \underline{\varepsilon}_F^{\mathrm{T}} \rangle = \langle \underline{\varepsilon}_O \varepsilon_B(\mathbf{r}_i) \rangle = \langle \underline{\varepsilon}_F \varepsilon_B(\mathbf{r}_i) \rangle = 0 \tag{5.6.6}$$

Then, square (5.6.3) and take expected values;

$$\langle \varepsilon_A^2(\mathbf{r}_i) \rangle = \langle \varepsilon_B^2(\mathbf{r}_i) \rangle - 2\underline{\mathrm{W}}_i^{\mathrm{T}}\underline{\underline{\Omega}}\langle \underline{\varepsilon}_B \varepsilon_B(\mathbf{r}_i) \rangle + \underline{\mathrm{W}}_i^{\mathrm{T}}[\langle \underline{\varepsilon}_O \underline{\varepsilon}_O^{\mathrm{T}} \rangle + \underline{\underline{\Omega}}\langle \underline{\varepsilon}_B \underline{\varepsilon}_B^{\mathrm{T}} \rangle \underline{\underline{\Omega}}^{\mathrm{T}} + \langle \underline{\varepsilon}_F \underline{\varepsilon}_F^{\mathrm{T}} \rangle]\underline{\mathrm{W}}_i \tag{5.6.7}$$

Minimize (5.6.7) with respect to $\underline{\underline{W}}_i$ using (5.1.12), giving

$$[\underline{\underline{O}} + \underline{\underline{\Omega}}\,\underline{\underline{B}}\,\underline{\underline{\Omega}}^{\mathrm{T}} + \underline{\underline{F}}]\underline{W}_i = \underline{\underline{\Omega}}\,\underline{B}_i \tag{5.6.8}$$

where $\underline{\underline{O}} = \langle \underline{\varepsilon}_{\mathrm{O}} \underline{\varepsilon}_{\mathrm{O}}^{\mathrm{T}} \rangle$ is the $K \times K$ observation error covariance matrix, $\underline{\underline{F}} = \langle \underline{\varepsilon}_{\mathrm{F}} \underline{\varepsilon}_{\mathrm{F}}^{\mathrm{T}} \rangle$ the $K \times K$ forward interpolation error covariance matrix, $\underline{\underline{B}} = \langle \underline{\varepsilon}_{\mathrm{B}} \underline{\varepsilon}_{\mathrm{B}}^{\mathrm{T}} \rangle$ the $I \times I$ background error covariance matrix, and $\underline{B}_i$ the background error covariance vector of length I. Equation (5.6.8) gives the weights $\underline{W}_i$ for the ith analysis gridpoint. The analysis weights for all I analysis gridpoints can then be written as

$$\underline{\underline{W}} = [\underline{\underline{O}} + \underline{\underline{\Omega}}\,\underline{\underline{B}}\,\underline{\underline{\Omega}}^{\mathrm{T}} + \underline{\underline{F}}]^{-1} \underline{\underline{\Omega}}\,\underline{\underline{B}} \tag{5.6.9}$$

and (5.6.3) becomes

$$\underline{f}_{\mathrm{A}} - \underline{f}_{\mathrm{B}} = \underline{\underline{W}}^{\mathrm{T}}[\underline{f}_{\mathrm{O}} - \underline{\underline{\Omega}}\,\underline{f}_{\mathrm{B}}] \tag{5.6.10}$$

where $\underline{f}_{\mathrm{A}}$ is the column vector of length I of analysis weights $f_{\mathrm{A}}(\mathbf{r}_i)$. Lorenc (1986) identifies $\underline{\underline{O}}$ with the instrument error only, $\underline{\underline{F}}$ with the errors of representativeness discussed in Section 1.3, and $\underline{\underline{B}}$ with the background error at the analysis gridpoints. Note that if the set of observation stations and analysis gridpoints coincide, $\underline{\underline{\Omega}} = \underline{\underline{I}}$ and (5.6.9–10) becomes equal to (4.5.3). Lorenc (1986) shows that a number of objective analysis algorithms discussed in this book are special or degenerate cases of (5.6.9–10). This is true of statistical interpolation, the method of successive corrections, and the constrained minimization algorithm discussed in Section 2.7.

Equations (5.6.9–10) can also be derived by minimizing the following functional with respect to $\underline{f}_{\mathrm{A}}$:

$$J = 0.5\{[\underline{\underline{\Omega}}\,\underline{f}_{\mathrm{A}} - \underline{f}_{\mathrm{O}}]^{\mathrm{T}}[\underline{\underline{O}} + \underline{\underline{F}}]^{-1}[\underline{\underline{\Omega}}\,\underline{f}_{\mathrm{A}} - \underline{f}_{\mathrm{O}}] + [\underline{f}_{\mathrm{A}} - \underline{f}_{\mathrm{B}}]^{\mathrm{T}}\underline{\underline{B}}^{-1}[\underline{f}_{\mathrm{A}} - \underline{f}_{\mathrm{B}}]\} \tag{5.6.11}$$

It will be seen that (5.6.11) is a generalization of (2.2.11) and becomes equal to it when the observation stations and analysis gridpoints coincide. The analyzed values $\underline{f}_{\mathrm{A}}$ of (5.6.10) minimize (5.6.7) and are consequently minimum variance estimates. If, in addition, the observation, background, and forward interpolation errors are normally distributed, then from (5.6.11), the analyzed values $\underline{f}_{\mathrm{A}}$ are also maximum likelihood estimates, and the analysis errors are normally distributed. Note that the algorithm (5.6.9–10), unlike (2.2.15), is an interpolation algorithm because it produces estimates at any desired location and not just at the observation stations.

An expression for the $I \times I$ analysis error covariance matrix $\underline{\underline{A}}$ for this algorithm will be useful in Chapter 13. Following (2.2.16–19),

$$\underline{\underline{A}} = [\underline{\underline{I}} - \underline{\underline{W}}^{\mathrm{T}} \underline{\underline{\Omega}}]\underline{\underline{B}}$$

or

$$\underline{\underline{A}} = \underline{\underline{B}} - \underline{\underline{B}}\,\underline{\underline{\Omega}}^{\mathrm{T}}[\underline{\underline{\Omega}}\,\underline{\underline{B}}\,\underline{\underline{\Omega}}^{\mathrm{T}} + \underline{\underline{O}} + \underline{\underline{F}}]^{-1}\underline{\underline{\Omega}}\,\underline{\underline{B}} \tag{5.6.12}$$

which degenerates to (2.2.19) when analysis gridpoints and observation stations coincide. The derivation of (5.6.11–12) is left as Exercise 5.5.

The algorithm (5.6.9–10) allows multivariate constraints to be included in the background error statistics in discrete rather than continuous form, as discussed by Ikawa (1984a). It is also possible explicitly to include a climatological background as well as a forecast background.

Of course, it is possible to use the algorithm (5.6.9–10) in a simplified or degenerate form by applying it locally to individual subdomains. When the forward interpolation is linear, then this form of the algorithm is not all that different from the operational use of statistical interpolation. In effect, statistical interpolation already accounts for both instrument error and error of representativeness though it does lump them together into a single observation error covariance matrix.

The strict application of an algorithm based on equations (5.6.9–10) over the whole analysis domains offers a number of potential advantages over other methods. The error of the forward interpolation is explicitly accounted for, the background error covariance matrix is not approximated, and there is no discretization error. This means that the expected analysis error variance over the whole domain is a minimum, and imposed constraints are satisfied exactly by the analysis increments.

However, a computational price must be paid. The most serious problem is the inversion of the $K \times K$ matrix in (5.6.9). Because K can be of $O(10^6)$ or more, direct inversion methods are out of the question. A more efficient method of solving (5.6.9) is based on serial processing of observations (Parrish and Cohn 1985). An alternate possibility is to directly minimize the variational form (5.6.11) by using a descent algorithm (see Section 13.1).

This concludes the discussion of purely spatial analysis. Our emphasis will now shift to the temporal aspects of the data assimilation problem.

Exercises

5.1 Derive the multivariate weight equations (5.1.13–16).

5.2 Derive (5.2.19–22) from (5.2.13–16).

5.3 For the continuous one-dimensional multivariate algorithm of Section 5.4, show that the analysis increments $h_A(x) - h_B(x)$ and $v_A(x) - v_B(x)$ are geostrophically coupled. Now suppose the *observation* increments are exactly geostrophically coupled, $\Delta\hat{v}_O(m) = im\,\Delta\hat{h}_O(m)$ in spectral form. Define $\zeta(m) = \Delta\hat{h}_A(m)/\Delta\hat{h}_O(m)$ as the height response in this case. Show that the wind response $\Delta\hat{v}_A(m)/\Delta\hat{v}_O(m) = \zeta(m)$. Show that $1 \geq \zeta(m) \geq R_h(m)$, $R_w(m)$ and that $\zeta(m) = R_h(m)$ for $m = 0$ and $\zeta(m) \to R_w(m)$ as $m \to \infty$.

5.4 For the collinear observation network of Exercise 4.6, assume there are three *wind* observations rather than height observations. Find the eigenvectors and eigenvalues of the background error correlation matrix. Using the one-dimensional form of the correlation model (4.3.21), plot the eigenvalues as a function of $\Delta x/L$. What happens at $\Delta x/L = 1$?

5.5 Minimize (5.6.11) to find an expression for the optimum weights. Derive the expression for the analysis error covariance matrix given by (5.6.12).

5.6 Consider the one-dimensional domain with analysis point x_2 and height observations at $x_1 = x_2 - \Delta x$ and $x_3 = x_2 + \Delta x$. Assume the height observation error is spatially uncorrelated with normalized error variance ε^2 at x_1 and x_3. Assume the background height error is homogeneous, and denote $\rho = \rho_{hh}(x_1 - x_3)$ and $\tilde{\rho} = \rho_{hh}(x_2 - x_3) = \rho_{hh}(x_2 - x_1)$. Assume the wind and height background errors are geostrophically related, so $\rho_{hv} = L\, d\rho_{hh}/dx$, where L is the scale given in (5.4.7). Denote $\rho_{hv}(x_2 - x_3) = -\rho_{hv}(x_2 - x_1) = \tilde{q}$. Determine the a posteriori weights for an analyzed height $h(x_2)$ and analyzed wind $v(x_2)$. Plot the a posteriori weights W_1 and W_3 for the analyzed height and LW_1 and LW_3 for the analyzed wind, as a function of $\Delta x/L$, using the correlation model (4.3.21) for ρ_{hh}. Find an expression for the normalized analysis error. What happens to the analysis error and the weights when the height observation error is spatially correlated?

6

The initialization problem

Chapters 2–5 discuss objective analysis procedures that use observations and background estimates to produce a representation of the state of the atmosphere at a fixed time. This representation is in the form of either point values on a uniform mesh or coefficients of a functional expansion, and its local accuracy is primarily a function of the local observation density and accuracy.

None of these objective analysis procedures are explicitly time dependent though time dependence is implicit in algorithms that use background fields derived from the time integration of numerical models. In Figures 1.1 and 1.3, it can be seen that atmospheric phenomena have a wide range of time scales. It goes without saying that an ideal four-dimensional assimilation system would have to take explicit account of both temporal and spatial characteristics of the atmosphere.

A more immediately compelling reason for considering time dependence is that objective analyses that are produced by the methods of Chapters 2–5 are often unsuitable when used as initial conditions for the time integration of numerical weather prediction (NWP) models. To see why this is so requires a closer examination of two important time scales of atmospheric motion.

6.1 Characterization of synoptic and planetary scale motion in the atmosphere

> We might say that the atmosphere is a musical instrument on which one can play many tunes. High notes are sound waves, low notes are long inertial waves, and nature is a musician more of the Beethoven than of the Chopin type. He much prefers the low notes and only occasionally plays arpeggios in the treble and then only with a light hand. The oceans and the continents are the elephants in Saint-Saens' animal suite, marching in a slow cumbrous rhythm, one step every day or so. Of course, there are overtones; sound waves, billow clouds (gravity waves), inertial oscillations, etc., but these are unimportant. (From an unpublished letter from Jule Charney to Phillip Thompson, 12 February 1947.)

Charney's characterization of the atmosphere in terms of a symphony orchestra illustrates the richness and diversity of atmospheric motion. In a later paper, Charney (1955) discusses the characteristics of atmospheric motion as follows. He presumes an atmosphere that is statically stable and in which the horizontal scale L_H far exceeds the vertical scale L_Z. If L_Z is assumed to be one atmospheric scale height (10 km), then the horizontal scale L_H considered here would be greater than 100 km. In other words, these atmospheric motions are in quasi-hydrostatic equilibrium and are of synoptic or planetary scale. In a hydrostatic atmosphere there are no sound waves, but this is not a serious restriction because sound waves are of extremely high frequency and are not thought to interact in any meaningful way with motions of meteorological interest. The equations governing such an atmosphere are called the hydrostatic or primitive equations.

Two time scales can be defined in this atmosphere:

$$\tau_1 = f^{-1}, \qquad \tau_2 = \frac{L_H}{V_H} \tag{6.1.1}$$

where f is the Coriolis parameter and V_H a characteristic horizontal velocity. τ_1 the inertial time scale and τ_2 the advective time scale. A Rossby number R_0 is defined to be the ratio τ_1 to τ_2:

$$R_0 = \frac{\tau_1}{\tau_2} = \frac{V_H}{L_H f} \tag{6.1.2}$$

In the atmosphere, R_0 is usually small, which means that the advective time scale is much longer than the inertial time scale. The time scale τ_1 is a few hours whereas τ_2 is longer than one day.

Two kinds of atmospheric motion can be identified as normal modes of the primitive equations, linearized about a simple basic state. (This will be discussed in detail in Section 6.4 and Chapter 9.) Motions of the first kind, also known as inertia–gravity waves, have time scales $\leq \tau_1$ and velocities of propagation considerably in excess of V_H. Motions of the second kind have time scales of order τ_2 and velocities of propagation similar to V_H. With some exceptions, it is motions of the second kind that are of primary meteorological significance.

From observations, we know that the bulk of the energy in the troposphere is confined to the second type of motion and is thus characterized by the advective time scale rather than the inertial time scale. This situation can be attributed to the following circumstances. First, the atmosphere is forced externally by the sun and the terrestrial surface. The characteristic time scales of the external forcing are longer than the inertial time scale and therefore excite few inertia–gravity oscillations. Second, the motions of the second kind are stable to any inertia–gravity waves that may be excited by the external forcing or through nonlinear interaction. In fact, Errico (1981) has recently shown that motions of the second type are *not* stable to inertia–gravity modes. However, the instability is so weak that it is more than counteracted by dissipative processes.

The preceding argument is most convincing when applied to synoptic scale extratropical flow in the troposphere. In the upper atmosphere (stratosphere and above), inertia–gravity waves play a more significant role; and at smaller spatial scales, inertia–gravity waves of substantial amplitude have been observed and studied (Savage, Weidner, and Stearns 1988). In the tropics (still assuming that $L_Z \ll L_H$), the preceding arguments are weakened because f is small, τ_1 may be of order τ_2, and R_0 can no longer be considered small. Thus, at low latitudes, inertia–gravity modes cannot easily be separated from other motions on the basis of a characteristic time scale. However, even in the tropics (with the possible exception of forced tidal motion), inertia–gravity waves with periods less than a day are a relatively minor component of the flow.

The consequence is that the synoptic and planetary scales of the atmosphere are dominated by motion with the advective time scale, and inertia–gravity waves are normally a minor component of the flow. Such flows are said to be *balanced*. Extending Charney's analogy, we can think of inertia–gravity waves as meteorological noise that is not loud enough to drown out the symphony orchestra though it does detract from the performance. In the atmosphere, this noise is an inconvenience; in numerical models, it can be disastrous.

6.2 Filtered models and initialized primitive equation models

> This leads us to the next problem, namely, how to filter out the noise. Pardon me, but let us again think metaphorically. The atmosphere is a transmitter. The computing machine is the receiver. The receiver is a very good one indeed, for it produces no appreciable noise itself, i.e., all noise comes from the input. (I am supposing that you can compute to any desired order of accuracy.) Now there are two ways to eliminate noise in the output. The first is to make sure that the input is free from objectionable noises, and the second is to employ a filtering system in the receiver. Translating, the first method implies that the unwanted harmonics shall be eliminated from the raw data by some type of harmonic analysis; the second that you transform the equations of motion and make approximations in such a way that the bad harmonics are automatically eliminated. (From an unpublished letter from Jule Charney to Phillip Thompson, 12 February 1947)

Charney's remarks suggest the concern of early numerical modelers for the problem of inertia–gravity wave noise in primitive equation models. In part this concern was based on the failure of Richardson's (1922) attempt to integrate these equations (Section 1.5). This noise problem in primitive equation models is illustrated in Figure 6.1, which shows the surface pressure in millibars as a function of time at a particular geographical location during the integration of an actual primitive equations model. The solid line indicates the surface pressure when the model is integrated from an unmodified objective analysis for 24 hours. It is evident that there are oscillations with amplitudes greater than 5 mb and with periods of only a few hours. These oscillations are due to the presence of large-amplitude inertia–gravity oscillations in the model. The amplitude of the inertia–gravity waves here is an order of magnitude larger than is observed for similar waves in the atmosphere (as evidenced

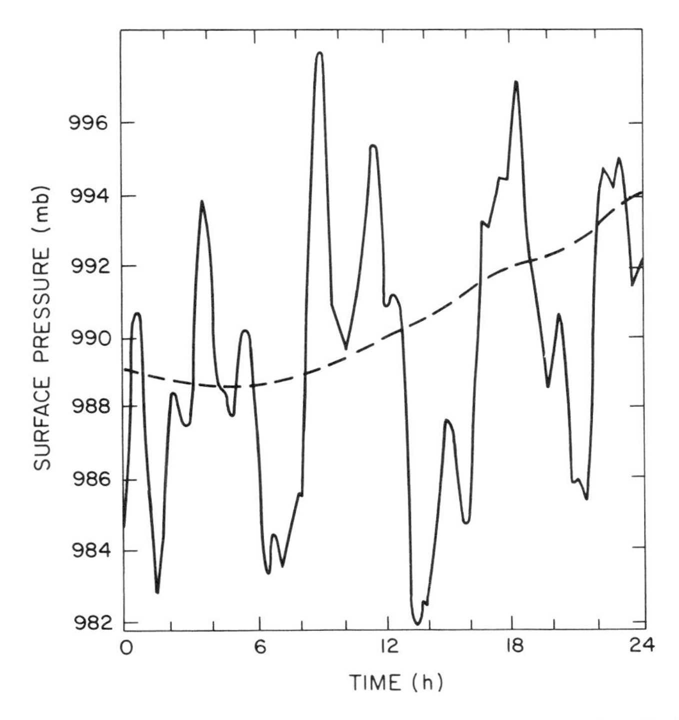

Figure 6.1 Surface pressure as a function of time during the integration of a primitive equations model. Uninitialized (solid), initialized (dashed). (After Williamson and Temperton, *Mon. Wea. Rev.* **109**: 745, 1981. The American Meteorological Society.)

in barograph traces, for example). The dashed line can be taken as an indication of the temporal evolution of the slowly varying or balanced motions of meteorological interest. The high-frequency oscillations in this primitive equations model are both spurious and undesirable.

Why do these high-frequency oscillations exist in primitive equation model integrations when they have virtually no counterpart in the atmosphere? Atmospheric observations are often poorly distributed in space and time and of dubious quality, and some state variables are not as well observed as others. Moreover, models only approximately simulate the atmosphere, and atmospheric multivariate relationships are not exactly duplicated in the model. The consequences are inconsistent, unbalanced initial conditions for the model, a large initial projection onto the model's inertia–gravity modes, and substantial high-frequency variance in subsequent model integrations. This subject is discussed at greater length in the next section.

Charney's letter (quoted earlier) suggests two solutions to this problem:

1 Integrate a model that does not permit inertia–gravity wave motions.
2 Integrate a primitive equations model but modify the initial state in such a way that the inertia–gravity waves are not excited.

The first solution was adopted by Charney and co-workers and led to the development of filtered or quasi-geostrophic models (see Chapter 7). These models filtered out inertia–gravity waves and hence had no spurious oscillations such as those in Figure 6.1 (Haltiner and Williams 1980).

In the 1960s, it became apparent that assumptions made in deriving the filtered models of the day not only eliminated inertia–gravity modes; they also adversely affected the meteorologically significant motions of the second kind. Moreover, these filtered models had difficulties in the tropics and with global scale motions. Consequently, there was a revival of interest in the integration of the less restrictive primitive equations. Integration of the primitive equations requires the modification of the initial state to prevent the excitation of inertia–gravity modes. This process is called *initialization*.

6.3 The benefits of initialization

An initialization process *can* produce balanced initial conditions that do *not* excite inertia–gravity oscillations in a model integration. This is illustrated by the dashed curve in Figure 6.1, which shows the surface pressure as a function of time after the initial state has been modified by an initialization process. The geographical location and model are the same for both solid and dashed curves. The surface pressure at $t = 0$ has been changed from 984.5 to 989 mb by the initialization process, and the initialized integration appears to have virtually no high-frequency oscillations. Examination of other variables and other geographical locations also indicates that the initialization process has virtually eliminated spurious high-frequency oscillations from the model integration. The example of Figure 6.1 will be considered again in Chapter 10, after the mechanics of initialization have been discussed in detail.

The terms *balanced* and *initialized* are often used interchangeably in the literature, but there are subtle distinctions. Thus, the atmosphere itself can be said to be balanced, but never initialized. Initialization is a process, and an initialized state is only a balanced state of the model, if the initialization process has been successful.

As Figure 6.1 demonstrates, spurious high-frequency oscillations do occur in primitive equal model integrations, and an initialization process can effectively remove these oscillations. Despite the apparent success of the initialization procedure, the experimental results of Figure 6.1 do suggest a number of questions:

1 Why don't multivariate objective analysis procedures produce balanced initial states?
2 Are these spurious high-frequency oscillations detrimental to the model integration?
3 Are there simple modifications that can be made to the model to damp these spurious oscillations?

With respect to the first question, consider objective analyses produced by multivariate statistical interpolation using a background field obtained by the time integration of a numerical forecast model. The background errors are assumed to satisfy certain *linear* multivariate relationships. Further assume that the background states are balanced and therefore satisfy complex *nonlinear* multivariate relationships, similar but not identical to those satisfied by the real atmosphere. (The case in which the background states are *not* balanced is discussed later in this section.)

Consider two extreme cases. When the observations are sparse, inaccurate, and infrequent, then the objective analysis should revert to the background state. By definition, the objective analysis would be balanced. Alternatively, when the observations are dense, accurate, and frequent, then the objective analysis should closely approximate the true atmospheric state. If this very accurate objective analysis were to be used as an initial condition for the forecast model, then high-frequency oscillations would be excited only to the extent of the incompatibility between the forecast model and the true atmosphere.

In reality the situation is more complex. Observations are often sporadic and mutually incompatible, and the observation density is highly variable, resulting in objective analyses that are close to the true atmospheric state in some regions, close to the background state in other regions, and elsewhere close to neither. Consequently, there can be no guarantee that the analyzed state would not excite high-frequency oscillations if used as an initial condition for a subsequent integration of the forecast model. In particular, the analysis increments may not actually satisfy the imposed linear constraints because of data selection considerations (Section 5.5). In other situations (the tropical flow is an obvious example), there are no easily exploitable linear multivariate relationships for the background error. Even when the analysis increments satisfy an appropriate linear constraint, the analyzed fields themselves may not be well balanced when the analysis increments are large. This is because balanced atmosphere and model states each satisfy (different) nonlinear multivariate relationships, and there is an implicit linearization in the application of linear multivariate relationships to background error statistics. This implied linearization may not be valid when the background error is large.

With respect to the second question, it can be argued that oscillations having the short inertial time scale τ_1 do not seriously interact with the meteorologically significant motion having the longer advective time scale τ_2. In other words, the time evolution of the meteorologically significant motion is virtually independent of the presence or absence of the high-frequency inertia–gravity oscillations. Experiments have shown (Daley 1980b, 1981) that this argument is essentially correct for simple models but is much less valid for complex models.

With respect to the third question, it is possible to define filters or artificial dissipation terms that damp out the spurious high-frequency oscillations (Section 12.2). These take up to 12 hours of integration to become really effective, however, so the early stages of an uninitialized integration still suffer from the spurious oscillations. Because the damping mechanisms that are customarily used are not very selective, the meteorologically significant motion can be adversely affected. Thus, after a sufficiently long integration (12–24 hours), the low-frequency flow would be

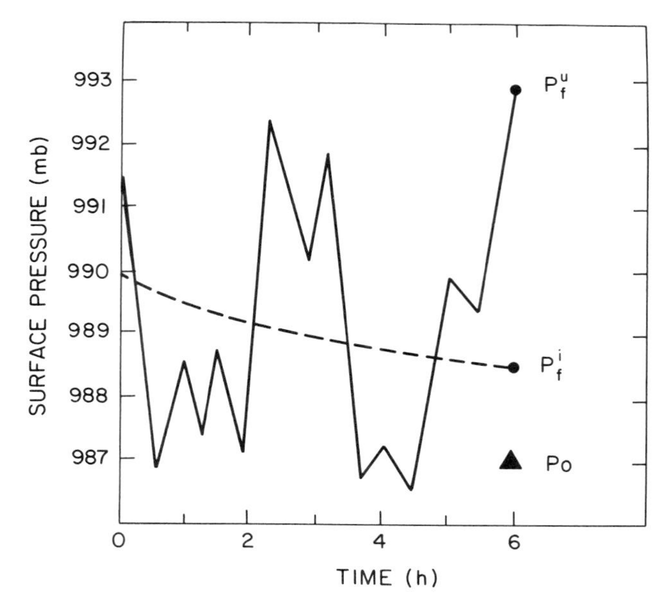

Figure 6.2 Initialization during forecast/analysis circle. Surface pressure as a function of time: ▲ observation, uninitialized (solid), and initialized (dashed).

similar (but not identical) in an uninitialized model, an initialized model, and an uninitialized but damped model. However, the short time evolution of the three models would be completely different. In particular, the short time evolution of sensitive quantities such as vertical motion, precipitation rate, and surface pressure tendency would be extremely unreliable in the uninitialized forecast whether damped or not.

There are even stronger arguments for the application of initialization procedures. In discussions earlier in this section, we assumed that the background states were balanced. Let us now relax this assumption and consider the whole data assimilation cycle (Section 1.6); a typical cycle is shown in Figure 1.11. In these cycles, a short model forecast is used to provide a background field for an objective analysis of the available observations. Figure 6.2 has been constructed to illustrate the effect of initialization on the data assimilation cycle. It is in a similar format to Figure 6.1, but the actual values are fictitious.

Suppose that the cycle has a period of 6 hours. That is, a 6 hour integration of a primitive equations model is used to provide a background field for the objective analysis at some geographical location. There are two values of the forecast surface pressure at 6 hours: $P_f^u = 993$ mb obtained from the uninitialized model (solid line in Figure 6.2) and $P_f^i = 988.5$ mb obtained from the initialized model (dashed line). Suppose that the observed surface $P_o = 987$ mb (solid triangle). Then, the two observation increments are $P_o - P_f^u = 6$ mb and $P_o - P_f^i = 1.5$ mb.

The observation increment for the initialized forecast is small and reasonable but is large and mostly spurious for the uninitialized forecast. The spurious observation increment from the uninitialized forecast could have detrimental effects on the subsequent objective analysis. In a statistical interpolation context (Chapter 4), several outcomes are possible, all unfortunate. Statistical interpolation requires the specification of background error variance and observation error variance. Suppose the specified background error variance of uninitialized forecasts was assumed to be small with respect to the observation error variance and thus *did not* reflect the poor quality of the uninitialized forecasts. Then the large observation increment, together with the supposedly accurate background field, might cause the outright rejection of a perfectly good observation. Now suppose that the specified background error variance of uninitialized forecasts was assumed to be large with respect to the observation error variance and thus *did* reflect the poor quality of the uninitialized forecasts. Then the background P_f^u would be given very little weight, and the objective analysis procedure would draw for the observation. If the background field is consistently given a very small weight, then it is not worth the trouble to make the 6 hour forecast. These arguments are consistent with those in Section 4.9.

The preceding arguments provide the justification for the initializaton step in the forecast–analysis cycle of Figure 1.6. However, other benefits can be obtained. The initialization process essentially provides powerful diagnostic constraints that can be used to generate approximate but model-consistent data that are not available from the observation network. In primitive equation models, initial fields of the mass (geopotential, temperature, or pressure), wind (rotational and divergent), moisture, and other variables are required to start the integrations. Not all of these variables are observed equally well. For example, over an otherwise unobserved region there could be a swath of temperature data obtained from the passage of an orbiting satellite radiometer. An initialization procedure could be used to generate an approximate but model-consistent windfield from the temperature in this region.

Some of the initialization procedures have other uses, peculiar to themselves. For instance, the normal-mode initialization procedure (Chapters 9 and 10) can provide useful information about the temporal aspects of the model forecast.

The forecast–analysis cycle has three stages: the model forecast, the objective analysis, and the initialization. As discussed in Section 1.6, a corresponding increment is associated with each stage in the cycle. For a state-of-the-art forecast–analysis cycle, the forecast increment is larger than the analysis increment, which is itself larger than the initialization increment. In most cases, the initialization increment is smaller than the assumed observation error. Thus, the benefits of the initialization step can generally be realized without making unacceptably large changes in the objectively analyzed fields.

Two more comments of a general nature are appropriate before beginning a detailed study of the initialization problem. First, there is no unique balanced model state corresponding to a given unbalanced model state; ideally, the initialization process should produce a balanced model state that is minimally different from the true atmospheric state. Second, the benefits of an initialization process are greatest when the observation network is poor and/or the objective analysis procedures are primitive; consequently, the need for initialization diminishes as the observation network is upgraded and more advanced objective analysis algorithms are implemented.

6.4 A linearized shallow water model

A simple model that can clarify many of the concepts introduced in the previous sections can be constructed from the shallow water equations. This model is complex enough to allow both time scales τ_1 and τ_2 yet simple enough to be solved analytically.

The primitive or hydrostatic equations mentioned in Section 6.1 describe the flow of a rotating, vertically stratified fluid in which the horizontal scales are much larger than the vertical. Because the fluid is vertically stratified, models based on these equations are called *baroclinic*. Much simpler, *barotropic* models can be constructed by vertically integrating the baroclinic equations. One such model, the *shallow water equations* model, is extremely useful in initialization theory. The shallow water equations describe a shallow rotating layer of homogeneous, incompressible, and inviscid fluid and are derived in standard textbooks on geophysical fluid dynamics such as Pedlosky (1979) or Gill (1982). In cartesian coordinates these equations can be written as

$$\frac{\partial u}{\partial t} + u\frac{\partial u}{\partial x} + v\frac{\partial u}{\partial y} - fv + g\frac{\partial h}{\partial x} = 0 \tag{6.4.1}$$

$$\frac{\partial v}{\partial t} + u\frac{\partial v}{\partial x} + v\frac{\partial v}{\partial y} + fu + g\frac{\partial h}{\partial y} = 0 \tag{6.4.2}$$

$$\frac{\partial h}{\partial t} + u\frac{\partial h}{\partial x} + v\frac{\partial h}{\partial y} + h\left(\frac{\partial u}{\partial x} + \frac{\partial v}{\partial y}\right) = 0 \tag{6.4.3}$$

where x is the eastward direction, y the northward direction, u the eastward wind component, v the northward wind component, h the height of the free surface of the fluid, g the gravitational constant, and f the Coriolis parameter. Equations (6.4.1) and (6.4.2) are the eastward and northward momentum equations and (6.4.3) is the continuity equation. It will be assumed that the Coriolis parameter f is a constant f_0 (the mid-latitude f-plane assumption).

The next step is to linearize the equations about a basic state which is at rest and has a free surface height $\bar{h}$ that is independent of space and time. The linearized equations are

$$\frac{\partial u}{\partial t} - f_0 v + \frac{\partial \Phi}{\partial x} = 0 \tag{6.4.4}$$

$$\frac{\partial v}{\partial t} + f_0 u + \frac{\partial \Phi}{\partial y} = 0 \tag{6.4.5}$$

$$\frac{\partial \Phi}{\partial t} + \tilde{\Phi}\left(\frac{\partial u}{\partial x} + \frac{\partial v}{\partial y}\right) = 0 \qquad (6.4.6)$$

where $\tilde{\Phi} = g\tilde{h}$ and $\Phi = gh$ is the deviation from $\tilde{\Phi}$.

The neglect of the nonlinear terms makes the initialization problem much more tractable but is not very realistic. The inclusion of nonlinear (and forcing) terms is considered in Chapters 7 and 9.

Equations (6.4.4–6) can be rewritten in terms of a streamfunction ψ and a velocity potential χ by use of Helmholtz's theorem,

$$u = -\frac{\partial \psi}{\partial y} + \frac{\partial \chi}{\partial x}, \qquad v = \frac{\partial \psi}{\partial x} + \frac{\partial \chi}{\partial y} \qquad (6.4.7)$$

Because of the *f*-plane assumption, the Coriolis parameter passes through any differentiation. Thus, differentiating (6.4.4) with respect to y and (6.4.5) with respect to x and subtracting the first equation from the second gives a vorticity equation:

$$\frac{\partial}{\partial t}\nabla^2\psi + f_0\,\nabla^2\chi = 0 \qquad (6.4.8)$$

Here $\nabla^2 = \partial^2/\partial x^2 + \partial^2/\partial y^2$ is the horizontal Laplacian operator in cartesian coordinates.

The *vorticity* $\zeta = \nabla^2\psi = \partial v/\partial x - \partial u/\partial y$ is, more strictly, the vertical component of the relative vorticity vector. The relative vorticity vector is defined with respect to a coordinate system that is rotating at the earth's angular velocity. The *divergence* $D = \nabla^2\chi = \partial u/\partial x + \partial v/\partial y$ is, more strictly, the horizontal divergence. A divergence equation is derived by differentiating (6.4.4) with respect to x and (6.4.5) with respect to y and adding the two equations. Thus,

$$\frac{\partial}{\partial t}\nabla^2\chi - f_0\,\nabla^2\psi + \nabla^2\Phi = 0 \qquad (6.4.9)$$

Equation (6.4.6) becomes

$$\frac{\partial \Phi}{\partial t} + \tilde{\Phi}\nabla^2\chi = 0 \qquad (6.4.10)$$

Equations (6.4.8–10) constitute a set of coupled linear partial differential equations. Assume the following functional dependence of ψ, χ, and Φ on x and y:

$$\begin{vmatrix} \psi(x, y, t) \\ \chi(x, y, t) \\ \Phi(x, y, t) \end{vmatrix} = \begin{vmatrix} \hat{\psi}_n^m(t) \\ i\hat{\chi}_n^m(t) \\ f_0\sqrt{K}\hat{\Phi}_n^m(t) \end{vmatrix} \exp\left[\frac{i(mx + ny)}{a}\right] \qquad (6.4.11)$$

where m is the east-west wavenumber, n is the north-south wavenumber, $i = \sqrt{-1}$, and a is the radius of the earth. $\hat{\psi}_n^m(t)$, $\hat{\chi}_n^m(t)$, and $\hat{\Phi}_n^m(t)$ are complex. The functional form (6.4.11) implies a periodic domain with dimensions $2\pi a$ in each of the x and y

directions. The introduction of i in the expression for χ represents a phase shift with respect to ψ. The constant K in (6.4.11) is given by

$$K = \frac{(m^2 + n^2)\bar{\Phi}}{a^2 f_0^2} \tag{6.4.12}$$

Inserting (6.4.11) into (6.4.8–10) yields

$$\begin{aligned} \frac{d\hat{\psi}}{dt} + if_0\hat{\chi} &= 0 \\ i\frac{d\hat{\chi}}{dt} - f_0\hat{\psi} + f_0\sqrt{K}\hat{\Phi} &= 0 \\ \frac{d\hat{\Phi}}{dt} - if_0\sqrt{K}\hat{\chi} &= 0 \end{aligned} \tag{6.4.13}$$

where $\hat{\psi}(t) = \hat{\psi}_n^m(t)$, $\hat{\chi}(t) = \hat{\chi}_n^m(t)$, and $\hat{\Phi}(t) = \hat{\Phi}_n^m(t)$.

Now assume the following temporal dependence for ψ, χ, and Φ:

$$\begin{vmatrix} \hat{\psi}(t) \\ \hat{\chi}(t) \\ \hat{\Phi}(t) \end{vmatrix} = \begin{vmatrix} \hat{\psi} \\ \hat{\chi} \\ \hat{\Phi} \end{vmatrix} \exp(-if_0\sigma t) \tag{6.4.14}$$

where σ is a nondimensional frequency.

Inserting expansion (6.4.14) into (6.4.13) yields, for each horizontal scale (m, n),

$$(\underline{\underline{L}} - \sigma\underline{\underline{I}}) \begin{vmatrix} \hat{\psi} \\ \hat{\chi} \\ \hat{\Phi} \end{vmatrix} = 0 \tag{6.4.15}$$

where

$$\underline{\underline{L}} = \begin{pmatrix} 0 & 1 & 0 \\ 1 & 0 & -\sqrt{K} \\ 0 & -\sqrt{K} & 0 \end{pmatrix}$$

and $\underline{\underline{I}}$ is the identity matrix. Note that the particular normalization used in (6.4.11) has been chosen to make $\underline{\underline{L}}$ real and symmetric. Equation (6.4.15) is an algebraic eigenvalue problem with eigenvalue σ. Because of the linearity of (6.4.8–10), each horizontal scale (m,n) does not interact with any other horizontal scale (m', n').

Matrix $\underline{\underline{L}}$ in (6.4.15) has three eigenvectors and three eigenvalues for a given horizontal scale (m, n). The three eigenvalues are obtained by setting the determinant of $\underline{\underline{L}} - \sigma\underline{\underline{I}}$ equal to zero, and the result is a cubic frequency equation,

$$\sigma^3 - \sigma K - \sigma = 0 \tag{6.4.16}$$

which has three real roots of $\sigma = 0$, $\pm\sqrt{1 + K}$.

The three eigenvectors are found by substituting each of the eigenvalues back into the matrix, $\underline{\underline{L}} - \sigma\underline{\underline{I}}$. Each eigenvector is defined up to a multiplicative constant, so by setting one element of the eigenvector equal to a constant (1 say), we can determine the other elements of the eigenvector.

The first eigenvalue is denoted σ_R, and the corresponding eigenvector is $|R_\psi R_\chi R_\Phi|^T$. Here T stands for transpose. Thus,

$$\sigma_R = 0, \qquad \begin{vmatrix} R_\psi \\ R_\chi \\ R_\Phi \end{vmatrix} = \frac{1}{\sqrt{1+K}} \begin{vmatrix} \sqrt{K} \\ 0 \\ 1 \end{vmatrix} \tag{6.4.17}$$

The R_ψ, R_χ, and R_Φ are the streamfunction, velocity potential, and geopotential elements of the eigenvector. The eigenvector is called the geostrophic or Rossby mode because it describes a stationary geostrophic solution, which can be seen by inserting this eigenvector back into (6.4.14). The normalization factor $(1+K)^{-1/2}$ has been chosen so that the vector magnitude is equal to 1. It will be seen that, for this eigenvector, $\hat{\psi} = \sqrt{K}\hat{\Phi}$, or in real-space terms (using 6.4.11 and 6.4.14), $f_0\psi = \Phi$. Making use of the Helmholtz theorem (6.4.7) implies that

$$u = -\frac{\partial\psi}{\partial y} = -\frac{1}{f_0}\frac{\partial\Phi}{\partial y}, \qquad v = \frac{\partial\psi}{\partial x} = \frac{1}{f_0}\frac{\partial\Phi}{\partial x}$$

which is the geostrophic relationship. In more complicated models (which are linearized about a nonzero mean wind or f is allowed to vary with latitude), this mode is no longer stationary (see Chapter 9). The geostrophic mode for this model has no divergence; in more realistic models (Chapter 9), the divergence of the geostrophic mode is small but nonzero.

The positive eigenvalue is denoted σ_G^1, with corresponding eigenvector $|G_\psi^1 \;\; G_\chi^1 \;\; G_\Phi^1|^T$:

$$\sigma_G^1 = \sqrt{1+K}, \qquad \begin{vmatrix} G_\psi^1 \\ G_\chi^1 \\ G_\Phi^1 \end{vmatrix} = \frac{1}{\sqrt{2(1+K)}} \begin{vmatrix} 1 \\ \sqrt{1+K} \\ -\sqrt{K} \end{vmatrix} \tag{6.4.18}$$

The negative eigenvalue is denoted σ_G^2, with corresponding eigenvector $|G_\psi^2 \;\; G_\chi^2 \;\; G_\Phi^2|^T$. The third eigensolution is similar to the second except that

$$\sigma_G^2 = -\sigma_G^1 = -\sqrt{1+K}, \qquad G_\psi^2 = G_\psi^1, \qquad G_\chi^2 = -G_\chi^1, \qquad G_\Phi^2 = G_\Phi^1$$

The two eigensolutions corresponding to σ_G^1 and σ_G^2 are inertia–gravity modes. If σ_G^1 is multiplied by f_0, the dimensional frequency becomes

$$f_0\sigma_G^1 = \sqrt{f_0^2 + \frac{(n^2+m^2)\tilde{\Phi}}{a^2}} \tag{6.4.19}$$

This frequency is always greater than f_0 and is thus characterized by the inertial time scale described in Section (6.1). The inertia–gravity modes have vorticity, divergence, and geopotential components. The ratio of the divergence to the vorticity $(G^1_\chi/G^1_\psi) = \sqrt{1+K}$ is greater than 1, indicating that the inertia–gravity modes are primarily irrotational. The eigenvalues σ_R, σ^1_G, and σ^2_G are all real, which follows from the fact that the matrix $\underline{\underline{L}}$ is real and symmetric.

The three eigenvectors (6.4.17–18) satisfy certain orthonormality conditions (Exercise 6.1). Let F, H stand for any of R, G^1, or G^2. Then,

$$F_\psi H_\psi + F_\chi H_\chi + F_\Phi H_\Phi = \delta^H_F \tag{6.4.20}$$

where δ^H_F is the Kronecker delta function $= 1$ if $F = H$ and 0 if $F \neq H$. In fact, (6.4.20) also implies orthonormality because of the normalization adopted in (6.4.17–18).

The $\hat{\psi}(t)$, $\hat{\chi}(t)$, and $\hat{\Phi}(t)$ defined in (6.4.11–14) can be expanded in the eigenfunctions defined by (6.4.17–18). Define $y(t)$, $z_1(t)$, and $z_2(t)$ to be complex expansion coefficients for the geostrophic model (6.4.17) and the pair of inertia–gravity modes (6.4.18), respectively. Because $\hat{\psi}(t)$, $\hat{\chi}(t)$, and $\hat{\Phi}(t)$ are complex and R_ψ, G^1_ψ, G^2_ψ, and so on, are real, then $y(t)$, $z_1(t)$, and $z_2(t)$ must be complex. Note that $y(t)$, the amplitude of the geostrophic model, should not be confused with the northward coordinate defined in (6.4.1–3). Then,

$$\begin{vmatrix} \hat{\psi}(t) \\ \hat{\chi}(t) \\ \hat{\Phi}(t) \end{vmatrix} = \underline{\underline{E}} \begin{vmatrix} y(t) \\ z_1(t) \\ z_2(t) \end{vmatrix} \tag{6.4.21}$$

where

$$\underline{\underline{E}} = \begin{pmatrix} R_\psi & G^1_\psi & G^2_\psi \\ R_\chi & G^1_\chi & G^2_\chi \\ R_\Phi & G^1_\Phi & G^2_\Phi \end{pmatrix}$$

is the matrix whose columns are the eigenvectors of $\underline{\underline{L}}$. Because of the orthogonality condition, the matrix of eigenvectors $\underline{\underline{E}}$ has the property that $\underline{\underline{E}}^{\mathrm{T}} = \underline{\underline{E}}^{-1}$:

$$|y(t)z_1(t)z_2(t)|^{\mathrm{T}} = \underline{\underline{E}}^{\mathrm{T}} |\hat{\psi}(t)\hat{\chi}(t)\hat{\Phi}(t)|^{\mathrm{T}} \tag{6.4.22}$$

Equation (6.4.13) can be written in the form

$$\begin{vmatrix} \dfrac{d\hat{\psi}}{dt} & \dfrac{d\hat{\chi}}{dt} & \dfrac{d\hat{\Phi}}{dt} \end{vmatrix}^{\mathrm{T}} = -if_0 \underline{\underline{L}} |\hat{\psi} \quad \hat{\chi} \quad \hat{\Phi}|^{\mathrm{T}} \tag{6.4.23}$$

The application of (6.4.21) to (6.4.23) gives

$$|\dot{y} \quad \dot{z}_1 \quad \dot{z}_2|^{\mathrm{T}} = -if_0 \underline{\underline{\Delta}} |y \quad z_1 \quad z_2|^{\mathrm{T}} \tag{6.4.24}$$

where $\underline{\underline{\Delta}} = \underline{\underline{E}}^{-1}\underline{\underline{L}}\,\underline{\underline{E}} = \underline{\underline{E}}^{\mathrm{T}}\underline{\underline{L}}\,\underline{\underline{E}}$ and $\dot{y} = dy/dt$, and so on. The columns of $\underline{\underline{E}}$ are eigenvectors of $\underline{\underline{L}}$. Application of the orthogonality condition (6.4.20) yields

$$\underline{\Lambda} = \begin{pmatrix} \sigma_R & & 0 \\ & \sigma_G^1 & \\ 0 & & \sigma_G^2 \end{pmatrix} \tag{6.4.25}$$

Thus, the matrix $\underline{\Lambda}$ is the diagonal matrix of eigenvalues, and (6.4.24) can be written as

$$\begin{aligned} \dot{y} &= -i\sigma_R f_0 y \\ \dot{z}_1 &= -i\sigma_G^1 f_0 z_1 \\ \dot{z}_2 &= -i\sigma_G^2 f_0 z_2 \end{aligned} \tag{6.4.26}$$

Equations (6.4.36) constitute a rewriting of (6.4.8–10) in terms of new dependent variables y, z_1, and z_2. In the form (6.4.26) the equations have been diagonalized or decoupled.

The equations (6.4.26) have the simple solutions

$$\begin{aligned} y(t) &= y(0)\exp(-i\sigma_R f_0 t) = y(0) \\ z_1(t) &= z_1(0)\exp(-i\sigma_G^1 f_0 t) \\ z_2(t) &= z_2(0)\exp(-i\sigma_G^2 f_0 t) = z_2(0)\exp(+i\sigma_G^1 f_0 t) \end{aligned} \tag{6.4.27}$$

From equation (6.4.21),

$$\hat{\psi}(t) = R_\psi y(t) + G_\psi^1 z_1(t) + G_\psi^2 z_2(t)$$

which can be rewritten as

$$\hat{\psi}(t) = R_\psi y(0) + G_\psi^1 z_1(0) e^{-i\alpha t} + G_\psi^2 z_2(0) e^{i\alpha t} \tag{6.4.28}$$

where $\alpha = f_0\sqrt{1+K}$.

R_ψ, G_ψ^1, and G_ψ^2 are defined in (6.4.17–18), and $y(0)$, $z_1(0)$, and $z_2(0)$ can be written in terms of $\hat{\psi}(0)$, $\hat{\chi}(0)$, and $\hat{\Phi}(0)$ using (6.4.22). Thus,

$$\hat{\psi}(t) = \frac{(K\hat{\psi}_0 + \sqrt{K}\hat{\Phi}_0)}{1+K} + \frac{(\hat{\psi}_0 - \sqrt{K}\hat{\Phi}_0)}{1+K}\cos\alpha t - \frac{i\hat{\chi}_0}{\sqrt{1+K}}\sin\alpha t \tag{6.4.29}$$

Similarly,

$$i\hat{\chi}(t) = \frac{(\hat{\psi}_0 - \sqrt{K}\hat{\Phi}_0)}{\sqrt{1+K}}\sin\alpha t + i\hat{\chi}_0\cos\alpha t \tag{6.4.30}$$

and

$$\sqrt{K}\hat{\Phi}(t) = \frac{(K\hat{\psi}_0 + \sqrt{K}\hat{\Phi}_0)}{1+K} - \frac{K(\hat{\psi}_0 - \sqrt{K}\hat{\Phi}_0)}{1+K}\cos\alpha t + \frac{iK\hat{\chi}_0}{\sqrt{1+K}}\sin\alpha t \tag{6.4.31}$$

where $\hat{\psi}_0 = \hat{\psi}(0)$, $\hat{\chi}_0 = \hat{\chi}(0)$, and $\hat{\Phi}_0 = \hat{\Phi}(0)$. It might be noted that (6.4.31) is derived using a completely different procedure in Section 13.4.

The simple model just developed is now used to investigate the initialization problem in Section 6.5 and geostrophic adjustment in Section 6.6.

6.5 Initialization of the linearized shallow water model

The linearized shallow water model introduced in the previous section is a simple analogue of the much more complex global baroclinic primitive equation models, which are used in numerical forecasting. In particular, this simple model permits inertia–gravity oscillations with the inertial time scale τ_1. It also permits stationary geostrophic modes that serve as analogues to the meteorologically significant motions of time scale τ_2 in more complex models. It should be noted, though, that an initialization procedure appropriate for the linear model (6.4.8–10) is not appropriate for the nonlinear model (6.4.1–3). The linear model is, however, sufficiently realistic to illustrate the initialization process.

An arbitrary initial state in this model would have initial amplitudes $y(0)$, $z_1(0)$, and $z_2(0)$ (for given m, n), all nonzero. The object of initialization is to modify the initial state so that the subsequent model integration has a minimum of high-frequency motions.

Equation (6.4.27) indicates that if both $z_1(0)$ and $z_2(0)$ are set to zero, then $z_1(t) = 0$ and $z_2(t) = 0$ for all t. This is the appropriate initialization condition for this model because it suppresses the inertia–gravity oscillations from the model integration for all time.

The condition $z_1(0) = z_2(0) = 0$ can be interpreted in terms of the streamfunction, velocity potential, and geopotential. For a given wavenumber m, n, consider observed/analyzed values of the streamfunction, velocity potential, and geopotential at time $t = 0$: $\hat{\psi}_A$, $\hat{\chi}_A$, $\hat{\Phi}_A$, The object is to produce values $\hat{\psi}_I$, $\hat{\chi}_I$, $\hat{\Phi}_I$ at time $t = 0$ (for the same wavenumber m, n) that have been adjusted in such a way that subsequent model integrations have no high-frequency oscillations. This requires $z_1(0) = z_2(0) = 0$ in the present model. Thus,

$$0 = z_1(0) = G^1_\psi \hat{\psi}_I + G^1_\chi \hat{\chi}_I + G^1_\Phi \hat{\Phi}_I$$
$$0 = z_2(0) = G^2_\psi \hat{\psi}_I + G^2_\chi \hat{\chi}_I + G^2_\Phi \hat{\Phi}_I$$

Eliminating $\hat{\chi}_I$ between these two equations gives

$$\hat{\psi}_I = -\frac{[G^2_\chi G^1_\Phi - G^1_\chi G^2_\Phi]}{[G^2_\chi G^1_\psi - G^1_\chi G^2_\psi]} \hat{\Phi}_I = \sqrt{K} \hat{\Phi}_I$$

using (6.4.18). Back substitution indicates $\hat{\chi}_I = 0$. Thus, the condition $z_1(0) = z_2(0) = 0$ implies

$$\hat{\psi}_I = \sqrt{K} \hat{\Phi}_I, \qquad \hat{\chi}_I = 0 \tag{6.5.1}$$

In real space, (6.5.1) becomes $f_0\psi_I = \Phi_I$, $\chi_I = 0$, which is the same condition satisfied by the geostrophic solution (6.4.17). Equation (6.5.1) describes initialization constraints or the diagnostic relationship among the three variables ψ, χ, and Φ, which must be satisfied in order that no inertia–gravity oscillations be excited in the model.

Equation (6.5.1) indicates the relationship among the various variables after the initialization process has occurred; it does not indicate how $\hat{\psi}_I$, $\hat{\chi}_I$, $\hat{\Phi}_I$ are to be related to $\hat{\psi}_A$, $\hat{\chi}_A$, $\hat{\Phi}_A$. In fact, there is no unique way of obtaining $\hat{\psi}_I$, $\hat{\chi}_I$, $\hat{\Phi}_I$ from $\hat{\Psi}_A$, $\hat{\chi}_A$, $\hat{\Phi}_A$, as noted in Section 6.3. We now consider four of the most straightforward procedures.

Method 1: Geopotential constrained

In this procedure it is assumed that $\hat{\Phi}_A$ is known, but $\hat{\psi}_A$ and $\hat{\chi}_A$ are not known or are known much less accurately than $\hat{\Phi}_A$. Then $\hat{\psi}_I$, $\hat{\chi}_I$, and $\hat{\Phi}_I$ are determined from $\hat{\Phi}_A$ as follows:

$$\hat{\Phi}_I = \hat{\Phi}_A, \qquad \hat{\psi}_I = \sqrt{K}\hat{\Phi}_A, \qquad \hat{\chi}_I = 0 \tag{6.5.2}$$

This method is called geopotential constrained because the geopotential is unchanged during the initialization process. The initialized values $\hat{\psi}_I$, $\hat{\chi}_I$, and $\hat{\Phi}_I$ also satisfy (6.5.1), as they must. This procedure allows the streamfunction field to be calculated from the observed/analyzed geopotential field.

Mehod 2: Rotational wind constrained

Here it is assumed that $\hat{\psi}_A$ is known, but $\hat{\chi}_A$ and $\hat{\Phi}_A$ are unknown or known much less accurately than $\hat{\psi}_A$. Thus,

$$\hat{\psi}_I = \hat{\psi}_A, \qquad \hat{\chi}_I = 0, \qquad \hat{\Phi}_I = \frac{\hat{\psi}_A}{\sqrt{K}} \tag{6.5.3}$$

Model 3: Divergent wind constrained

Here it is assumed that $\hat{\chi}_A$ is known, but $\hat{\psi}_A$ and $\hat{\Phi}_A$ are unknown or known much less accurately than $\hat{\chi}_A$. Clearly, if only the χ field is known, it is impossible to determine $\hat{\psi}_I$ and $\hat{\Phi}_I$ from $\hat{\chi}_A$, so there is no way to integrate this model with no inertia–gravity oscillations. In other words, an observed/analyzed divergent wind supplies no information about the balanced state in this simple model.

Method 4: Slow mode/potential vorticity constrained

Here it is assumed that $\hat{\psi}_A$, $\hat{\chi}_A$, and $\hat{\Phi}_A$ are all available and that there is no means to choose one as significantly better than the others. The object is to produce $\hat{\psi}_I$, $\hat{\chi}_I$, and $\hat{\Phi}_I$ such that the geostrophic mode amplitude is unchanged during the initialization process. Thus,

$$\begin{aligned} y_I &= y_A \\ z_1(0) &= z_2(0) = 0 \end{aligned} \tag{6.5.4}$$

Equation (6.5.4) filters out the gravity modes while leaving the geostrophic mode untouched. Substituting into (6.5.4) from (6.4.21) gives

$$y_{\rm I} - y_{\rm A} = 0 = R_\psi(\hat{\psi}_{\rm I} - \hat{\psi}_{\rm A}) + R_\chi(\hat{\chi}_{\rm I} - \hat{\chi}_{\rm A}) + R_\Phi(\hat{\Phi}_{\rm I} - \hat{\Phi}_{\rm A})$$
$$z_1(0) = 0 = G^1_\psi \hat{\psi}_{\rm I} + G^1_\chi \hat{\chi}_{\rm I} + G^1_\Phi \hat{\Phi}_{\rm I}$$
$$z_2(0) = 0 = G^2_\psi \hat{\psi}_{\rm I} + G^2_\chi \hat{\chi}_{\rm I} + G^2_\Phi \hat{\Phi}_{\rm I}$$

Substituting in from (6.4.17–18) gives

$$\hat{\psi}_{\rm I} = \sqrt{K}\hat{\Phi}_{\rm I} = \frac{K\hat{\psi}_{\rm A} + \sqrt{K}\hat{\Phi}_{\rm A}}{1+K}, \qquad \hat{\chi}_{\rm I} = 0 \tag{6.5.5}$$

It might be noted that if $\hat{\psi}_{\rm A}$ and $\hat{\Phi}_{\rm A}$ happen to be geostrophically related ($\hat{\psi}_{\rm A} = \sqrt{K}\hat{\Phi}_{\rm A}$), then

$$\hat{\psi}_{\rm I} = \sqrt{K}\hat{\Phi}_{\rm I} = \sqrt{K}\hat{\Phi}_{\rm A} = \hat{\psi}_{\rm A}$$

Equation (6.5.5) can be rewritten as

$$K(\hat{\psi}_{\rm I} - \hat{\psi}_{\rm A}) + \sqrt{K}(\hat{\Phi}_{\rm I} - \hat{\Phi}_{\rm A}) = 0 \tag{6.5.6}$$

In other words, $K\hat{\psi} + \sqrt{K}\hat{\Phi}$ is conserved during the slow mode constrained initialization process. Condition (6.5.6) can be interpreted in real space (using 6.4.11) as

$$Q_{\rm I} - Q_{\rm A} = 0 \tag{6.5.7}$$

where $Q = \nabla^2\psi + f_0 - f_0\Phi/\tilde{\Phi}$ is the quasi-geostrophic *potential vorticity*.

The quasi-geostrophic potential vorticity is conserved during the slow mode constrained initialization process for this model. It might also be noted that the potential vorticity is time independent in (6.4.8–10), which can easily be verified by substituting (6.4.10) into (6.4.8). The potential vorticity of the gravity modes is $KG^1_\psi + \sqrt{K}G^1_\Phi = KG^2_\psi + \sqrt{K}G^2_\Phi = 0$ from (6.4.18).

Methods 1 and 2 preserve either the geopotential or the rotational wind during the initialization process. However, the geostrophic component of the flow is modified in these methods ($y_{\rm I} \neq y_{\rm A}$). In contrast, the potential vorticity constrained method (4) does not alter the geostrophic component y, but the geopotential and rotational wind are both modified ($\psi_{\rm I} \neq \psi_{\rm A}$, $\Phi_{\rm I} \neq \Phi_{\rm A}$).

The general properties of initialization procedures for complex global baroclinic primitive equation models are similar to those of this simple linearized shallow water model. However, the details of the initialization process are quite different.

6.6 Geostrophic adjustment

The linearized shallow water model of Section 6.4 can also be used to investigate the problem of geostrophic adjustment. Many practical results of initialization and continuous data assimilation experiments can be simply explained by the theory of geostrophic adjustment. The theory was first formulated by Rossby (1938) and is described in detail by Blumen (1972) and Okland (1970). It is also covered in dynamical meteorology textbooks such as Gill (1982).

Geostrophic adjustment is usually concerned with localized perturbations and is most easily examined in plane polar coordinates. In these coordinates, the nonlinear shallow water equations (6.4.1–3) can be written (Godske et al. 1957) as

$$\frac{du}{dt} - \frac{v^2}{r} - f_0 v + \frac{\partial \Phi}{\partial r} = 0$$

$$\frac{dv}{dt} + \frac{uv}{r} + f_0 u + \frac{1}{r}\frac{\partial \Phi}{\partial \phi} = 0 \qquad (6.6.1)$$

$$\frac{d\Phi}{dt} + \frac{\Phi}{r}\left(\frac{\partial ru}{\partial r} + \frac{\partial v}{\partial \phi}\right) = 0$$

where r is the radial coordinate, ϕ the tangential coordinate, and u and v are the radial and tangential velocity components, respectively. Here,

$$\frac{d}{dt} = \frac{\partial}{\partial t} + u\frac{\partial}{\partial r} + \frac{v}{r}\frac{\partial}{\partial \phi}$$

Introducing a streamfunction ψ and velocity potential χ, using the plane polar coordinate version of Helmholtz's theorem, gives

$$u = \frac{\partial \chi}{\partial r} - \frac{1}{r}\frac{\partial \psi}{\partial \phi}, \qquad v = \frac{\partial \psi}{\partial r} + \frac{1}{r}\frac{\partial \chi}{\partial \phi} \qquad (6.6.2)$$

Linearization of (6.6.1) about a state of rest with mean geopotential $\tilde{\Phi}$ and substitution of (6.6.2) yields (6.4.8–10), except that the Laplacian operator is

$$\nabla^2 = \frac{1}{r}\frac{\partial}{\partial r} r \frac{\partial}{\partial r} + \frac{1}{r^2}\frac{\partial^2}{\partial \phi^2} \qquad (6.6.3)$$

Assume solutions of the form

$$e^{im\phi} J_m(kr), \qquad \begin{matrix} -\infty \leq m \leq \infty \\ 0 \leq k \leq \infty \end{matrix}, \qquad J_{-m}(kr) = (-1)^m J_m(kr) \qquad (6.6.4)$$

where m is the tangential wavenumber, k a radial wavenumber with dimensions of $(\text{length})^{-1}$, and J the Bessel function of the first kind of integer order m (which is bounded at $r = 0$). General information on Bessel functions is given in Appendix G, and the $J_m(r)$ for $m = 0$, 1 are plotted in Figure G1. From Equation (G16), functions of the form (6.6.4) are eigenfunctions of the Laplace operator (6.6.3) with eigenvalues equal to $-k^2$, and they are orthogonal over the domain (G12).

Assume an expansion of the form (G15) for arbitrary fields ψ, χ, and Φ and insert into (6.4.8–10):

$$\begin{vmatrix} \psi(r,\phi,t) \\ \chi(r,\phi,t) \\ \Phi(r,\phi,t) \end{vmatrix} = \sum_{m=-\infty}^{\infty} \int_0^{\infty} \begin{vmatrix} \hat{\psi}_m(k,t) \\ i\hat{\chi}_m(k,t) \\ f_0\sqrt{K}\hat{\Phi}_m(k,t) \end{vmatrix} J_m(kr)e^{im\phi}k\,dk, \tag{6.6.5}$$

where $K = k^2\bar{\Phi}/f_0^2$. The result is (6.4.13) with $\hat{\psi}(t) = \hat{\psi}_m(k,t)$, $\hat{\chi}(t) = \hat{\chi}_m(k,t)$ and $\hat{\Phi}(t) = \hat{\Phi}_m(k,t)$. $\hat{\psi}_m(k,t)$, $\hat{\chi}_m(k,t)$, and $\hat{\Phi}_m(k,t)$ can be recovered from $\psi(r,\phi,t)$, $\chi(r,\phi,t)$, and $\Phi(r,\phi,t)$ by the reverse transform,

$$\begin{vmatrix} \hat{\psi}_m(k,t) \\ i\hat{\chi}_m(k,t) \\ f_0\sqrt{K}\hat{\Phi}_m(k,t) \end{vmatrix} = \frac{1}{2\pi}\int_{-\pi}^{\pi}\int_0^{\infty} \begin{vmatrix} \psi(r,\phi,t) \\ \chi(r,\phi,t) \\ \Phi(r,\phi,t) \end{vmatrix} e^{-im\phi}J_m(kr)r\,dr\,d\phi \tag{6.6.6}$$

making use of (2.4.6) and (G12).

It is now possible to find the general solution to (6.4.8–10) in plane polar coordinates. Suppose there exist initial conditions $\psi(r,\phi,0)$, $\chi(r,\phi,0)$, and $\Phi(r,\phi,0)$. Equation (6.6.6) is used to find $\hat{\psi}_m(k,0)$, $\hat{\chi}_m(k,0)$, and $\hat{\Phi}_m(k,0)$. The solution to (6.4.13) – that is, (6.4.29–31) – is used to find $\hat{\psi}_m(k,t)$, $\hat{\chi}_m(k,t)$, and $\hat{\Phi}_m(k,t)$ for any t. These are then inserted back into (6.6.5) to yield $\psi(r,\phi,t)$, $\chi(r,\phi,t)$, and $\Phi(r,\phi,t)$.

Figures 6.3 and 6.4 demonstrate the geostrophic adjustment process for two localized initial perturbations. Figure 6.3(a) illustrates an initial local axisymmetric geopotential perturbation; the initial windfield is zero. Figure 6.3(b–e) illustrates the evolution of the geopotential and windfields – using (6.6.2) to derive u, v from ψ, χ – at $t = 1$, 2, 3, and 6 hours. The geopotential field is contoured, and the windspeed is indicated on the diagram. In both Figures 6.3 and 6.4, the mean height $\bar{h}$ is 8485 m, and f_0 is appropriate for 38.25 degrees north. The horizontal scale is defined by Figure 6.3(a). The contour at which the initial geopotential falls to $1/\sqrt{e} = 0.607$ of its value at $r = 0$ has a radius of 400 km.

Examination of Figure 6.3 illustrates that inertia–gravity waves propagate outward from the original perturbation. Note the almost completely divergent windfield at 1 hour. At later times, a small, clockwise rotational wind component is apparent. After 6 hours, nothing is left but a small geopotential perturbation (only 3% of its initial value) and a rotational windfield that is in geostrophic balance with it. The rest of the perturbation has propagated off toward infinity as inertia–gravity waves. The inertia–gravity waves carry energy away from the initial region of imbalance. Calculation of the group velocities of these waves (Gal-Chen 1983) demonstrates that the small-scale energy moves out more rapidly than the large-scale energy.

Figure 6.4(a) has a local initial wind perturbation with zero initial geopotential field. The wind perturbation is completely nondivergent and isotachs (dotted) are shown every 2 ms^{-1}. Figure 6.4(b–d) shows the evolution at 1, 2, and 4 hours. A geopotential field is quickly generated; part of the geopotential field (the banana-shaped contours) is associated with propagating inertia–gravity modes, and part is in geostrophic balance [see (d)] with the largely unchanged initial wind perturbation.

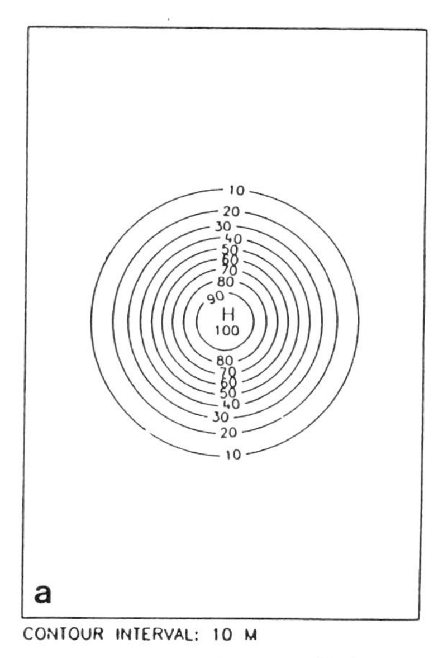

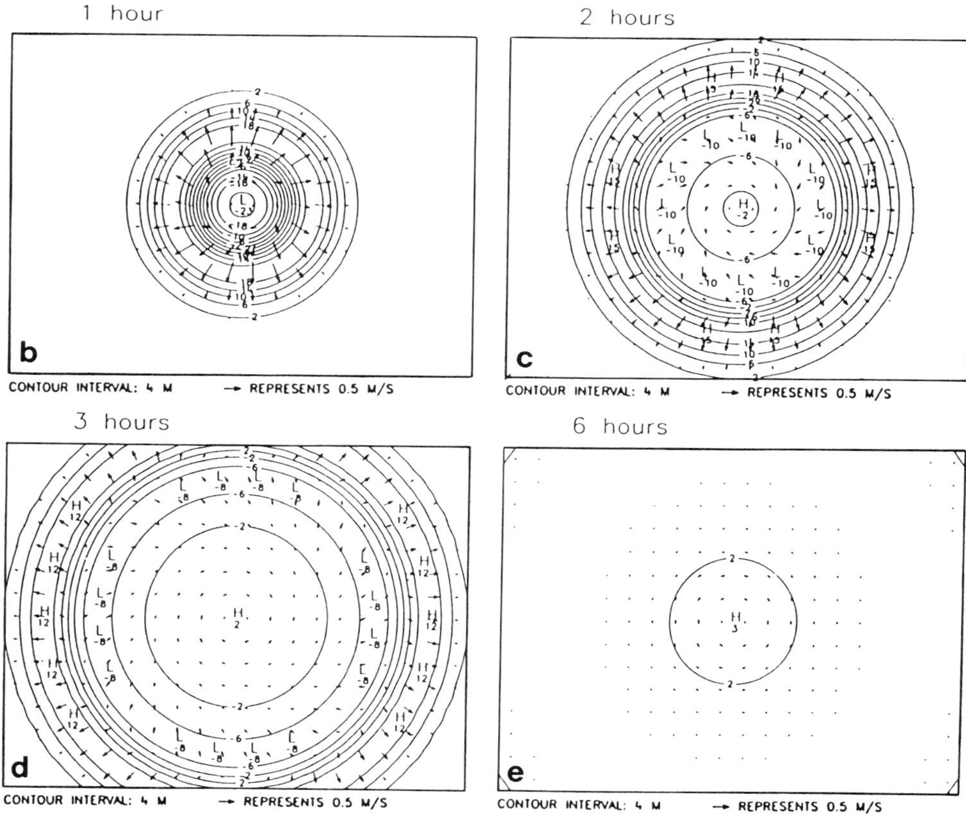

Figure 6.3 (a) Geostrophic adjustment of initial geopotential perturbation. (b–e) Solutions at 1, 2, 3, and 6 hours. Contoured field is geopotential, and wind arrows indicate speed and direction of windfield. (After Barwell and Bromley, 1988)

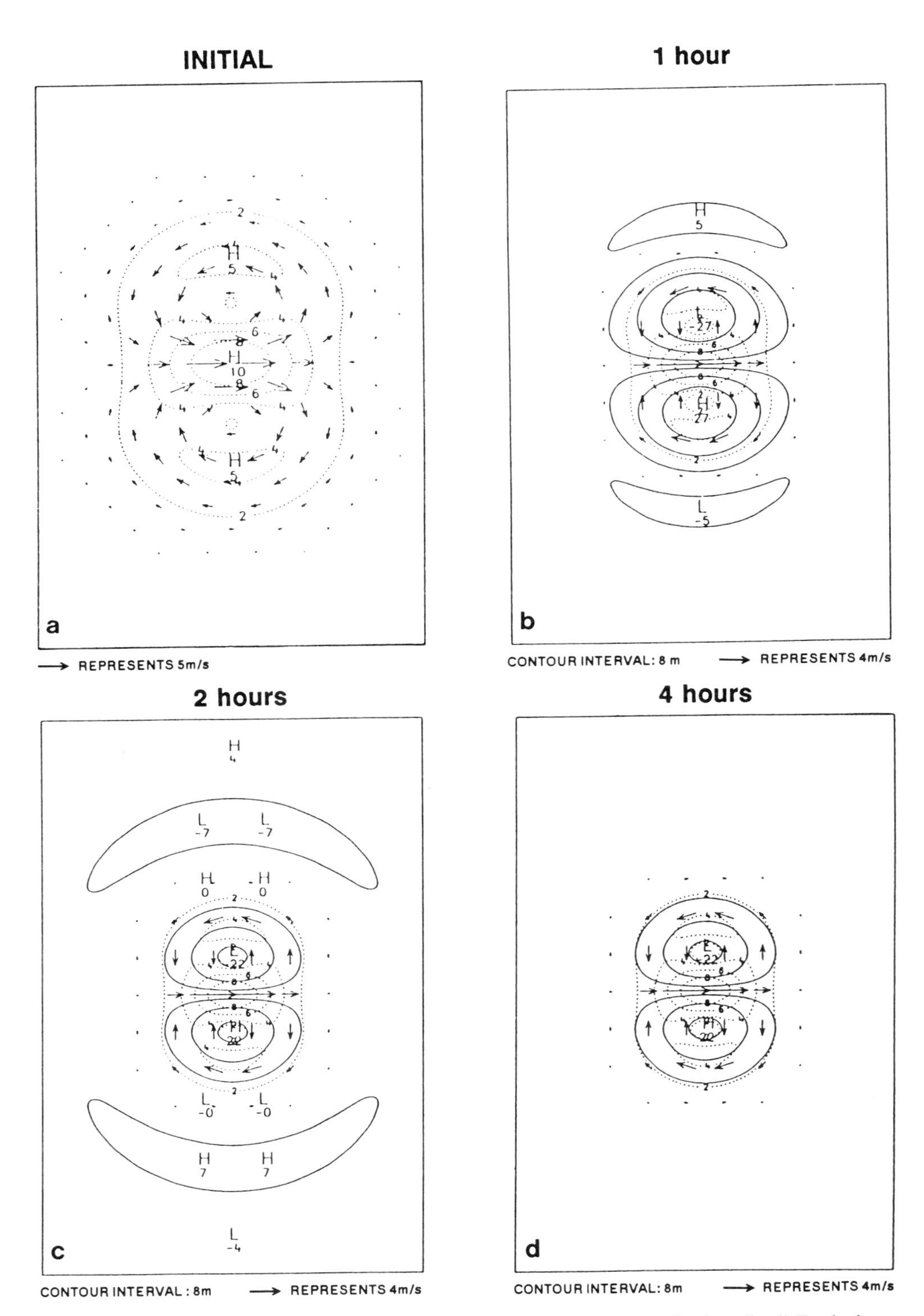

Figure 6.4 (a) Geostrophic adjustment for initial rotational wind perturbation. (b–d) Evolution of solution is shown at 1, 2, and 4 hours in the same format as Figure 6.3. (After Barwell and Bromley, 1988)

Figures 6.3 and 6.4 illustrate a number of aspects of geostrophic adjustment. An unbalanced (nongeostrophic in this simple model) local perturbation generates inertia–gravity waves that propagate away in a few hours. Remaining behind are time-independent geostrophically balanced fields, which can be quite similar to the initial perturbation in some cases (Figure 6.4) or quite different in others (Figure 6.3). Both figures illustrate the transient and the stationary components of the solutions to (6.4.29–31). Analysis of the transient solutions (Gill 1982) shows that the characteristic time scale for geostrophic adjustment is the inertial time scale, $\tau_1 = f_0^{-1}$. Of primary interest in the initialization problem is the stationary component of (6.4.29–31), which remains after the transients have propagated elsewhere or have been damped out by the numerical procedures used in the assimilating model (Sections 11.1 and 12.2). The stationary components of (6.4.29–31) are given by

$$\hat{\psi}_s = \sqrt{K}\hat{\Phi}_s = \frac{K\hat{\psi}_0 + \sqrt{K}\hat{\Phi}_0}{1+K}, \qquad \hat{\chi}_s = 0 \tag{6.6.7}$$

which is, of course, the geostrophic solution. (Subscript s stands for stationary.) The steady-state solution can be rewritten as follows:

$$\hat{\psi}_s = \frac{L_R}{L_H}\hat{\Phi}_s = \frac{\hat{\psi}_0 + (L_H/L_R)\hat{\Phi}_0}{1 + L_H^2/L_R^2} \tag{6.6.8}$$

where $L_H = \sqrt{a^2/(n^2+m^2)}$ defines a horizontal scale, and $L_R = \sqrt{\tilde{\Phi}}/f_0$ is called the Rossby radius of deformation. K can be written as L_R^2/L_H^2.

The Rossby radius of deformation is a horizontal scale at which rotation effects become as important as buoyancy effects. Consider the geostrophic potential vorticity defined in (6.5.7). ζ/f_0 ($\zeta = \nabla^2\psi$ is the vorticity) is a nondimensional ratio that measures the rotation of the fluid with respect to the rotation rate of the earth. $\Phi/\tilde{\Phi}$ is the nondimensional ratio of the amplitude of the free surface perturbation with respect to the mean depth of the fluid. From (6.5.7) and (6.6.8), it follows that

$$\frac{\zeta f_0^{-1}}{\Phi\tilde{\Phi}^{-1}} = \frac{L_R^2}{L_H^2} \tag{6.6.9}$$

Thus, for $L_H < L_R$, the vorticity term dominates; for $L_H > L_R$, the surface elevation term dominates. In energetic terms, Gill (1982) demonstrates that short-wavelength geostrophic flow contains mainly kinetic energy whereas long-wavelength geostrophic flow contains most of the energy in potential form.

Equation (6.6.8) describes the final state $[\hat{\psi}_s, \hat{\Phi}_s]$ in terms of the initial state $[\hat{\psi}_0, \hat{\Phi}_0]$ and the ratio of the horizontal scale L_H to the Rossby radius of deformation L_R. Two asymptotic cases can be considered, $L_H \ll L_R$ and $L_H \gg L_R$.

Case 1: $L_H \ll L_R$

In this case $\hat{\psi}_s = (L_R/L_H)\hat{\Phi}_s = \hat{\psi}_0$. In real space form this is equivalent to

$$\psi_s = \frac{1}{f_0}\Phi_s = \psi_0 \tag{6.6.10}$$

The final rotational windfield is equal to the initial rotational windfield, and the final geopotential is related to the initial rotational windfield. $L_H \ll L_R$ implies $K \gg 1$, which occurs if f_0 is small, $\tilde{\Phi}$ is large, or L_H is small. Now, $\tilde{h} = g^{-1}\tilde{\Phi}$ defines a vertical scale. Thus, the geopotential adjusts to the wind at low latitudes, for large vertical scales, and for small horizontal scales.

Case 2: $L_H \gg L_R$
This time $\hat{\psi}_s = (L_R/L_H)\hat{\Phi}_s = (L_R/L_H)\hat{\Phi}_0$. In real space form this becomes

$$\psi_s = \frac{1}{f_0}\Phi_s = \frac{1}{f_0}\Phi_0 \tag{6.6.11}$$

The final geopotential is equal to the initial geopotential, and the final windfield is geostrophically related to the initial geopotential. $L_H \gg L_R$ implies $K \ll 1$, which occurs if f_0 is large, $\tilde{\Phi}$ is small, or L_H is large. Thus, the wind adjusts to the geopotential at high latitudes, for small vertical scales, and for large horizontal scales. Cases 1 and 2 are further illustrated in Exercise 6.4 and 6.5.

The shallow water equations (6.4.1–3) define an atmosphere with no vertical stratification and with mean depth $\tilde{h}$. A stratified atmosphere may have a number of vertical scales. Equation (6.6.10) and the discussion following it apply in the stratified case as well, except that $\tilde{h}$ is replaced by the appropriate vertical scales (see Okland 1970). The connection between $\tilde{h}$ and the vertical scales in a baroclinic atmosphere is shown in Sections 9.1 and 9.2.

The Rossby radius L_R defined in (6.6.8) is not completely appropriate in the tropics. An equatorial Rossby radius of deformation can be defined (Gill 1982) but it does not change the preceding conclusion. For the purposes of initialization, (6.6.8) leads to the following general statement: When the horizontal scale of the initial field L_H is small compared with the Rossby radius of deformation L_R, the final state is determined by the initial rotational windfield whereas when L_H is large compared with L_R, the final state is determined by the initial mass field.

6.7 A brief history of initialization

Most operational analysis forecast systems have both objective analysis and initialization components. The objective analysis is required to transform the initial data into some type of regular representation, and the initialization is required to suppress high-frequency oscillations in the forecast model integration. As noted in Chapter 1, objective analysis dates from the late 1940s when the first successful NWP models were developed. These early models were filtered and did not require initialization. In the mid-1950s, however, there was increasing interest in primitive equation models and consequently in the initialization problem.

The earliest initialization procedures were based on the quasi-geostrophic theory. Charney (1955) demonstrated that primitive equation models had high-frequency oscillations not present in the atmosphere. He proposed using the nonlinear balance equation to calculate the streamfunction ψ from an observed/analyzed geopotential Φ. It was thought that this balanced initial state would be sufficient to suppress inertia–gravity waves. Hinkelmann (1959) and Phillips (1960) demonstrated that the use of the nonlinear balance equation alone was not sufficient to suppress inertia–gravity waves. Their results suggested that an initial velocity potential field χ also had to be determined from ψ and Φ using the quasi-geostrophic omega equation.

These early quasi-geostrophic initialization procedures could be loosely characterized as geopotential constrained, in the terminology of Section 6.5. There were a number of technical problems with the solution of the quasi-geostrophic omega equation and particularly the nonlinear balance equation. The most difficult was the *ellipticity problem* in the nonlinear balance equation. These difficulties generated considerable interest in the late 1950s (Bolin 1955, 1956; Miyakoda 1956; Bushby and Huckle 1956; Shuman 1957; Arnason 1958). The quasi-geostrophic techniques were reasonably successful at high latitudes, but the geopotential constrained formulation doomed them to failure at lower latitudes.

In the 1960s, Sasaki (1958, 1969, 1970a), Thompson (1969), and Stephens (1970) formulated initialization as a *variational problem*, in which the final initialized analysis must (a) fit the observations as closely as possible and (b) satisfy exactly or approximately some imposed dynamical constraints. In principle, at least this procedure could perform the functions of both objective analysis (a) and initialization (b). The dynamic constraints to be imposed were usually of the quasi-geostrophic form: that is, the nonlinear balance equation or the quasi-geostrophic omega equation.

The variational generalization permitted different types of data to be used in the initialization when appropriate. For example, consistent with Section 6.6, it is more appropriate to use wind information at low latitudes and geopotential information at high latitudes. This could be handled elegantly and straightforwardly in the variational formulation, which was an advantage over purely quasi-geostrophic initialization. However, variational formulations were computationally expensive; and because they usually used quasi-geostrophic constraints, they suffered from many of the problems of quasi-geostrophic initialization.

In the late 1960s a new initialization technique was introduced by Miyakoda and Moyer (1968) and Nitta and Hovermale (1969). This technique was called *dynamic initialization* and had the unusual feature that it actually made use of the primitive equations forecast model. As mentioned in Section 6.3, it is possible to eventually damp inertia–gravity oscillations in a primitive equations model by integrating the model with artificial damping terms. Such a procedure requires 12–24 hours of model integration to be effective. However, by integrating the same damped model forward and backward in time around $t = 0$, using a numerical scheme that damps in both directions, it is possible, in principle, to arrive at an initial model state that is free of high-frequency oscillations. The early dynamic initialization schemes required many hours of integration before the high-frequency oscillations were removed and were thus prohibitively expensive. Later schemes of Okamura (Temperton 1976) tended to converge more rapidly.

The most widely used initialization technique is *normal mode initialization*. It was first contemplated in the early 1970s by Dickinson and Williamson (1972) and Williamson and Dickinson (1976). They developed procedures for finding the normal modes and associated eigenfrequencies of discretized models. Modes with high frequencies could be identified as inertia–gravity waves, and the normal mode structures corresponding to these eigenfrequencies could be filtered from the objective analyses. This initialization procedure was not very successful because it was a linear procedure whereas the forecast model equation was nonlinear.

Machenhauer (1977) and Baer (1977b) demonstrated how the normal mode procedure could be generalized to include nonlinear terms. The nonlinear schemes were very successful in suppressing inertia–gravity oscillations. These were basically slow mode constrained schemes in the nomenclature of Section 6.5, but they were extended to the data constrained cases by Daley (1978). The nonlinear normal mode techniques were the most general initialization techniques of all. They made use of the full primitive equation forecast model as in dynamic initialization. They could also be used in a variational context, and Leith (1980) demonstrated that the quasi-geostrophic methods were simply special cases of the normal mode procedure.

Since 1980 there have been a number of new techniques such as the bounded derivative method and initialization by Laplace transform. The most intense research activity has shifted from synoptic scale initialization to mesoscale initialization (Section 13.7).

In the following chapters, four initialization procedures are described in detail: the static initialization procedures first followed by the dynamic method. Thus, quasi-geostrophic and variational methods are discussed in Chapters 7 and 8, and normal mode methods are discussed in Chapters 9 and 10. The normal mode methods are the most widely used at the present time and are discussed in more detail than the other methods. Dynamic initialization is discussed in Chapter 11 and the newer techniques in Sections 13.4 and 13.5.

Exercises

6.1 Show that the eigenvectors of matrix $\underset{\sim}{L}$ of (6.4.15) are orthonormal. Show that the matrix $\underset{\approx}{E}$ of eigenvectors in (6.4.21) has the property that $\underset{\approx}{E}^{T} = \underset{\approx}{E}^{-1}$, where T stands for matrix transpose.

6.2 In Section 6.5, geopotential, rotational wind, and slow mode constrained initialization can be written in the form

$$|\hat{\psi}_I \quad \hat{\chi}_I \quad \hat{\Phi}_I|^T = \underset{\approx}{\Pi}|\hat{\psi}_A \quad \hat{\chi}_A \quad \hat{\Phi}_A|^T$$

where $\underset{\approx}{\Pi}$ is a 3×3 matrix known as the *slow mode projection matrix*. Find $\underset{\approx}{\Pi}$ for each of these three types of initialization. The matrix $(\underset{\approx}{I} - \underset{\approx}{\Pi})$ is the fast mode projection. Show that

$$\underset{\approx}{\Pi}^n = \underset{\approx}{\Pi} \qquad \text{and} \qquad (\underset{\approx}{I} - \underset{\approx}{\Pi})\underset{\approx}{\Pi} = 0$$

6.3 Consider the quadratic form

$$J = 0.5[w_\psi(\hat{\psi}_A - \hat{\psi}_I)^2 + w_\chi(\hat{\chi}_A - \hat{\chi}_I)^2 + w_\Phi(\hat{\Phi}_A - \hat{\Phi}_I)^2]$$

where w_ψ, w_χ, and w_Φ are specified (a priori) weights and

$$\hat{\psi}_I = R_\psi y, \qquad \hat{\chi}_I = R_\chi y, \qquad \hat{\Phi}_I = R_\Phi y$$

Here y is the slow mode amplitude.
Show that the minimization of J produces the slow mode projection matrix:

$$\underline{\underline{\Pi}} = \frac{1}{K + w_\Phi/w_\psi}\begin{pmatrix} K & 0 & \sqrt{K}\dfrac{w_\Phi}{w_\psi} \\ 0 & 0 & 0 \\ \sqrt{K} & 0 & \dfrac{w_\Phi}{w_\psi} \end{pmatrix}$$

Show that this slow mode projection matrix reduces to the geopotential, rotational wind, and slow mode constrained cases with appropriate choices of w_ψ, w_χ, and w_Φ.

6.4 Two-dimensional time dependent solutions to the linearized shallow water equations on an infinite f-plane can be found using (6.6.6) followed by (6.4.29–31) and (6.6.5). Consider the same equations on the one-dimensional domain $-\infty \le x \le \infty$, with Fourier transforms instead of Hankel transforms.

Define a Rossby radius $L_R = \sqrt{\bar{\Phi}}/f_0$. Suppose the initial geopotential distribution is given by,

$$\Phi(x, t = 0) = \Phi_0 \exp\left(\frac{-|x|}{L}\right)$$

where L is a characteristic scale and $\psi(x, t = 0) = \chi(x, t = 0) = 0$. (This is a continuous one-dimensional analogue for the insertion of a single geopotential observation, as discussed in Chapter 12.) Show that the stationary component of the geopotential field is

$$\Phi_s(x) = \Phi_0 L\left[\frac{Le^{-|x|/L} - L_R e^{-|x|/L_R}}{L^2 - L_R^2}\right]$$

Use integral tables if necessary. Show that when $L = L_R$, the stationary geopotential has the functional form (3.3.15). Plot $\Phi_s(x)$ as a function of x/L for several values of L_R.

6.5 Consider a similar situation to that in Exercise 6.4, except that $\Phi(x, t = 0) = \chi(x, t = 0) = 0$, and the initial velocity distribution is given by

$$v(x, t = 0) = \frac{\partial\psi(x, t = 0)}{\partial x} = v_0 \exp\left(\frac{-|x|}{L}\right)$$

(This is a continuous one-dimensional analogue for the insertion of a single wind observation as discussed in Chapter 12.) Find the stationary components of the windfield $v_s(x)$ and geopotential $\Phi_s(x)$. Find the solutions in the case $L = L_R$. Plot $v_s(x)$ and $\Phi_s(x)$ as a function of x/L for $L_R = 0$, L, and ∞.

7

Quasi-geostrophic constraints

The earliest initialization procedures were based on the quasi-geostrophic theory, which was developed in the 1940s and 1950s by Rossby, Charney, and others. The theory has been reviewed by Phillips (1963) and is covered extensively in most atmospheric dynamics textbooks; see, for example, Holton (1972) or Pedlosky (1979). Quasi-geostrophic theory has been extremely useful in numerical modeling, instability theory, and in a variety of other theoretical contexts.

Quasi-geostrophic initialization is a rather special application of the general quasi-geostrophic theory. Consequently, this treatment of the theory emphasizes the temporal aspects of the problem. Quasi-geostrophic initialization will be approached through the analysis of a simple model, and the conclusions drawn from this analysis will be used as a guide in deriving the quasi-geostrophic constraints from the primitive equations.

7.1 The Hinkelmann–Phillips model

Hinkelmann (1951) and Phillips (1960) developed and analyzed a very simple model based on the shallow water equations linearized about a constant basic current. They obtained the quasi-geostrophic initialization constraints appropriate for this model, which could then be applied to much more realistic models.

The nonlinear shallow water equations are given by (6.4.1–3). Hinkelmann (1951) and Phillips (1960) assumed that the equations are linearized about a constant wind U, which is independent of x, y, and t, and a basic state geopotential $\bar{\Phi} = g\bar{h}$, which is a function only of y (the northward coordinate). In particular, U is related geostrophically to $\bar{\Phi}$:

$$U = -\frac{1}{f_0}\frac{\partial\bar{\Phi}}{\partial y} \tag{7.1.1}$$

where f_0 is the Coriolis parameter (which is assumed constant). The perturbation velocity and geopotential components are assumed to be function of x only. With these assumptions, (6.4.1–3) reduce to

$$\frac{\partial u}{\partial t} + U\frac{\partial u}{\partial x} - f_0 v + \frac{\partial \Phi}{\partial x} = 0 \tag{7.1.2}$$

$$\frac{\partial v}{\partial t} + U\frac{\partial v}{\partial x} + f_0 u = 0 \tag{7.1.3}$$

$$\frac{\partial \Phi}{\partial t} + U\frac{\partial \Phi}{\partial x} + v\frac{\partial \bar{\Phi}}{\partial y} + \tilde{\Phi}\frac{\partial u}{\partial x} = 0 \tag{7.1.4}$$

where $\tilde{\Phi} = \int_y \bar{\Phi}(y)\, dy / \int_y dy$.

Equations (7.1.2–4) can be rewritten in terms of a streamfunction ψ and a velocity potential χ using Helmholtz's theorem (6.4.7):

$$\frac{\partial}{\partial t}\frac{\partial^2 \psi}{\partial x^2} + U\frac{\partial}{\partial x}\frac{\partial^2 \psi}{\partial x^2} + f_0\frac{\partial^2 \chi}{\partial x^2} = 0 \tag{7.1.5}$$

$$\frac{\partial}{\partial t}\frac{\partial^2 \chi}{\partial x^2} + U\frac{\partial}{\partial x}\frac{\partial^2 \chi}{\partial x^2} - f_0\frac{\partial^2 \psi}{\partial x^2} + \frac{\partial^2 \Phi}{\partial x^2} = 0 \tag{7.1.6}$$

$$\frac{\partial \Phi}{\partial t} + U\frac{\partial \Phi}{\partial x} - f_0 U\frac{\partial \psi}{\partial x} + \tilde{\Phi}\frac{\partial^2 \chi}{\partial x^2} = 0 \tag{7.1.7}$$

where (7.1.1) has been used in deriving (7.1.7).

The solution to (7.1.5–7) can be found by assuming the following form:

$$\begin{vmatrix} \psi(x,t) \\ \chi(x,t) \\ \Phi(x,t) \end{vmatrix} = \begin{vmatrix} \hat{\psi}(t) \\ \hat{\chi}(t) \\ \hat{\Phi}(t) \end{vmatrix} \exp\left[\frac{imx}{a} - \frac{iUmt}{a}\right] \tag{7.1.8}$$

where m is the x wavenumber and a the earth's radius. Inserting (7.1.8) into (7.1.5–7) yields for each wavenumber m,

$$\frac{d\hat{\psi}}{dt} + f_0\hat{\chi} = 0 \tag{7.1.9}$$

$$\frac{d\hat{\chi}}{dt} - f_0\hat{\psi} + \hat{\Phi} = 0 \tag{7.1.10}$$

$$\frac{d\hat{\Phi}}{dt} - \frac{if_0 Um\hat{\psi}}{a} - \frac{m^2\tilde{\Phi}\hat{\chi}}{a^2} = 0 \tag{7.1.11}$$

The normalization of $\hat{\psi}$, $\hat{\chi}$, and $\hat{\Phi}$ in (7.1.8) is purposely different from that in (6.4.11). The terms of the form $U(\partial/\partial x)$ in (7.1.5–7) have been absorbed by the form of the time dependence assumed in expansion (7.1.8).

At this point, it is possible to proceed by the normal mode methods developed in Chapter 6. In fact, a normal mode expansion was used by Phillips (1960) in his examination of this model. An alternative procedure (which will turn out to be very useful in Chapter 13) is the Laplace transform method.

Laplace transforms are widely used in electrical engineering and signal processing for the purpose of solving coupled, linear ordinary differential equations with constant coefficients; see, for example, Holbrook (1959) or Kuhfittig (1977). Equations (7.1.9–11) fit this description exactly and can be solved by the Laplace transform method. The Laplace transform of the function $\hat{\psi}(t)$ is defined as follows:

$$L(\hat{\psi}) = \int_0^\infty \hat{\psi}(t) e^{-st}\, dt \tag{7.1.12}$$

where the variable s can be thought of as a complex frequency.

Some simple properties of the Laplace transform follow immediately from (7.1.12). Suppose C is a constant, then

$$\begin{aligned} L(C\hat{\psi}) &= CL(\hat{\psi}) \\ L(\hat{\psi} + \hat{\chi}) &= L(\hat{\psi}) + L(\hat{\chi}) \\ L(Ct^n) &= \frac{n!C}{s^{n+1}} \\ L\left(\frac{d\hat{\psi}}{dt}\right) &= -\hat{\psi}(0) + sL(\hat{\psi}) \end{aligned} \tag{7.1.13}$$

The application of the Laplace transform to (7.1.9–11) yields

$$sL(\hat{\psi}) + f_0 L(\hat{\chi}) = \hat{\psi}_0 \tag{7.1.14}$$

$$sL(\hat{\chi}) - f_0 L(\hat{\psi}) + L(\hat{\Phi}) = \hat{\chi}_0 \tag{7.1.15}$$

$$sL(\hat{\Phi}) - \frac{if_0 Um}{a} L(\hat{\psi}) - \frac{m^2 \tilde{\Phi} L(\hat{\chi})}{a^2} = \hat{\Phi}_0 \tag{7.1.16}$$

where $\hat{\psi}_0 = \hat{\psi}(0)$, $\hat{\chi}_0 = \hat{\chi}(0)$, and $\hat{\Phi}_0 = \hat{\Phi}(0)$. Equations (7.1.14–16) constitute a set of three algebraic equations for the three unknowns $L(\hat{\psi})$, $L(\hat{\chi})$, and $L(\hat{\Phi})$. Eliminating $L(\hat{\psi})$ and $L(\hat{\Phi})$ yields a single equation for $L(\hat{\chi})$:

$$L(\hat{\chi}) = \frac{\hat{\chi}_0[s^2 + s(f_0\hat{\psi}_0 - \hat{\Phi}_0)/\hat{\chi}_0 - if_0^2\sigma_2\hat{\psi}_0/\hat{\chi}_0]}{s^3 + sf_0^2\sigma_1^2 - if_0^3\sigma_2} \tag{7.1.17}$$

where $\sigma_1 = \sqrt{1 + m^2\tilde{\Phi}/a^2 f_0^2}$ and $\sigma_2 = Um/af_0$. The σ_1 and σ_2 are nondimensional frequencies: σ_1 is the inertia–gravity wave frequency of (6.4.18), which has characteristic time scale τ_1 in the notation of Chapter 6, and σ_2 corresponds to the advective time scale τ_2.

To find $\hat{\chi}$ from (7.1.17), it is necessary to invert the Laplace transform $L(\hat{\chi})$. The inversion of (7.1.17) involves factoring the denominator, which is a fairly general cubic polynomial. This yields a complicated expression for $\hat{\chi}(t)$. The problem can be made much more tractible by approximating the denominator of (7.1.17) by

$$s^3 + sf_0^2\sigma_1^2 \tag{7.1.18}$$

In other words, the term $if_0^3\sigma_2$ is assumed to be small. Working backward from (7.1.17), we can show that the consequences of the assumption (7.1.18) is that (7.1.4) is replaced by

$$\frac{\partial \Phi}{\partial t} + U\frac{\partial \Phi}{\partial x} + \tilde{\Phi}\frac{\partial u}{\partial x} = -v(0)\frac{\partial \bar{\Phi}}{\partial y} \tag{7.1.19}$$

Thus, in the continuity equation, the term $v(\partial\bar{\Phi}/\partial y)$ has been replaced by its initial value and becomes a forcing function. This is likely to be a reasonable approximation if v or ψ do not vary rapidly in time. This will be verified after the fact.

With assumption (7.1.18), expression (7.1.17) can be written in a standard form:

$$L(\hat{\chi}) = \hat{\chi}_0\left[\frac{s^2 + a_1 s + a_0}{s(s^2 + a_2^2)}\right] \tag{7.1.20}$$

which has the solution

$$\frac{\hat{\chi}(t)}{\hat{\chi}_0} = \left[\frac{a_0}{a_2^2} + \frac{a_2^2 - a_0}{a_2^2}\cos(a_2 t) + \frac{a_1}{a_2}\sin(a_2 t)\right] \tag{7.1.21}$$

This solution can be obtained from Laplace transform tables (Holbrook 1959). In Section 13.4, (7.1.21) will be derived from (7.1.20) using the method of residues. The solution $\chi(t)$ can be obtained from (7.1.21). To obtain the complete time-dependent solution, multiply by the factor $\exp(-iUmt/a)$, which was transformed out in expansion (7.1.8). Thus,

$$\hat{\chi}(t)e^{-if_0\sigma_2 t} = \left[\frac{-i\sigma_2\hat{\psi}_0}{\sigma_1^2} + \left(\hat{\chi}_0 + \frac{i\sigma_2\hat{\psi}_0}{\sigma_1^2}\right)\cos f_0\sigma_1 t + \left(\frac{f_0\hat{\psi}_0 - \hat{\Phi}_0}{f_0\sigma_1}\right)\sin f_0\sigma_1 t\right]e^{-if_0\sigma_2 t} \tag{7.1.22}$$

Solution (7.1.22) for the time dependence takes the form of inertia–gravity wave solutions with characteristic frequency σ_1 modulated by the slower advective frequency σ_2. Equation (7.1.22) is more complex than (6.4.30). The amplitude of the sine factor is proportional to the geostrophic departure $f_0\hat{\psi}_0 - \hat{\Phi}_0$ as before, but the amplitude of the cosine factor contains an extra term.

7.2 Initialization of the Hinkelmann–Phillips model

We can now discuss the initialization of the Hinkelmann–Phillips model. First consider the degenerate case when the terms $v(\partial\bar{\Phi}/\partial y)$ in (7.1.4) or $if_0Um\hat{\psi}/a$ in (7.1.11) are set equal to zero. From (7.1.22), there will be no high-frequency oscillations when the coefficients of the sine and cosine terms are set equal to zero initially. That is,

$$\hat{\chi}_0 = 0 \quad \text{and} \quad f_0\hat{\psi}_0 - \hat{\Phi}_0 = 0 \tag{7.2.1}$$

which is equivalent to (6.5.1).

Now suppose that the term $v(\partial\bar{\Phi}/\partial y)$ in (7.1.4) is nonzero, but make the same approximation (7.1.19) as in the previous section. Then (7.1.22) implies that there will be no high-frequency oscillations if

$$\hat{\chi}_0 + \frac{i\sigma_2\hat{\psi}_0}{\sigma_1^2} = 0 \qquad \text{and} \qquad f_0\hat{\psi}_0 - \hat{\Phi}_0 = 0 \tag{7.2.2}$$

In real space, the first equation of (7.2.2) can be written as

$$\tilde{\Phi}\frac{\partial^2\chi}{\partial x^2} - f_0^2\chi = U\frac{\partial\Phi}{\partial x} = -v\frac{\partial\bar{\Phi}}{\partial y} \tag{7.2.3}$$

which is a form of the quasi-geostrophic omega equation (see Section 7.4). The initialization conditions (7.2.2) can also be derived by setting

$$\frac{d}{dt}[f_0\hat{\psi} - \hat{\Phi}] = 0 \qquad \text{and} \qquad \frac{d\hat{\chi}}{dt} = 0 \tag{7.2.4}$$

at $t = 0$ in (7.1.9–11).

It can be seen that conditions (7.2.4) are the time derivatives of conditions (7.2.1). In fact, (7.2.1) can be regarded as a zeroth-order initialization condition and (7.2.4) as a first-order condition; the second-order condition is left as Exercise 7.2. The initialization conditions (7.2.2) imply a geostrophic balance between ψ and Φ, but they also imply that there should be a nonzero initial velocity potential field. The primary contribution of Hinkelmann (1959) and Phillips (1960) was to demonstrate that the initial velocity potential or divergence fields should be nonzero and that (for this simple model) they should be given approximately by (7.2.3).

The relative magnitude of the divergent wind to the rotational wind for the initialized model can be obtained from (7.2.2). Thus,

$$\left|\frac{\hat{\chi}_0}{\hat{\psi}_0}\right| = \frac{\sigma_2}{\sigma_1^2} \ll 1 \tag{7.2.5}$$

The most important observations to be made about the initialization constraints (7.2.2) for this model are as follows:

1 The streamfunction and geopotential are in geostrophic balance. This requires that the time tendency term in the divergence equation (7.1.6) be neglected.
2 The divergence obtained from the quasi-geostrophic omega equation (7.2.3) is nonzero but is much smaller than the vorticity.

Before we leave the Hinkelmann–Phillips model, it is appropriate to discuss the effects of approximation (7.1.19). Note that approximations of this type are commonly made in initialization procedures, for example, the Machenhauer procedure discussed in Chapter 9. Rather than compare the approximate initial conditions (7.2.2) with the exact initial conditions obtainable from the complete solution of (7.1.2–4), we take a more practical approach.

Assume that a forecast model is described by (7.1.9–11). This model is integrated in time using a centered difference approximation for the time derivatives $d\hat{\psi}/dt$, $d\hat{\chi}/dt$, and $d\hat{\Phi}/dt$. Thus, (7.1.9) can be approximated as

$$\frac{\hat{\psi}(t+\Delta t)-\hat{\psi}(t-\Delta t)}{2\,\Delta t}+f_0\hat{\chi}(t)=0 \tag{7.2.6}$$

where Δt is some finite increment in time. Similar expressions can be written for (7.1.10–11). In this way, expressions for $\hat{\psi}(t+\Delta t)$, $\hat{\chi}(t+\Delta t)$, and $\hat{\Phi}(t+\Delta t)$ in terms of $\hat{\psi}(t)$, $\hat{\chi}(t)$, $\hat{\Phi}(t)$, $\hat{\psi}(t-\Delta t)$, $\hat{\chi}(t-\Delta t)$, and $\hat{\Phi}(t-\Delta t)$ can be obtained. Provided Δt is sufficiently small, accurate solutions to the exact equations (7.1.9–11) can be obtained for finite times. (For more information on numerical time integration schemes, consult Haltiner and Williams 1980 or Richtmeyer and Morton 1967.) This model is integrated using two initial conditions:

$$1\ \hat{\Phi}_0=f_0\hat{\psi}_0,\qquad \hat{\chi}_0=0\qquad \text{(nondivergent)} \tag{7.2.7}$$

$$2\ \hat{\Phi}_0=f_0\hat{\psi}_0,\qquad \hat{\chi}_0=\frac{-i\sigma_2\hat{\Phi}_0}{f_0\sigma_1^2}\qquad \text{(divergent)} \tag{7.2.8}$$

The constants in (7.1.9–11) are chosen to be representative midlatitude values. Thus,

$$\begin{aligned} U &= 10\ \mathrm{ms}^{-1} & a &= 6.37\times 10^6\ \mathrm{m}\\ f_0 &= 10^{-4}\ \mathrm{s}^{-1} & g &= 9.8\ \mathrm{ms}^{-2}\\ L_\mathrm{H} &= \frac{2\pi a}{m}=2\times 10^6\ \mathrm{m} & \tilde{h} &= g^{-1}\tilde{\Phi}=10\ \mathrm{km} \end{aligned} \tag{7.2.9}$$

From these constants, $\sigma_1=9.888$, $\sigma_2=0.314$, $\tau_1=2\pi/f_0\sigma_1=1.76$ hours, and $\tau_2=2\pi/f_0\sigma_2=55.55$ hours.

The next step is to choose initial values for $\hat{\psi}$ or $\hat{\Phi}$. $\hat{\psi}_0$ has been specified to be the complex number (1, 1). Thus, $\hat{\Phi}_0=f_0\psi_0=(10^{-4},\ 10^{-4})$. $\hat{\chi}_0$ is specified using (7.2.7) or (7.2.8).

Figure 7.1(a–c) shows the results of a 24 hour integration with this model. Shown are the real parts of $\hat{\psi}(t)\exp(-if_0\sigma_2 t)$, $\hat{\chi}(t)\exp(-if_0\sigma_2 t)$, and $\hat{\Phi}(t)\exp(-if_0\sigma_2 t)$. Thus, the full time dependence of the solutions is plotted, including the slow variation that was transformed out in expansion (7.1.8). Part (a) shows the streamfunction, (b) the velocity potential, and (c) the geopotential. The solid curve shows the integration initialized with the nondivergent initial conditions (7.2.7), and the dashed curve is the same except for the divergent initial conditions (7.2.8). The two curves for the streamfunction (a) are indistinguishable.

Examining Figure 7.1(a–c), we see that time scales τ_1 and τ_2 are both present. The advective time scale of approximately 55 hours shows up in all plots, but the inertia–gravity time scale of approximately 1.7 hours shows up only in the velocity potential and geopotential plots. It is apparent that the divergent initial condition (7.2.8) is very successful in suppressing inertia–gravity oscillations despite the approximation made in (7.1.19). That the streamfunction varies smoothly with time

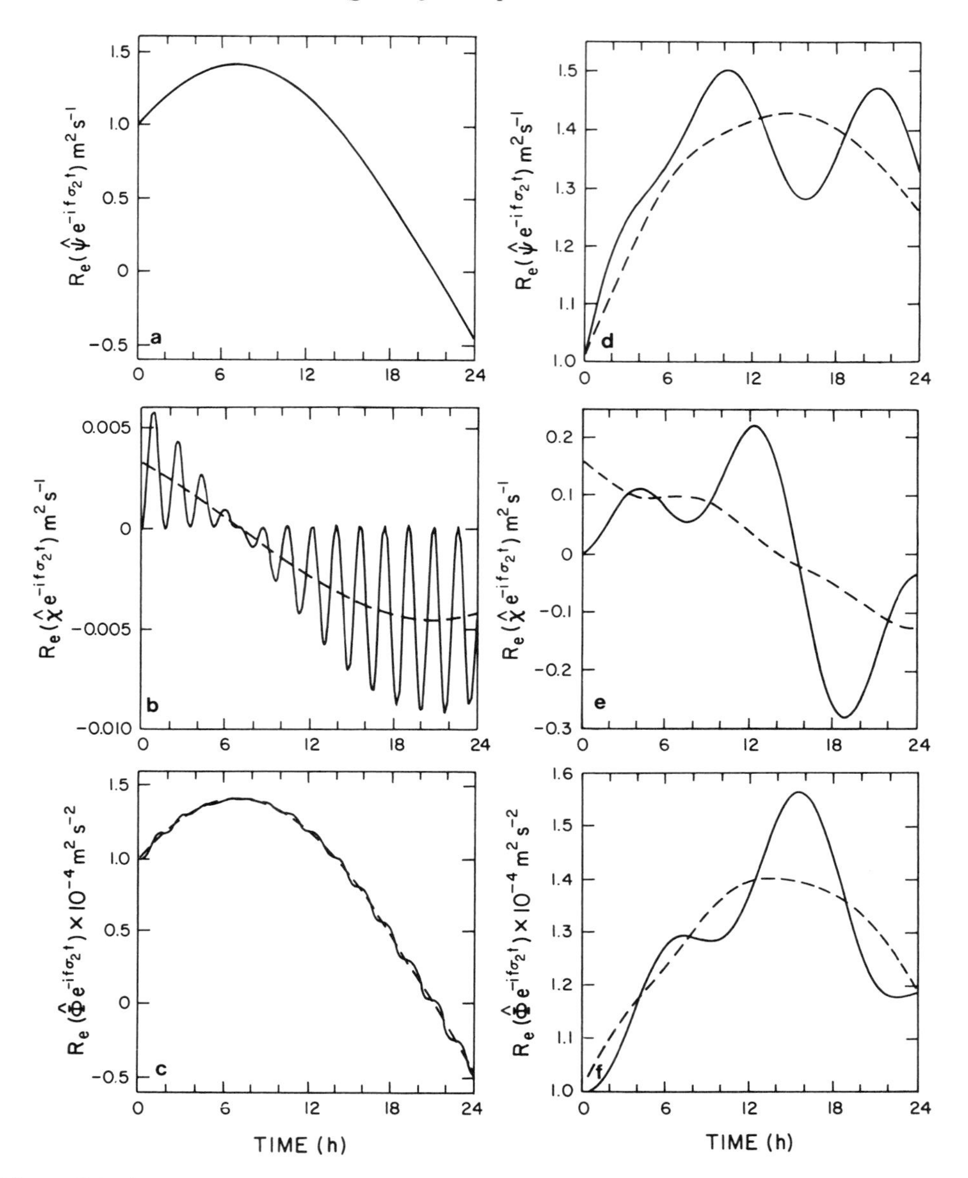

Figure 7.1 The 24 hour integrations with the Hinkelmann–Phillips model: (a–c) equivalent depth 10 k (d–f) equivalent depth 100 m. The streamfunction, velocity potential, and geopotential components are shown in (a,d), (b,e), and (c,f), respectively.

also indicates the reasonableness of approximation (7.1.19), and the assumption of holding it constant appears justified. It might be noted, in agreement with (7.2.5), that the velocity potential is much smaller than the streamfunction.

In Figure 7.1(a–c), the inertia–gravity oscillations have a very short time scale τ_1, only 1.76 hours. It is also interesting to examine a case with slower inertia–gravity waves. From the definition of σ_1 (7.1.17), it is apparent that reducing the mean height of the fluid $\tilde{h}$ reduces the inertia–gravity wave frequency. In Figure 7.1(a–c), the mean height of the fluid was 10 km, which corresponds to a large vertical scale.

The results for a mean fluid height of 100 m are shown in Figure 7.1(d–f). These plots are the same as in Figure 7.1(a–c) except that $\tilde{h}$ is 100 m. The corresponding values of σ_1 and τ_1 are 1.402 and 12.44 hours, respectively. It is apparent that the solutions for the streamfunction, velocity potential, and geopotential for this case are now quite different. In the case of the nondivergent initialization (solid curves), the inertia–gravity oscillations are visible in the streamfunction, velocity potential, and geopotential plots. The divergent initialization (dashed curves) largely removes the inertia–gravity oscillations from the streamfunction and geopotential, but they are still discernible in the velocity potential. Note also that the amplitudes of both the inertia–gravity oscillations and the velocity potential are much larger than in the 10 km case.

The inertia–gravity wave frequency σ_1 can also be written as follows:

$$\sigma_1 = \sqrt{1 + \frac{L_R^2}{L_H^2}}$$

where $L_H = 2\pi a/m$ and where $L_R = 2\pi\sqrt{\tilde{\Phi}}/f_0$ is the Rossby radius of deformation. Thus, when L_R becomes small compared with L_H, σ_1 decreases toward 1. At the same time, it is apparent in Figure 7.1 that the amplitude of the inertia–gravity oscillations increases substantially. The initialized divergence in (7.2.2) also increases as σ_1 decreases. Figure 7.1 suggests that when L_R becomes smaller than L_H, the initialization condition (7.2.8) does not completely suppress inertia–gravity waves.

Integrations with the Hinkelmann–Phillips model indicate that for a given latitude and horizontal scale, the quasi-geostrophic initialization procedure is successful for large vertical scales but is less successful for small vertical scales.

7.3 Scale analysis of the primitive equations

The simple diagnostic relationships (7.2.2) are sufficient to largely suppress inertia–gravity modes from the Hinkelmann–Phillips model. In other words, if the model variables ψ, χ, and Φ are initially related in an appropriate way, the resulting model integration will have little or no high-frequency variance. Thus, an initialization procedure can be thought of as the application at initial time of an appropriate set of diagnostic relationships or constraints among the dependent variables of the model.

It would be expected that as the complexity of the model increases, so does the complexity of the appropriate initialization conditions. However, initialization is both a practical and theoretical problem, and it is possible to have diagnostic relationships that are too complex to be of any value. Thus, highly implicit or highly nonlinear diagnostic relationships may not be useful. For example, an approximate but tractable diagnostic relationship may be more useful than a more accurate but highly implicit relationship. The simpler relationship may be less successful in suppressing inertia–gravity modes, but this could be compensated by ease of application.

Quasi-geostrophic initialization conditions can be derived for both the primitive equations and the shallow water equations. In fact, quasi-geostrophic initial conditions for the shallow water equations are just a special case of the quasi-geostrophic initial conditions for the primitive equations. Consequently, the quasi-geostrophic initial conditions will be derived for the (baroclinic) primitive equations and the corresponding (barotropic) shallow water results will be quoted where appropriate. The following development of the quasi-geostrophic initial conditions is based on the midlatitude beta-plane derivation of Kasahara (1982a). Alternative derivations of the quasi-geostrophic equations can be seen in Pedlosky (1979) or Gill (1982), and the general quasi-geostrophic theory is reviewed in Phillips (1963).

The primitive equations are derived in most textbooks on atmospheric dynamics (see, for example, Holton 1972). They are valid in an atmosphere that is shallow (thickness of atmosphere is small compared with the earth's radius) and hydrostatic (horizontal scales are much larger than vertical). They are written here in a vector form in pressure coordinates (P), pressure increasing downward; the horizontal coordinates are curvilinear (x and y directed eastward and northward, respectively):

$$\frac{\partial \mathbf{v}}{\partial t} + \mathbf{v} \cdot \nabla \mathbf{v} + \omega \frac{\partial \mathbf{v}}{\partial p} + f\mathbf{k} \times \mathbf{v} + \nabla \Phi = \mathbf{F} \tag{7.3.1}$$

$$\frac{\partial \Phi}{\partial p} + \frac{RT}{P} = 0 \tag{7.3.2}$$

$$\nabla \cdot \mathbf{v} + \frac{\partial \omega}{\partial p} = 0 \tag{7.3.3}$$

$$\left(\frac{\partial}{\partial t} + \mathbf{v} \cdot \nabla\right) \frac{\partial \Phi}{\partial p} + \omega \Gamma = -\frac{RQ}{C_p P} \tag{7.3.4}$$

where $\mathbf{v}$ is the horizontal wind $= u\mathbf{i} + v\mathbf{j}$, $\nabla = \mathbf{i}(\partial/\partial x) + \mathbf{j}(\partial/\partial y)$, $\mathbf{k}$ is the upward directed unit vector, and $\omega = dP/dt$ is the vertical velocity in pressure coordinates. R is the gas constant, C_p the specific heat at constant pressure, T the temperature, and Φ the geopotential. $\mathbf{F}$ is the frictional force per unit mass, Q is the time rate of heating per unit mass, and

$$\Gamma = \frac{1}{P} \frac{\partial}{\partial P} \left(P \frac{\partial \Phi}{\partial P} - \frac{R\Phi}{C_p} \right)$$

is the static stability.

Equation (7.3.1) is the equation of motion, (7.3.2) is the hydrostatic equation, (7.3.3) the equation of continuity, and (7.3.4) the thermodynamic equation. Equation (7.3.4) is derived from the first law of thermodynamics:

$$\frac{d\theta}{dt} = \frac{\theta}{C_p} \frac{Q}{T} \tag{7.3.5}$$

where $\theta = T(P_0/P)^{\kappa}$ is the potential temperature, P_0 is 1000 mb, and $\kappa = R/C_p$. Equation (7.3.4) is obtained from (7.3.5) by logarithmic differentiation of θ and use of (7.3.2).

The static stability Γ is a measure of the vertical stratification of the atmosphere. It can also be written as

$$\Gamma = \frac{N^2}{g^2}\left(\frac{\partial \Phi}{\partial P}\right)^2 \tag{7.3.6}$$

where

$$N = \left(\frac{g}{\theta}\frac{\partial \theta}{\partial z}\right)^{1/2} = g\left[\frac{1}{\theta}\frac{\partial \theta/\partial P}{\partial \Phi/\partial P}\right]^{1/2}$$

is known as the Brunt–Väisälä frequency. In a stably stratified atmosphere, Γ is positive, and a parcel displaced vertically will oscillate with frequency N about its equilibrium state.

For the sake of simplicity, the following vertical boundary conditions are assumed:

$$\omega = \frac{dP}{dt} = 0 \quad \text{at} \quad \begin{matrix} P = 1000 \text{ mb} \\ P = 0 \end{matrix} \tag{7.3.7}$$

The lower boundary condition assumes that the earth's surface is flat with no orography.

The first step in deriving the quasi-geostrophic initial conditions is to scale the primitive equations (7.3.1–4). The following characteristic scales are defined:

$$\begin{array}{llll} L_{\mathrm{H}} = \text{horizontal scale (m)} & \dfrac{L_{\mathrm{H}}}{V_{\mathrm{H}}} & = \text{advective time scale (s)} & \\ L_{\mathrm{Z}} = \text{vertical scale (m)} & g & = \text{gravitational constant } (\mathrm{ms}^{-2}) & \\ \Pi \ = \text{vertical pressure scale (mb)} & \Phi & = \text{earth's rotation rate } (\mathrm{s}^{-1}) & (7.3.8) \\ V_{\mathrm{H}} = \text{horizontal wind speed } (\mathrm{ms}^{-1}) & a & = \text{earth radius (m)} & \\ N_0 = \text{Brunt–Vaisala frequency } (\mathrm{s}^{-1}) & & & \end{array}$$

The variables in (7.3.1–4) are then nondimensionalized, using the scales defined in (7.3.8), and the (*) is used to indicate a nondimensionalized variable. Thus,

$$\begin{array}{llll} \mathbf{v}^* = V_{\mathrm{H}}^{-1}\mathbf{v}, & \nabla^* & = L_{\mathrm{H}}\nabla & \\ t^* = V_{\mathrm{H}}L_{\mathrm{H}}^{-1}t & (x^*, y^*) & = L_{\mathrm{H}}^{-1}(x, y) & (7.3.9) \\ P^* = \Pi^{-1}P & \omega^* & = L_{\mathrm{H}}\Pi^{-1}V_{\mathrm{H}}^{-1}\omega & \end{array}$$

The characteristic vertical scale L_{Z} and the Brunt–Väisälä frequency N_0 are obtained from a characteristic vertical stratification. Definition (7.3.6) implies the relation $N_0^2 L_{\mathrm{Z}}^2 = gL_{\mathrm{Z}}$. Define $(\tilde{\ })$ to be the horizontal averaging operator. From (7.3.6), the horizontally averaged static stability $\tilde{\Gamma}$ is nondimensionalized as

$$\tilde{\Gamma}^* = \frac{\Pi^2 \tilde{\Gamma}}{L_{\mathrm{Z}}^2 N_0^2}$$

The geopotential Φ is expressed as

$$\Phi = N_0^2 L_Z^2 \tilde{\Phi}^*(P^*) + 2\Omega V_H L_H \Phi^* \tag{7.3.10}$$

where $\tilde{\Phi}^*$ is the nondimensionalized horizontally averaged geopotential, which is only a function of pressure, and Φ^* indicates a deviation from $\tilde{\Phi}^*$. The scaling for Φ is not immediately obvious and will be justified subsequently. In scaling the Coriolis parameter, the midlatitude beta-plane approximation is assumed (see Pedlosky 1979, chap. 3). Thus,

$$f^* = (2\Omega)^{-1} f = [f_0^* + L_H a^{-1} \beta^* y^*] \tag{7.3.11}$$

where $f_0^* = \text{sine(latitude)}$ and $\beta^* = \text{cosine(latitude)}$.

If (7.3.9–11) are inserted into (7.3.1–4), the following scaled equations result:

$$\frac{\partial \mathbf{v}^*}{\partial t^*} + \mathbf{v}^* \cdot \nabla \mathbf{v}^* + \omega^* \frac{\partial \mathbf{v}^*}{\partial P^*} + R_0^{-1}[(f_0^* + L_H a^{-1} \beta^* y^*)\mathbf{k} \times \mathbf{v}^* + \nabla^* \Phi^*] = 0 \tag{7.3.12}$$

$$\nabla^* \cdot \mathbf{v}^* + \frac{\partial \omega^*}{\partial P^*} = 0 \tag{7.3.13}$$

$$\left[\frac{\partial}{\partial t^*} + \mathbf{v}^* \cdot \nabla^*\right] \frac{\partial \Phi^*}{\partial P^*} + \frac{L_R^2}{R_0 L_H^2} \omega^* \tilde{\Gamma}^* + \omega^* \Gamma^* = 0 \tag{7.3.14}$$

where

$$\Gamma^* = \frac{1}{P^*} \frac{\partial}{\partial P^*} \left[P^* \frac{\partial \Phi^*}{\partial P^*} - \kappa \Phi^*\right]$$

and Q and $\mathbf{F}$ have been neglected for simplicity.

$$R_0 = \frac{V_H}{2\Omega L_H} \tag{7.3.15}$$

is the Rossby number, and the Rossby radius of deformation is

$$L_R = \frac{N_0 L_Z}{2\Omega} \tag{7.3.16}$$

which is the analogue of (6.6.8) for a stratified (baroclinic) atmosphere. All starred variables in (7.3.12–14) are nondimensional; there are also three dimensionless numbers: R_0, L_H/a, and $R_0 L_H^2/L_R^2$. The quasi-geostrophic approximation, which will be formally derived in the next section, is valid if the following three conditions hold:

$$R_0 \ll 1, \qquad \frac{L_H}{a} \ll 1, \qquad \frac{R_0 L_H^2}{L_R^2} \ll 1 \tag{7.3.17}$$

Note that the last inequality is strengthened for $L_H < L_R$.

For midlatitudes, representative values of the characteristic scales of (7.3.8) are as follows:

$L_{\mathrm{H}} = 10^6$ m, $L_{\mathrm{Z}} = 10^4$ m, $V_{\mathrm{H}} = 10$ ms^{-1}, $g = 10$ ms^{-2}
$\Omega = 10^{-4}$ s^{-1}, $a = 10^7$ m, $N_0 = 10^{-2}$ s^{-1}, which implies $L_{\mathrm{R}} = 10^6$ m

In midlatitudes R_0, L_{H}/a, and $R_0 L_{\mathrm{H}}^2/L_{\mathrm{R}}^2$ are all clearly much less than one, and so the quasi-geostrophic approximation is valid.

Define ε to be 0.1. In scaling terms, R_0, L_{H}/a, and $R_0 L_{\mathrm{H}}^2/L_{\mathrm{R}}^2$ are $\mathrm{O}(\varepsilon)$ (i.e., of order ε). Equations (7.3.12–14) can be rewritten as follows if the three nondimensional numbers R_0, L_{H}/a, and $R_0 L_{\mathrm{H}}^2/L_{\mathrm{R}}^2$ are replaced by ε:

$$\frac{\partial u}{\partial t} + \varepsilon^{-1}\left[\frac{\partial \Phi}{\partial x} - f_0 v\right] = -\mathbf{v}\cdot\nabla u - \omega\frac{\partial u}{\partial P} + \beta y v \tag{7.3.18}$$

$$\frac{\partial v}{\partial t} + \varepsilon^{-1}\left[\frac{\partial \Phi}{\partial y} + f_0 u\right] = -\mathbf{v}\cdot\nabla v - \omega\frac{\partial v}{\partial P} - \beta y u \tag{7.3.19}$$

$$\nabla\cdot\mathbf{v} + \frac{\partial \omega}{\partial P} = 0 \tag{7.3.20}$$

$$\frac{\partial^2 \Phi}{\partial t\,\partial P} + \varepsilon^{-1}\omega\tilde{\Gamma} = -\mathbf{v}\cdot\nabla\frac{\partial \Phi}{\partial P} - \frac{\omega}{P}\frac{\partial}{\partial P}\left(P\frac{\partial \Phi}{\partial P} - \kappa\Phi\right) \tag{7.3.21}$$

where (7.3.18–19) are the eastward and northward components, respectively, of (7.3.12). The * notation has been dropped for the sake of simplicity, but all quantities in (7.3.18–21) are assumed to be scaled as in (7.3.12–14). Note that Φ in (7.3.21) is the nondimensional geopotential deviation (Φ^*) defined in (7.3.10).

Before considering the derivation of the quasi-geostrophic initial conditions for (7.3.18–21), we must justify the scaling adopted for the geopotential (7.3.10). Equation (7.3.10) can also be written as

$$\Phi = 4\Omega^2 L_{\mathrm{R}}^2\left[\tilde{\Phi}^* + \frac{L_{\mathrm{H}}^2}{L_{\mathrm{R}}^2} R_0 \Phi^*\right]$$

using the definition of the Rossby radius of deformation (L_{R}). As mentioned earlier $R_0 L_{\mathrm{H}}^2/L_{\mathrm{R}}^2$ is $\mathrm{O}(\varepsilon)$, so the assumed deviation geopotential is an order of magnitude smaller than the mean geopotential. In (7.3.10), the scaling for the mean term is consistent with (7.3.6). The scaling for the deviation term Φ^* is chosen so that the two terms $f\mathbf{k}\times\mathbf{v}$ and $\nabla\Phi$ in (7.3.1) have the same scaling. This ensures that these two terms are of the same magnitude. As noted in Pedlosky (1979), if the Coriolis term is large compared with the time derivative term, then $\mathbf{v}$ can be nonzero only if the pressure gradient term is large enough to approximately balance the Coriolis term.

7.4 Quasi-geostrophic initialization of the primitive equations

The quasi-geostrophic equations are often derived from (7.3.18–21) by expanding the dependent variables u, v, Φ, and so on, in an asymptotic series in ε (see Pedlosky

1979, for example). An alternative procedure, more relevant to the initialization problem, is the bounded derivative method of Kreiss (1980) and Browning, Kasahara, and Kreiss (1980). The bounded derivative procedure will also be discussed in Chapter 13.

The idea behind the bounded derivative method is as follows. The time scale chosen in (7.3.8) was the advective time scale τ_2. That is, terms that are O(1) in the scaled equations (7.3.18–21) can be considered to have the advective time scale. It would be desirable to find solutions to (7.3.18–21) that have only time scales of O(1), that is only the advective time scale. Such solutions would naturally have time derivatives $\partial u/\partial t$, $\partial v/\partial t$, $\partial \Phi/\partial t$, and so on, that are of O(1). Moreover, all higher time derivatives $\partial^2 u/\partial t^2$, $\partial^2 v/\partial t^2$, and so on, would be of O(1). If any higher time derivatives were not O(1), they would introduce other time scales into the problem, and so the solution would no longer be O(1). The fact that the higher time derivatives must be of O(1) can be used to generate initialization constraints. The procedure develops a hierarchy of higher-order initialization constraints as higher time derivatives are invoked.

For (7.3.18–21), the bounded derivative method proceeds as follows. First, the first-order time derivatives $\partial u/\partial t$, $\partial v/\partial t$, and $\partial \Phi/\partial t$ are used to obtain the lowest-order initialization constraints. For $\partial u/\partial t$, $\partial v/\partial t$, and $\partial \Phi/\partial t$ to be of O(1) in (7.3.18–21), the following conditions must be true:

$$\frac{\partial \Phi}{\partial x} - f_0 v = \varepsilon G_u(x, y, P, t) \tag{7.4.1}$$

$$\frac{\partial \Phi}{\partial y} + f_0 u = \varepsilon G_v(x, y, P, t) \tag{7.4.2}$$

$$\omega = \varepsilon G_\Phi(x, y, P, t) \tag{7.4.3}$$

where G_u, G_v, and G_Φ are as yet unspecified functions of O(1).

From (7.4.3), it is clear that ω is O(ε). Thus, from the continuity equation (7.3.20), the divergence $D = \nabla \cdot \mathbf{v}$ is also of O(ε). This result can also be obtained by differentiating (7.4.2) with respect to x and subtracting (7.4.1) differentiated with respect to y. If the horizontal windfield is written in terms of its rotational and divergent components, then

$$\mathbf{v} = \mathbf{v}_\psi + \varepsilon \mathbf{v}_\chi \tag{7.4.4}$$

where $\mathbf{v}_\psi = \mathbf{k} \times \nabla\psi$ is the rotational wind and ψ the streamfunction; $\mathbf{v}_\chi = \nabla\chi$ is the divergent wind and χ the velocity potential. Note that in (7.4.4), $\mathbf{v}$, $\mathbf{v}_\psi$, and $\mathbf{v}_\chi$ are all assumed to be scaled as in (7.3.18–21). The fact that the divergent wind is an order of magnitude smaller than the rotational wind is consistent with the result obtained from the Hinkelmann–Phillips model (7.2.5).

The simplest way to ensure that the conditions (7.4.1–3) are satisfied at $t = 0$ is to neglect terms of O(ε). This implies that at $t = 0$,

$$v = \frac{1}{f_0}\frac{\partial \Phi}{\partial x}, \qquad u = -\frac{1}{f_0}\frac{\partial \Phi}{\partial y}, \qquad \omega = 0 \tag{7.4.5}$$

These conditions are the geostrophic initial conditions derived in Chapter 6. Because $\omega = 0$, $\mathbf{v}_\chi = 0$, and the initial windfields are purely rotational,

$$u = \frac{\partial \psi}{\partial y}, \qquad v = \frac{\partial \psi}{\partial x}, \qquad \Phi = f_0 \psi \tag{7.4.6}$$

These are the lowest-order quasi-geostrophic initial conditions, and, as demonstrated in Section 7.2, they are far from satisfactory in the initialization of the Hinkelmann–Phillips model.

The next-order quasi-geostrophic initial conditions can be derived as follows. This time we demand that the second time derivatives $\partial^2 u/\partial t^2$, $\partial^2 v/\partial t^2$, and $\partial^2 \Phi/\partial t^2$ are also of O(1). Substitute into (7.3.18–21) for the functions G_u, G_v, and G_Φ defined in (7.4.1–3). Thus,

$$\frac{\partial u}{\partial t} + G_u + \mathbf{v}\cdot\nabla u + \varepsilon G_\Phi \frac{\partial u}{\partial P} - \beta y v = 0 \tag{7.4.7}$$

$$\frac{\partial v}{\partial t} + G_v + \mathbf{v}\cdot\nabla v + \varepsilon G_\Phi \frac{\partial v}{\partial P} + \beta y u = 0 \tag{7.4.8}$$

$$D + \varepsilon \frac{\partial G_\Phi}{\partial P} = 0 \tag{7.4.9}$$

$$\frac{\partial^2 \Phi}{\partial P\, \partial t} + G_\Phi \tilde{\Gamma} + \mathbf{v}\cdot\nabla \frac{\partial \Phi}{\partial P} + \frac{\varepsilon G_\Phi}{P}\frac{\partial}{\partial P}\left(P\frac{\partial \Phi}{\partial P} - \kappa\Phi\right) = 0 \tag{7.4.10}$$

Now, if (7.4.7–10) are differentiated with respect to time, expressions for $\partial^2 u/\partial t^2$, $\partial^2 v/\partial t^2$, and $\partial^2 \Phi/\partial t^2$ will be obtained. These expressions will contain $\partial G_u/\partial t$, $\partial G_v/\partial t$, and $\partial G_\Phi/\partial t$. If $\partial^2 u/\partial t^2$, $\partial^2 v/\partial t^2$, and $\partial^2 \Phi/\partial t^2$ are to be of O(1), then $\partial G_u/\partial t$, $\partial G_v/\partial t$, and $\partial G_\Phi/\partial t$ also have to be of O(1). Expressions for the time derivatives of G_u, G_v, and G_Φ can be obtained by differentiating (7.4.1–3) with respect to time. Thus,

$$\varepsilon \frac{\partial^2 G_u}{\partial P\, \partial t} = \frac{\partial^3 \Phi}{\partial x\, \partial P\, \partial t} - f_0 \frac{\partial^2 v}{\partial P\, \partial t} = -\frac{\partial}{\partial x}\left[G_\Phi \tilde{\Gamma} + \mathbf{v}\cdot\nabla \frac{\partial \Phi}{\partial P} + \varepsilon \frac{G_\Phi}{P}\frac{\partial}{\partial P}\left(P\frac{\partial \Phi}{\partial P} - \kappa\Phi\right)\right]$$
$$+ f_0 \frac{\partial}{\partial P}\left[G_v + \mathbf{v}\cdot\nabla v + \varepsilon G_\Phi \frac{\partial v}{\partial P} + \beta y u\right] \tag{7.4.11}$$

$$\varepsilon \frac{\partial^2 G_v}{\partial P\, \partial t} = \frac{\partial^3 \Phi}{\partial y\, \partial P\, \partial t} + f_0 \frac{\partial^2 u}{\partial P\, \partial t} = -\frac{\partial}{\partial y}\left[G_\Phi \tilde{\Gamma} + \mathbf{v}\cdot\nabla \frac{\partial \Phi}{\partial P} + \varepsilon \frac{G_\Phi}{P}\frac{\partial}{\partial P}\left(P\frac{\partial \Phi}{\partial \Phi} - \kappa\Phi\right)\right]$$
$$- f_0 \frac{\partial}{\partial P}\left[G_u + \mathbf{v}\cdot\nabla u + \varepsilon G_\Phi \frac{\partial u}{\partial P} - \beta y v\right] \tag{7.4.12}$$

$$\varepsilon\frac{\partial^2 G_\Phi}{\partial P\,\partial t}=\frac{\partial^2\omega}{\partial P\,\partial t}=-\frac{\partial D}{\partial t}=-\frac{\partial^2 u}{\partial t\,\partial x}-\frac{\partial^2 v}{\partial t\,\partial y}=\frac{\partial}{\partial x}\left[G_u+\mathbf{v}\cdot\nabla u+\varepsilon G_\Phi\frac{\partial u}{\partial P}-\beta yv\right]$$
$$+\frac{\partial}{\partial y}\left[G_v+\mathbf{v}\cdot\nabla v+\varepsilon G_\Phi\frac{\partial v}{\partial P}+\beta yu\right] \qquad (7.4.13)$$

The simplest way to ensure that $\partial G_u/\partial t$, $\partial G_v/\partial t$, and $\partial G_\Phi/\partial t$ are of O(1) is to neglect terms of O(ε) in (7.4.11–13). Now,

$$\mathbf{v}\cdot\nabla\frac{\partial\Phi}{\partial P}=\mathbf{v}_\psi\cdot\nabla\frac{\partial\Phi}{\partial P}+\varepsilon\mathbf{v}_\chi\cdot\nabla\frac{\partial\Phi}{\partial P}$$

$$\mathbf{v}\cdot\nabla u=\mathbf{v}_\psi\cdot\nabla u_\psi+\varepsilon(\mathbf{v}_\psi\cdot\nabla u_\chi+\mathbf{v}_\chi\cdot\nabla u_\psi)+\varepsilon^2\mathbf{v}_\chi\cdot\nabla u_\chi$$

If all the terms in (7.4.11–13) that are of O(ε), O(ε^2), ... are neglected, then these equations become

$$f_0\frac{\partial}{\partial P}[G_v+\mathbf{v}_\psi\cdot\nabla v_\psi+\beta yu_\psi]-\frac{\partial}{\partial x}\left[G_\Phi\tilde{\Gamma}+\mathbf{v}_\psi\cdot\nabla\frac{\partial\Phi}{\partial P}\right]=0 \qquad (7.4.14)$$

$$f_0\frac{\partial}{\partial P}[G_u+\mathbf{v}_\psi\cdot\nabla u_\psi-\beta yv_\psi]+\frac{\partial}{\partial y}\left[G_\Phi\tilde{\Gamma}+\mathbf{v}_\psi\cdot\nabla\frac{\partial\Phi}{\partial P}\right]=0 \qquad (7.4.15)$$

$$\frac{\partial}{\partial x}[G_u+\mathbf{v}_\psi\cdot\nabla u_\psi-\beta yv_\psi]+\frac{\partial}{\partial y}[G_v+\mathbf{v}_\psi\cdot\nabla v_\psi+\beta yu_\psi]=0 \qquad (7.4.16)$$

These three conditions are not independent. Take $\partial/\partial y$ of (7.4.14) and add $\partial/\partial x$ of (7.4.15). The result is $\partial/\partial P$ of (7.4.16). This result could have been anticipated from the definitions of G_u, G_v, and G_Φ in (7.4.1–3).

Substituting into (7.4.16) for G_u and G_v from (7.4.1–3) gives

$$\frac{\partial}{\partial x}\left[\frac{\partial\Phi}{\partial x}-f_0v_\psi\right]+\frac{\partial}{\partial y}\left[\frac{\partial\Phi}{\partial y}+f_0u_\psi\right]=-\varepsilon\left[\frac{\partial}{\partial x}(\mathbf{v}_\psi\cdot\nabla u_\psi)+\frac{\partial}{\partial y}(\mathbf{v}_\psi\cdot\nabla v_\psi)\right.$$
$$\left.-\beta y\left(\frac{\partial v_\psi}{\partial x}-\frac{\partial u_\psi}{\partial y}\right)+\beta u_\psi\right] \qquad (7.4.17)$$

Now, $\partial v_\psi/\partial x-\partial u_\psi/\partial y=\nabla^2\psi=\zeta$, the vorticity. Also, $f=f_0+\beta y$. Thus, (7.4.17) becomes

$$\nabla^2\Phi-f\zeta=-\varepsilon[\beta u_\psi+\nabla\cdot(\mathbf{v}_\psi\cdot\nabla\mathbf{v}_\psi)] \qquad (7.4.18)$$

This equation is a form of the nonlinear balance equation, and terms of O(ε^2) have been neglected in its derivation. The terms of O(ε) on the right-hand side of (7.4.18), together with the higher-order terms, were the terms neglected in deriving the geostrophic initial conditions (7.4.6). It is clear from (7.4.18) that the use of the geostrophic initial conditions has errors of O(ε).

The unscaled form of the nonlinear balance equation can be obtained by taking the divergence of (7.3.1) and dropping the frictional forcing **F** and all terms involving the divergence D, the vertical motion ω, or the divergent wind $\mathbf{v}_\chi$.

The nonlinear balance equation (7.4.18) relates ψ to Φ. A relationship among ω, ψ, and Φ also exists. This initial condition can be derived by taking $\partial/\partial y$ of (7.4.15) and subtracting $\partial/\partial x$ of (7.4.14). Substitution into the resulting equation for G_u, G_v, and G_Φ from (7.4.1–3) gives

$$\varepsilon^{-1}\tilde{\Gamma}\nabla^2\omega + \nabla^2\left(\mathbf{v}_\psi\cdot\nabla\frac{\partial\Phi}{\partial P}\right) - \varepsilon^{-1}f_0^2\frac{\partial D}{\partial P}$$
$$+ f_0\frac{\partial}{\partial P}\left[\frac{\partial}{\partial y}\left(\mathbf{v}_\psi\cdot\nabla u_\psi - \beta y v_\psi\right) - \frac{\partial}{\partial x}\left(\mathbf{v}_\psi\cdot\nabla v_\psi + \beta y u_\psi\right)\right] = 0 \quad (7.4.19)$$

But

$$\frac{\partial}{\partial y}(\mathbf{v}_\psi\cdot\nabla u_\psi) - \frac{\partial}{\partial x}(\mathbf{v}_\psi\cdot\nabla v_\psi) = -\mathbf{v}_\psi\cdot\nabla\zeta$$
$$\frac{\partial}{\partial y}(\beta y v_\psi) + \frac{\partial}{\partial x}(\beta y u_\psi) = \beta v_\psi + \mathbf{v}_\psi\cdot\nabla\beta y \quad (7.4.20)$$

Thus,

$$\tilde{\Gamma}\nabla^2\omega + f_0^2\frac{\partial^2\omega}{\partial P^2} = \varepsilon\left\{-\nabla^2\left[\mathbf{v}_\psi\cdot\nabla\frac{\partial\Phi}{\partial P}\right] + f_0\frac{\partial}{\partial P}[\mathbf{v}_\psi\cdot\nabla(\zeta + \beta y)]\right\} \quad (7.4.21)$$

Equation (7.4.21) is the scaled form of the quasi-geostrophic omega equation and expresses a relationship among ω, Φ, and ψ. Through the use of (7.3.20), the divergent wind $\mathbf{v}_\chi$ can thus be related to ψ and Φ. Higher-order quasi-geostrophic initial conditions can be derived by considering the third time derivative. In general, this procedure leads to more accurate but less tractable constraints.

Quasi-geostrophic initial conditions can also be obtained for the shallow water equations (6.4.1–3) using the bounded derivative procedure. The initial conditions equivalent to (7.4.18) and (7.4.21) for the shallow water system are

$$\nabla^2\Phi - f\zeta = -\varepsilon[\beta u_\psi + \nabla\cdot(\mathbf{v}_\psi\cdot\nabla\mathbf{v}_\psi)] \quad (7.4.22)$$
$$\tilde{\Phi}\,\nabla^2 D - f_0^2 D = \varepsilon[-\nabla^2(\mathbf{v}_\psi\cdot\nabla\Phi) + f_0\mathbf{v}_\psi\cdot\nabla(\zeta + \beta y)] \quad (7.4.23)$$

Equations (7.4.22–23) can be deduced quite easily by noting that in the shallow water equations $\tilde{\Phi}$ plays the role of $\tilde{\Gamma}$ and that D plays the role of ω in the primitive equations. Equation (7.4.22) is exactly the same as (7.4.18) because the equations of motion in the two systems are the same except for terms such as $\omega(\partial v/\partial P)$ in the primitive equations that are of $O(\varepsilon^2)$ or greater and thus do not appear in (7.4.18). Equation (7.2.3) is a simplified form of (7.4.23) appropriate for the Hinkelmann–Phillips model.

7.5 The linear balance equation

The simplest quasi-geostrophic initialization constraints are given by (7.4.5–6) which have errors of $O(\varepsilon)$. These linear equations could be used in a variety of ways to initialize a primitive equations model. Geopotential constrained, rotational wind constrained, or potential vorticity constrained initialization (as defined in Chapter 6) are all possible. Traditionally, however, only the geopotential constrained or rotational wind constrained methods have been in common use.

In the geopotential constrained procedure, (7.4.5) is used directly to obtain rotational winds from the geopotential field:

$$u_\psi = -\frac{1}{f_0}\frac{\partial \Phi}{\partial y}, \qquad v_\psi = \frac{1}{f_0}\frac{\partial \Phi}{\partial x} \tag{7.5.1}$$

The rotational wind constrained procedure is a little more complex. The method proceeds by appropriately differentiating (7.4.6). Thus,

$$\nabla^2 \Phi = f_0\, \nabla^2 \psi = f_0 \zeta = f_0\left[\frac{\partial v}{\partial x} - \frac{\partial u}{\partial y}\right] \tag{7.5.2}$$

In (7.5.2), u and v are presumed known, and ζ can then be calculated by differentiation; f_0 is regarded as a forcing function or inhomogeneous term. Equation (7.5.2) is an example of an elliptic partial differential equation or elliptic boundary-value problem. The particular form of (7.5.2), $\nabla^2\Phi = G$, where G is known, is called a Poisson equation. To solve (7.5.2) for Φ, it is necessary to know G everywhere. In addition, information is required about Φ on the boundaries of the domain.

Suppose c is a closed curve that defines the boundary of the domain. Then, three possible boundary conditions can be applied for the solution of (7.5.2):

1 Φ is known everywhere on c (Dirichlet condition)
2 $\partial\Phi/\partial n$ is known everywhere on c (Neumann condition)
3 A linear combination of conditions 1 and 2 is known on c (mixed condition)

For condition 2, $\partial\Phi/\partial n$ is the spatial derivative of Φ that is normal to the boundary curve c. The solution of a Poisson equation can be obtained only up to an additive constant, so the horizontal average of Φ over the domain cannot be obtained from (7.5.2). There are a variety of numerical procedures for the solution of (7.5.2) subject to the boundary conditions, and there are many excellent texts on the numerical solution of elliptic boundary-value problems. See, for example, Forsythe and Wasaw (1960) or Schulz (1981).

It is clear that (7.5.1) becomes singular at the equator where f_0 vanishes. Of course, midlatitude scaling was assumed in deriving (7.5.1), so the equations are strictly valid only in the extratropics. From (7.4.18), however, it is possible to derive a linear relationship between Φ and ψ that appears not to suffer from this equatorial singularity. In particular, retaining the β term in (7.4.18) but eliminating the nonlinear term gives a tractable linear relationship between Φ and ψ. The resulting equation (in dimensional form) can be written (using 6.4.7) as

$$\nabla^2\Phi = f\zeta - u_\psi \frac{\partial f}{\partial y} = \nabla \cdot (f\, \nabla\psi) \tag{7.5.3}$$

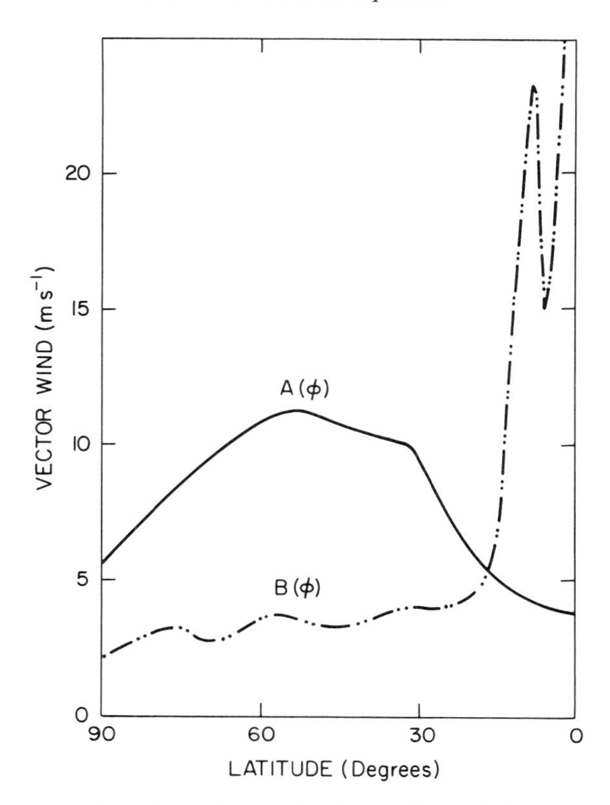

Figure 7.2 The rms error in determining wind using the linear balance equation as a function of latitude, $B(\phi)$. $A(\phi)$ is the square root of the wind variance. (After Daley 1983a)

This equation, known as the linear balance equation, can be used in either geopotential constrained or rotational wind constrained form.

If u and v are known, ζ, ψ, and the right-hand side of (7.5.3) can be calculated and Φ can be determined by solving the Poisson equation. If Φ is known, it appears that (7.5.3) is not singular at the equator because $\partial f/\partial y = 2\Omega\cos(\text{latitude})$ is a maximum at the equator. It is tempting to use (7.5.3) to generate a global nondivergent windfield from a global geopotential field. However, as shown by Eliasen and Machenhauer (1965), Merilees (1968), and Baer (1977a), this is not possible. Their analyses of the linear balance equation demonstrate that any arbitrary geopotential field generates a rotational windfield whose kinetic energy is infinite. Here, the kinetic energy is defined to be the integral of $0.5(u^2 + v^2)$ over the globe.

This result can be seen more graphically in Figure 7.2. Daley (1983a) used the linear balance equation (7.5.3) to calculate wind components (u and v) from a global objective analysis of the geopotential. These estimates of u and v were then compared with the objectively analyzed u and v. The high-quality GWE (see Chapter 1) objective analyses were used for the comparison. In Figure 7.2, two quantities are plotted as a function of latitude (only the northern hemisphere is shown):

$$A(\phi) = \left\{\frac{1}{2\pi}\int_0^{2\pi} [u_0^2 + v_0^2]\, d\lambda\right\}^{1/2}$$

$$B(\phi) = \left\{\frac{1}{2\pi}\int_0^{2\pi} [(u_0 - u_1)^2 + (v_0 - v_1)^2]\, d\lambda\right\}^{1/2}$$

where $u_0(\lambda, \phi)$ and $v_0(\lambda, \phi)$ are the observed wind components, $u_1(\lambda, \phi)$ and $v_1(\lambda, \phi)$ the corresponding wind components calculated from (7.5.3), λ is the longitude, and ϕ the latitude. Where (7.5.3) is a good approximation, $B(\phi) \ll A(\phi)$. It is apparent in Figure 7.2 that the linear balance equation is a good approximation poleward of 20 degrees, but the results are meaningless in the tropics. This suggests that the linear balance equation is of little value in the tropics and cannot be used on the global scale.

7.6 The nonlinear balance equation

The nonlinear balance equation given in scaled form by (7.4.18) has been a widely used initialization constraint relating the streamfunction and the geopotential. On an f-plane ($\beta = 0$), (7.4.18) can be written in dimensional form as

$$\nabla^2\Phi = f_0\, \nabla^2\psi + 2[\psi_{xx}\psi_{yy} - \psi_{xy}^2] \tag{7.6.1}$$

where $\psi_{xx} = \partial^2\psi/\partial x^2$, $\psi_{xy} = \partial^2\psi/\partial x\, \partial y$, and so on. For rotational wind constrained initialization (ψ specified), the right-hand side of (7.6.1) can be calculated, resulting in a Poisson equation for Φ. This equation can be solved for Φ by the numerical methods discussed in Section 7.5.

The more interesting case is when Φ is specified (geopotential constrained case), resulting in a nonlinear equation in the unknown ψ. Equation (7.6.1) is then known as a Monge–Ampere equation.

Rather than analyze (7.6.1), it is helpful to follow the procedure of Kasahara (1982b), which uses a more general equation. Consider the equation of motion (7.3.1) in pressure coordinates on a midlatitude f-plane. Taking the horizontal divergence, we obtain

$$f_0\zeta - \nabla\cdot[\mathbf{v}_\psi\cdot\nabla\mathbf{v}_\psi] = \nabla^2\Phi + R \tag{7.6.2}$$

where

$$R = \left[\frac{\partial D}{\partial t} + \nabla\cdot(\mathbf{v}_\psi\cdot\nabla\mathbf{v}_\chi + \mathbf{v}_\chi\cdot\nabla\mathbf{v}_\psi + \mathbf{v}_\chi\cdot\nabla\mathbf{v}_\chi) + \nabla\cdot\omega\left(\frac{\partial\mathbf{v}_\psi}{\partial P} + \frac{\partial\mathbf{v}_\chi}{\partial P}\right) + \omega\frac{\partial D}{\partial P}\right] - \nabla\cdot\mathbf{F}$$

Then (7.6.2) can be written as

$$f_0\, \nabla^2\psi + 2[\psi_{xx}\psi_{yy} - \psi_{xy}^2] = \nabla^2\Phi + R \tag{7.6.3}$$

which is (7.6.1) with a residual term R. As mentioned earlier, (7.6.3) is a Monge–

Ampere equation, which can be written in the general form

$$a\psi_{xx} + 2b\psi_{xy} + c\psi_{yy} + \psi_{xx}\psi_{yy} - \psi_{xy}^2 = e \tag{7.6.4}$$

where

$$a = c = \frac{f_0}{2}, \qquad b = 0, \qquad e = \frac{\nabla^2\Phi + R}{2}$$

From the theory of Monge–Ampere equations (see Courant and Hilbert 1962), there exist at most two solutions of (7.6.4) as a boundary-value problem if a, b, c, and e are continuous functions of x and y in a closed domain and satisfy the inequality

$$ac - b^2 + e > 0$$

From the definitions of a, b, c, and e, this implies that

$$\nabla^2\Phi + \frac{f_0^2}{2} + R > 0 \tag{7.6.5}$$

or alternatively, using Equation (7.6.3),

$$\left(\frac{f_0}{2} + \psi_{xx}\right)\left(\frac{f_0}{2} + \psi_{yy}\right) - \psi_{xy}^2 > 0 \tag{7.6.6}$$

Consider first the case $R = 0$, that is, (7.6.1). Then,

$$M = \nabla^2\Phi + 0.5f_0^2 > 0 \tag{7.6.7}$$

Inequality (7.6.7), known as the ellipticity condition, is a condition on the specified geopotential that must be satisfied before a solution ψ can be obtained from (7.6.1). M is a measure of the ellipticity, which is positive in elliptic regions and negative in nonelliptic regions. Because $0.5f_0^2 > 0$, $\nabla^2\Phi$ cannot be large and negative. In other words, in a very intense high-pressure region ($\nabla^2\Phi < 0$), it is possible (particularly at low latitudes where f_0 is small) that (7.6.1) will have no solution for ψ.

Inequality (7.6.6) is an a posteriori condition on the solution ψ rather than on the specified geopotential, and is obviously less useful. However, the first term in (7.6.6) must be positive, so $(0.5f_0 + \psi_{xx})$ and $(0.5f_0 + \psi_{yy})$ must both be positive or both negative. This implies that $\psi_{xx} + \psi_{yy} + f_0 = \nabla^2\psi + f_0$ (the absolute vorticity) must be positive everywhere or negative everywhere (see Bolin 1955 or Arnason 1958). For synoptic scale motions in the northern hemisphere, it is appropriate to choose $\nabla^2\psi + f_0 > 0$ and in the southern hemisphere $\nabla^2\psi + f_0 < 0$.

The ellipticity condition (7.6.6) is more easily understood in a simpler context. Consider the equations of motion (7.3.1) in cylindrical coordinates (r, ϕ, P), where r is the radial coordinate pointing outwards, ϕ the tangential coordinate pointing counterclockwise, and P the pressure coordinate. Assume axisymmetry, that is, all variables are independent of ϕ. Also make the midlatitude f-plane assumption and neglect the forcing term **F**. In this case, (7.3.1) can be written (see Godske et al. 1957 and Section 6.6) as

$$\frac{\partial u}{\partial t} + u\frac{\partial u}{\partial r} - \frac{v^2}{r} + \frac{\partial \Phi}{\partial r} - f_0 v + \omega\frac{\partial u}{\partial P} = 0 \qquad \text{(radial)} \tag{7.6.8}$$

$$\frac{\partial v}{\partial t} + u\frac{\partial v}{\partial r} + \frac{uv}{r} + f_0 u + \omega\frac{\partial v}{\partial P} = 0 \qquad \text{(tangential)} \tag{7.6.9}$$

where u is the radial wind component, v the tangential wind component, and ω, Φ, and so on are defined as in (7.3.1).

A streamfunction ψ and velocity potential χ can be defined that are independent of ϕ. From (6.6.2–3),

$$u = \frac{\partial \chi}{\partial r}, \qquad v = \frac{\partial \psi}{\partial r}$$

$$\begin{aligned} D = \nabla\cdot\mathbf{v} &= \frac{1}{r}\frac{\partial}{\partial r}(ru) + \frac{1}{r}\frac{\partial v}{\partial \phi} = \frac{1}{r}\frac{\partial}{\partial r}\left(r\frac{\partial \chi}{\partial r}\right) = \nabla^2\chi \\ \zeta = \mathbf{k}\cdot\nabla\times\mathbf{v} &= \frac{1}{r}\frac{\partial}{\partial r}(rv) - \frac{1}{u}\frac{\partial u}{\partial \phi} = \frac{1}{r}\frac{\partial}{\partial r}\left(r\frac{\partial \psi}{\partial r}\right) = \nabla^2\psi \end{aligned} \tag{7.6.10}$$

Taking the horizontal divergence of (7.6.8–9) yields

$$f_0\,\nabla^2\psi + \frac{1}{r}\frac{\partial}{\partial r}\left(\frac{\partial \psi}{\partial r}\right)^2 - \nabla^2\Phi = R \tag{7.6.11}$$

where

$$R = \frac{\partial D}{\partial t} + \frac{1}{r}\frac{\partial}{\partial r}\left(ru\frac{\partial u}{\partial r}\right) + \frac{1}{r}\frac{\partial}{\partial r}\left(r\omega\frac{\partial u}{\partial P}\right)$$

Now R consists entirely of terms that depend on the divergent wind (D, or ω or u). Setting $R = 0$ in (7.6.11) yields the analogue of (7.6.1) appropriate for this geometry. Equation (7.6.11) can then be rewritten as

$$\frac{1}{r}\frac{\partial}{\partial r}r\left[f_0 v + \frac{v^2}{r} - \Phi_r\right] = 0 \tag{7.6.12}$$

where $\Phi_r = \partial\Phi/\partial r$. The term inside the square brackets will be recognized as the gradient wind equation which is derived and discussed in Holton (1972). Thus, in this simple geometry, the nonlinear balance equation (7.6.1) is reduced to the gradient wind equation. The gradient wind equation can be treated as a quadratic algebraic equation, which can be solved for v:

$$v = \frac{rf_0}{2} \pm \sqrt{\frac{r^2 f_0^2}{4} + r\Phi_r} \tag{7.6.13}$$

For (7.6.13) to have a real solution, the term under the radical must be positive:

$$\frac{r^2 f_0^2}{4} + r\Phi_r > 0 \tag{7.6.14}$$

Taking $(1/r)(\partial/\partial r)$ of inequality (7.6.14) gives

$$\frac{1}{r}\frac{\partial}{\partial r} r \frac{\partial \Phi}{\partial r} + \frac{f_0^2}{2} = \nabla^2 \Phi + \frac{f_0^2}{2} > 0 \tag{7.6.15}$$

which is the ellipticity condition (7.6.7). Thus, the ellipticity condition is related to the existence of a real solution to the gradient wind equation. The gradient wind equation has no solution for the windfield in regions of intense high pressure, and neither does the nonlinear balance equation.

Traditionally, for actual applications, given a geopotential field Φ that violates (7.6.7) in some region, the geopotential field was smoothed locally until (7.6.7) was satisfied everywhere in the domain. A streamfunction ψ could then be obtained from (7.6.1) by an iterative process (see, e.g., Arnason 1958). Figure 7.3 shows an example of solving the nonlinear balance equation for the streamfunction, taken from Bolin (1956). Part (a) shows the 500 mb geopotential field in dekameters; part (b) shows the streamfunction ψ in units of $10^5 \mathrm{m}^2 \mathrm{s}^{-1}$ calculated from a nonlinear balance equation similar to (7.6.1). (The beta term was included in Bolin's calculation.) Also plotted in (b) are a few observed wind arrows. The observed wind directions and speeds are consistent with the calculated streamfunction fields.

The violation of the ellipticity condition (7.6.7) is fairly frequent in the atmosphere (Paegle and Paegle 1976; Kasahara 1982b). Figure 7.4 (Kasahara 1982b) plots the ellipticity measure M at 200 mb as a function of latitude and longitude for 00 GMT 13 January 1979 (from the GWE data set described in Chapter 1). Stippled areas indicate $M < 0$. It is apparent that the nonelliptic regions are mostly confined to the tropics and subtropics, but some appear even in midlatitudes.

Because the nonlinear balance equation (7.6.1) is derived for a midlatitude f-plane, it is not surprising that it is incapable of describing some observed tropical flows. Kasahara (1982b) has investigated this aspect as follows. Using the high-quality GWE analyses, it is possible to estimate the quantity R defined by (7.6.2) as a function of latitude, longitude, and pressure. The same is done with M (7.6.7), and the results are averaged zonally over 14 tropical latitudes between 20° south and 20° north. The averaged values of M and R are defined by

$$\bar{M} = \frac{1}{2\pi}\int_0^{2\pi} M \, d\lambda \qquad \text{and} \qquad \bar{R} = \frac{1}{2\pi}\int_0^{2\pi} R \, d\lambda$$

$\bar{M}$ and $\bar{R}$ are plotted in Figure 7.5. It is apparent that $\bar{M}$ is negative at a number of latitudes and/or pressure levels. However, $\bar{M} + \bar{R}$ is virtually always positive. Thus, condition (7.6.5), which is more general than (7.6.7), is not violated in the atmosphere, at least in the zonal mean. This result suggests that nonelliptic areas ($M < 0$) exist

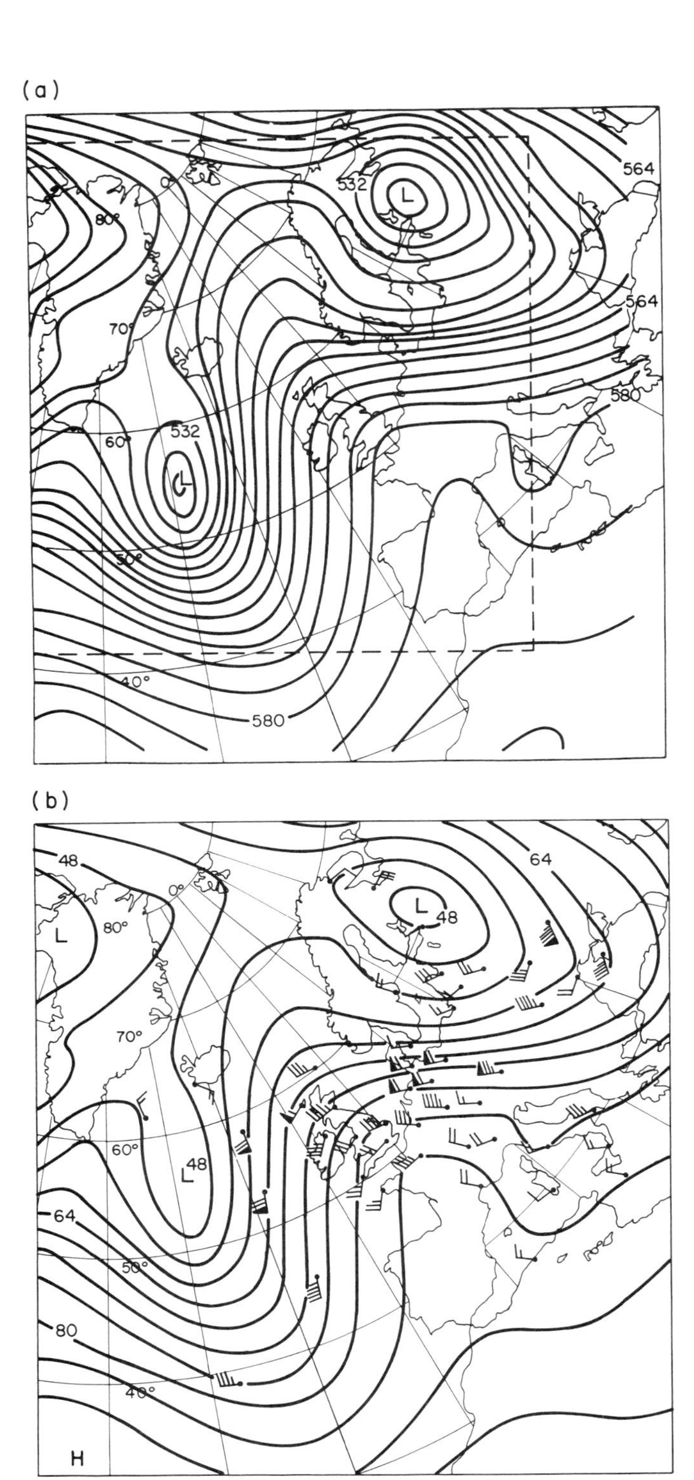

Figure 7.3 Determination of the streamfunction from the geopotential using the nonlinear balance equation: (a) the 500 mb geopotential and (b) the streamfunction. A barb on the wind arrows in (b) denotes 10 knots, and a triangle denotes 50 knots. (After Bolin 1956)

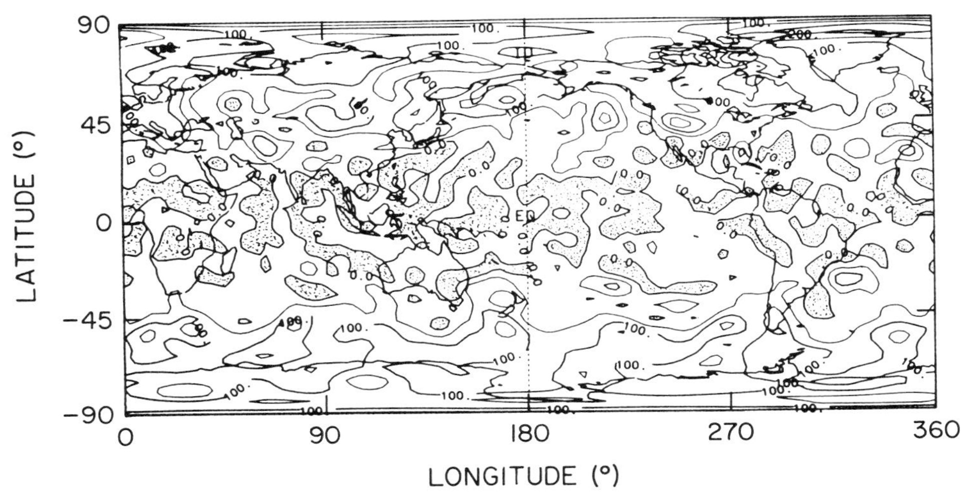

Figure 7.4 Ellipticity measure M as a function of latitude and longitude. Stippled areas indicated $M < 0$. (After Kasahara, *Mon. Wea. Rev.* **110**: 1956, 1982. The American Meteorological Society.)

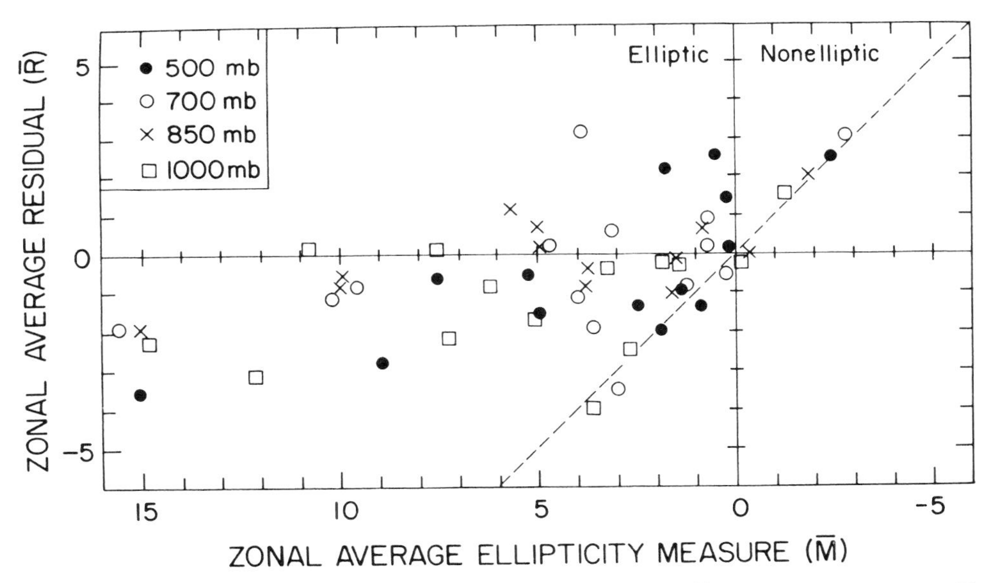

Figure 7.5 Distribution of zonally averaged ellipticity measure $\bar{M}$ and residual measure $\bar{R}$. (After Kasahara, *Mon. Wea. Rev.* **110**: 1956, 1982. The American Meteorological Society.)

in the atmosphere because the terms neglected in deriving the nonlinear balance equation (7.6.7) are not always negligible.

7.7 The quasi-geostrophic omega equation

The quasi-geostrophic omega equation, written in scaled form in (7.4.21), can be written in dimensional form as follows:

$$\tilde{\Gamma}\,\nabla^2\omega + f_0^2\frac{\partial^2\omega}{\partial P^2} = -\nabla^2\left[\mathbf{v}_\psi\cdot\nabla\frac{\partial\Phi}{\partial P}\right] + f_0\frac{\partial}{\partial P}[\mathbf{v}_\psi\cdot\nabla(\zeta+\beta y)] - f_0\mathbf{k}\cdot\nabla\times\frac{\partial\mathbf{F}}{\partial P} - \frac{\kappa}{P}\nabla^2 Q \tag{7.7.1}$$

where $\mathbf{F}$ is the frictional force and Q the time rate of heating per unit mass, which were defined in (7.3.1) and (7.3.4). The terms $\mathbf{F}$ and Q can be carried through the scaling analysis of Sections 7.3 and 7.4 if suitable scaling assumptions are made (see Kasahara 1982a for representative midlatitude scalings for these terms).

Equation (7.7.1) is an elliptic boundary-value problem in the unknown ω. It requires lateral boundary conditions of either the Dirichlet, Neumann, or mixed type. It also requires boundary conditions on ω at $P = 1000$ mb and $P = 0$. Equation (7.3.7) suggests that $\omega = 0$ at $P = 0$ and $P = 1000$ mb, which correspond to Dirichlet conditions. More general boundary conditions are also possible. The right-hand side of (7.7.1) can be calculated if three-dimensional distributions of ψ and Φ are available. In geopotential constrained initialization, ψ would be calculated from Φ using the linear or nonlinear balance equations before calculating the right-hand side of (7.7.1). The reverse would be true for rotational wind constrained initialization. The form of (7.7.1) is similar to a three-dimensional Poisson equation, and a number of numerical procedures exist for solving (7.7.1) for ω, including both iterative and direct methods. See Haltiner and Williams (1980), Forsythe and Wasaw (1960), and Schulz (1981).

Various physical processes can be parametrized in the terms $\mathbf{F}$ and Q of (7.7.1). In particular, planetary boundary layer forcing (Ekman pumping) and internal dissipation can be parametrized in the frictional term $\mathbf{F}$. Latent heat release in moist processes can be parametrized in the term Q. In addition, topographic effects can be introduced by modifying the lower boundary condition (7.3.7). Detailed discussion of the parametrization of physical processes is not within the scope of this book, but the subject is covered in Haltiner and Williams (1980).

Figure 7.6 demonstrates the use of the quasi-geostrophic omega equation. It is derived from Daley (1966) and shows the solution of the quasi-geostrophic omega equation over a midlatitude limited domain covering North America for 1200 GMT 26 February 1961. Part (a) shows the 1000 mb geopotential field m for this date. Schematic frontal positions are also shown on this chart. A version of the quasi-geostrophic omega equation (7.7.1) was used to obtain vertical motion ω for this date. The approach was basically geopotential constrained and included parametrizations of latent heat release (Smagorinsky 1956), topographic effects, and Ekman

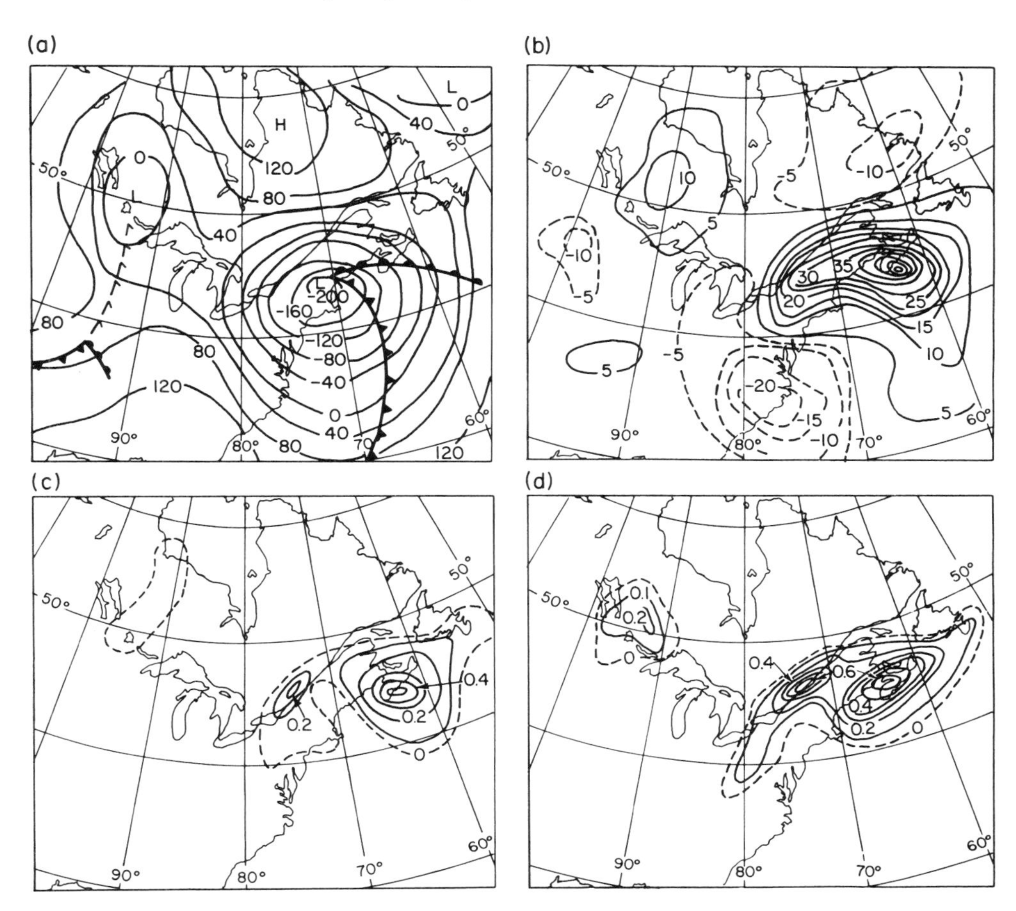

Figure 7.6 Vertical motion (mm/hour) fields (b) and precipitation (in./6 hours) fields (c), calculated using the quasi-geostrophic omega equation; (a) shows the 1000 mb geopotential and (d) the observed precipitation. (After Daley 1966)

pumping (Cressman 1960). The resulting pressure vertical motion ($\omega = dP/dt$) was then converted into true vertical motion ($w = dz/dt$), which is plotted in part (b). It can be seen that the upward vertical motion ($w > 0$, solid contours) is a maximum ahead of the warm front in the Massachusetts low and that there is a maximum descending motion ($w < 0$, dashed contours) behind the cold front.

It is not easy to verify the accuracy of vertical motion fields because of the difficulty of objectively analyzing the divergence field. However, the vertical motion field w can be used to diagnose an instantaneous precipitation rate (Bannon 1948), which can be verified against the observed precipitation rate. Figure 7.6(c) shows the precipitation rate (in the operational units of inches/6 hours) calculated from the w of (b); and (d) shows the corresponding observed precipitation. It is evident that the calculated and observed precipitation rates are in substantial agreement.

The results in Figure 7.6 are probably better than can usually be expected from the quasi-geostrophic omega equation. (Other cases in Daley 1966 are less impressive.) However, this figure does demonstrate that the quasi-geostrophic omega equation is capable of producing synoptically reasonable vertical motion fields (and divergent winds) in midlatitudes. Another perspective on the omega equation is given by Hoskins, Draghici, and Davies (1978).

7.8 The quasi-geostrophic potential vorticity equation

Equations (7.4.18) and (7.4.21) are diagnostic equations that explicitly relate the mass field and the rotational and divergent components of the wind. Mass and wind fields that satisfy these constraints can then be used as initial conditions for the integration of a model based on the primitive equations (7.3.1–4).

The bounded derivative procedure can also be used to produce a prognostic equation that allows the advective time scale τ_2 but not the inertial time scale τ_1. This equation is known as the quasi-geostrophic potential vorticity equation and is derived as follows. Take $\partial/\partial x$ of (7.4.8), subtract $\partial/\partial y$ of (7.4.7), and drop all β terms and all terms of order $\varepsilon, \varepsilon^2, \ldots$. This gives

$$\frac{\partial \zeta}{\partial t} + \frac{\partial G_v}{\partial x} - \frac{\partial G_u}{\partial y} + \frac{\partial}{\partial x}[\mathbf{v}_\psi \cdot \nabla v_\psi] - \frac{\partial}{\partial y}[\mathbf{v}_\psi \cdot \nabla u_\psi] = 0 \tag{7.8.1}$$

where ζ is the vorticity. From (7.4.1–3),

$$\frac{\partial G_v}{\partial x} - \frac{\partial G_u}{\partial y} + f_0 \frac{\partial G_\Phi}{\partial P} = 0 \tag{7.8.2}$$

Using (7.8.2), we can write (7.8.1) as

$$\frac{\partial \zeta}{\partial t} - f_0 \frac{\partial G_\Phi}{\partial P} + \mathbf{v}_\psi \cdot \nabla \zeta = 0 \tag{7.8.3}$$

Take $f_0(\partial/\partial P)\tilde{\Gamma}^{-1}$ of (7.4.10) and drop all terms of order $\varepsilon, \varepsilon^2, \ldots$, then

$$f_0 \frac{\partial^2}{\partial t\, \partial P} \tilde{\Gamma}^{-1} \frac{\partial \Phi}{\partial P} + f_0 \frac{\partial G_\Phi}{\partial P} + f_0 \frac{\partial}{\partial P} \tilde{\Gamma}^{-1}\left[\mathbf{v}_\psi \cdot \nabla \frac{\partial \Phi}{\partial P}\right] = 0 \tag{7.8.4}$$

Adding (7.8.3) to (7.8.4) gives

$$\frac{\partial}{\partial t}\left[\zeta + f_0 \frac{\partial}{\partial P} \tilde{\Gamma}^{-1} \frac{\partial \Phi}{\partial P}\right] + \mathbf{v}_\psi \cdot \nabla \zeta + f_0 \frac{\partial}{\partial P} \tilde{\Gamma}^{-1}\left[\mathbf{v}_\psi \cdot \nabla \frac{\partial \Phi}{\partial P}\right] = 0 \tag{7.8.5}$$

Now apply (7.4.18), dropping all β terms and all terms of order ε and noting that

$$\mathbf{k} \times \nabla \frac{\partial \psi}{\partial P} \cdot \nabla \frac{\partial \psi}{\partial P} = 0$$

which gives

$$\frac{\partial Q}{\partial t} + \mathbf{v}_\psi \cdot \nabla Q = 0 \tag{7.8.6}$$

where

$$Q = \nabla^2\psi + f_0 + f_0^2 \frac{\partial}{\partial P}\left(\tilde{\Gamma}^{-1}\frac{\partial\psi}{\partial P}\right) \tag{7.8.7}$$

is known as the quasi-geostrophic potential vorticity. The symbol for quasi-geostrophic potential vorticity is not to be confused with that used for the heating in (7.3.4).

The quasi-geostrophic potential vorticity is conserved following the horizontal geostrophic flow (ignoring all vertical motion) in the absence of heating and friction. The scaled and real space forms of (7.8.6) are identical because all terms are O(1).

Equation (7.8.6) is a prognostic equation in only one variable: Q. If Q is known everywhere at some time, it can be obtained at some future time using (7.8.6). ψ can be obtained from Q by inverting (7.8.7) subject to boundary conditions at the periphery of the domain, and then Φ and ω (or χ) can be obtained by using (7.4.18) and (7.4.21). Equation (7.8.6) does not permit inertia–gravity oscillations because there is only a single prognostic equation, and Φ and χ are always related diagnostically to ψ. Equation (7.8.6) is a prognostic equation for the geostrophic or Rossby component of the flow and is thus a real-space nonlinear baroclinic analogue to the first equation of (6.4.26).

The shallow water equivalent of (7.8.6–7) is

$$\left[\frac{\partial}{\partial t} + \mathbf{v}_\psi \cdot \boldsymbol{\nabla}\right]\left[\nabla^2\psi + f_0 - \tilde{\Phi}^{-1} f_0^2 \psi\right] = 0 \tag{7.8.8}$$

where $g^{-1}\tilde{\Phi}$ is the height of the free surface. The quasi-geostrophic potential vorticity in this case is the same as that defined in (6.5.7).

A more general potential vorticity equation can be constructed from the shallow water equations (6.4.1–3). A vorticity equation derived from the horizontal motion equations (6.4.1–2) is

$$\frac{\partial\zeta}{\partial t} + \mathbf{v}\cdot\boldsymbol{\nabla}(\zeta + f) + (\zeta + f)\,\boldsymbol{\nabla}\cdot\mathbf{v} = 0 \tag{7.8.9}$$

Combining (7.8.9) with the continuity equation (6.4.3) yields

$$\frac{d}{dt}\left[\frac{\zeta + f}{\Phi}\right] = \left[\frac{\partial}{\partial t} + \mathbf{v}\cdot\boldsymbol{\nabla}\right]\left[\frac{\zeta + f}{\Phi}\right] = 0 \tag{7.8.10}$$

$(\zeta + f)/\Phi$ is a generalized potential vorticity for the shallow water equations and is conserved following the motion in two-dimensional flow. In principle, complete details of the flow field can be obtained by inverting the spatial distribution of $(\nabla^2\psi + f)/\Phi$. However, unlike the geostrophic potential vorticity, the generalized potential vorticity involves both the mass field and the motion field. Moreover, (7.8.10) is derived from the complete shallow water equations and thus involves motion on both advective and inertia–gravity time scales, unlike (7.8.8) which involves only slow motion. To obtain the slowly varying fields from the spatial distribution of $(\nabla^2\psi + f)/\Phi$ requires the imposition of additional constraints such as geostrophic balance (7.4.5), the linear balance equation (7.5.3), the nonlinear balance equation (7.6.1), or the divergence equation (7.4.23).

There is a three-dimensional analogue of (7.8.10) called the conservation of Ertel potential vorticity. It can be shown (Hoskins, McIntyre, and Robertson 1985) that in the absence of heating or friction,

$$\frac{dQ_\varepsilon}{dt} = 0 \tag{7.8.11}$$

where

$$Q_\varepsilon = \rho^{-1}(\boldsymbol{\zeta} + 2\boldsymbol{\Omega})\cdot\boldsymbol{\nabla}_3\theta \tag{7.8.12}$$

is the Ertel potential vorticity. d/dt is a three-dimensional total time derivative (following an air parcel), ρ the density, and θ the potential temperature. $\boldsymbol{\nabla}_3$ is the three-dimensional gradient operator, ζ is the relative vorticity vector, and $\boldsymbol{\Omega}$ the earth's angular rotation vector. The vertical component of $\zeta + 2\boldsymbol{\Omega}$ (see Section 6.4) is simply

$$\mathbf{k}\cdot(\zeta + 2\boldsymbol{\Omega}) = \nabla^2\psi + f$$

There are no approximations in (7.8.11); Q_ε is conserved even for fully three-dimensional, nonhydrostatic motion on the sphere. Charney and Stern (1962) have demonstrated that the variation of quasi-geostrophic potential vorticity Q on a constant pressure surface is approximately proportional to the variation of Q_ε on an isentropic (constant θ) surface. Equation (7.8.11) was derived with no approximations and contains both slow and fast time scales. Thus, the slowly varying fields can be derived from the three-dimensional distribution of Q_ε only with the imposition of side constraints as noted.

7.9 The limitations of quasi-geostrophic initialization

The scale analysis of Section 7.4 assumed normal midlatitude scaling. That is,

$$R_0 = \frac{V_H}{2\Omega L_H}, \qquad \frac{L_H}{a}, \qquad \frac{R_0 L_H^2}{L_R^2} = \frac{2\Omega V_H L_H}{N_0^2 L_Z^2} \tag{7.9.1}$$

are $O(\varepsilon)$, where $\varepsilon = O(0.1)$. The quasi-geostrophic procedures described in this chapter are capable, in principle, of successfully initializing a primitive equation model when the (7.9.1) are satisfied. That is, a primitive equation model integrated from the quasi-geostrophically constrained initial fields would exhibit little high-frequency variance under these conditions. This is most likely to happen in mid or high latitudes for horizontal scales that are close to or smaller than the Rossby radius of deformation. Primitive equation models contain a number of vertical scales, however, and it is clear from (7.9.1) and also from results with the Hinkelmann–Phillips model (Section

7.2) that even in midlatitudes the quasi-geostrophic initialization procedures are less successful with the smaller vertical scales.

Can the quasi-geostrophic initialization procedures be applied when the (7.9.1) are not strictly satisfied? What about the cases of tropical flow (f small) or global scales (L_H large)? Sections 7.5 and 7.6 indicated that the linear and nonlinear balance equations both had serious problems at low latitudes. The rotational wind constrained forms of these equations, though solvable, do not give very accurate results in the tropics. Thus, the quasi-geostrophic initialization procedures have the best chance of success in midlatitudes. This naturally requires that the primitive equation model be run over a bounded domain, and lateral boundary conditions are required. If the lateral boundary conditions in the primitive equation model are inconsistent with boundary conditions used in the linear or nonlinear equations and/or the omega equation, then the initialization may be unsuccessful.

One of the main limitations of the quasi-geostrophic initialization procedures, as traditionally applied, are that they are either geopotential or rotational wind constrained. In fact, most applications of these procedures have tended to be geopotential constrained. Geopotential or rotational wind constrained initialization procedures do not necessarily make best use of the available information, which could well be a mixture of wind and mass data. Procedures that do make use of both wind and mass information are described in the next chapter.

Exercises

7.1 Show that the initial conditions (7.2.2) are equivalent to demanding $dD/dt = d^2D/dt^2 = 0$ at $t = 0$, for (7.1.2–4). Here D is the (horizontal) divergence and d/dt the total time derivative. Apply the initial condition $dD/dt = 0$ to (6.4.1–2) and find a relation among h, u, and v. Compare this relation with the nonlinear balance equation (7.6.1). Are they exactly equivalent?

7.2 To derive solution (7.1.22), $\hat{\psi}(t)$ in (7.1.11) was expanded in a Taylor series in time about $t = 0$, and only the first term $\hat{\psi}(0)$ was retained. Now include the next term in the series, $\hat{\psi}(t) = \hat{\psi}(0) + t(d\hat{\psi}/dt)|_{t=0}$, and derive a more accurate solution of the form (7.1.22) using Laplace transforms. (The problem is solvable by the method of residues, Section 13.4.) Find the initial conditions that generate no high-frequency oscillations and show that these initial conditions are also obtained by setting $d^2\hat{\chi}/dt^2 = 0$ and $(d^2/dt^2)[f_0\hat{\psi} - \hat{\Phi}] = 0$ at $t = 0$ in (7.1.9–11). Derive the corresponding real space relationships among ψ, χ, and Φ that must be satisfied at $t = 0$.

8

Variational procedures

The development of the variational calculus in the seventeenth and eighteenth century was motivated by the need to find the largest or smallest values of rapidly varying quantities. One problem was to find the maximum range of a projectile under the influence of gravity and air resistance, as a function of the elevation angle of launch. Another early problem was to find the least water resistance that can be achieved by varying the shape of an object propelled through the water. Many early applications of this calculus were to the problems of classical mechanics in which it provided an attractive alternative to the procedures of Leibniz and Newton. The appeal of variational procedures is that they consider a system as a whole and do not deal explicitly with the individual components of the system. Thus, in principle, it is possible to derive the behavior of a system without knowing the details of all the interactions among the various subcomponents.

The variational calculus has been employed to advantage in several branches of atmospheric science. In particular, it has stimulated both the theory and the practice of objective analysis and initialization. The variational procedures discussed in this chapter are closely related to the least squares estimation theory of Chapter 2. The emphasis in Chapter 2 was on spatial interpolation of data from an irregular observation network to a regular analysis grid, subject to (at most) simple constraints. In this chapter, the primary concern is initialization and four-dimensional data assimilation; consequently, there is more interest in temporal aspects and the imposition of more complex dynamic constraints. Spatial aspects will usually be treated in continuous rather than discrete fashion.

The emphasis here is on the development of the basic tools of the variational calculus and on straightforward applications to meteorological problems. In this way, we construct a firm foundation for the more sophisticated discussions of Chapter 13. Sections 8.1 and 8.2 briefly review the concept of stationary values of functions and definite integrals. Simple meteorological applications are discussed in Section 8.3 and the theory of weak constraints in Section 8.4. Section 8.5 covers inequality constraints, and Section 8.6 introduces four-dimensional variational analysis.

8.1 The stationary value of a function

The variational calculus involves the determination of stationary points (extrema) of integral expressions known as *functionals*. Before introducing functionals, we briefly review the stationary values of functions. For a more rigorous and complete discussion of the general theory, the reader is referred to Lanczos (1970), Courant and Hilbert (1962), or any standard treatise in mathematical physics.

Consider the function $f(x, y)$, where x and y are the independent variables. This function f is said to have a *stationary* value at the point (x_0, y_0) if, in the infinitesimal neighborhood surrounding the point, the rate of change of the function in every possible direction from that point is zero. Thus, if we proceed an infinitesimal distance in any arbitrary direction on the x-y plane away from (x_0, y_0), the rate of change of the function f must be zero in order that $f(x_0, y_0)$ be a stationary value.

The concept of the stationary value can be examined by using the variational operator δ, first introduced by Lagrange. This operator is in many ways similar to the ordinary differential operator d, but there are subtle differences. The differential operator d refers to *actual* infinitesimal displacements whereas the variational operator δ refers to *virtual* infinitesimal displacements. The distinction between actual and virtual displacements is subtle but is made clearer by the following simple example. Consider a marble that is at rest at the lowest point of a bowl. The actual displacement of the marble is zero. However, we want to find out how the potential energy of the marble changes when the marble is brought to a neighboring position. An exploratory displacement of this nature is virtual and is called a *variation* of the position.

The operator δ is used to explore the neighborhood of the point (x_0, y_0). Suppose a small virtual displacement is made from the point (x_0, y_0). This will be called $(\delta x, \delta y)$. The change of the function δf can be expressed as

$$\delta f = \frac{\partial f}{\partial x}\,\delta x + \frac{\partial f}{\partial y}\,\delta y \tag{8.1.1}$$

δf is called the first variation of the function f. δx and δy can be written in terms of their directional cosines:

$$\delta x = \varepsilon\alpha_x, \qquad \delta y = \varepsilon\alpha_y$$

where α_x and α_y are the directional cosines of the virtual direction in which we have proceeded, and ε is a small parameter that tends to zero. The rate of change of the function in the specified direction now becomes

$$\frac{\delta f}{\varepsilon} = \frac{\partial f}{\partial x}\alpha_x + \frac{\partial f}{\partial y}\alpha_y \tag{8.1.2}$$

For (x_0, y_0) to be a stationary value, $\delta f/\varepsilon$ must vanish for any virtual displacement.

$\delta f/\varepsilon$ must vanish regardless of the choice of the direction of the virtual displacement (i.e., independently of α_x and α_y). Thus,

$$\frac{\partial f}{\partial x} = 0, \qquad \frac{\partial f}{\partial y} = 0 \tag{8.1.3}$$

at a stationary point. The condition that all partial derivatives of the function f vanish at a point is a necessary and sufficient condition for the function f to have a stationary value at that point.

A stationary value of the function f may also be an extremum of f. If the value $f(x_0, y_0)$ is a stationary value of f, then this is a necessary but not sufficient condition for $f(x_0, y_0)$ to be an extremum of f. If, in an infinitesimal neighborhood around (x_0, y_0), f is everywhere greater than $f(x_0, y_0)$, then $f(x_0, y_0)$ is said to be a local minimum. Conversely, if in an infinitesimal neighborhood around (x_0, y_0), f is everywhere less than $f(x_0, y_0)$, then $f(x_0, y_0)$ is said to be a local maximum. In these two cases, the stationary value is also an extremum; in any other case, $f(x_0, y_0)$ is a stationary value but not an extremum. An example of this latter situation is a saddle point, where $\partial f/\partial x = \partial f/\partial y = 0$, but some values of f in the neighborhood of (x_0, y_0) are greater than $f(x_0, y_0)$ and others are smaller.

Second derivatives can be used to determine whether a stationary point is a maximum, minimum, or neither. Thus, in one dimension the stationary point $x = x_0$ is a maximum/minimum if

$$\left.\frac{d^2 f}{dx^2}\right|_{x=x_0}$$

is less/greater than zero.

The concept of stationary values and extrema requires some qualification. First of all, it is necessary to specify the domain in which the stationary values or extrema are to be found. The function $\sin(x)$, for example, has an infinite number of stationary values and extrema. In the domain $x_x < x < x_b$, where $x_b = x_a + \pi$, there is never more than one stationary value of $\sin(x)$.

Second, to have a stationary value, the derivatives $\partial f/\partial x$, $\partial f/\partial y$, and so on, must exist everywhere in the domain. Consider the domain $0 \leq x \leq 2a$ where $f(x)$ is defined as

$$\begin{aligned} f(x) &= x, && 0 \leq x \leq a \\ f(x) &= 2a - x, && a \leq x \leq 2a \end{aligned} \tag{8.1.4}$$

Clearly, df/dx is not equal to zero anywhere in the domain, and yet $f(a)$ appears to be a maximum. Continuous derivatives are required for the concept of a stationary value to have any meaning.

Many of the problems arising in the variational calculus have constraints. Suppose a stationary value is sought for the function $f(x, y)$ subject to the auxiliary or side condition that

$$g(x, y) = 0 \tag{8.1.5}$$

In this constrained problem, the stationary value is sought in the subspace of the x-y plane where $g(x, y) = 0$. It might be possible to rewrite the constraint (8.1.5) as

$$y = h(x) \tag{8.1.6}$$

In this case, finding a stationary value of $f(x, y)$ is equivalent to finding a stationary value of $f(x, h(x))$. Thus, because the constraint can be written in the form (8.1.6), the constrained two-dimensional problem can be reduced to an unconstrained one-dimensional problem.

It is not always possible to rewrite a constraint (8.1.5) in the form (8.1.6), so an alternative procedure is required. Lagrange's method of undetermined multipliers has already been encountered in Section 4.1. Take the variation of the function $f(x, y)$ and constraint (8.1.5):

$$\delta f = \frac{\partial f}{\partial x}\,\delta x + \frac{\partial f}{\partial y}\,\delta y \tag{8.1.7}$$

$$\delta g = \frac{\partial g}{\partial x}\,\delta x + \frac{\partial g}{\partial y}\,\delta y = 0 \tag{8.1.8}$$

Form the new function $f_1 = f + \lambda g$, where λ is an as yet undetermined function. Take the variation of f_1,

$$\delta f_1 = \delta(f + \lambda g) = \delta f + \lambda\,\delta g + g\,\delta\lambda = \delta f \tag{8.1.9}$$

because the operator δ obeys the same rules of multiplication and addition as the differential operator d. The function $g = 0$ from (8.1.5) and $\delta g = 0$ from (8.1.8). δf_1, considered as a function of x, y, and λ, can be written by (8.1.1) as

$$\delta f_1 = \frac{\partial f_1}{\partial x}\,\delta x + \frac{\partial f_1}{\partial y}\,\delta y + \frac{\partial f_1}{\partial \lambda}\,\delta\lambda = \left(\frac{\partial f}{\partial x} + \lambda\frac{\partial g}{\partial x}\right)\delta x + \left(\frac{\partial f}{\partial y} + \lambda\frac{\partial g}{\partial y}\right)\delta y \tag{8.1.10}$$

as $\partial f_1/\partial\lambda = g = 0$.

A stationary point of f occurs when $\delta f = 0$. From (8.1.9), a stationary point of f_1 is also a stationary point of f. Thus, the condition for a stationary point of f_1, subject to the constraint (8.1.5), is

$$\frac{\partial f}{\partial x} + \lambda\frac{\partial g}{\partial x} = 0, \qquad \frac{\partial f}{\partial y} + \lambda\frac{\partial g}{\partial y} = 0 \tag{8.1.11}$$

The two equations of (8.1.11) plus the constraint (8.1.5) can then be solved to find the stationary value. The Lagrange multiplier λ can be interpreted as a measure of the sensitivity of the value of the function f at the stationary point to changes in the constraint (8.1.5).

In N dimensions $(x_1, \ldots, x_N)$, a stationary point of the function $f(x_1, \ldots, x_N)$ subject to the M constraints $g_1(x_1, \ldots, x_N) = 0, \ldots, g_M(x_1, \ldots, x_N) = 0$ can be found in exactly the same manner. The result equivalent to (8.1.11) in this case is

$$\frac{\partial}{\partial x_n}\left(f + \sum_{m=1}^{M} \lambda_m g_m\right) = 0, \qquad 1 \le n \le N \tag{8.1.12}$$

The Lagrange multiplier λ_m measures the sensitivity of the value of f at the stationary point to changes in the mth constraint, all other constraints being held fixed.

Occasionally, the constraint $g(x, y) = 0$ is actually a differential equation. This is known as a *nonholonomic constraint*, but the method of Lagrange multipliers can be applied in this case as well.

All the elementary calculus concepts reviewed in this section have counterparts in the variational calculus to be introduced in the next section.

8.2 The stationary value of a definite integral

I is said to be a functional of the function $u(x)$ in the interval (x_a, x_b) when it depends on all the values $u(x)$, $x_a \leq x \leq x_b$. Two examples of functionals are

$$I\{u(x)\} = \int_{x_a}^{x_b} F(u(x))\,dx$$

$$I\{u(x)\} = \int_{x_a}^{x_b} F(x, u, u'\ u'', \ldots, u^{(n)})\,dx$$

where

$$u' = \frac{du}{dx}, \ldots, u^{(n)} = \frac{d^n u}{dx^n}$$

The domain of a functional is a set of admissible functions rather than a region of coordinate space. For example, the domain of the functional could consist of the set of all positive continuous functions in the interval $x_a \leq x \leq x_b$.

The fundamental problem of the variational calculus can be stated as follows. *In a given domain of admissible functions, find the function $u(x)$ of a given functional I for which $I\{u(x)\}$ is a stationary value.* The result is not necessarily an extremum.

Consider first the so-called simplest problem of the variational calculus. Find the function $u(x)$ that makes the functional

$$I\{u(x)\} = \int_{x_a}^{x_b} F(x, u, u')\,dx \tag{8.2.1}$$

stationary, subject to the boundary conditions $u(x_a) = \alpha$ and $u(x_b) = \beta$. Here α and β are prescribed, and u is assumed to be continuous and have continuous derivatives to the second order. This problem was first solved by Euler using a double limit process. A more elegant and instructive solution to the problem (8.2.1) is due to Lagrange.

Suppose the function $u(x)$ gives a stationary value to the functional $I\{u(x)\}$ of (8.2.1). Consider a slightly modified function,

$$\tilde{u}(x) = u(x) + \varepsilon\eta(x) \tag{8.2.2}$$

where $\eta(x)$ is some arbitrary continuous and differentiable function with $\eta(x_a) = \eta(x_b) = 0$ and ε is a small parameter that tends to zero as in (8.1.2). Figure 8.1 depicts

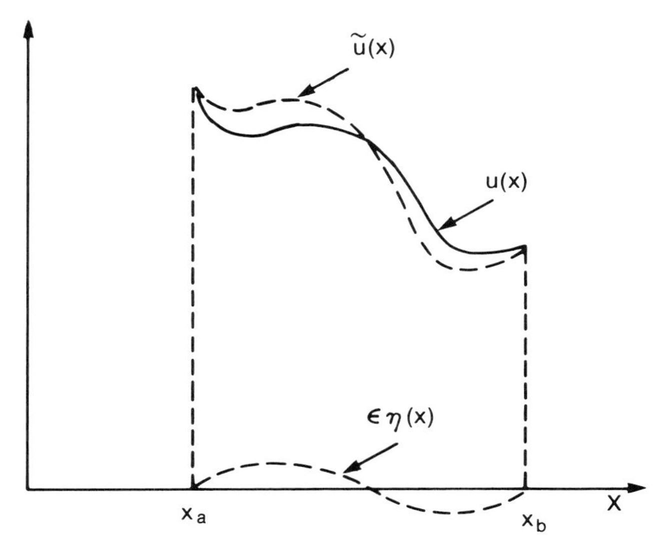

Figure 8.1 Illustration of $u(x)$, $\tilde{u}(x)$, and $\eta(x)$ used in the derivation of the Euler–Lagrange equations.

schematically the functions $u(x)$, $u(\tilde{x})$, and $\varepsilon\eta(x)$. The difference between $u(x)$ and $\tilde{u}(x)$ is called the variation of the function u:

$$\delta u = \tilde{u}(x) - u(x) = \varepsilon\eta(x) \tag{8.2.3}$$

The variation of a function, in full analogy with the variation of the position of a point that was examined in the previous section, is infinitesimal and virtual. That is, the parameter $\varepsilon \to 0$ and the change can be made in an arbitrary manner.

The difference between the variational operator δ and the differential operator d can now be seen more clearly; δu and du are both infinitesimal changes of the function u. However, du refers to the infinitesimal change of the function $u(x)$ caused by the infinitesimal change of the independent variable dx whereas δu is an infinitesimal change of $u(x)$ that produces a new function $u(x) + \delta u(x)$. In the process of variation on the dependent variable, u is varied while the independent variable x is not varied. Thus, δx is always set to zero. Moreover, if $u(x_a)$ and $u(x_b)$ are prescribed, then these two boundary values cannot be varied:

$$\delta x = 0, \qquad \delta u\big|_{x_a} = 0, \qquad \delta u\big|_{x_b} = 0$$

The δ operator commutes with both the normal derivative operator and the integral operator:

$$\frac{d}{dx}(\delta u) = \frac{d}{dx}(\varepsilon\eta(x)) = \varepsilon\frac{d\eta}{dx} = \frac{d\tilde{u}}{dx} - \frac{du}{dx} = \delta\left(\frac{du}{dx}\right) \tag{8.2.4}$$

Similarly, it is straightforward to show that

$$\delta\int_{x_a}^{x_b} F(x, u, u')\,dx = \int_{x_a}^{x_b} \delta F(x, u, u')\,dx \qquad (\text{'}$$

Now consider the variation of the integrand $F(x, u, u')$ of (8.2.1):

$$\delta F(x, u, u') = F(x, u + \delta u, u' + \delta u') - F(x, u, u')$$
$$= F(x, u + \varepsilon\eta, u' + \varepsilon\eta') - F(x, u, u') \tag{8.2.6}$$

Expand F in a Taylor series around u, u', but ignore higher-order terms because ε is small. Then,

$$\delta F(x, u, u') = \varepsilon\left(\frac{\partial F}{\partial u}\eta + \frac{\partial F}{\partial u'}\eta'\right) \tag{8.2.7}$$

Now consider the variation of the integral I of (8.2.1):

$$\delta I = \delta\int_{x_a}^{x_b} F\,dx = \int_{x_a}^{x_b} \delta F\,dx = \varepsilon\int_{x_a}^{x_b}\left(\frac{\partial F}{\partial u}\eta + \frac{\partial F}{\partial u'}\eta'\right)dx$$

Integrate the second term by parts:

$$\delta I = \varepsilon\int_{x_a}^{x_b}\eta\left(\frac{\partial F}{\partial u} - \frac{d}{dx}\frac{\partial F}{\partial u'}\right)dx + \varepsilon\eta(x)\frac{\partial F}{\partial u'}\bigg|_{x_a}^{x_b} \tag{8.2.8}$$

The δI is called the variation or, more precisely, the first variation of I. To get a stationary value of I, by analogy with (8.1.2), $\delta I/\varepsilon$ must vanish. Now $\eta(x) = 0$ at $x = x_a$ and x_b, and $\eta(x)$ is arbitrary, so the integral in (8.2.8) must vanish for any choice of $\eta(x)$. This can only occur if

$$\frac{\partial F}{\partial u} - \frac{d}{dx}\frac{\partial F}{\partial u'} = 0, \qquad x_a \le x \le x_b \tag{8.2.9}$$

Condition (8.2.9) is necessary and sufficient for the vanishing of δI and is commonly called the *Euler–Lagrange* equation. The vanishing of δI is a necessary but not sufficient condition for an extremum of I. A function $u(x)$ that satisfies the Euler–Lagrange equation is called a *stationary function.*

The simplest problem of the variational calculus (8.2.1) can be generalized in a number of ways. Suppose I is a functional of the two functions $u(x)$ and $v(x)$:

$$I\{u(x), v(x)\} = \int_{x_a}^{x_b} F(x, u, u', v, v')\,dx \tag{8.2.10}$$

It is straightforward to show that if I is stationary, there are two Euler–Lagrange equations for $u(x)$ and $v(x)$:

$$\frac{\partial F}{\partial u} - \frac{d}{dx}\frac{\partial F}{\partial u'} = 0 \quad \text{and} \quad \frac{\partial F}{\partial v} - \frac{d}{dx}\frac{\partial F}{\partial v'} = 0 \tag{8.2.11}$$

If the integrand contains higher derivatives of u,

$$I\{u(x)\} = \int_{x_a}^{x_b} F(x, u, u', \ldots, u^{(n)})\,dx \tag{8.2.12}$$

then the Euler–Lagrange equation becomes

$$\frac{\partial F}{\partial u} - \frac{d}{dx}\frac{\partial F}{\partial u'} + \frac{d^2}{dx^2}\frac{\partial F}{\partial u''} - \cdots (-1)^n \frac{d^n}{dx^n}\frac{\partial F}{\partial u^{(n)}} = 0 \tag{8.2.13}$$

If there are two independent variables x and y, then I might take the form

$$I\{u(x, y)\} = \iint_S F(x, y, u, u_x, u_y)\, dx\, dy \tag{8.2.14}$$

where

$$u_x = \frac{\partial u}{\partial x}, \qquad u_y = \frac{\partial u}{\partial y}$$

S is a given domain on the (x, y) plane where u is continuous and has continuous derivatives up to the second order. u is assumed to take on prescribed values on the boundary of the domain S. Then, the Euler–Lagrange equation for (8.2.14) is

$$\frac{\partial F}{\partial u} - \frac{\partial}{\partial x}\left(\frac{\partial F}{\partial u_x}\right) - \frac{\partial}{\partial y}\left(\frac{\partial F}{\partial u_y}\right) = 0 \tag{8.2.15}$$

Analogous expressions to (8.2.15) can be derived for any number of independent variables.

Consider the problem of finding a stationary value of I given in (8.2.1) subject to the auxiliary condition,

$$\int_{x_a}^{x_b} g(x, u, u')\, dx = 0 \tag{8.2.16}$$

Form the new functional,

$$I_1\{u, g, \lambda\} = \int_{x_a}^{x_b} (F + \lambda g)\, dx \tag{8.2.17}$$

where λ is an as yet undetermined Lagrange multiplier. Take the variation of I_1:

$$\delta I_1 = \int_{x_a}^{x_b} (\delta F + \lambda\, \delta g + g\, \delta\lambda)\, dx \tag{8.2.18}$$

Following through the procedure of (8.2.7–9), it is simple to show that the Euler–Lagrange equation of this system is

$$\frac{\partial}{\partial u}(F + \lambda g) - \frac{d}{dx}\frac{\partial}{\partial u'}(F + \lambda g) = 0 \tag{8.2.19}$$

Equation (8.2.19), together with the constraint condition (8.2.16), is sufficient to determine the stationary value of the integral. Note that the multipliers λ, in this case, are constants. If there is more than one constraint, then (8.2.19) involves several multipliers, analogous to (8.1.12). As in Section 8.1, the Lagrange multiplier λ can be interpreted as a measure of the sensitivity of the stationary value of I to change in the constraint (8.2.16).

A slightly different constrained problem is to find a stationary value of I in (8.2.1) subject to the side condition

$$g(x, u, u') = 0 \tag{8.2.20}$$

If the function g depends on u', then (8.2.20) is a differential equation and is thus a nonholonomic constraint. It can be shown that the Euler–Lagrange equation of this system is

$$\frac{\partial}{\partial u}[F + \lambda(x)g] - \frac{d}{dx}\frac{\partial}{\partial u'}[F + \lambda(x)g] = 0 \tag{8.2.21}$$

Note that the Lagrange multiplier λ is now a function of the independent variable.

Up to this point, the boundary conditions of the problem have been largely ignored. One of the features of the variational calculus is that it includes boundary conditions in an elegant fashion. Consider the functional

$$I = \int_{x_a}^{x_b} F(x, u, u', u'')\, dx \tag{8.2.22}$$

Take the first variation,

$$\delta I = \int_{x_a}^{x_b} \delta F\, dx = \int_{x_a}^{x_b} \left(\frac{\partial F}{\partial u''}\delta u'' + \frac{\partial F}{\partial u'}\delta u' + \frac{\partial F}{\partial u}\delta u\right) dx$$

or

$$\delta I = \int_{x_a}^{x_b} \delta u\left(\frac{d^2}{dx^2}\frac{\partial F}{\partial u''} - \frac{d}{dx}\frac{\partial F}{\partial u'} + \frac{\partial F}{\partial u}\right) dx + \left[\delta u'\frac{\partial F}{\partial u''} - \delta u\frac{d}{dx}\frac{\partial F}{\partial u''} + \delta u\frac{\partial F}{\partial u'}\right]\Bigg|_{x_a}^{x_b} \tag{8.2.23}$$

The integrand in (8.2.23) corresponds to (8.2.13) with $n = 2$. The boundary terms vanish if $\delta u = \delta u' = 0$ at $x = x_a$, x_b. This corresponds to specifying u and u' on the boundaries (*imposed boundary conditions*).

However, δI can vanish even if δu and $\delta u'$ are not both equal to zero on the boundaries. If,

$$\frac{\partial F}{\partial u'} - \frac{d}{dx}\frac{\partial F}{\partial u''} = 0 \quad \text{and} \quad \frac{\partial F}{\partial u''} = 0 \quad \text{at} \quad x = x_a, x_b \tag{8.2.24}$$

then the boundary terms also vanish. Conditions (8.2.24) are known as the *natural boundary conditions* of the problem. The natural boundary conditions do not require any specification on the boundaries and are internally consistent with the solution inside the domain. If there is no second derivative term in F, then the natural boundary conditions become

$$\frac{\partial F}{\partial u'} = 0 \quad \text{at} \quad x = x_a, x_b \tag{8.2.25}$$

The natural boundary conditions can be generalized to higher derivatives or more independent variables.

Finally, it should be noted that it is possible to determine if a stationary value of I is also an extremum by calculating the second variation of I. This is denoted symbolically as $\delta^2 I$ and is analogous to the second derivative tests discussed in Section 8.1. This completes the background material on the variational calculus.

8.3 Application to atmospheric analysis problems

The variational calculus provides a powerful conceptual framework for examination of atmospheric analysis and initialization problems. In this section, simple analysis/initialization problems are examined using the variational tools developed in the previous sections.

The examples to be discussed are continuous formulations of time-independent problems. Consider first a three-dimensional problem with two dependent variables. Suppose $u(x, y, P)$ and $v(x, y, P)$ are atmospheric variables (e.g., temperature, geopotential, and wind components) where x, y, and P are the independent variables. Assume that there exist analyzed values of $u_A(x, y, P)$ and $v_A(x, y, P)$. The object is to produce initialized values $u_I(x, y, P)$ and $v_I(x, y, P)$ subject to imposed constraints. That is, u and v must satisfy exactly some condition such as the geostrophic relation (6.5.1), the hydrostatic relation (7.3.2), the continuity equation (7.3.3), or the nonlinear balance equation (7.4.18). The spirit of this approach is to keep the initialized fields close to the observations while exactly satisfying the constraints. Mathematically, the problem can be expressed as follows. Find the stationary values (usually only the minima are of interest) of

$$I = \iint_{PS} \{w_u(u_I - u_A)^2 + w_v(v_I - v_A)^2\}\, dS\, dP \tag{8.3.1}$$

subject to the constraint

$$f_1(u, v) = 0 \tag{8.3.2}$$

Here the integral is over the domain defined by P (pressure) and S (the horizontal domain); w_u and w_v are specified weights that are continuous functions of the independent variables, and f_1 is the externally imposed constraint.

In practice, (8.3.1) would be replaced by a discrete form in which u_A and v_A are specified at the gridpoints of a regular three-dimensional grid $\mathbf{r}_j$, $1 \leq j \leq J$, where $\mathbf{r}_j = (x_j, y_j, P_j)$:

$$\begin{aligned} I &= \sum_{j=1}^{J} \{w_u(\mathbf{r}_j)[u_I(\mathbf{r}_j) - u_A(\mathbf{r}_j)]^2 + w_v(\mathbf{r}_j)[v_I(\mathbf{r}_j) - v_A(\mathbf{r}_j)]^2\} \\ &= [\underline{u}_I - \underline{u}_A]^T \underline{\underline{w}}_u [\underline{u}_I - \underline{u}_A] + [\underline{v}_I - \underline{v}_A]^T \underline{\underline{w}}_v [\underline{v}_I - \underline{v}_A] \end{aligned} \tag{8.3.3}$$

where $\underline{u}_I$ is a column vector of length J with elements $u_I(\mathbf{r}_j)$, and $\underline{\underline{w}}_u$ a diagonal $J \times J$ matrix with elements $w_u(\mathbf{r}_j)$, and with corresponding definitions for $\underline{u}_A$, $\underline{v}_I$, $\underline{v}_A$, and $\underline{\underline{w}}_v$. Equation (8.3.2) would also be replaced by an appropriate discrete form.

The weight functions $\underline{\underline{w}}_u$ and $\underline{\underline{w}}_v$ can be specified arbitrarily. Following the arguments of Section 2.2, a reasonable choice might be

$$w_u(\mathbf{r}_j) = 0.5\langle \varepsilon_u^A(\mathbf{r}_j)^2 \rangle^{-1} \quad \text{and} \quad w_v(\mathbf{r}_j) = 0.5\langle \varepsilon_v^A(\mathbf{r}_j)^2 \rangle^{-1}$$

where $\varepsilon_u^A(\mathbf{r}_j)$ and $\varepsilon_v^A(\mathbf{r}_j)$ are the analysis errors of the variables u and v at the jth gridpoint. Note that functionals of the form (8.3.1) or (8.3.3) allow neither for spatially correlated analysis error nor for correlations between the analysis errors of u and v.

Consider now the formulation and solution of a simple one-dimensional problem. The vertical velocity field is not routinely measured directly, but it can be estimated indirectly from the horizontal divergence via the continuity equation (7.3.3). This so-called kinematic method is simpler than the application of the quasi-geostrophic omega equation (7.7.1). Suppose, in a vertical column, estimates of the divergence field $D_A(P)$ derived from observations or an objective analysis are available. The problem can be posed in variational form (O'Brien 1970) as follows. Minimize

$$I = \int_{P_T}^{P_B} w_D(P)\{D_I(P) - D_A(P)\}^2 \, dP \tag{8.3.4}$$

subject to the constraint

$$D_I + \frac{d\omega_I}{dP} = 0 \tag{8.3.5}$$

$D_I(P)$ is the initialized divergence field, (8.3.5) is the continuity equation (7.3.3), P_B and P_T are the pressure at the bottom and top of the column, respectively, $w_D(P) = 0.5\langle \varepsilon_D^2(P) \rangle^{-1}$ is the weight, and ω_I is the initialized vertical motion. The constraint (8.3.5) can be rewritten as

$$\int_{P_T}^{P_B} D_I \, dP = \omega_T - \omega_B \tag{8.3.6}$$

where ω_T and ω_B are the vertical motion dP/dt at the top and bottom of the column and are assumed to be known.

The constraint (8.3.6) is now in integral form, suggesting the use of a constant (independent of P) Lagrange multiplier as in (8.2.16). Form the new functional:

$$I_1 = \int_{P_T}^{P_B} \left\{ w_D(D_I - D_A)^2 + \lambda\left(D_I + \frac{\omega_B - \omega_T}{P_B - P_T} \right) \right\} dP \tag{8.3.7}$$

λ is the unknown constant Lagrange multiplier; $D_I(P)$ is also unknown, but everything else in (8.3.7) is known. Taking the first variation of I and noting that $\delta\lambda$ is independent of P yields

$$\delta I_1 = \int_{P_T}^{P_B} \delta D_I\{2w_D(P)(D_I - D_A) + \lambda\} \, dP + \delta\lambda \int_{P_T}^{P_B} \left\{ D_I + \frac{\omega_B - \omega_T}{P_B - P_T} \right\} dP \tag{8.3.8}$$

Variation on λ yields the constraint (8.3.6) whereas variation on D_I yields

$$D_I(P) - D_A(P) = -\frac{\lambda}{2w_D(P)} \tag{8.3.9}$$

Note that (8.3.7) contains no derivative of D with respect to pressure, and therefore there are no boundary terms. Integrating (8.3.9) vertically, and using the constraint (8.3.6) gives

$$\lambda = \frac{2\{\int_{P_T}^{P_B} D_A(P)\,dP + \omega_B - \omega_T\}}{\int_{P_T}^{P_B} \frac{dP}{w_D(P)}} \tag{8.3.10}$$

λ is a measure of how well $D_A(P)$ satisfies the condition (8.3.6). Substitute (8.3.10) into (8.3.9), giving

$$D_I(P) = D_A(P) + \alpha(P)\left\{\omega_T - \omega_B - \int_{P_T}^{P_B} D_A(P')\,dP'\right\} \tag{8.3.11}$$

where

$$\alpha(P) = \frac{\langle\varepsilon_D^2(P)\rangle}{\int_{P_T}^{P_B} \langle\varepsilon_D^2(P')\rangle\,dP'}$$

The dummy variable in the integrals of (8.3.11) have been changed from P to P' to avoid confusion. The term in curly brackets in (8.3.11) is a measure of the degree to which the observed divergence field fails to satisfy the continuity equation (7.3.3), and $\alpha(P)$ spreads out the discrepancy through the vertical column. If $\langle\varepsilon_D^2(P)\rangle$ is independent of P,

$$D_I(P) = D_A(P) + \frac{1}{P_B - P_T}\left\{\omega_T - \omega_B - \int_{P_T}^{P_B} D_A(P')\,dP'\right\} \tag{8.3.12}$$

Thus, it is easy to see that the second term in (8.3.12) provides a correction term to the observed divergence $D_A(P)$ so that the initialized divergence satisfies the continuity equation (8.3.6). If $\omega_T = \omega_B = 0$, then (8.3.12) implies that the divergence profile is shifted by a constant value so that the vertical integral of the divergence $D_I(P)$ vanishes. The vertical velocity is obtained by integrating (8.3.5). Exercise 8.1 discusses a discrete version of this problem.

Now consider a two-dimensional problem with three dependent variables and two constraints. On a domain (x, y) with analyzed geopotentials Φ_A and winds $(u_A\ v_A)$, obtain initialized values Φ_I, u_I, and v_I that exactly satisfy the geostrophic relation and are as close as possible to the analyzed fields. The aim is to produce geostrophically balanced winds and geopotentials that are minimally different from the analyzed fields. A continuous form of the problem can be formulated as follows. Minimize the functional

$$I = \int_S \{w_v(u_I - u_A)^2 + w_v(v_I - v_A)^2 + w_\Phi(\Phi_I - \Phi_A)^2\}\,dS \tag{8.3.13}$$

subject to the constraints

$$\frac{\partial \Phi_I}{\partial x} - f_0 v_I = 0, \qquad \frac{\partial \Phi_I}{\partial y} + f_0 u_I = 0 \tag{8.3.14}$$

The weights w_v and w_Φ are assumed to be specified. Now (8.3.13–14) can be solved by Lagrange multipliers using a two-dimensional generalization of (8.2.21). In the present case, however, it is simpler to proceed by direct substitution of (8.3.14) into (8.3.13):

$$I = \int_S \left\{ w_v \left(\frac{1}{f_0} \frac{\partial \Phi_I}{\partial y} + u_A \right)^2 + w_v \left(\frac{1}{f_0} \frac{\partial \Phi_I}{\partial x} - v_A \right)^2 + w_\Phi (\Phi_I - \Phi_A)^2 \right\} dS \tag{8.3.15}$$

Take the first variation δ of (8.3.15) with respect to Φ_I:

$$\delta I = 2 \int_S \left\{ \frac{w_v}{f_0} \left(\frac{1}{f_0} \frac{\partial \Phi_I}{\partial y} + u_A \right) \delta \left(\frac{\partial \Phi_I}{\partial y} \right) + \frac{w_v}{f_0} \left(\frac{1}{f_0} \frac{\partial \Phi_I}{\partial x} - v_A \right) \delta \left(\frac{\partial \Phi_I}{\partial x} \right) + w_\Phi (\Phi_I - \Phi_A)\, \delta \Phi_I \right\} dS \tag{8.3.16}$$

Set (8.3.16) to zero and integrate by parts:

$$\int_S \left\{ w_\Phi (\Phi_I - \Phi_A) - \frac{\partial}{\partial y} \frac{w_v}{f_0} \left(\frac{1}{f_0} \frac{\partial \Phi_I}{\partial y} + u_A \right) - \frac{\partial}{\partial x} \frac{w_v}{f_0} \left(\frac{1}{f_0} \frac{\partial \Phi_I}{\partial x} - v_A \right) \right\} \delta \Phi_I \, dS$$

$$+ \oint_\tau \left\{ \frac{1}{f_0} \frac{\partial \Phi_I}{\partial n} - \boldsymbol{\tau} \cdot \mathbf{v}_A \right\} \delta \Phi_I \, d\tau = 0 \tag{8.3.17}$$

where $\oint_\tau (\ ) \, d\tau$ is a contour integral around the boundaries of the domain S, $\boldsymbol{\tau} \cdot \mathbf{v}_A$ is the tangential component of $\mathbf{v}_A$, and the direction of integration is counterclockwise. $\partial/\partial n$ indicates differentiation in the direction normal to the boundary curve (positive outward).

Now assume that w_Φ and w_v are spatially independent. Then the vanishing of (8.3.17) requires

$$w_\Phi \Phi_I - \frac{w_v}{f_0^2} \nabla^2 \Phi_I = w_\Phi \Phi_A - \frac{w_v}{f_0} \zeta_A \tag{8.3.18}$$

in the interior, and either

$$\delta \Phi_I = 0 \qquad \text{or} \qquad \frac{1}{f_0} \frac{\partial \Phi_I}{\partial n} = \boldsymbol{\tau} \cdot \mathbf{v}_A \tag{8.3.19}$$

on the boundaries. Here, $\zeta_A = \partial v_A / \partial x - \partial u_A / \partial y$ is the analyzed vorticity.

The boundary condition $\delta \Phi_I = 0$ requires Φ_I to be specified on the boundaries. This is, in effect, a Dirichlet condition (Section 7.5). The other boundary condition is the natural boundary condition and corresponds to a Neumann condition (specifying the normal gradient geostrophically). Equation (8.3.18), with boundary condition (8.3.19), is a second-order equation in Φ_I, similar but slightly more complicated than the Poisson equation discussed in Section 7.5.

If $w_\Phi = 0$ and $w_v = 1$, then (8.3.18) becomes identical to (7.5.2). If $w_\Phi = 1$ and $w_v = 0$, then $\Phi_I = \Phi_A$. When $w_\Phi \gg w_v$, the initialized geopotential and windfields adjust to the analyzed geopotential. When $w_v \gg w_\Phi$, the initialized geopotential and windfields adjust to the analyzed windfields.

The constraints discussed in this section have been simple and linear. However, the real goal of variational initialization is to apply more realistic constraints such as the nonlinear balance equation (7.4.18). This has been discussed by Stephens (1970) and Barker, Haltiner, and Sasaki (1977). Achtemeier (1975) used the complete primitive equations as strong constraints.

8.4 Weak constraint formulations

In the classical variational procedures introduced in Section 8.2 and illustrated in Section 8.3, all the imposed constraints were satisfied exactly. That is, the final initialized variables were required to satisfy identically the imposed constraints whether they be the geostrophic relation, hydrostatic relation, nonlinear balance equation, or other. When constraints are satisfied exactly, they are known as *strong* constraints. Sasaki (1969, 1970a,b) introduced a number of variational techniques whereby the imposed constraints are satisfied only approximately, not exactly. In this case, the constraints are known as *weak* constraints.

Consider again the system (8.3.1–2). The strong constraint approach would be to introduce Lagrange multipliers and attempt to find the stationary value of an augmented functional:

$$I_1 = \iint_{PS} \{w_u(u_I - u_A)^2 + w_v(v_I - v_A)^2 + \lambda f_1\}\, dS\, dP \qquad (8.4.1)$$

where λ is the (unknown) Lagrange multiplier.

Sasaki's weak constraint approach introduces the constraints with prespecified weights. Thus, a functional is constructed as follows:

$$I_2 = \iint_{PS} \{w_u(u_I - u_A)^2 + w_v(v_I - v_A)^2 + \gamma(f_1)^2\}\, dS\, dP \qquad (8.4.2)$$

Here γ is a prespecified weight that is a function of x, y, and P. If we want to satisfy the condition $f_1 = 0$ very precisely, then we should specify γ to be very large. Conversely, the constraint $f_1 = 0$ is only approximately satisfied if γ is small.

The weak constraint formulation is related to the penalty methods of Section 2.7. Thus, the first two terms of the integral (8.4.2) constitute a continuous cost function, and the remaining term is the penalty function. The cost function is minimized while controlling the constraint violations by penalizing them. Although the examples in this section are in continuous form, in practice, discrete formulations analogous to (8.3.3) would be used.

Sasaki's weak constraint variational formulation is first illustrated for the simple spatial filtering problem described by (2.7.1–4) of Chapter 2. Suppose $\Phi_A(x)$ is an analyzed geopotential field and consider the functional

$$I = \int_{x_a}^{x_b} \left\{ w_\Phi(\Phi_I - \Phi_A)^2 + \gamma\left(\frac{d^2\Phi_I}{dx^2}\right)^2 \right\} dx \tag{8.4.3}$$

where w_Φ and γ are prespecified and assumed to be independent of x. $\Phi_I(x)$ is the geopotential field resulting from the minimization of (8.4.3). Taking the first variation of (8.4.3) yields the Euler–Lagrange equation:

$$w_\Phi \Phi_I + \gamma \frac{d^4\Phi_I}{dx^4} = w_\Phi \Phi_A \tag{8.4.4}$$

where it is assumed that $\Phi_I(x_a)$ and $\Phi_I(x_b)$ are specified. Suppose $\Phi_A(x)$ is given as

$$\Phi_A(x) = \Phi_0 + \Phi_1 \exp(ikx)$$

Then,

$$\Phi_I(x) = \Phi_0 + \frac{w_\Phi \Phi_1 e^{ikx}}{w_\Phi + \gamma k^4} \tag{8.4.5}$$

In the limit where $k \to 0$, then $\Phi_I(x)$ approaches $\Phi_A(x)$. As k increases, the x varying component $\Phi_1 \exp(ikx)$ is increasingly damped in Φ_I, and the weakly constrained functional (8.4.3) effectively damps small-scale variance. The spatial filtering implied by (8.4.5) is the same as that implied by (2.7.4) for $p = 2$.

The functional (8.4.3) damps smaller spatial scales. The essence of the initialization problem is to suppress small time scales. A functional similar to (8.4.3) that is suitable for damping high frequencies is

$$I = \int_{x_a}^{x_b} \left\{ w_\Phi(\Phi_I - \Phi_A)^2 + \gamma\left(\frac{\partial\Phi_I}{\partial t}\right)^2 \right\} dx \tag{8.4.6}$$

Here, the first temporal derivative rather than the second spatial derivative is used as a weak constraint. From the result (8.4.5), it is reasonable to suppose that (8.4.6) would suppress high frequencies if γ were sufficiently large. The following example, based on Lewis and Grayson (1972), illustrates this procedure.

In the planetary boundary layer on the f-plane, a linearized form of (7.3.1) can be written in component form:

$$\frac{\partial u}{\partial t} = f_0 v - \frac{\partial \Phi}{\partial x} - Ku \tag{8.4.7}$$

$$\frac{\partial v}{\partial t} = -f_0 u - \frac{\partial \Phi}{\partial y} - Kv \tag{8.4.8}$$

Here K is a constant estimated from the mean surface wind and surface drag coefficients (see Lewis and Grayson 1972). The variational problem can be posed as follows. Find the stationary value of

$$I = \int_S \left\{ w_v(u_I - u_A)^2 + w_v(v_I - v_A)^2 + w_\Phi(\Phi_I - \Phi_A)^2 + \gamma\left(\frac{\partial u}{\partial t}\right)^2 + \gamma\left(\frac{\partial v}{\partial t}\right)^2 \right\} dS \qquad (8.4.9)$$

where $\partial u/\partial t$ and $\partial v/\partial t$ are given by (8.4.7–8). The weights w_v and w_Φ are defined as in (8.3.13), and Φ_I, Φ_A, and so on, are the initialized and analyzed values as in Section 8.3. Assume that Φ_I, u_I, and v_I are specified on the boundary of the domain. Functional (8.4.9) is similar to (8.3.15) except that the constraints are now weak and nongeostrophic.

Take the first variation of I. Because there are three dependent variables u, v, and Φ, there are three Euler–Lagrange equations. There are no boundary integrals because u, v, and Φ are specified on the boundaries. Thus,

$$w_v(u_I - u_A) + \gamma(K^2 + f_0^2)u_I + \gamma\left(K\frac{\partial \Phi_I}{\partial x} + f_0\frac{\partial \Phi_I}{\partial y}\right) = 0 \qquad (8.4.10)$$

$$w_v(v_I - v_A) + \gamma(K^2 + f_0^2)v_I + \gamma\left(-f_0\frac{\partial \Phi_I}{\partial x} + K\frac{\partial \Phi_I}{\partial y}\right) = 0 \qquad (8.4.11)$$

$$w_\Phi(\Phi_I - \Phi_A) - \gamma\,\nabla^2\Phi_I + \gamma\frac{\partial}{\partial x}(f_0 v_I - K u_I) + \gamma\frac{\partial}{\partial y}(-f_0 u_I - K v_I) = 0 \qquad (8.4.12)$$

Assume γ, w_Φ, and w_v are independent of x and y. Solving (8.4.10) for u_I and (8.4.11) for v_I and substituting into (8.4.12) gives

$$w_\Phi\Phi_I - \beta\,\nabla^2\Phi_I = w_\Phi\Phi_A - \beta f_0\zeta_A + \beta K D_A \qquad (8.4.13)$$

where ζ_A is the vorticity of the analyzed field as in (8.3.21), D_A is the analyzed divergence, and

$$\beta = \frac{\gamma w_v}{w_v + \gamma(K^2 + f_0^2)}$$

If $\gamma \gg w_v$ and $f_0 \gg K$, then (8.4.13) becomes identical to (8.3.18). This corresponds to the case of a very strongly specified geostrophic constraint. Sections 8.3 and 8.4 have discussed continuous variational problems. The extension of these ideas to discrete problems is straightforward, and discrete variational problems are left as Exercises 8.1 and 8.3.

8.5 Inequality constraints

Inequality constraints can also be imposed in the variational formalism. Valentine (1937) showed that certain problems with inequality constraints are equivalent to problems with equality constraints. The following simple meteorological application was discussed by Sasaki and McGinley (1980).

Suppose the estimated potential temperature at some location is given by $\theta_A(z)$, where z is the height above the earth's surface. In general, the atmosphere is stable, and $d\theta_A/dz$ is positive. Occasionally, however, superadiabatic layers occur, and the values of $d\theta_A/dz$ are negative for some layers. Usually this phenomenon is short-lived or due to instrument error. The potential temperature profile can be adjusted so that $d\theta/dz$ is greater than or equal to zero everywhere using the following technique. The variational problem is the minimization of

$$I = \int_0^{z_T} w_\theta(z)[\theta_I(z) - \theta_A(z)]^2 \, dz \tag{8.5.1}$$

subject to the constraint that $d\theta_I/dz \geq 0$ for $0 \leq z \leq z_T$. Here $w_\theta(z)$ is the specified weight function, and $\theta_I(z)$ is the final initialized potential temperature.

Define the function $F(z)$, called the *slack function*, such that

$$\frac{d\theta_I}{dz} - F^2(z) = 0 \tag{8.5.2}$$

If F is real valued, F^2 is always positive. Equation (8.5.2) now becomes the constraint for the problem, but a new dependent variable has been introduced. Define a new functional I_1 that includes a Lagrange multiplier λ:

$$I_1 = \int_0^{z_T} \left[w_\theta(\theta_I - \theta_A)^2 + \lambda\left(\frac{d\theta_I}{dz} - F^2\right)\right] dz \tag{8.5.3}$$

The constraint (8.5.2) is of the form (8.2.20) except that there are two dependent variables, θ_I and F. Variation with respect to θ_I and F, together with the constraint (8.5.2), gives

$$2w_\theta(z)(\theta_I - \theta_A) - \frac{d\lambda}{dz} = 0 \tag{8.5.4}$$

$$\lambda F = 0 \tag{8.5.5}$$

$$\frac{d\theta_I}{dz} - F^2 = 0 \tag{8.5.6}$$

The system of Equations (8.5.4–6) can then be solved subject to the imposed boundary condition, $\theta_I(0) = \theta_A(0)$ and $\theta_I(z_T) = \theta_A(z_T)$. In practice, this problem would be solved in discrete form (Exercise 8.3).

8.6 Four-dimensional variational analysis: Thompson's scheme

The least squares estimation algorithms of Chapter 2 were developed for spatial interpolation of irregularly distributed observations to a regular analysis grid. In this chapter, the variational calculus has been used to minimize functionals subject to initialization constraints. In both cases, the cost functions or functionals are not explicitly time dependent, and neither are the constraints.

It is now time to introduce the temporal dimension and consider variational formulations of four-dimensional data assimilation. This concept was first advanced in the meteorological literature by Thompson (1969) and explored further by Sasaki (1969, 1970a), Lewis and Bloom (1978), Lewis (1980), and Lewis and Panetta (1983). In its simplest form, the idea is variationally to minimize the differences between time series of objective analyses and model forecasts over a finite time period. In recent years, the concept has been expanded to include the whole analysis/forecast cycle. This section introduces Thompson's original scheme, and Section 13.2 will discuss the more general algorithm.

For the arbitrary variables $u(x, y, P, t)$ and $v(x, y, P, t)$ discussed in Section 8.3, the functional would take the form

$$I = \int_{t_1}^{t_2} \iint_{PS} \{w_u(u_{\mathrm{I}} - u_{\mathrm{A}})^2 + w_v(v_{\mathrm{I}} - v_{\mathrm{A}})^2\}\, dS\, dP\, dt \tag{8.6.1}$$

Constraints (8.3.2) could be included in either a strong constraint formulation such as (8.4.1) or a weak constraint formulation (8.4.2). Thompson's (1969) formulation is temporally discrete and spatially continuous. The completely discrete formulation is left until Section 13.2.

The determination of the stationary value of (8.6.1) subject to either strong or weak constraints yields values of u_{I} and v_{I} that exactly or approximately satisfy the imposed constraints and have internal temporal consistency over the period t_1 to t_2.

In principle, many choices of constraint are possible, depending on the particular characteristics of the problem. For example, the primitive equations themselves (7.3.1–4) could be used as a constraint in (8.6.1). In this case, the initialized fields would be approximately or exactly consistent with the primitive equations over the period t_1 to t_2. The primitive equations allow both the advective time scale τ_1 and the inertial time scale τ_2. Perhaps a better choice of constraint would be one that allowed only the advective time scale τ_2. In this way, the initialized values would not only be temporally consistent, they would also contain no high frequencies. Constraints that have the last property are the geostrophic potential vorticity equation (7.8.6), its shallow water counterpart (7.8.8), and the slow manifold equations of Section 12.9.

We consider the minimization of a functional such as (8.6.1) using (7.8.8) as a strong constraint. This is a slight modification of the original problem discussed by Thompson (1969). Define the functional

$$I = \int_{t_1}^{t_2} \int_S \left\{(Q_{\mathrm{I}} - Q_{\mathrm{A}})^2 + \lambda\left(\frac{\partial Q_{\mathrm{I}}}{\partial t} + \mathbf{v}_\psi \cdot \boldsymbol{\nabla} Q_{\mathrm{I}}\right)\right\} dS\, dt \tag{8.6.2}$$

where

$$Q = \nabla^2\psi + f_0 - \frac{f_0^2\psi}{\tilde{\Phi}}$$

Here ψ is the streamfunction, $g^{-1}\tilde{\Phi}$ the depth of the fluid, and subscripts I and A indicate initialized and analyzed values, respectively. S is a two-dimensional (horizontal) domain.

We consider a particularly simple form of (8.6.2). Suppose complete analyses of Q_A are available at times t_1 and t_2. In practice, t_1 and t_2 might be 6–12 hours apart. Assume that $\mathbf{v}_\psi$ is known and time dependent over this relatively brief period. The constraint (7.8.8) is then discretized as follows, using the centered time-difference approximation discussed briefly in Section 7.2:

$$G = \frac{Q_I^2 - Q_I^1}{2\,\Delta t} + \mathbf{v}_\psi \cdot \nabla\left(\frac{Q_I^2 + Q_I^1}{2}\right) = 0 \tag{8.6.3}$$

where $Q_I^2 = Q_I(t_2)$, $Q_A^1 = Q_A(t_1)$, and so on, and $\Delta t = 0.5(t_2 - t_1)$.

Approximation (8.6.3) is equivalent to approximating the time derivative in (7.8.8) by centered differences and taking the time mean value of the advected potential vorticity. An approximate form of the functional (8.6.2) can then be written as

$$I_1 = \int_S \{(Q_I^2 - Q_A^2)^2 + (Q_I^1 - Q_A^1)^2 + \lambda G\}\, dS \tag{8.6.4}$$

The quadratic terms in (8.6.4) are equivalent to those in (8.6.2) if $(Q_I - Q_A)^2$ is assumed to vary linearly in t over the interval $t_1 \leqq t \leqq t_2$. Functional (8.6.2) had a single time-dependent variable Q_I whereas (8.6.4) has two time-independent variables Q_I^1 and Q_I^2. Taking the first variation of (8.6.4) yields two Euler–Lagrange equations:

$$2(Q_I^1 - Q_A^1) - \frac{\lambda}{2\,\Delta t} - \frac{\mathbf{v}_\psi}{2} \cdot \nabla\lambda = 0 \tag{8.6.5}$$

$$2(Q_I^2 - Q_A^2) + \frac{\lambda}{2\,\Delta t} - \frac{\mathbf{v}_\psi}{2} \cdot \nabla\lambda = 0 \tag{8.6.6}$$

as $\nabla \cdot \mathbf{v}_\psi = 0$. It has been assumed that Q_I^1 and Q_I^2 are specified on the boundaries of the domain, so that the boundary integrals vanish. The elimination of the Lagrange multiplier λ leads to

$$(Q_I^2 + Q_I^1) - (\Delta t)^2 \mathbf{v}_\psi \cdot \nabla[\mathbf{v}_\psi \cdot \nabla(Q_I^2 + Q_I^1)] = Q_A^2 + Q_A^1 + \Delta t\, \mathbf{v}_\psi \cdot \nabla(Q_A^2 - Q_A^1) \tag{8.6.7}$$

Equation (8.6.7) can be solved for $(Q_I^2 + Q_I^1)$ because all the other quantities are known. Back substitution of $(Q_I^2 + Q_I^1)$ into (8.6.3) yields $(Q_I^2 - Q_I^1)$, from which can be obtained Q_I^2 and Q_I^1 in isolation. Streamfunctions ψ_I^1 and ψ_I^2 can be obtained from Q_I^1 and Q_I^2 using the definition of Q in (8.6.2).

This procedure produces potential vorticities and streamfunctions that contain only the advective time scale and are temporally consistent over the period $t_1 \leq t \leq t_2$. Thompson (1969) discussed this approach theoretically, and it was tested by Lewis (1980) in a fairly simple context.

The variational formalism is a powerful and elegant tool for attacking a number of problems in atmospheric data analysis. The idea of minimizing a quadratic cost function while simultaneously satisfying (exactly or approximately) imposed physical or mathematical constraints is inherently appealing and has been applied in a variety of contexts.

Variational concepts, in the form of least squares estimation, were first encountered in Chapter 2. In spatial objective analysis, observations on an irregular network are interpolated to a regular analysis grid. In this case, the cost function measures the fit to the observations, and the applied constraints can be physical (Section 2.6) or mathematical (Section 2.7). In Chapter 5, another quadratic form for three-dimensional objective analysis was derived, Equation (5.6.11). The minimization of this cost function will be discussed at greater length in Section 13.1.

The variational formalism has also been introduced into the initialization problem. As noted in Section 6.5, there is no unique initialized state corresponding to an uninitialized state. Consequently, gridded objective analyses can be variationally adjusted using cost functions with weights determined from the expected analysis error, and constraints can be determined from an initialization theory.

Lewis (1972), Barker et al. (1977), and Phillips (1977) have performed variational initialization using quasi-geostrophic constraints. The variational formalism permits a generalization of the geopotential or wind constrained initialization of Chapter 7. For example, the nonlinear balance equation (7.4.18) does not converge at low latitude for the geopotential constrained case. In a variational formulation, it is possible to specify latitudinally varying weights that force toward the analyzed geopotential at mid and high latitudes and toward the analyzed wind at low latitudes. With an appropriate choice of weights, the ellipticity problems of Section 7.6 can be avoided. The constraints used in this chapter were derived from the quasi-geostrophic theory. As noted in Section 7.7, the quasi-geostrophic theory is really limited to mid and high latitudes and is not applicable on global scales or in the tropics. Extensions of quasi-geostrophic theory to global scales will be discussed in Chapters 9 and 10, and the variational formulation will also be extended to this case (Section 10.5).

In this section, we have introduced yet another variational approach. The idea was to minimize a four-dimensional cost function while exactly satisfying a four-dimensional dynamic constraint (the forecast model itself). If the forecast model excludes high frequencies (slow manifold model), then this approach solves the initialization problem as well. In Thompson's scheme, the inputs were gridded analyses, which implies prior objective analysis. In Section 13.2, Thompson's scheme is generalized so that the input is raw data from the irregular observation network. In this way, objective analysis and initialization are both incorporated, resulting in a complete four-dimensional data assimilation system.

Exercises

8.1 Consider a discrete form of the problem (8.3.4–5). Suppose there are $N + 1$ pressure levels between $P_B = P_{N+1}$ and $P_T = P_1$. Assume that $D_A(P_n)$, $1 \le n \le N + 1$, and $\omega_B = \omega_I(P_{N+1})$ and $\omega_T = \omega_I(P_1)$ are known. Replace the integral (8.3.4) and the constraint (8.3.5) with the following simple discrete approximations:

$$I = \sum_{n=1}^{N} w_n[D_I(P_n) - D_A(P_n)]^2 \delta_n$$

$$\frac{D_I(P_{n+1}) + D_I(P_n)}{2} + \frac{\omega_I(P_{n+1}) - \omega_I(P_n)}{\delta_n} = 0, \qquad 1 \le n \le N$$

where $\delta_n = P_{n+1} - P_n$ and $w_n = 0.5\langle \varepsilon_D^2(P_n)\rangle^{-1}$ are the specified weights. Minimize I subject to the N constraints, and compare the results with the continuous counterpart (8.3.11).

8.2 Consider a three-dimensional domain in (x, y, P) and analyzed geopotential Φ_A, temperature T_A, and wind components u_A, v_A. Minimize

$$I = \iint_{PS} \{w_v(u_I - u_A)^2 + w_v(v_I - v_A)^2 + w_\Phi(\Phi_I - \Phi_A)^2 + w_T(T_I - T_A)^2\}\, dS\, dP$$

using the geostrophic and hydrostatic relations as strong constraints on the unknowns u_I, v_I, Φ_I, and T_I. Assume that w_v, w_Φ, and w_T are spatially independent and u_I, v_I, Φ_I, and T_I are specified on the boundaries of the domain. Repeat using a weak constraint formulation.

8.3 Consider a discrete form of the problem discussed in Section 8.5. Suppose there are $N + 1$ levels in z between $z = 0 = z_0$ and $z = z_T = z_N$. Assume that $\theta_A(z_n)$, $0 \le n \le N$, is known. Define

$$\bar{\theta}_n = 0.5[\theta(z_n) + \theta(z_{n+1})], \qquad 0 \le n \le N - 1$$

as the layer mean average of the potential temperature and denote $\Delta\theta(z) = \theta_I(z) - \theta_A(z)$. Replace the integral (8.5.1) and the constraint (8.5.2) with the discrete approximations

$$\sum_{n=0}^{N-1} w_n[\Delta\bar{\theta}_n]^2\, \delta_n \qquad \text{and} \qquad \frac{\theta_I(z_{n+1}) - \theta_I(z_n)}{\delta_n} - F_n^2 = 0, \qquad 0 \le n \le N - 1$$

where w_n is the a priori weight w_θ, and $\delta_n = z_{n+1} - z_n$. Find the discrete counterparts of (8.5.4–6). If θ remains constant on the boundaries, and w_n and δ_n are independent of n, show that

$$\sum_{n=1}^{N-1} [\theta(z_{n+1}) + 2\theta(z_n) + \theta(z_{n-1})] + \theta(z_{N-1}) + \theta(z_1)$$

is conserved during the adjustment.

8.4 The super-obbing technique discussed in Section 4.6 is a procedure for combining almost coincident observations to produce a single superobservation. The super-obbing procedure is based on statistical interpolation, and the expected analysis error of the super-obbing procedure can be regarded as the expected observation error of the superobservation. Because the superobservation is to be used in a subsequent objective analysis, it is desirable that the error of the superobservation is uncorrelated with the background error, $\langle (A_i - T_i)(B_i - T_i)\rangle = 0$ in the nomenclature of Section 4.2. Define $\langle (B_i - T_i)^2\rangle = E_B^2$ and background and observation error covariances $\underline{B}$, $\underline{B}_i$, and $\underline{O}$ and a posteriori weights $\underline{W}_i$.

(a) Show $\underline{W}_i^T \underline{B}_i = E_B^2$.

(b) Using (a) as a constraint, minimize a modified version of (4.2.5) to produce an appropriate set of weights $\underline{W}_i$.

(c) Show that if the observations are exactly co-located, the background error is homogeneous, and the observation errors are uncorrelated, then the super-obbing algorithm produces a weighted average of the observations as in Case 6 of Section 4.6.

9

Normal mode initialization: theory

In Chapter 6, a linearized shallow water f-plane model was initialized by zeroing the initial amplitudes of the inertia–gravity modes. The simple example contains the essence of normal mode initialization. That is, a forecast model is linearized and the normal modes determined. The normal modes are divided into a set of high-frequency (fast) modes and a set of low-frequency (slow) modes. The model equations are then projected onto the normal modes, yielding sets of coupled fast mode and slow mode predictive equations. The fast (and perhaps the slow) normal mode expansion coefficients are then modified by applying appropriate initialization conditions. Transformation of the initialized normal mode expansion coefficients back into real space yields initial conditions for a model integration that should not excite high-frequency oscillations.

In practice, of course, forecast models are much more complicated than the example of Chapter 6. In particular, they are usually baroclinic rather than barotropic, they contain parametrizations of complicated physical processes, they are nonlinear rather than linear, they may have boundary conditions, and they do not make the f-plane assumption. Consequently, each of the steps of the normal mode initialization process is much more complex in a realistic model. Normal mode initialization has had its greatest success with the baroclinic primitive equations on the sphere, and it is this model that we discuss in detail. The first step is to linearize the equations and find the normal modes.

9.1 The linearized baroclinic primitive equations

The baroclinic primitive equations in pressure coordinates are given in (7.3.1–4). They are rewritten as follows:

$$\frac{\partial \mathbf{v}}{\partial t} + \mathbf{k} \times f\mathbf{v} + \nabla\Phi = \mathbf{R}_{\mathbf{v}} \tag{9.1.1}$$

$$\nabla \cdot \mathbf{v} + \frac{\partial \omega}{\partial P} = 0 \tag{9.1.2}$$

$$\frac{\partial}{\partial t}\frac{\partial \Phi}{\partial P} + \omega \tilde{\Gamma} = R_{\Phi} \tag{9.1.3}$$

where

$$\mathbf{R}_{\mathrm{v}} = -\mathbf{v} \cdot \nabla \mathbf{v} - \omega \frac{\partial \mathbf{v}}{\partial P} + \mathbf{F} \tag{9.1.4}$$

$$R_{\Phi} = -\mathbf{v} \cdot \nabla \frac{\partial \Phi}{\partial P} - \omega \Gamma' - \frac{\kappa Q}{P} \tag{9.1.5}$$

$$\Gamma = \frac{1}{P}\frac{\partial}{\partial P}\left\{P \frac{\partial \Phi}{\partial P} - \kappa \Phi\right\}$$

and $\partial \Phi / \partial P = -RT/P$, $(\tilde{\ })$ indicates global horizontal average, and $(')$ indicates deviation from the average. All other symbols are as in Section 7.3, and we assume that $\tilde{\Gamma}(P)$ is always positive and finite.

Forecast models frequently use a terrain-following coordinate system such as the sigma coordinate system of Phillips (1957). The governing equations in such models are similar to those in (9.1.1–5), but they contain additional coordinate transformation terms. It turns out (Kasahara and Puri 1981) that the normal mode procedure is virtually identical in sigma and pressure coordinates. Consequently, pressure coordinates are used in the following development.

The boundary conditions that we assume at the earth's surface ($P = P_{\mathrm{S}}$) and the top of the model atmosphere ($P = P_{\mathrm{T}}$) are

$$w = \frac{dh}{dt} = 0 \tag{9.1.6}$$

These boundary conditions differ slightly from those in (7.3.7) and do not necessarily correspond exactly to present modeling practices.

The left-hand sides of (9.1.1–3) contain all the terms that have been linearized about a state of rest with a mean static stability $\tilde{\Gamma}$ that is a function of P only. All the remaining terms in the equation have been grouped into $\mathbf{R}_{\mathrm{v}}$ and R_{Φ}. Linearization consists in setting $\mathbf{R}_{\mathrm{v}}$ and R_{Φ} equal to zero.

At this point, it is appropriate to justify the linearization about a state of rest. Equations (9.1.1–3) can be linearized about a more complex basic state. In fact, they can be linearized about a state that contains a nonzero mean flow field. For the purpose of normal mode initialization, however, only the inertia–gravity modes are of interest. The inertia–gravity modes that are initialized have high frequencies, and these structures and frequencies are relatively unaffected by the use of a more realistic basic state that is not at rest (Dickinson and Williamson 1972; Kasahara 1980).

After setting $\mathbf{R}_{\mathrm{v}}$ and R_{Φ} to zero, we can eliminate the vertical velocity ω between (9.1.1–3). Thus, dividing (9.1.3) by $\tilde{\Gamma}$, taking $\partial/\partial P$ of the resulting equation, and substituting from (9.1.2) yields

$$\frac{\partial}{\partial t}\left(\frac{\partial}{\partial P}\frac{1}{\tilde{\Gamma}}\frac{\partial \Phi}{\partial P}\right) - \nabla \cdot \mathbf{v} = 0 \tag{9.1.7}$$

Now (9.1.1) has no explicit pressure dependence; all explicit pressure dependence is contained in (9.1.7). Suppose that $\mathbf{r}$ is the (as yet unspecified) horizontal coordinate, then the horizontal dependence on $\mathbf{r}$ and the vertical dependence on P can be separated out by considering (9.1.7) only. Suppose that

$$\begin{aligned} \Phi(\mathbf{r}, P, t) &= \Phi^*(\mathbf{r}, t) Z(P) \\ \mathbf{v}(\mathbf{r}, P, t) &= \mathbf{v}^*(\mathbf{r}, t) Z(P) \end{aligned} \tag{9.1.8}$$

where Φ^* and $\mathbf{v}^*$ are geopotential and velocity functions that depend only on $\mathbf{r}$ and t.

Inserting (9.1.8) into (9.1.7) and separating variables yields

$$\frac{1}{Z}\frac{\partial}{\partial P}\left(\frac{1}{\tilde{\Gamma}}\frac{\partial Z}{\partial P}\right) = \frac{\nabla \cdot \mathbf{v}^*}{\partial \Phi^*/\partial t} = -\frac{1}{g\tilde{h}} \tag{9.1.9}$$

The left-hand term is a function of P only, the center term is a function of $\mathbf{r}$ and t, and $-1/g\tilde{h}$ is the separation constant. Here g is the gravitational constant, and $\tilde{h}$ is known as the equivalent depth and has units of length. Note the dimensional consistency of the separation constant. Equation (9.1.9) can then be written as

$$\frac{\partial}{\partial t}\Phi^* + g\tilde{h}\,\nabla \cdot \mathbf{v}^* = 0 \tag{9.1.10}$$

$$\frac{d}{dP}\left(\frac{1}{\tilde{\Gamma}}\frac{dZ}{dP}\right) + \frac{Z}{g\tilde{h}} = 0 \tag{9.1.11}$$

where the partial derivative notation in (9.1.11) has been dropped because Z is a function of P only.

Equation (9.1.11) is known as the vertical structure equation. It describes an eigenvalue problem with a second-order differential operator in P and an eigenvalue (unknown at this point) that is the separation constant $[g\tilde{h}]^{-1}$. A second-order operator requires two boundary conditions, which can be derived from (9.1.6).

At the top and bottom boundaries, $w = dh/dt = 0$. Thus, expanding the total time derivative dh/dt in terms of its derivatives with respect to $\mathbf{r}$, P, and t gives

$$g\frac{dh}{dt} = \frac{d\Phi}{dt} = \frac{\partial \Phi}{\partial t} + \omega\frac{\partial \tilde{\Phi}}{\partial P} - R_b = 0, \qquad \text{at } P = P_S, P_T$$

where

$$R_b = -\mathbf{v}\cdot\nabla\Phi - \omega\frac{\partial}{\partial P}(\Phi - \tilde{\Phi}) \tag{9.1.12}$$

R_b is a nonlinear term, which is neglected for now, consistent with the linearization adopted in (9.1.1–3). With the hydrostatic equation (7.3.2), the linearized boundary conditions at $P = P_S$ and P_T become

$$\frac{\partial \Phi}{\partial t} - \frac{R\tilde{T}\omega}{P} = 0 \tag{9.1.13}$$

ω can be eliminated between (9.1.13) and (9.1.3) with $R_\Phi = 0$. Substitution from (9.1.8) gives

$$\frac{dZ}{dP} + \frac{P\tilde{\Gamma}Z}{R\tilde{T}} = 0 \qquad \text{at } P = P_S, P_T \tag{9.1.14}$$

Equation (9.1.11) with boundary conditions (9.1.14) constitutes the complete vertical structure equation.

Now set $\mathbf{R}_v = 0$ in (9.1.1) and introduce (9.1.8). The resulting equation, together with (9.1.10), constitute the horizontal equations:

$$\frac{\partial \mathbf{v}}{\partial t} + \mathbf{k} \times f\mathbf{v} + \nabla\Phi = 0 \tag{9.1.15}$$

$$\frac{\partial \Phi}{\partial t} + g\tilde{h}\,\nabla\cdot\mathbf{v} = 0 \tag{9.1.16}$$

where the * notation has been dropped.

Equations (9.1.15–16) are identical to the linearized shallow water equations with fluid depth $\tilde{h}$. That is why the specific form $[-g\tilde{h}]^{-1}$ was chosen for the separation constant. The equivalent depth $\tilde{h}$ plays the same role in the baroclinic primitive equations as the fluid depth plays in the shallow water equations. Taylor (1936) was the first to demonstrate that the linearized primitive equations could be vertically decomposed in this way. Equations (9.1.15–16) are purely equations of time t and the horizontal coordinate $\mathbf{r}$ and are known as the horizontal structure equations.

To find the free normal modes of the linearized equations (9.1.1–3), we must first solve the vertical structure equation (9.1.11) with boundary conditions (9.1.14) as an eigenvalue problem in which $[g\tilde{h}]^{-1}$ is the (unknown) eigenvalue. This yields a series of eigenvalues (equivalent depths) and corresponding eigenvectors (characteristic vertical structures). Each equivalent depth $\tilde{h}$ obtained in this way is then inserted into the horizontal structure equations (9.1.15–16). The horizontal structure equations are solved as a normal mode or eigenvalue problem for each specified equivalent depth. The eigenvalues of the horizontal structure equations are the normal mode frequencies, as in Section 6.4, and each corresponds to a particular horizontal structure. The complete three-dimensional structure corresponding to each normal model frequency is the product of the appropriate vertical and horizontal eigenstructures (9.1.8).

Before going on to the horizontal problem, we first describe the properties and solution of the vertical problem.

9.2 The vertical structure equation

The eigenvalues and eigenvectors of (9.1.11) with boundary conditions (9.1.14) are orthogonal. This can be shown as follows. Suppose $Z_i(p)$ and $Z_j(p)$ are two different eigenfunctions of (9.1.11) with corresponding eigenvalues $[g\tilde{h}_i]^{-1}$ and $[g\tilde{h}_j]^{-1}$, respectively. Then,

$$\frac{d}{dP}\left(\frac{1}{\tilde{\Gamma}}\frac{dZ_i}{dP}\right)+\frac{Z_i}{g\tilde{h}_i}=0 \qquad (9.2.1)$$

$$\frac{d}{dP}\left(\frac{1}{\tilde{\Gamma}}\frac{dZ_j}{dP}\right)+\frac{Z_j}{g\tilde{h}_j}=0 \qquad (9.2.2)$$

Multiply (9.2.1) by Z_j and (9.2.2) by Z_i and subtract. Then integrate the resulting expression from $P = P_S$ to $P = P_T$:

$$\left(\frac{1}{g\tilde{h}_j}-\frac{1}{g\tilde{h}_i}\right)\int_{P_T}^{P_S} Z_iZ_j\,dP+\int_{P_T}^{P_S}\left\{Z_i\frac{d}{dP}\left(\frac{1}{\tilde{\Gamma}}\frac{dZ_j}{dP}\right)-Z_j\frac{d}{dP}\left(\frac{1}{\tilde{\Gamma}}\frac{dZ_i}{dP}\right)\right\}dP=0 \qquad (9.2.3)$$

The second integral in (9.2.3) equals zero. This can be shown by addition and subtraction of the term $(1/\tilde{\Gamma})(dZ_i/dP)(dZ_j/dP)$ in the integrand and application of the boundary condition (9.1.14). Consequently,

$$\int_{P_T}^{P_S} Z_i(P)Z_j(P)\,dP=0 \qquad \text{if } \tilde{h}_j\neq\tilde{h}_i \qquad (9.2.4)$$

Thus, if the eigenvalues are distinct, the eigenfunctions are orthogonal. In Exercise 9.1, it is shown that the eigenvalues are real and positive.

The eigenfunctions of (9.1.11) can be determined only up to a multiplicative constant because the problem is linear. Thus, they can be normalized in a variety of ways. An appropriate normalization, consistent with (9.2.4), is

$$\int_{P_T}^{P_S}\{Z_i(P)\}^2\,dP=1 \qquad (9.2.5)$$

The vertical structure functions of (9.1.11) with boundary values (9.1.14) can be derived straightforwardly for an isothermal atmosphere. Suppose that

$$\frac{d\tilde{\Phi}}{dP}=-\frac{RT_0}{P}$$

where T_0 is a constant. Then,

$$\tilde{\Gamma}=\frac{R\kappa T_0}{P^2}$$

Equation (9.1.11) becomes

$$\frac{d}{dP}\left(P^2\frac{dZ}{dP}\right)+\frac{R\kappa T_0 Z}{g\tilde{h}}=0 \qquad (9.2.6)$$

with boundary conditions

$$\frac{dZ}{dP} + \frac{\kappa}{P} Z = 0, \qquad \text{at } P = P_S, P_T \tag{9.2.7}$$

Now make the following transformation of the independent variable:

$$P = P_S \exp\left(\frac{-gz}{RT_0}\right)$$

Equation (9.2.6) becomes

$$\frac{d^2Z}{dz^2} - \frac{g}{RT_0}\frac{dZ}{dz} + \frac{g\kappa Z}{RT_0\tilde{h}} = 0 \tag{9.2.8}$$

while the boundary conditions (9.2.7) become

$$\frac{dZ}{dz} - \frac{\kappa g Z}{RT_0} = 0 \tag{9.2.9}$$

Here $z = 0$ when $P = P_S$ and $z = z_T = [RT_0/g] \log_e(P_S/P_T)$ when $P = P_T$.

Equation (9.2.8) has a solution called the external or Lamb mode. It has the following functional form:

$$Z_0(z) = \exp\left(\frac{\kappa g z}{RT_0}\right) \tag{9.2.10}$$

with the equivalent depth

$$\tilde{h}_0 = \frac{1}{g}\frac{RT_0}{1-\kappa}$$

The external mode (which satisfies the boundary conditions 9.2.9) increases monotonically with height with no zero crossings. In terms of pressure, the external mode structure becomes

$$Z_0(P) = \left(\frac{P_S}{P}\right)^\kappa$$

Equation (9.2.8) also has a set of solutions called internal modes, which are most easily found by making the following transformation in the dependent variable:

$$Z(z) = F(z) \exp\left(\frac{gz}{2RT_0}\right)$$

Equation (9.2.8) becomes

$$\frac{d^2F}{dz^2} + \frac{g}{RT_0}\left(\frac{\kappa}{\tilde{h}} - \frac{g}{4RT_0}\right)F = 0 \tag{9.2.11}$$

while the boundary condition (9.2.9) becomes

$$\frac{dF}{dz}+\frac{g}{RT_0}\left(\frac{1}{2}-\kappa\right)F=0, \qquad \text{at } z=0, z_T \tag{9.2.12}$$

Equation (9.2.11) has oscillatory solutions that satisfy boundary conditions (9.2.12) provided

$$0<\tilde{h}\leq\frac{4R\kappa T_0}{g} \tag{9.2.13}$$

Those solutions are called internal modes. For the case $P_T \to 0$, $z_T \to \infty$, there is an eigensolution of (9.2.11) corresponding to any equivalent depth $\tilde{h}$ that satisfies (9.2.13). Equation (9.2.11) is said to have a continuous spectrum of eigenvalues in this case (see Kasahara 1976). Note that the maximum equivalent depth for internal modes is less than $\tilde{h}_0$.

For $P_T \neq 0$, $z_T \neq \infty$, the solution to (9.2.11) with boundary conditions (9.2.12) can be written as

$$F_j=\cos\frac{j\pi z}{z_T}+\frac{gz_T}{RT_0 j\pi}\left(\kappa-\frac{1}{2}\right)\sin\frac{j\pi z}{z_T} \tag{9.2.14}$$

for $j=1, 2, 3, \ldots$ with equivalent depths $\tilde{h}_j$ given by

$$\tilde{h}_j=\frac{g\kappa}{RT_0(g^2/4R^2T_0^2+j^2\pi^2/z_T^2)} \tag{9.2.15}$$

The Z_j can be constructed from F_j by multiplying by $\exp(gz/2RT_0)$. The modes F_j with equivalent depths $\tilde{h}_j$ constitute a discrete spectrum of internal modes. The structures and eigenvalues of the modes are governed, in part, by the position of the top boundary. Setting $w=0$ at $P_T \neq 0$ is effectively a lid condition that artificially reflects outward propagating oscillations instead of letting them propagate out of the model (Lindzen, Batten, and Kim 1968). In operational models, a boundary condition similar to (7.3.7) is generally used at the top of the model whereas boundary condition (9.1.6) is used at the bottom. Boundary condition (7.3.7) also acts as an effective lid to the model and artificially reflects outward propagating waves. It is appropriate to assume the model's upper boundary conditions when attempting to initialize these models. If, at some time in the future, more realistic upper boundary conditions are used in forecast models, then the initialization procedure will have to be modified accordingly.

Discretization of the vertical structure equation (as is the practice with operational models) also causes modification of the internal mode structures. For example, if (9.2.6–7) were discretized using finite differences, the vertical structures of the discrete system would be a modification of the corresponding vertical structures of the continuous system. The differences between the continuous and discrete vertical structures would be most noticeable for small equivalent depths. An example of a

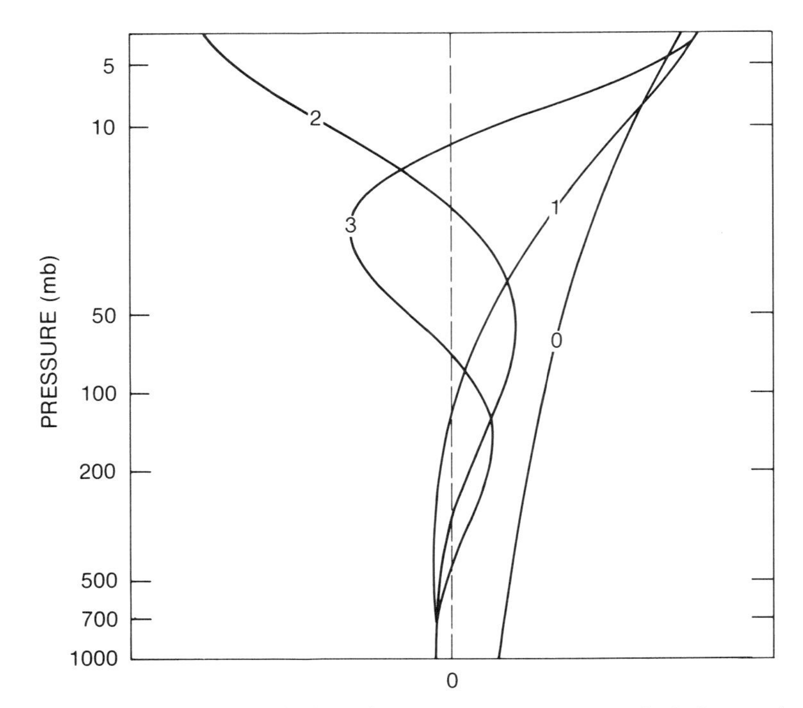

Figure 9.1 External mode (0) and three internal modes (1, 2, and 3) for an isothermal atmosphere as a function of pressure. (After Daley, "The Application of Variational Methods of Initialization on the Sphere," in Sasaki (ed): *Variational Methods in Geoscience*, New York: Elsevier Science Publishing Co.)

discrete vertical structure equation is discussed in Exercise 9.2. (See also Kashara and Puri 1981.)

The external mode and first three internal modes corresponding to (9.2.10) and (9.2.14) are plotted in Figure 9.1. Here $T_0 = 250$ K, $P_S = 1013$ mb, and $P_T = 4.3$ mb. This choice implies $z_T = 40$ km. The equivalent depths in this case are as follows: For the external mode $\tilde{h}_0 = 10.25$ km, and for the internal modes $\tilde{h}_1 = 3.61$ km, $\tilde{h}_2 = 1.33$ km, and $\tilde{h}_3 = 650$ m. By way of comparison, the maximum equivalent depth for an internal mode in an unbounded isothermal atmosphere at 250 K is 8.37 km (9.2.13).

The vertical structures corresponding to $\tilde{h}_0$, $\tilde{h}_1$, $\tilde{h}_2$, and $\tilde{h}_3$ are plotted in Figure 9.1 as a linear function of the natural logarithm of pressure. The increasing number of zero crossings with decreasing equivalent depth is evident. The vertical structure equation for arbitrary basic state temperature profiles is discussed by Cohn and Dee (1989).

9.3 The horizontal structure equation

Equations (9.1.15–16) constitute the horizontal structure equations. $\tilde{h}$ is assumed to have been determined in the solution of the vertical structure equation (9.1.11) and is thus considered a known constant in (9.1.15–16). In spherical polar coordinates (Godske et al. 1957), Equations (9.1.15–16) become

$$\frac{\partial u}{\partial t} - 2\Omega v \sin\phi + \frac{1}{a\cos\phi}\frac{\partial \Phi}{\partial \lambda} = 0 \tag{9.3.1}$$

$$\frac{\partial v}{\partial t} + 2\Omega u \sin\phi + \frac{1}{a}\frac{\partial \Phi}{\partial \phi} = 0 \tag{9.3.2}$$

$$\frac{\partial \Phi}{\partial t} + \frac{\tilde{\Phi}}{a\cos\phi}\left(\frac{\partial u}{\partial \lambda} + \frac{\partial v\cos\phi}{\partial \phi}\right) = 0 \tag{9.3.3}$$

where ϕ is the latitude (increasing northward), λ the longitude (increasing eastward), v the northward wind component, u the eastward wind component, $\tilde{\Phi} = g\tilde{h}$, and a is the earth's radius; Ω is the earth's rotation rate, so $f = 2\Omega\sin\phi$.

Equations (9.3.1–3) are the spherical polar equivalents of the shallow water f-plane equations (6.4.4–6). As in Chapter 6, the next step is to determine the normal modes of (9.3.1–3). Because of the spherical geometry, the problem is somewhat more complicated than in Chapter 6, but many of the useful properties of the normal modes can be inferred directly from (9.3.1–3).

The first property of (9.3.1–3) is that the normal modes are orthogonal. Assume u, v, and Φ are expanded as follows:

$$\begin{bmatrix} u \\ v \\ \Phi \end{bmatrix} = \sum_p \begin{bmatrix} u_p(\lambda, \phi) \\ v_p(\lambda, \phi) \\ \Phi_p(\lambda, \phi) \end{bmatrix} \exp(-2\Omega i\sigma_p t) \tag{9.3.4}$$

where u_p, v_p, and Φ_p are complex expansion coefficients, and σ_p is a nondimensional complex frequency. The normal modes are assumed to be ordered in some fashion, and subscript p indicates the pth normal mode; $[u_p \quad v_p \quad \Phi_p]^{\mathrm{T}}$ indicates the latitudinal and longitudinal structures of the u, v, and Φ components of the pth normal mode of (9.3.1–3), with frequency σ_p. The normal modes are assumed to have the form (9.3.4), though the actual spatial structures of u_p, v_p, and Φ_p are unknown at this point.

Now, u_p, v_p, and Φ_p are all complex numbers whereas u, v, and Φ are real. Thus,

$$u = \sum_p u_p \exp(-2\Omega i\sigma_p t) = \sum_p u_p^* \exp(2\Omega i\sigma_p^* t) \tag{9.3.5}$$

where * indicates complex conjugate, and similarly for v and Φ. Inserting (9.3.4) into (9.3.1–3) gives

$$-2\Omega i\sigma_p u_p - 2\Omega v_p \sin\phi + \frac{1}{a\cos\phi}\frac{\partial \Phi_p}{\partial \lambda} = 0 \tag{9.3.6}$$

$$-2\Omega i\sigma_p v_p + 2\Omega u_p \sin\phi + \frac{1}{a}\frac{\partial \Phi_p}{\partial \phi} = 0 \tag{9.3.7}$$

$$-2\Omega i\sigma_p\Phi_p + \frac{\tilde{\Phi}}{a\cos\phi}\left\{\frac{\partial u_p}{\partial\lambda} + \frac{\partial}{\partial\phi}(v_p\cos\phi)\right\} = 0 \tag{9.3.8}$$

for any particular normal mode $[u_p \quad v_p \quad \Phi_p]^{\mathrm{T}}$. Suppose $[u_q \quad v_q \quad \Phi_q]^{\mathrm{T}}$ is another normal mode of (9.3.1–3) with nondimensional frequency σ_q. From (9.3.5), there is also a complex conjugate form of this mode: that is, $[u_q^* \quad v_q^* \quad \Phi_q^*]^{\mathrm{T}}$ with frequency σ_q^*. From (9.3.1–3), the model equations for the complex conjugate form of mode q are

$$2\Omega i\sigma_q^* u_q^* - 2\Omega v_q^* \sin\phi + \frac{1}{a\cos\phi}\frac{\partial\Phi_q^*}{\partial\lambda} = 0 \tag{9.3.9}$$

$$2\Omega i\sigma_q^* v_q^* + 2\Omega u_q^* \sin\phi + \frac{1}{a}\frac{\partial\Phi_q^*}{\partial\phi} = 0 \tag{9.3.10}$$

$$2\Omega i\sigma_q^*\Phi_q^* + \frac{\tilde{\Phi}}{a\cos\phi}\left\{\frac{\partial u_q^*}{\partial\lambda} + \frac{\partial(v_q^*\cos\phi)}{\partial\phi}\right\} = 0 \tag{9.3.11}$$

Multiply (9.3.6) by $\tilde{\Phi}u_q^*$, (9.3.7) by $\tilde{\Phi}v_q^*$, (9.3.8) by Φ_q^*, (9.3.9) by $\tilde{\Phi}u_p$, (9.3.10) by $\tilde{\Phi}v_p$, (9.3.11) by Φ_p, and then add. This gives

$$\begin{aligned} &2\Omega i[\sigma_p - \sigma_q^*][\tilde{\Phi}(u_p u_q^* + v_p v_q^*) + \Phi_p\Phi_q^*] \\ &\quad - \frac{\tilde{\Phi}}{a\cos\phi}\left\{\frac{\partial}{\partial\lambda}[u_p\Phi_q^* + u_q^*\Phi_p] + \frac{\partial}{\partial\phi}[(v_p\Phi_q^* + v_q^*\Phi_p)\cos\phi]\right\} = 0 \end{aligned} \tag{9.3.12}$$

Now integrate (9.3.12) over the sphere:

$$2\Omega i[\sigma_p - \sigma_q^*]\int_S [\tilde{\Phi}(u_p u_q^* + v_p v_q^*) + \Phi_p\Phi_q^*]\, dS = 0 \tag{9.3.13}$$

where

$$\int_S (\quad)\, dS = \int_0^{2\pi}\int_{-\pi/2}^{\pi/2} (\quad) a^2\cos\phi\, d\phi\, d\lambda$$

The last term in (9.3.12) integrates to zero on the sphere because of the periodicity in λ and because $\cos(\pi/2) = \cos(-\pi/2) = 0$ while v_p and Φ_p remain bounded at the poles. Now if $p = q$, then

$$\int_S [\tilde{\Phi}(|u_p|^2 + |v_p|^2) + |\Phi_p|^2]\, dS \tag{9.3.14}$$

is a real positive number, where $|u_p^2| = u_p u_p^*$. Thus, from (9.3.13), $\sigma_p - \sigma_p^* = 0$, which means that the imaginary part of σ_p must equal zero. Consequently, σ_p is purely real and

$$\int_S [\tilde{\Phi}(u_p u_q^* + v_p v_q^*) + \Phi_q^*\Phi_p]\, dS = 0 \qquad \text{for } \sigma_p \neq \sigma_q \tag{9.3.15}$$

The eigenstructures $[u_p \quad v_p \quad \Phi_p]^{\mathrm{T}}$ and $[u_q^* \quad v_q^* \quad \Phi_q^*]^{\mathrm{T}}$ are orthogonal if the two eigenvalues σ_p and σ_q are distinct. Because σ_p is purely real, the normal modes of (9.3.1–3) correspond to free oscillations in time with no damped or growing components. This is exactly the same situation as in Chapter 6. Note that (9.3.14) divided by $\tilde{\Phi}$ has the form of kinetic plus potential energy per unit mass.

The eigenvalues of (9.3.1–3) are usually distinct, and the normal modes satisfy (9.3.15). However, some multiple eigenvalues exist where (9.3.15) is not satisfied. This will be discussed further in the next section.

Another property of (9.3.1–3) is important on the sphere. ϕ (the latitude) is positive north of the equator and negative south of the equator. A function that is symmetric about the equator has the property $f_S(-\phi) = f_S(\phi)$ whereas an antisymmetric function has $f_A(-\phi) = -f_A(\phi)$. It is left as Exercise 9.3 to show that (9.3.1–3) can be divided into two independent normal model problems: a symmetric problem involving u_S, v_A, and Φ_S and an antisymmetric problem involving u_A, v_S, and Φ_A.

Equations (9.3.1–3) can also be written in terms of a streamfunction ψ and a velocity potential χ, using the spherical form of Helmholtz's equation,

$$u = -\frac{1}{a}\frac{\partial \psi}{\partial \phi} + \frac{1}{a \cos \phi}\frac{\partial \chi}{\partial \lambda}$$

$$v = \frac{1}{a \cos \phi}\frac{\partial \psi}{\partial \lambda} + \frac{1}{a}\frac{\partial \chi}{\partial \phi} \tag{9.3.16}$$

The streamfunction and velocity potential form of (9.3.1–3) then becomes

$$\frac{\partial}{\partial t}\nabla^2\psi + 2\Omega \sin \phi\, \nabla^2\chi + \frac{2\Omega}{a^2}\left\{\frac{\partial \psi}{\partial \lambda} + \cos \phi \frac{\partial \chi}{\partial \phi}\right\} = 0 \tag{9.3.17}$$

$$\frac{\partial}{\partial t}\nabla^2\chi - 2\Omega \sin \phi\, \nabla^2\psi + \frac{2\Omega}{a^2}\left\{\frac{\partial \chi}{\partial \lambda} - \cos \phi \frac{\partial \psi}{\partial \phi}\right\} + \nabla^2\Phi = 0 \tag{9.3.18}$$

$$\frac{\partial \Phi}{\partial t} + \tilde{\Phi}\, \nabla^2\chi = 0 \tag{9.3.19}$$

where

$$\nabla^2 = \frac{1}{a^2}\left\{\frac{1}{\cos \phi}\frac{\partial}{\partial \phi}\left(\cos \phi \frac{\partial}{\partial \phi}\right) + \frac{1}{\cos^2 \phi}\frac{\partial^2}{\partial \lambda^2}\right\} \tag{9.3.20}$$

is the Laplacian operator in spherical polar coordinates. The linearized divergence equation (9.3.18) becomes the linear balance equation (7.5.3) if terms involving χ are dropped ($\zeta = \nabla^2\psi$, $D = \nabla^2\chi$). From the Helmholtz equation (9.3.16), u and χ have the same symmetry and v and ψ have the same symmetry with respect to the equator. Thus, the symmetric set includes u, χ, and Φ symmetric with ψ and v antisymmetric. The antisymmetric set includes u, χ, and Φ antisymmetric with ψ and v symmetric.

The structures of the normal modes and the associated eigenfrequencies can be found from either (9.3.1–3) or (9.3.17–19). In either case, some form of discretization is required to solve the equations because analytic solutions to the continuous equations have never been found. As with the vertical structure equation discussed in Section 9.2, the discretization of the horizontal structure equations should be completely consistent with the form of discretization that is used in the forecast model itself. A sample discretization of (9.3.17–19) is discussed in Appendix A.

Note that the horizontal structure equations (9.3.1–3), together with the vertical structure equation (9.1.11), are known as the Laplace tidal equations. They are widely used in tidal theory (Chapman and Lindzen 1970). The tidal problem uses the same equations as the initialization problem (except for a radiation rather than a lid condition at the upper boundary), but it solves them in a different order. In tidal theory, a forcing function and forcing frequency are imposed. In this case, the horizontal structure equations (9.3.6–8) with the frequency σ specified are solved to find the equivalent depth, and then $\tilde{h}$ is inserted in the vertical structure equation (9.1.11) to find the vertical structure consistent with the imposed forcing function. The tidal problem is a forced mode problem whereas the initialization problem is a free mode problem.

9.4 Horizontal structure functions (Hough modes)

A normal mode of (9.3.1–3) can be written in the form

$$\begin{bmatrix} u_{mn}(\lambda, \phi, \tilde{\Phi}) \\ v_{mn}(\lambda, \phi, \tilde{\Phi}) \\ \Phi_{mn}(\lambda, \phi, \tilde{\Phi}) \end{bmatrix} \exp(-2\Omega i \sigma_{mn} t) \tag{9.4.1}$$

where u_{mn}, v_{mn}, and Φ_{mn} are the horizontal structure functions of the mode and σ_{mn} is the real (see previous section) nondimensional frequency of the mode; u_{mn}, v_{mn}, and Φ_{mn} can be separated into functions of latitude ϕ and longitude λ. Thus,

$$\begin{bmatrix} u_{mn}(\lambda, \phi, \tilde{\Phi}) \\ v_{mn}(\lambda, \phi, \tilde{\Phi}) \\ \Phi_{mn}(\lambda, \phi, \tilde{\Phi}) \end{bmatrix} = \begin{bmatrix} u_{n}(\lambda, \phi, \tilde{\Phi}) \\ v_{n}(\lambda, \phi, \tilde{\Phi}) \\ \Phi_{n}(\lambda, \phi, \tilde{\Phi}) \end{bmatrix} \exp(im\lambda) \tag{9.4.2}$$

The decomposition into the latitudinal and meridional structures is discussed in Appendix A. As can be seen, m is a longitudinal wavenumber and plays exactly the same role as the x wavenumber in the f-plane (Chapter 6); n is defined to be the latitudinal mode number in analogy with the y wavenumber of Chapter 6. The latitudinal structures are more complex than in the f-plane case, being functions of both the equivalent depths $\tilde{\Phi}$ and the zonal wavenumber m. This is evident from Appendix A and will also become apparent in illustration shown later in this section. It is also possible, by using Helmholtz's equation (9.3.16), to write the normal mode structures in terms of streamfunction $\psi_{mn}(\lambda, \phi, \tilde{\Phi})$ and velocity potential $\chi_{mn}(\lambda, \phi, \tilde{\Phi})$.

The normal modes (horizontal structure functions or Hough functions) of (9.3.17–19) together with their associated eigenfrequencies were obtained using the techniques of Appendix A. The value of L in Equation (A11) was taken to be

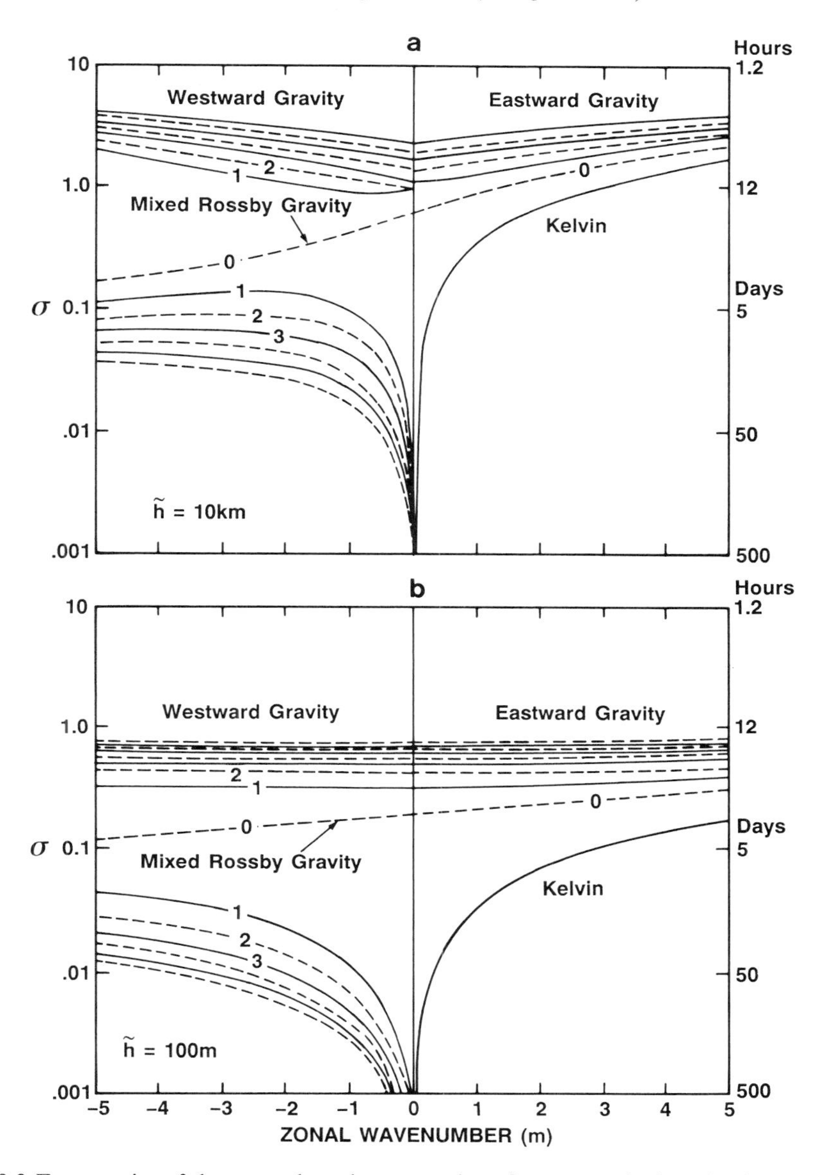

Figure 9.2 Frequencies of the normal modes on a sphere for two equivalent depths: (a) 10 km and (b) 100 m. Frequencies are shown in both dimensional and nondimensional form as a function of zonal wavenumber m.

sufficiently large to properly resolve any mode of interest. The frequencies of the normal modes are shown in Figure 9.2 for two equivalent depths ($\tilde{h} = g^{-1}\tilde{\Phi}$): (a) is for an equivalent depth of 10 km, which corresponds to the external mode; and (b) is the same except for an equivalent depth of 100 m, which is an internal mode.

Because λ is positive eastward, we see from definition (9.4.1–2) that for $m > 0$,

$\sigma > 0$ implies eastward propagation and $\sigma < 0$ implies westward propagation. Conversely, for $m < 0$, $\sigma < 0$ implies eastward propagation and $\sigma > 0$ implies westward propagation. Figure 9.2 shows only positive frequencies, so eastward propagating modes are for positive m and westward propagating modes for negative m. The frequency σ_{mn} is indicated on the left-hand side with the corresponding period $T = 2\pi/2\Omega\sigma_{mn}$ in hours or days indicated on the right-hand side. The diagram is logarithmic in frequency.

In Chapter 6, the linearized shallow water f-plane model had three normal modes for each horizontal wavenumber. Two of the three were inertia–gravity modes, and the third was a stationary Rossby or geostrophic mode. In the same way. Figure 9.2 indicates a natural separation into three groups of modes for each equivalent depth: eastward propagating and westward propagating inertia–gravity modes and westward propagating Rossby modes. In general, for each eastward propagating inertia–gravity mode there is a westward propagating inertia–gravity mode with approximately equal and opposite frequency; thus, the curves in (a) are almost symmetric about $m = 0$. Usually, the inertia–gravity modes of a given equivalent depth have much higher frequencies than the Rossby modes of the same equivalent depth. Note, however, that unlike the f-plane case (6.4.19), the magnitudes of inertia–gravity wave frequencies for the spherical case are not bounded below. In particular, for the equivalent depth of 100 m, Figure 9.2(b), the inertia–gravity wave freqencies σ_{mn} are considerably smaller than 1 (i.e., less than 2Ω in dimensional terms). This is because the minimum value of $|f|$ on the sphere is zero (at the equator).

The Rossby waves in Figure 9.2 do not have zero frequency as in the f-plane case of Chapter 6. Because of the variation of f with latitude (beta effect), the spherical Rossby waves have a small but nonzero westward propagation. The time scales of the Rossby modes are of the order of days and are thus comparable to the advective time scale τ_2. Remember, however, that though the Rossby mode time scale and the advective time scale are comparable, the Rossby mode time scale is due to beta effects, not advective effects.

Two special modes on the sphere, which do not exist on the f-plane, are denoted on Figure 9.2 as the Kelvin wave and the mixed Rossby–gravity wave. The character of these modes can be that of either a Rossby mode or an inertia–gravity mode depending on $\tilde{\Phi}$ or m.

As mentioned in Section 9.3, both symmetric and antisymmetric modes exist. Symmetric modes are identified by solid lines in Figure 9.2 and antisymmetric modes by dashed lines. For a fixed value of m and $\tilde{\Phi}$, the modes can be ordered by frequency. The convention adopted here for ordering the modes is as follows:

E_m^n, an eastward inertia–gravity mode
W_m^n, a westward inertia–gravity mode
R_m^n, a Rossby mode
K_m, a Kelvin mode

where m is the zonal wavenumber and n the latitudinal mode number of (9.4.1–2).

For eastward and westward propagating inertia–gravity modes, $n \geq 0$ and is defined to increase as the magnitude of the frequency increases (for given m and $\tilde{\Phi}$). The mixed Rossby–gravity modes are designated W_m^0. For Rossby modes, $n \geq 1$ and is defined to increase as the magnitude of the frequency decreases. The latitudinal mode number n for each type of mode is shown in Figure 9.2 for the lowest few modes of each type.

Figure 9.2 also indicates frequencies for $0 \leq m \leq 5$. Although only the modes with m integer are of interest on the sphere, values of σ_{mn} for m noninteger have been calculated and are shown in the figure for the sake of clarity.

Zonally averaged ($m = 0$) modes are degenerate. The analysis of this case is left for Exercise 9.4, but the important features are as follows. The frequencies of all the Rossby modes and the Kelvin mode tend to zero as $m \to 0$. When $m = 0$ ($\partial/\partial\lambda = 0$) and $\sigma = 0$ ($\partial/\partial t = 0$), then (9.3.17–18) degenerate to the linear balance equation (7.5.3). Thus, the Rossby modes and the Kelvin mode K_0 satisfy the linear balance equation for $m = 0$. These degenerate modes all have the same eigenvalue ($\sigma = 0$) and thus do not naturally satisfy the orthogonality condition (9.3.15). Kasahara (1978) has constructed orthogonal Rossby modes for $m = 0$ using a Gram–Schmidt procedure (Section 2.3). Shigehisa (1983) constructed orthogonal Rossby modes in this case by considering a limiting process as $m \to 0$. The eigensolutions for $m = 0$ shown in Figures 9.2 and 9.4 were also obtained by a limiting process. The eigenvalues (Figure 9.2) and eigenvectors (Figure 9.4; R_0^1, W_0^0, E_0^0) for $m = 0$ were obtained numerically by setting $|m| = 10^{-6}$.

For each value of $\tilde{\Phi}$, two modes exist whose frequencies tend to the value of 1 (i.e., 2Ω in dimensional form) as $m \to 0$. This can be seen in Figure 9.2 for $\tilde{h} = 10$ km, but it occurs at a much higher value of n for the $\tilde{h} = 100$ m case and is not shown. These two modes correspond to the globally averaged streamfunction (antisymmetric mode) and globally averaged velocity potential (symmetric mode), respectively. These are degenerate modes (for $m = 0$) that have no physical meaning because, from Helmholtz's equation (9.3.16), a globally averaged streamfunction or velocity potential field corresponds to zero windfield.

In Chapter 6, we demonstrated that there are two characteristic frequencies: the inertial frequency τ_1 and the advective frequency τ_2. In the initialization problem, we assumed that the high frequencies corresponding to τ_1 were undesirable and must be suppressed. This implies that the time scales τ_1 and τ_2 can be easily separated. If τ_1 is regarded as characterizing the eastward and westward propagating inertia–gravity modes and τ_2 as characterizing the Rossby mode frequencies, it is apparent from Figure 9.2 that for a given equivalent depth $\tilde{h}$ there is a reasonably clear separation between τ_1 and τ_2. However, the inertia–gravity oscillations for an equivalent depth of 100 m have frequencies that are not very different from the Rossby mode frequencies for an equivalent depth of 10 km. Thus, in the spherical baroclinic problem, separation of the fast and slow time scales is not straightforward.

The horizontal structures of the normal modes are shown in Figures 9.3 and 9.4 in six latitude–longitude plots. The particular mode designator (E_m^n, W_m^n, R_m^n, or K_m)

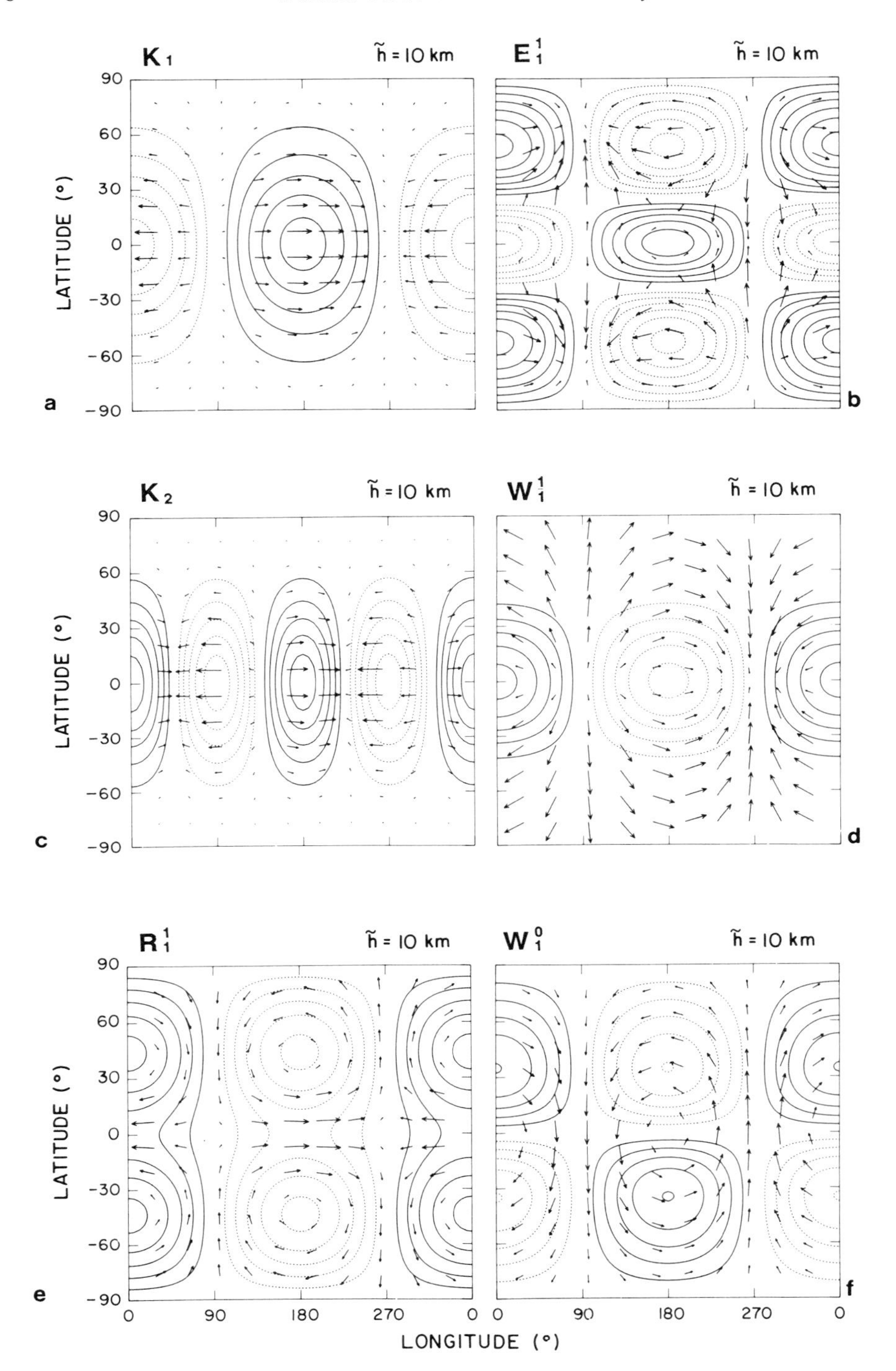

Figure 9.3 Structures of geopotential (contours) and windfield (arrows) as a function of longitude and latitude for six normal modes.

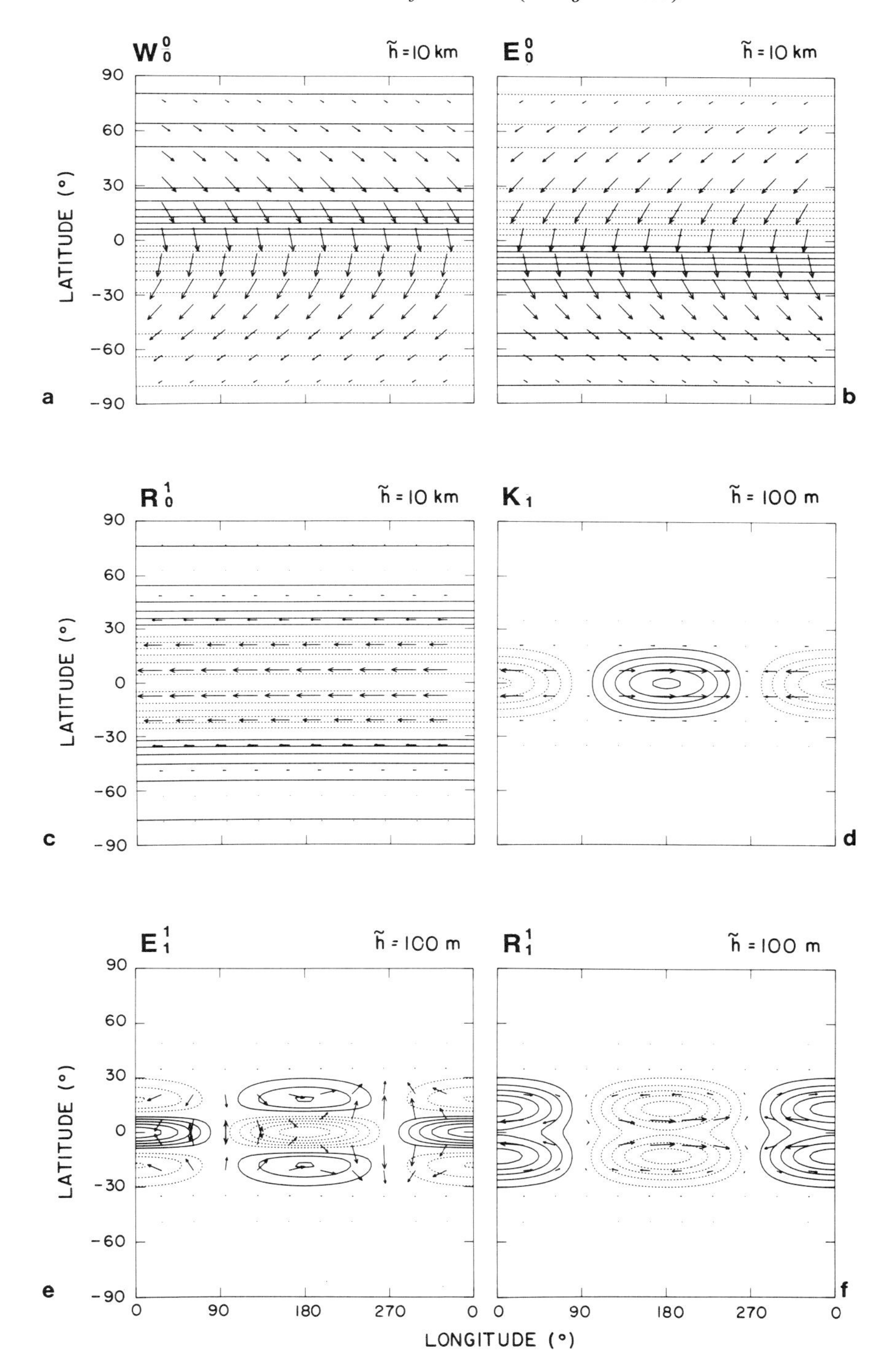

Figure 9.4 Similar to Figure 9.3 except for different modes.

is shown above each mode together with the equivalent depth $\tilde{h}$. All six modes in Figure 9.3 have an equivalent depth of 10 km whereas in Figure 9.4 three are 10 km modes and three are 100 m modes. The geopotentials (solid > 0, dashed < 0) and wind vectors are shown. The geopotential fields are contoured so that approximately 10 contour intervals occur between maximum and minimum. The length of a wind arrow indicates the relative wind speed.

Figure 9.3(a–f) shows the following:

(a) The Kelvin mode K_1 is a symmetric mode with nondimensional frequency $\sigma = 0.37$. In extratropical latitudes, the flow is approximately geostrophic, but at low latitudes the meridional wind v vanishes.

(b) E_1^1 ($\sigma = 1.28$) is the gravest symmetric eastward inertia–gravity mode. It is apparent that it is highly divergent and the flow is mostly ageostrophic at higher latitudes. It is, in fact, a pure inertia–gravity mode. It has two latitudinal zero crossings in the geopotential field and significant amplitude at high latitude. In general, for modes E_m^n, W_m^n, or R_m^n, m fixed, increasing n implies an increasing number of latitudinal zero crossings and more amplitude at high latitude.

(c) K_2 ($\sigma = 0.73$) is the Kelvin mode of zonal wavenumber 2. It is similar in latitudinal structure to K_1 (geostrophic at high latitudes, zonal winds in the tropics).

(d) W_1^1 ($\sigma = -0.91$) is a westward propagating inertia–gravity mode (highly divergent and ageostrophic).

(e) R_1^1 ($\sigma = -0.099$) is a Rossby mode for $\tilde{h} = 10$ km. It is symmetric with vanishing meridional wind at the equator and is geostrophic at high latitudes.

(f) W_1^0 ($\sigma = -0.421$) is the mixed Rossby–gravity wave of zonal wavenumber 1. It is antisymmetric, geostrophic at high latitudes, and has purely meridional flow at the equator.

Figures 9.4(a–f) show the following:

(a,b) These are zonal modes ($m = 0$) of equivalent depth $\tilde{h} = 10$ km. W_0^0 ($\sigma = -0.629$) and E_0^0 ($\sigma = +0.629$) are a pair of antisymmetric inertia–gravity modes with exactly equal and opposite frequencies. The v fields are identical, but the Φ and u fields have opposite signs. W_0^0 and E_0^0 correspond to a pair of standing oscillations; there is no longitudinal propagation.

(c) R_0^1 ($\sigma = 0.0$, $\tilde{h} = 10$ km) is a symmetric Rossby mode that satisfies the linear balance equation (7.5.3). It was obtained by the limiting process ($m \to 0$) discussed earlier.

(d) K_1 ($\sigma = 0.03$, $\tilde{h} = 100$ km) can be compared with Figure 9.3(a). It is a Kelvin mode but very clearly tropically trapped.

(e) E_1^1 ($\sigma = 0.32$, $\tilde{h} = 100$ km) is the gravest symmetric eastward gravity mode of zonal wavenumber 1 and can be compared with Figure 9.3(b). It is clearly a tropically trapped inertia–gravity mode.

(f) R_1^1 ($\sigma = -0.01$, $\tilde{h} = 100$ km) corresponds to Figure 9.3(e). It is a symmetric Rossby mode, tropically trapped but with a tendency to be geostrophic away from the equator. In general, with decreasing equivalent depth, both Rossby and inertia–gravity modes tend to become increasingly tropically trapped.

For more information on the latitudinal structures of the normal modes of (9.3.1–3), see Longuet–Higgins (1968), Kasahara (1976), and Kasahara (1978). As noted in Appendix A, the normal modes of a discretized model would differ from the Hough modes discussed here. The greatest differences would be in modes with small horizontal scales.

9.5 Normal mode form of the model equations

The next step is to project the model equations onto the normal modes determined in Sections 9.1–9.4.

The pressure coordinate baroclinic primitive equations (9.1.1–3) in spherical polar coordinates can be written as

$$\frac{\partial u}{\partial t} - 2\Omega v \sin\phi + \frac{1}{a\cos\phi}\frac{\partial \Phi}{\partial \lambda} = R_u \tag{9.5.1}$$

$$\frac{\partial v}{\partial t} + 2\Omega u \sin\phi + \frac{1}{a}\frac{\partial \Phi}{\partial \phi} = R_v \tag{9.5.2}$$

$$\frac{\partial}{\partial t}\frac{\partial}{\partial P}\frac{1}{\tilde{\Gamma}}\frac{\partial \Phi}{\partial P} - \frac{1}{a\cos\phi}\left\{\frac{\partial u}{\partial \lambda} + \frac{\partial(v\cos\phi)}{\partial \phi}\right\} = \frac{\partial}{\partial P}\left(\frac{R_\Phi}{\tilde{\Gamma}}\right) \tag{9.5.3}$$

where (9.1.2–3) have been used to eliminate ω. R_u and R_v are the eastward and northward components of $\mathbf{R}_\mathbf{v}$.

First consider the linearized form of (9.5.1–3), that is, $R_u = R_v = R_\Phi = 0$. A normal mode of the linearized equations has the following form:

$$\begin{bmatrix} u_{mn}(\lambda, \phi, \tilde{\Phi}_j) \\ v_{mn}(\lambda, \phi, \tilde{\Phi}_j) \\ \Phi_{mn}(\lambda, \phi, \tilde{\Phi}_j) \end{bmatrix} Z_j(P)\exp(-2\Omega i \sigma^j_{mn} t) \tag{9.5.4}$$

Three indices govern this mode: j the vertical index, m the zonal wavenumber, and n the latitudinal index. u_{mn}, v_{mn}, and Φ_{mn} constitute the horizontal structures of the u, v, and Φ components of the normal mode. They are complex and depend on the equivalent depths $\tilde{h}_j$ of the vertical structure functions Z_j, as shown in Section 9.3. σ^j_{mn} is the real nondimensional frequency of the mode. In expansion (9.5.4), u_{mn}, v_{mn}, Φ_{mn}, and σ^j_{mn} are analogous to u_p, v_p, Φ_p, and σ_p in expansion (9.3.4).

Suppose u, v, Φ denotes an arbitrary set of fields of the wind components and geopotential. Then u, v, Φ can be expanded in the normal modes of the linearized form of (9.5.1–3):

$$\begin{bmatrix} u(\lambda, \phi, P, t) \\ v(\lambda, \phi, P, t) \\ \Phi(\lambda, \phi, P, t) \end{bmatrix} = \sum_m \sum_n \sum_j x^j_{mn}(t) \begin{bmatrix} u_{mn}(\lambda, \phi, \tilde{\Phi}_j) \\ v_{mn}(\lambda, \phi, \tilde{\Phi}_j) \\ \Phi_{mn}(\lambda, \phi, \tilde{\Phi}_j) \end{bmatrix} Z_j(P) \tag{9.5.5}$$

where x^j_{mn} is a complex expansion coefficient given by

$$x^j_{mn}(t) = \int_S \int_{P_T}^{P_S} \{\tilde{\Phi}_j(uu^*_{mn} + vv^*_{mn}) + \Phi\Phi^*_{mn}\} Z_j \, dP \, dS \tag{9.5.6}$$

and $\int(\quad)\,dS$ is defined by (9.3.13). Relationship (9.5.6) can be easily derived by noting the vertical orthogonality condition (9.2.4) and the horizontal orthogonality condition (9.3.15). Note that expansions (9.5.5–6) are analogous to a three-dimensional Fourier expansion.

Now consider the nonlinear equations (9.5.1–3) with $R_u \neq 0$, $R_v \neq 0$, and $R_\Phi \neq 0$. For each of these equations, multiply by $Z_j(P)$ and integrate from $P = P_T$ to $P = P_S$. The result is

$$\dot{u}_j - 2\Omega v_j \sin\phi + \frac{1}{a\cos\phi}\frac{\partial\Phi_j}{\partial\lambda} = R^j_u \tag{9.5.7}$$

$$\dot{v}_j + 2\Omega u_j \sin\phi + \frac{1}{a}\frac{\partial\Phi_j}{\partial\phi} = R^j_v \tag{9.5.8}$$

$$\dot{\Phi}_j + \frac{\tilde{\Phi}_j}{a\cos\phi}\left\{\frac{\partial u_j}{\partial\lambda} + \frac{\partial(v_j\cos\phi)}{\partial\phi}\right\} = R^j_\Phi \tag{9.5.9}$$

where

$$g_j = \int_{P_T}^{P_S} gZ_j \, dP, \qquad R^j_g = \int_{P_T}^{P_S} R_g Z_j \, dP, \qquad \dot{g}_j = \frac{\partial g_j}{\partial t}$$

except that

$$R^j_\Phi = \tilde{\Phi}_j \int_{P_T}^{P_S} \frac{R_\Phi}{\tilde{\Gamma}}\frac{dZ_j}{dP}\,dP + \left.\frac{\tilde{\Phi}_j P R_b Z_j}{R\tilde{T}}\right|_{P_T}^{P_S}$$

and R_b is defined in (9.1.12).

Equations (9.5.7–8) are derived in a straightforward manner from (9.5.1–3). The derivation of (9.5.9) is more complicated. It is necessary to perform integration by parts and make use of (9.1.11) with boundary conditions (9.1.14) on $Z_j(p)$. It is also necessary to invoke upper and lower boundary conditions on $\dot{\Phi}$. These are obtained by the elimination of ω between (9.1.12) and (9.1.3) applied at $P = P_S$ and P_T:

$$\frac{P\dot{\Phi}}{R\tilde{T}} + \frac{1}{\tilde{\Gamma}}\frac{\partial\dot{\Phi}}{\partial P} = \frac{R_\Phi}{\tilde{\Gamma}} + \frac{PR_b}{R\tilde{T}} \qquad \text{at } P = P_S, P_T \tag{9.5.10}$$

The left-hand sides of (9.5.7–9) have the same form as the shallow water equations with equivalent depth $\tilde{\Phi}_j$.

Now multiply (9.5.7) by $\tilde{\Phi}_j u^*_{mn}$, (9.5.8) by $\tilde{\Phi}_j v^*_{mn}$, and (9.5.9) by Φ^*_{mn} defined in

(9.5.4); then add and integrate over the sphere. The result is

$$\frac{\partial}{\partial t}\int_S \{\tilde{\Phi}_j(u_j u^*_{mn} + v_j v^*_{mn}) + \Phi_j \Phi^*_{mn}\}\, dS$$
$$+ 2\Omega i\sigma^j_{mn}\int_S \{\tilde{\Phi}_j(u_j u^*_{mn} + v_j v^*_{mn}) + \Phi_j \Phi^*_{mn}\}\, dS = R^j_{mn} \quad (9.5.11)$$

where

$$R^j_{mn} = \int_S \{\tilde{\Phi}_j(R^j_u u^*_{mn} + R^j_v v^*_{mn}) + R^j_\Phi \Phi^*_{mn}\}\, dS$$

Here * indicates complex conjugation. The first left-hand term in (9.5.11) and the expression for R^j_{mn} are straightforwardly derived from (9.6.7–9). The second term on the left-hand side of (9.5.11) is derived by integrating by parts and using the conditions of periodicity in λ and $\cos(\pi/2) = \cos(-\pi/2) = 0$ that were used in Section 9.3. Use has also been made of all (9.3.9–11) with u^*_q, v^*_q, Φ^*_q, and σ^*_q replaced by u^*_{mn}, v^*_{mn}, Φ^*_{mn}, and $\sigma^*_{mn} = \sigma_{mn}$.

Equation (9.5.11) can then be written for each (j, m, n) as

$$\frac{\partial}{\partial t} x^j_{mn} + 2\Omega i\sigma^j_{mn} x^j_{mn} = R^j_{mn} \quad (9.5.12)$$

where x^j_{mn} is defined by (9.5.6), and the definitions of u_j, v_j, and Φ_j in (9.5.7–9) have been used. Equation (9.5.12) describes the time evolution of a single normal mode coefficient.

In deriving (9.5.12), we have assumed that $Z_j(P)$ and u_{mn}, v_{mn}, and Φ_{mn} are analytic functions and that they obey the orthogonality conditions (9.2.4) and (9.3.15). In practice, when dealing with a particular atmospheric model, the vertical and horizontal structure functions are derived from discretized equations and obey discrete orthogonality conditions. Equations of the form (9.5.12) are also valid for discretized equations, as can be demonstrated symbolically following Section 6.4. Consider a discrete version of (9.5.1–3) with u, v, Φ and R_u, R_v, R_Φ defined at the gridpoints of a regular mesh. Define $\underset{\sim}{s}$ as the column vector of all values of u, v, Φ at the gridpoints of the mesh and define $\underset{\sim}{R}_s$ in a similar fashion. The number of elements (N) in vector $\underset{\sim}{s}$ is equal to the number of dependent variables times the number of gridpoints in the mesh. Then, any discrete form of (9.5.1–3) can be written symbolically as

$$\dot{\underset{\sim}{s}} + \underset{\approx}{L}\,\underset{\sim}{s} = \underset{\sim}{R}_s \quad (9.5.13)$$

where $\underset{\approx}{L}$ is the discrete form of all operators (except the time derivatives) on the left-hand side of (9.5.1–3). $\underset{\approx}{L}$ is a real matrix of order N, and we assume that it is symmetric. The discrete form of the boundary conditions is assumed to be included in $\underset{\sim}{s}$, $\underset{\sim}{R}_s$, and $\underset{\approx}{L}$. Define $\underset{\approx}{E}$ as the matrix whose columns are the eigenvectors of $\underset{\approx}{L}$. Then, the eigenvectors are orthogonal, and $\underset{\approx}{E}^T = \underset{\approx}{E}^{-1}$. Denote $\underset{\sim}{X}$ as the column vector of all normal mode amplitudes, and

$$\underline{s} = \underline{\underline{E}}\,\underline{X} \tag{9.5.14}$$

Substitution of (9.5.14) into (9.5.13) and left multiplication by $\underline{\underline{E}}^{\mathrm{T}}$ yields the very simple form

$$\dot{\underline{X}} + 2\Omega i\,\underline{\underline{\Lambda}}\,\underline{X} = \underline{R}_x \tag{9.5.15}$$

where $2\Omega i\underline{\underline{\Lambda}} = \underline{\underline{E}}^{\mathrm{T}}\,\underline{\underline{L}}\,\underline{\underline{E}}$ and $\underline{R}_x = \underline{\underline{E}}^{\mathrm{T}}\,\underline{R}_s$. $\underline{\underline{\Lambda}}$ is a real diagonal matrix whose elements are the nondimensional normal mode frequencies σ. The normal mode coefficient x^j_{mn} of (9.5.12) is an element of the column vector $\underline{X}$; thus, (9.5.15) is simply a vector form of (9.5.12). In a formal sense, the normal mode form (9.5.15) is the same for both continuous equations and discrete approximations to them, though the elements of $\underline{\underline{\Lambda}}$ and $\underline{R}_x$ differ in the two cases. Other model equations can also be put in the simple form (9.5.15). Note that the form of (9.5.15) can be derived in the discrete case when the matrix $\underline{\underline{L}}$ is not symmetric.

In Chapter 6, the normal modes were broken into two groups: the inertia–gravity modes with the fast time scale τ_1 and the geostrophic or Rossby modes with the slow time scale τ_2. An initialization process was applied to the modes with the fast time scale τ_1. In the same way, the normal modes of the linearized version of (9.5.1–3) can be broken into two groups. For the spherical baroclinic case (unlike the shallow water f-plane case), the division of modes into slow and fast is not completely straightforward. Nonetheless, on the basis of considerations to be discussed in Section 10.2, we can choose a cutoff frequency σ_c. Any mode x^j_{mn} with $|\sigma^j_{mn}| < \sigma_c$ is designated a slow mode. If $|\sigma^j_{mn}| > \sigma_c$, the mode is designated a fast mode.

The vector of slow mode amplitudes is designated $\underline{Y}$ with individual elements y. The vector of fast mode amplitudes is designated $\underline{Z}$ with individual elements z. The slow and fast mode equations corresponding to (9.5.15) are

$$\dot{\underline{Y}} + 2\Omega i\,\underline{\underline{\Lambda}}_y\,\underline{Y} = \underline{R}_y(\underline{Y}, \underline{Z}) \qquad \text{slow} \tag{9.5.16}$$

$$\dot{\underline{Z}} + 2\Omega i\,\underline{\underline{\Lambda}}_z\,\underline{Z} = \underline{R}_z(\underline{Y}, \underline{Z}) \qquad \text{fast} \tag{9.5.17}$$

where $\underline{R}_y$ and $\underline{R}_z$ have been written as functions of both the slow and fast modes. $\underline{\underline{\Lambda}}_y$ and $\underline{\underline{\Lambda}}_z$ are the real diagonal matrices of slow and fast frequencies, respectively. The corresponding equations for individual elements of $\underline{Y}$ and $\underline{Z}$ are

$$\dot{y} + 2\Omega i\sigma_y y = R_y(\underline{Y}, \underline{Z}) \tag{9.5.18}$$

$$\dot{z} + 2\Omega i\sigma_z z = R_z(\underline{Y}, \underline{Z}) \tag{9.5.19}$$

where σ_y is an element of $\underline{\underline{\Lambda}}_y$ and σ_z an element of $\underline{\underline{\Lambda}}_z$. R_y and R_z are nonlinear and therefore functions of all fast modes $\underline{Z}$ and slow modes $\underline{Y}$.

9.6 The Machenhauer balance condition

We derived (9.5.18–19) for a baroclinic primitive equations model in spherical and pressure coordinates. Similar equations can be derived for any model with sufficiently simple boundary conditions. At this point we can ignore the origins of (9.5.18–19) and treat them as being given.

We consider two approaches to the initialization of (9.5.18–19). The Machenhauer approach is discussed in this section and the Baer–Tribbia procedure in Section 9.7.

Consider first the linear problem $R_y(\underline{Y}, \underline{Z}) = R_z(\underline{Y}, \underline{Z}) = 0$ in (9.5.18–19). The fast mode equation then becomes

$$\dot{z} + 2\Omega i\sigma_z z = 0 \tag{9.6.1}$$

The appropriate initialization procedure, as in Section 6.5, is to set z at time $t = 0$ equal to zero. Thus,

$$z(0) = 0, \qquad z(t) = z(0)\exp(-2\Omega i\sigma_z t) = 0 \tag{9.6.2}$$

Suppose, however, that the model is nonlinear and $R_z(\underline{Y}, \underline{Z}) \neq 0$. Clearly, (9.6.2) is not the appropriate initial condition in this case. Machenhauer (1977) demonstrated what can happen in a nonlinear model. In an experiment performed with a nonlinear shallow water model (6.4.1–3), he plotted separately the linear and nonlinear contributions to the time tendencies of certain fast modes during the integration of the model. Thus, for a particular fast mode z, $-2\Omega i\sigma_z z$ and $R_z(\underline{Y}, \underline{Z})$ were plotted as functions of time.

Figure 9.5, based on Machenhauer's results, shows the time behavior of a fast mode whose natural period $2\pi/(2\Omega\sigma_z)$ is approximately 12 hours. Both real and imaginary parts are shown. The curves in (a) indicate the nonlinear term and those in (b) the linear term. It can be seen that the linear oscillations are large and, in fact, have exactly the same frequency and period as would be given by the linearized equations. The nonlinear term R_z has only a very low amplitude high-frequency oscillation. However, the time averages of R_z and $-2\Omega i\sigma_z z$ are approximately equal and opposite (particularly noticeable in the imaginary parts).

From these results, Machenhauer (1977) derived a condition for balancing the linear and nonlinear terms at initial time. As can be seen from Figure 9.5, R_z has only a relatively small time variation. Suppose R_z were assumed to be independent of time. Then, the time-dependent solution to (9.5.19) would be

$$z(t) = \frac{R_z(0)}{2\Omega i\sigma_z} + \left[z(0) - \frac{R_z(0)}{2\Omega i\sigma_z}\right]e^{-2\Omega i\sigma_z t} \tag{9.6.3}$$

where $R_z(0)$ is the value of $R_z(\underline{Y}, \underline{Z})$ at time $t = 0$ and is assumed independent of time. Equation (9.6.3) consists of a constant term, plus a term oscillating at the gravity-wave frequency σ_z. To eliminate high-frequency oscillations, we simply require that the second term vanish initially. Then, for elements z of $\underline{Z}$,

$$z(0) = \frac{R_z(0)}{2\Omega i\sigma_z} \tag{9.6.4}$$

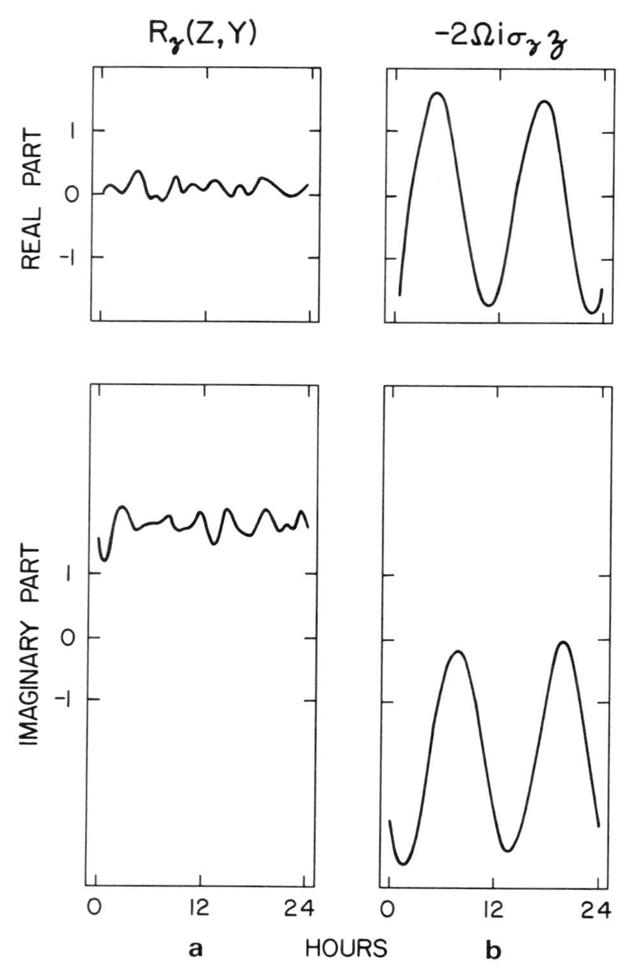

Figure 9.5 Time behavior of a particular fast mode in a nonlinear shallow water model. The real and imaginary parts are shown (a) with the projection on the nonlinear terms and (b) with the projection on the linear terms. (After Machenhauer 1977)

or, equivalently,

$$\left.\frac{dz}{dt}\right|_{t=0} = 0 \tag{9.6.5}$$

Machenhauer (1977) suggested that the balance condition (9.6.4) or (9.6.5) could be used as an initial condition for the nonlinear equation (9.5.19). The assumption of the time independence of R_z resembles assumption (7.1.19) made in deriving initial conditions for the Hinkelmann–Phillips model.

Initial conditions (9.6.4–5) are exact only if R_z is time independent. Of course, R_z is not really time independent, but balance condition (9.6.5) can still be used in deriving an initialization scheme. First note that R_z is a function of z and thus z appears on both sides of (9.6.4). If (9.6.4) is to be satisfied, the z that appears on the left-hand side must be equal to the z that is used in the calculation of R_z. Because this is an implicit relation in z, Machenhauer (1977) suggested that (9.6.4) be iterated until (9.6.5) was satisfied.

The iterative scheme tested by Machenhauer (1977) is a form of Picard iteration (see Section 10.7). The slow mode amplitudes $\underline{Y}$ are not changed as the iteration proceeds and would correspond to method 4 or slow mode constrained initialization in the nomenclature of Section 6.5. One such Picard iteration procedure is as follows:

Step 1 Linear initialization,

$$z_0(0) = 0 \tag{9.6.6}$$

Step 2 First nonlinear iteration,

$$z_1(0) = \frac{R_z^0(0)}{2\Omega i\sigma_z} \tag{9.6.7}$$

Step 3 Second nonlinear iteration,

$$z_2(0) = \frac{R_z^1(0)}{2\Omega i\sigma_z} \tag{9.6.8}$$

Step $k+1$

$$z_k(0) = \frac{R_z^{k-1}(0)}{2\Omega i\sigma_z} \tag{9.6.9}$$

Here $R_z^k(0) = R_z(\underline{Y}(0), \underline{Z}_k(0))$ for all elements z of $\underline{Z}$.

This procedure is repeated until convergence. In each step, R_z is recalculated for all elements z of $\underline{Z}$ using the values of z calculated in the previous step. To avoid confusion it should be noted that condition (9.6.5) is called the *Machenhauer condition*. However, a number of iteration procedures like (9.6.7–9) could lead to the same result (9.6.4); see Section 10.7 for a discussion of advanced iteration schemes.

Linear initialization $z(0) = 0$ and the Machenhauer condition $\dot{z}(0) = 0$ are really the first two orders of a very general initialization procedure. If R_z is infinitely differentiable in time, then a solution to (9.5.19) can be written formally as

$$z(t) = \left\{ z(0) + \sum_{s=1}^{\infty} \frac{1}{(-2\Omega i\sigma_z)^s} \left. \frac{d^{s-1} R_z(t)}{dt^{s-1}} \right|_{t=0} \right\} e^{-2\Omega i\sigma_z t} - \sum_{s=1}^{\infty} \frac{1}{(-2\Omega i\sigma_z)^s} \frac{d^{s-1} R_z(t)}{dt^{s-1}} \tag{9.6.10}$$

The condition for no high frequencies in (9.6.10) is simply

$$z(0) + \sum_{s=1}^{\infty} \frac{1}{(-2\Omega i\sigma_z)^s} \left. \frac{d^{s-1} R_z(t)}{dt^{s-1}} \right|_{t=0} = 0 \tag{9.6.11}$$

It is clear that linear initialization corresponds to truncation of (9.6.11) at $s=0$ whereas the Machenhauer scheme corresponds to truncation at $s=1$. Substitution from (9.5.19) for $R_z(0)$ into (9.6.11) gives

$$z(0)+\sum_{s=1}^{\infty}\frac{1}{(-2\Omega i\sigma_z)^s}\left.\frac{d^s z}{dt^s}\right|_{t=0}-\sum_{s=1}^{\infty}\frac{1}{(-2\Omega i\sigma_z)^{s-1}}\left.\frac{d^{s-1}z}{dt^{s-1}}\right|_{t=0}=0$$

which is

$$\lim_{p\to\infty}\frac{1}{(-2\Omega i\sigma_z)^p}\left.\frac{d^p z}{dt^p}\right|_{t=0}=0 \tag{9.6.12}$$

In other words, the condition of no high frequencies is equivalent to setting a very high time derivative of z equal to zero at $t=0$. Equation (9.6.12) is known as the *superbalance condition* (Lorenz 1980). Thus, the Machenhauer condition, $\dot{z}(0)=0$, is a first-order initial condition. The second-order condition, $\ddot{z}(0)=0$, implies that

$$z(0)=\frac{R_z(0)}{2\Omega i\sigma_z}+\frac{\dot{R}_z(0)}{4\Omega^2\sigma_z^2} \tag{9.6.13}$$

Equation (9.6.13) is similar to (9.6.4), but with a correction term that depends on the time derivative of R_z at $t=0$.

$R_z(0)$ is the projection onto the fast normal modes of the nonlinear and forcing terms $\mathbf{R}_v$ and R_Φ in (9.1.1–3). These terms appear naturally in the model, and $R_z(0)$ can be simply calculated using the actual model software (see Section 10.7). Time derivative terms such as $\dot{R}_z(0)$ involve time derivatives $\dot{\mathbf{R}}_v$ and $\dot{R}_\Phi$, which are not usually calculated during the model integration. Thus, in practice, time derivatives of R_z are not easy to calculate. However, Tribbia (1984) has developed a procedure in which time derivatives of R_z at $t=0$ can be estimated from short time integrations of the model, thus making the application of higher-order conditions such as (9.6.13) possible in practice.

If a low-order balance condition of order p is satisfied, then (9.6.10) implies that the amplitude of the remaining inertia–gravity wave oscillations is

$$\sum_{s=p+1}^{\infty}\frac{1}{(-2\Omega i\sigma_z)^s}\left.\frac{d^{s-1}R_z(t)}{dt^{s-1}}\right|_{t=0} \tag{9.6.14}$$

If the time derivatives of $R_z(0)$ are bounded, then the factor $(-2\Omega i\sigma_z)^s$ in the denominator of (9.6.14) causes decreasing inertia–gravity wave amplitude with increasing order of the scheme (p).

9.7 The Baer–Tribbia scheme

Baer (1977b) and Baer and Tribbia (1977) developed a normal mode initialization procedure that is conceptually different from the Machenhauer (1977) scheme discussed in the previous section. The Baer–Tribbia scheme is based on the scaled equations (Section 7.3) and was originally derived using an asymptotic expansion procedure. It was also derived by Kasahara (1982a) using the bounded derivative method (Section 7.4). To derive the procedure rigorously, we must first follow all the steps of Sections 9.1–9.5 using scaled equations. This results in scaled versions of the normal mode equations (9.5.16–17). Rederiving the normal mode equations from the scaled form of (9.1.1–3) is lengthy and tedious and is omitted here. The scaled fast and slow equations are given by (see Kasahara 1982a)

$$\dot{y} + i\sigma_y y = R_y(\underline{Y}, \underline{Z}) \tag{9.7.1}$$

$$\dot{z} + i\varepsilon^{-1}\sigma_z z = R_z(\underline{Y}, \underline{Z}) \tag{9.7.2}$$

Here y and z are the slow and fast mode amplitudes, and σ_y and σ_z are the two frequencies. All variables are assumed to be O(1), and ε is assumed to be O(0.1).

In Equation (9.7.1), all quantities are O(1), which means the time tendencies $\dot{y}$ of the slow modes are also O(1). The situation is different in (9.7.2). Here the linear term is multiplied by ε^{-1}, and the possibility exists that $\dot{z}$ will not be O(1).

Equation (9.7.2) can be initialized using the bounded derivative method discussed in Section 7.4. Thus for the time derivative $\dot{z}$ to be O(1), $\varepsilon^{-1}z(t)$ must be O(1). Following the arguments of Section 7.4, this requires that

$$z(t) = \varepsilon c(t) \tag{9.7.3}$$

where $c(t)$ is O(1) and is a smooth function of time. The simplest way to ensure that (9.7.3) is satisfied is to set $c(t)$ equal to zero at $t = 0$. This implies $z(0) = 0$, which is linear initialization and is equivalent to (9.6.2). It is possible to proceed to the next order as in Section 7.4 and demand that $\dot{c} = 0$ at $t = 0$. From (9.7.3), this implies $\dot{z}(0) = 0$, which is the Machenhauer condition (9.6.5). Clearly,

$$\left.\frac{d^p c}{dt^p}\right|_{t=0} = 0 \quad \text{implies} \quad \left.\frac{d^p z}{dt^p}\right|_{t=0} = 0$$

for any value of p, so the application of the bounded derivative principle to (9.7.3) is identical to the procedure of (9.6.12).

The Baer–Tribbia procedure approaches the initialization of (9.7.2) from a slightly different perspective. It is assumed that $z(t)$ can be expanded in a power series in the small parameter ε. Thus,

$$z(0) = \sum_{k=0}^{\infty} z_k(t)\varepsilon^k \tag{9.7.4}$$

where $z_k(t)$ is the kth expansion coefficient of the power series. Denote

$$R_z^k(t) = R_z(\underline{Y}, \underline{Z}_k, t) \tag{9.7.5}$$

Substitute (9.7.4) into (9.7.2) and collect the terms multiplying different powers of ε. Evaluating the result at $t = 0$ leads to the following sequence of equations:

$$\varepsilon^0[i\sigma_z z_0(0)] = 0 \tag{9.7.6}$$

$$\varepsilon^1[\dot{z}_0(0) + i\sigma_z z_1(0) - R_z^0(0)] = 0 \tag{9.7.7}$$

$$\varepsilon^k[\dot{z}_{k-1}(0) + i\sigma_z z_k(0) - R_z^{k-1}(0)] = 0 \qquad (9.7.8)$$

From (9.7.6), it follows that $z_0(0) = 0$. This implies that $\dot{z}_0(0) = 0$ in (9.7.7). Thus,

$$z_0(0) = 0 \qquad (9.7.9)$$

$$z_1(0) = \frac{R_z^0(0)}{i\sigma_z} \qquad (9.7.10)$$

$$z_2(0) = \frac{R_z^1(0)}{i\sigma_z} - \frac{\dot{z}_1(0)}{i\sigma_z} = \frac{R_z^1(0)}{i\sigma_z} + \frac{1}{(\sigma_z)^2}\frac{dR_z^0(0)}{dt} \qquad (9.7.11)$$

and the general term is given by

$$z_k(0) = -\sum_{s=1}^{k} \frac{1}{(-i\sigma_z)^s} \frac{d^{s-1}\, R_z^{k-s}(0)}{dt^{s-1}} \qquad (9.7.12)$$

The sequence $z_0(0), z_1(0), \ldots, z_k(0)$ is noniterative, unlike the Machenhauer scheme. The unscaled form of (9.7.9–12) is an approximation to (9.6.11). Equation (9.6.11) is an implicit relationship because $R_z(0)$ is a function of z. The sequence (9.7.9–12) is easier to use because $R_z^{k-s}(0)$ is a function of $z_{k-1}, z_{k-2}, \ldots, z_0$, all of which have been previously calculated. Thus, no iteration is necessary.

The Baer–Tribbia scheme is used less than the Machenhauer scheme for two reasons. First, it involves the scaled equations whereas most models are written in unscaled form. Thus, it is not simple to use the existing model software directly in the initialization process. Second, it requires generating time derivatives of R_z. However, the procedure of Tribbia (1984) discussed at the end of Section 9.6 is used to circumvent this problem.

Exercises

9.1 Show that the eigenvalues of (9.1.11) with boundary conditions (9.1.14) are real and positive if $P_T = 0$.

9.2 Suppose that (9.1.11) with boundary conditions (9.1.14) applied at P_S and $P_T = 0$ is discretized in the following way. Define $K + 1$ equally spaced pressure levels P_k between $P_S = P_{K+1}$ and $P_T = P_1$. The distance between pressure levels $\Delta P = P_{k+1} - P_k$. Define $\tilde{\Gamma}_k = \tilde{\Gamma}(P_k + \Delta P/2)$ and $Z_k = Z(P_k)$, $1 \le k \le K$, with $Z_S = Z_{K+1} = Z(P_S)$ and approximate (9.1.11) by

$$\frac{1}{\Delta P}\left[\frac{1}{\tilde{\Gamma}_k}\frac{Z_{k+1} - Z_k}{\Delta P} - \frac{1}{\tilde{\Gamma}_{k-1}}\frac{Z_k - Z_{k-1}}{\Delta P}\right] + \frac{Z_k}{g\tilde{h}} = 0$$

for $2 \le k \le K$ and (9.1.14) by

$$\frac{Z_2 - Z_1}{\Delta P} = 0 \quad \text{and} \quad \frac{Z_S - Z_K}{\Delta P} + \frac{P_S \tilde{\Gamma}_K Z_S}{R\tilde{T}_S} = 0$$

Show that the eigenvalues $g\tilde{h}$ of this discrete approximation are real and positive.

9.3 On the sphere, a symmetric function has the property $f_S(-\phi) = f_S(\phi)$ whereas an antisymmetric function has $f_A(-\phi) = -f_A(\phi)$, where ϕ is latitude. Thus, the velocity and geopotential components u, v, and Φ can be written as the sum of symmetric and antisymmetric components: $u = u_S + u_A$, $v = v_S + v_A$, and $\Phi = \Phi_S + \Phi_A$. Show that the horizontal structure equations (9.3.1–3) constitute two separate sets of equations: a symmetric set involving Φ_S, u_S, and v_A and an antisymmetric set involving Φ_A, u_A, and v_S.

9.4 Consider the horizontal structure equations (9.3.1–3). Expand u, v, and Φ using (9.3.4) and eliminate u and v to derive a single equation in Φ. Consider the special case when u, v, and Φ are independent of longitude λ. Show that there must be zero eigenvalues in this case and that the eigenvectors corresponding to these eigenvalues must have $v = \chi = 0$ and u, ψ, and Φ must satisfy the linear balance equation (7.5.3). Show that the remaining eigenvalues must come in pairs, say, σ_+ and σ_-, with $\sigma_- = -\sigma_+$. Define the eigenvector associated with σ_+ as $[\Phi_+(\phi) \quad u_+(\phi) \quad v_+(\phi)]^T$ and similarly for σ_-. Show that $\Phi_+ = -\Phi_-$, $u_+ = -u_-$, and $v_+ = v_-$.

10

Normal mode initialization: applications

The normal mode initialization procedure, described in detail in the previous chapter, is widely used to initialize global numerical models. This method is successful because it deals explicitly with the essence of the initialization problem: the separation between fast and slow time scales. In addition to its practical value in the data assimilation cycle, normal mode initialization has a more theoretical dimension as an extension to quasi-geostrophic theory. Leith (1980) has developed a useful conceptual framework called the *slow manifold*, in which we can examine the whole data assimilation cycle in a straightforward manner. This chapter demonstrates the practical utility of the normal mode procedure and then examines the slow manifold concept and its ramifications.

10.1 Some results from normal mode initialization experiments

The normal mode initialization procedure is effective in eliminating high-frequency oscillations from integrations of nonlinear global barotropic and baroclinic models. Temperton and Williamson (1981) and Williamson and Temperton (1981) used the Machenhauer balance condition (Section 9.6) to initialize a global baroclinic forecast model. This is demonstrated in Figure 10.1, which shows a plot of the surface pressure as a function of time for 24 hours. When there is no initialization (solid curves), surface pressure oscillations with amplitudes of almost 12 mb are present. In (a), the dashed curve shows the surface pressure trace after linear initialization, Equation (9.6.6); the amplitude of the oscillation is slightly reduced, but the results are clearly far from satisfactory. Apparently, the nonlinear terms in (9.5.19) are reexciting the high-frequency oscillations. In (b), the curve shows the results of two nonlinear iterations of the initialization scheme (9.6.7–8). Now all high-frequency oscillations have been eliminated. Because Figure 10.1 is for the same case as Figure 6.1, we now see how the high-frequency oscillations in Figure 6.1 were eliminated.

The convergence of iterative schemes of the form (9.6.6–9) was also demonstrated

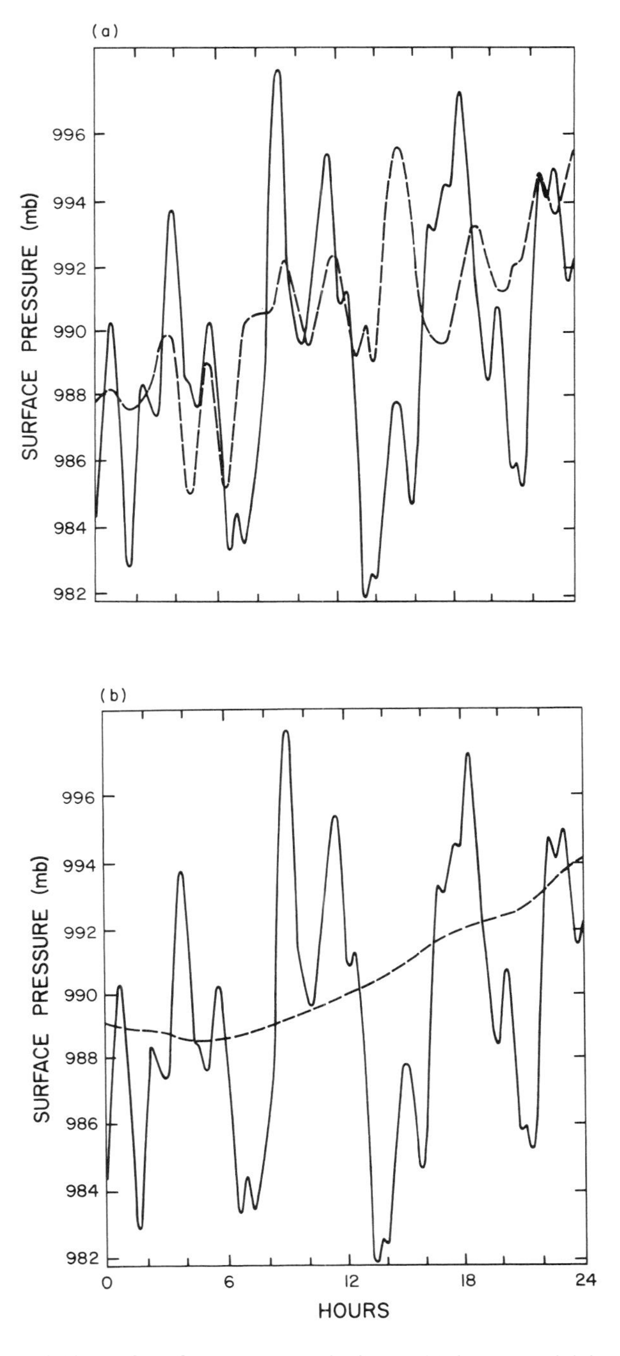

Figure 10.1 Time evolution of surface pressure during a 24 hour model integration for (a) linear and (b) nonlinear normal mode initialization. Solid curves, uninitialized; dashed curves, initialized. (After Williamson and Temperton, *Mon. Wea. Rev.* **109**: 745, 1981. The American Meteorological Society.)

by Temperton and Williamson (1981) and Williamson and Temperton (1981). They defined an objective measure of the convergence as follows. Suppose y_j is any slow mode with equivalent depth $\tilde{h}_j$. Correspondingly, z_j is a fast mode of equivalent depth $\tilde{h}_j$. Then the model equations (9.5.18–19) for these modes can be written as

$$\dot{y}_j + 2\Omega i\sigma_y^j y_j = R_y^j$$

$$\dot{z}_j + 2\Omega i\sigma_z^j z_j = R_z^j$$

where j is the vertical mode number and σ_y^j, σ_z^j, R_y^j, and R_z^j correspond to σ_y, σ_z, R_y, and R_z of (9.5.18–19). A measure of the balance achieved by the initialization scheme is written as

$$\mathrm{BAL}_j(y) = \sum_{y_j} (\dot{y}_j)(\dot{y}_j^*) \tag{10.1.1}$$

$$\mathrm{BAL}_j(z) = \sum_{z_j} (\dot{z}_j)(\dot{z}_j^*) \tag{10.1.2}$$

where * indicates complex conjugate. $\mathrm{BAL}_j(y)$ indicates the normal level of slow mode time tendencies and remains virtually unchanged throughout the initialization process. $\mathrm{BAL}_j(z)$ would be identically zero if the Machenhauer condition (9.6.5) were perfectly satisfied for all fast modes z_j with equivalent depth $\tilde{h}_j$. Presumably, $\mathrm{BAL}_j(z)$ would be large for an uninitialized model and would decrease after each iteration of the initialization procedure.

Figure 10.2 plots $\mathrm{BAL}_j(y)$ and $\mathrm{BAL}_j(z)$ for successive iterations of the initialization procedure. The external mode is $j = 1$, and decreasing equivalent depth is indicated by j increasing. The Rossby curve indicates the slow mode balance $\mathrm{BAL}_j(y)$ and is plotted as a reference. The uninitialized curve indicates $\mathrm{BAL}_j(z)$ before initialization whereas the L curve indicates linear initialization (9.6.6); N1, N2, and N3 indicate successive nonlinear iterations (9.6.9). As is apparent, the model is initially unbalanced, but the balance is improved slightly by linear initialization. Two nonlinear iterations of the scheme substantially improve the balance, but the third iteration causes little change. Most improvement in the balance comes with the external and graver internal modes. This result is consistent with the results of the Hinkelmann–Phillips model of Section 7.2.

The synoptic implications of the Machenhauer balance condition (9.6.5) have been examined by Temperton and Williamson (1981), Williamson and Temperton (1981), and Daley (1979). Of particular interest is the generation of a divergent windfield. Daley (1979) considered the balance obtained by the application of (9.6.9) to a hemispheric baroclinic model. Figure 10.3 demonstrates the vertical motion generated by this procedure. (a) shows the surface synoptic situation (surface pressure and a schematic frontal contour) over a portion of the model domain. (b) shows the vertical motion ω (dP/dt in millibars/hour) generated by the application of normal mode initialization to this model. U indicates upward vertical motion ($\omega \leq 0$); D indicates downward vertical motion ($\omega \geq 0$). It is apparent that upward vertical motion is

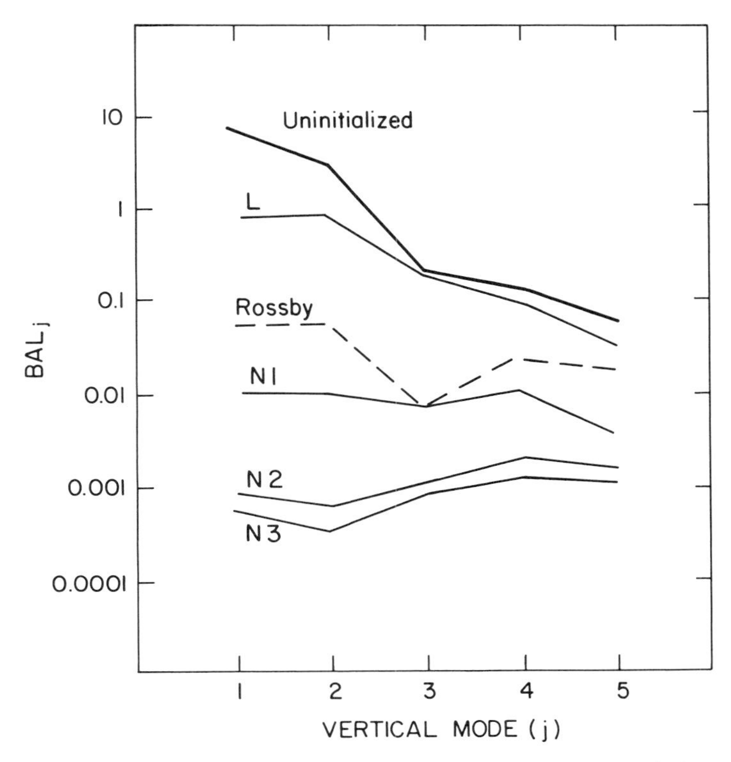

Figure 10.2 Convergence of normal mode initialization as a function of vertical mode number j. BAL statistics are plotted for uninitialized fast modes, Rossby modes, linear initialization (L), and three nonlinear iterations (N1, N2, and N3). (After Williamson and Temperton, *Mon. Wea. Rev.* **109**: 745, 1981. The American Meteorological Society.)

generated ahead of the warm front, with downward vertical motion behind the cold front. This is consistent with the quasi-geostrophic result (Figure 7.6).

The evolution of a vertical profile of ω with time is shown in Figure 10.4 for the spot marked X (Labrador coast) in Figure 10.3(b). Figure 10.4(a) is the uninitialized case and (b) the initialized case. We can see that in the uninitialized case, the vertical motion profile (essentially zero at initial time) oscillates before settling down by 12 hours. In the initialized case, the initial ω profile is consistent with the other variables of the model (rotational wind and mass); thus, there is little oscillation. The ω profiles at 12 hours in (a) and (b) are similar, consistent with the discussion of Section 6.3.

Vertical motions can also be induced in the atmosphere by orographic forcing or heating. In (9.1.1), the $\mathbf{R}_v$ term would contain the orographic forcing and the R_Φ term of Equation (9.1.3) would contain the heating. Figure 10.5 depicts the orographically induced vertical motion as generated by the normal mode initialization procedure. In general, the orographically induced vertical motion is smaller than the

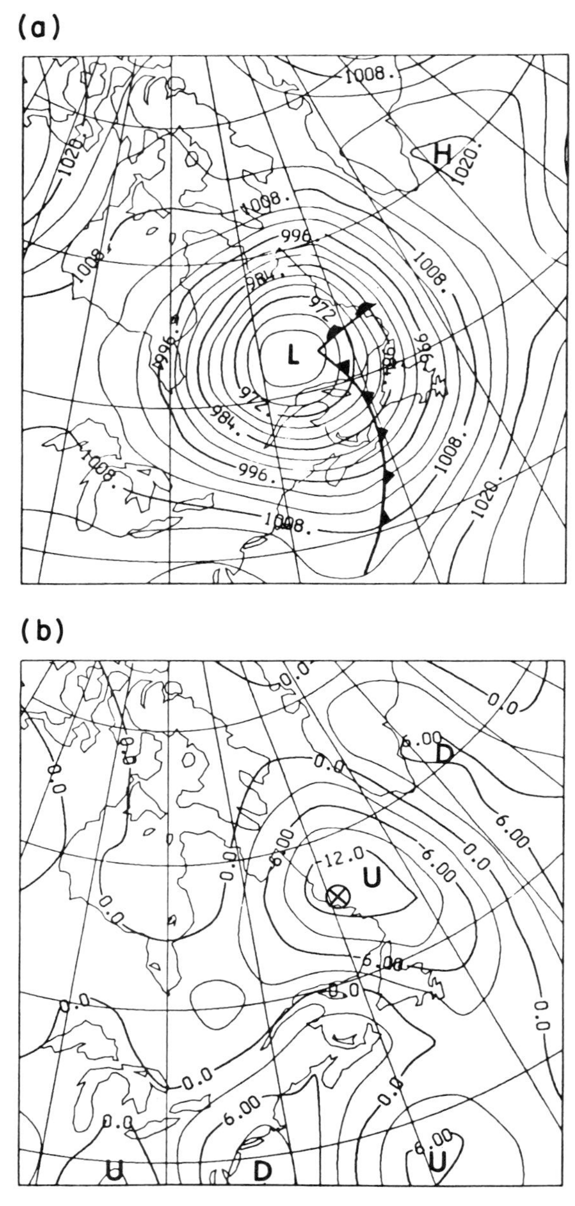

Figure 10.3 Vertical motion fields ω generated by nonlinear normal mode initialization: (a) surface pressure (millibars) and schematic frontal positions; (b) corresponding vertical motions (millibars/hour). (After Daley 1979)

synoptically induced vertical motion and is hard to identify in normal synoptic situations. Daley (1979) demonstrated that the procedure (9.6.6–9) generates realistic orographically induced vertical motions by the use of idealized data. Objective analyses were zonally averaged before being initialized. The zonally averaged 500 mb

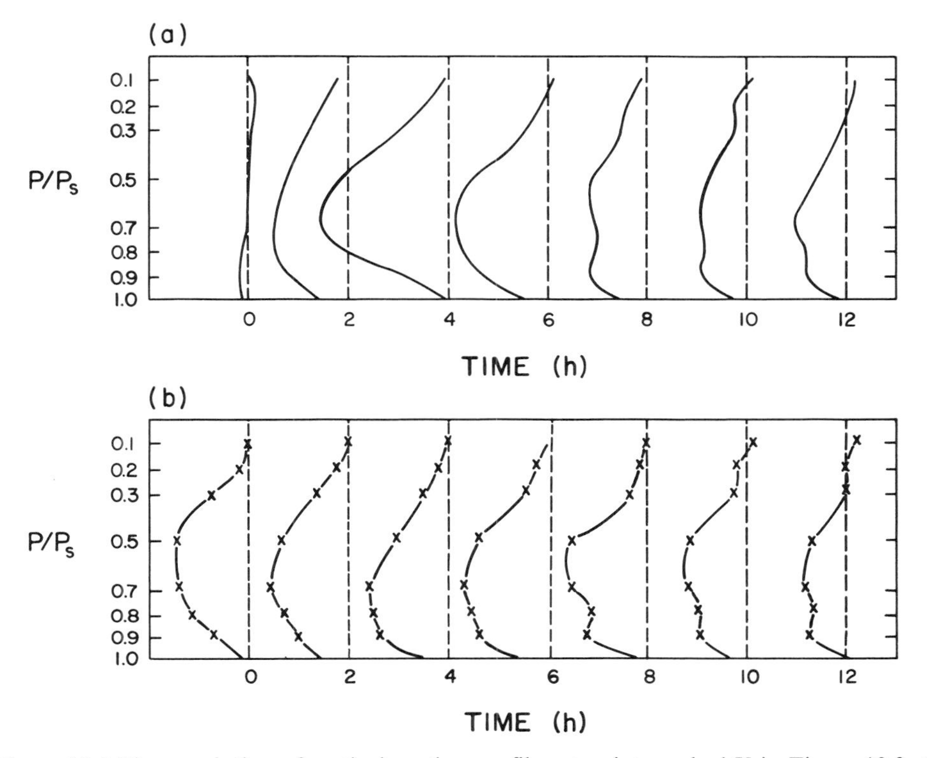

Figure 10.4 Time evolution of vertical motion profiles at point marked X in Figure 10.3: (a) uninitialized; (b) initialized. (After Daley 1979)

geopotential field (dekameters) is shown over a portion of the domain in Figure 10.5(a); (b) shows the orography (dekameters) – essentially the Rocky Mountain chain, which is perpendicular to the zonal flow; (c) shows the vertical motion (millibars/hour) generated by the initialization process. The initialization procedure indicates ascending air upstream of the barrier and descending air downstream of the barrier.

The normal mode procedure can generate a consistent vertical motion field. This is a useful by-product because the divergent wind is difficult to measure, as indicated in earlier chapters.

10.2 Separation of slow and fast time scales

In Section 9.5, we divided the model equations into fast mode equations (9.5.19) which required initialization and slow mode equations (9.5.18) which did not. At that point, we assumed there was a clear distinction between the slow modes and the fast modes. For a midlatitude f-plane (see Chapter 6), this distinction is always clear because the inertia–gravity modes never have a frequency smaller than $|f|$.

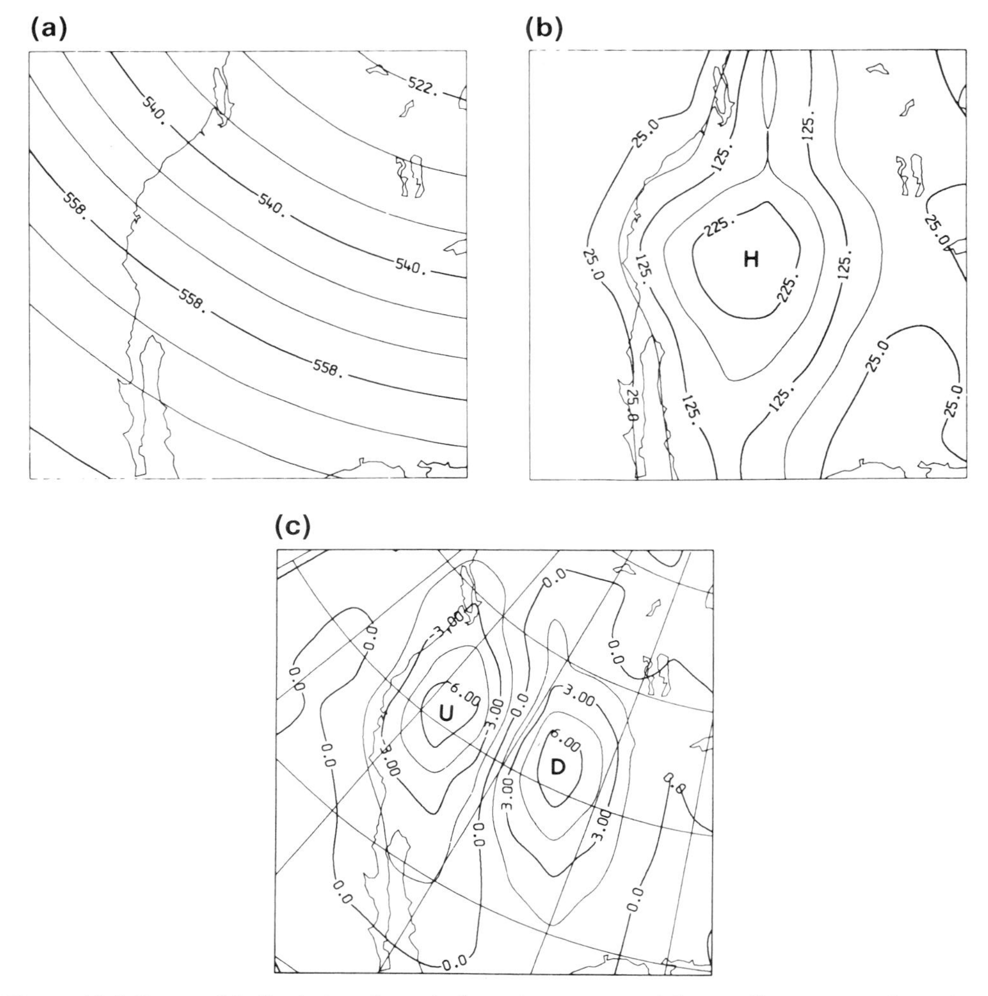

Figure 10.5 Orographically induced vertical motion generated by nonlinear normal mode initialization: (a) 500 millibar geopotential (dekameters); (b) orography (dekameters); (c) vertical motion (millibars/hours). U and D are upward and downward vertical motions. (After Daley 1979)

Thus, the time scales of the inertia–gravity modes are less than about 12 hours and are easily distinguished from the advective time scales of a day or more.

This simple frequency separation does not occur on the sphere (Figure 9.2). Inertia–gravity modes with small equivalent depths can have lower frequencies than Rossby waves with large equivalent depths. Thus, on the sphere, slow and fast modes cannot be separated merely on the basis of being characterized as Rossby modes or inertia–gravity modes. All Rossby modes on the sphere can be characterized as slow, but not all inertia–gravity modes can be characterized as fast.

We can classify modes as slow or fast on the basis of kind (Rossby, Kelvin, inertia–gravity), horizontal scale, vertical scale (equivalent depth), linear frequency σ, or even by examining the temporal characteristics of each mode in a numerical integration. There are many possibilities, and the normal mode theory outlined in Chapter 9 does not give us any guidance. Investigators have often used equivalent depth to classify modes operationally, though linear frequency seems a more natural choice.

Errico (1984b) approached this problem by considering the actual balance in a time integration of the model that was to be initialized – a spherical baroclinic model with a comprehensive physical parametrization that was normally used for the simulation of the earth's climate. Because this model could be integrated indefinitely, stable statistics could be collected. During the integration, the model variables were projected into normal mode space (see Section 9.5) to produce equations in the normal mode form (9.5.12).

Equation (9.5.12) can be written as

$$\dot{x} + 2\Omega i \sigma_x x - R_x = 0 \tag{10.2.1}$$

which has a time tendency term, a linear term, and a nonlinear (and forcing) term. Following the Machenhauer condition (9.6.5), Errico could say that the mode x is in balance if $\dot{x}$ is small compared with either $2\Omega i \sigma_x x$ or R_x. From the long time integration, he could collect statistics on R_x, $2\Omega i \sigma_x x$, and $\dot{x}$. The results were then summarized as follows. Each discrete frequency band $\Delta\sigma_k$, $1 < k < K$, contains frequencies corresponding to several different normal modes. From (10.2.1), the following quantities are calculated for each frequency band:

$$\begin{aligned} A_k^2 &= \frac{1}{N_k} \sum \dot{x}\dot{x}^*, \qquad & B_k^2 &= \frac{4\Omega^2}{N_k} \sum \sigma_x^2 x x^* \\ C_k^2 &= \frac{1}{N_k} \sum R_x R_x^*, \qquad & N_k &= \sum x x^* \end{aligned} \tag{10.2.2}$$

where $\sum$ indicates the sum over all σ_x in the frequency band $\Delta\sigma_k$, and (*) indicates complex conjugation. Thus, A_k, B_k, and C_k were normalized magnitudes (s^{-1}) of the time tendency, linear, and nonlinear terms for frequencies in the band.

These results are illustrated in Figure 10.6 for the inertia–gravity modes of the model. The nondimensional frequency σ is plotted logarithmically against normalized magnitudes, and the periods corresponding to σ are at the top of the diagram. The values of A_k, B_k, and C_k in the various frequency intervals $\Delta\sigma_k$ have been plotted and then smooth curves drawn through them. It is clear that curve B should be nearly a straight line. (It would be a straight line if there were one mode per frequency band.) In general, the tendency term (curve A) is small compared with the nonlinear term (curve C) and the linear term (curve B) for the higher frequencies; but it is not small for the lower frequencies. It would appear that the Machenhauer balance ($\dot{x}$ negligible) is well maintained for time scales of 12 hours or less. The balance begins to deteriorate for time scales between 12 and 24 hours; and for time scales greater than 24 hours, there is clearly no Machenhauer balance.

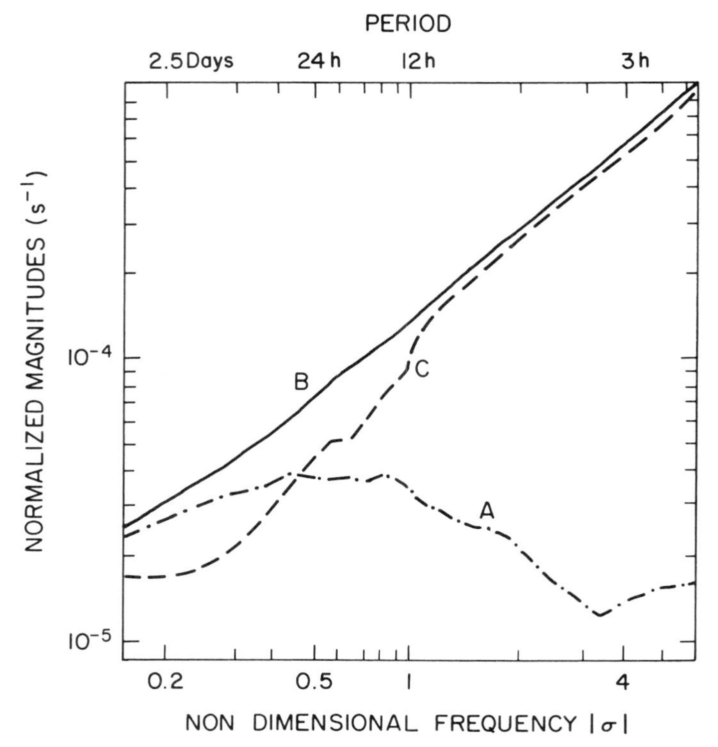

Figure 10.6 Normalized time tendencies in the model as a function of the linear frequency of the normal modes: curve A, time tendencies; curve B, linear terms, and curve C, nonlinear and forcing terms. (After Errico, *Mon. Wea. Rev.* **112**: 2439, 1984. The American Meteorological Society.)

This experiment makes it clear that there is no point in trying to apply the Machenhauer condition (9.6.5) to inertia–gravity modes with time scales longer than 24 hours. Even inertia–gravity modes with time scales between 12 and 24 hours have to be examined carefully. Thus, for initialization, a cutoff frequency σ_c corresponding to time scales of 12–24 hours should be chosen to separate the slow and fast modes. Figure 10.6 presents the balance information in a condensed form. In principle, the balance in each mode could be examined in turn and a decision made as to whether that mode was balanced or not in the model integration. Thus, only modes that were balanced in the model would be initialized.

Note that this procedure for determining the separation between slow and fast modes depends on the balance that exists in the model to be integrated. This is not necessarily the same balance that exists in the real atmosphere, but the true atmospheric balance is difficult to determine accurately with the present observation network. Moreover, the aim of initialization is to suppress oscillations in the model itself, so it is the model's own balance that is important here.

10.3 The slow manifold

In Section 6.4, we introduced a simple linear shallow water model that illustrated many of the basic initialization concepts. In reality, as noted in Chapters 7 and 9, initialization is really concerned with nonlinear equations. In general, it is difficult to examine the nonlinear initialization process analytically because few nonlinear fluid dynamics equations have analytic solutions. One exception is the isolated barotropic vortex model of Tribbia (1981). We use this simple model here to introduce the *slow manifold* concept and in Section 10.4 to explore the relation between normal mode and quasi-geostrophic initialization.

The shallow water equations on an f-plane in plane polar coordinates are given by (6.6.1). Assume the flow is axisymmetric, that is, independent of the tangential coordinate ϕ. Now introduce a streamfunction ψ and a velocity potential χ, using Helmholtz's theorem (6.6.2). This gives

$$\frac{\partial}{\partial t}\nabla^2\psi + f_0\nabla^2\chi = \frac{1}{r}\frac{\partial}{\partial r}rR_v \tag{10.3.1}$$

$$\frac{\partial}{\partial t}\nabla^2\chi - f_0\nabla^2\psi + \nabla^2\Phi = \frac{1}{r}\frac{\partial}{\partial r}rR_u \tag{10.3.2}$$

$$\frac{\partial}{\partial t}\Phi + \tilde{\Phi}\nabla^2\chi = R_\Phi \tag{10.3.3}$$

where

$$u = \frac{\partial\chi}{\partial r}, \qquad v = \frac{\partial\psi}{\partial r}, \qquad R_v = \left(-u\frac{\partial v}{\partial r} - \frac{uv}{r}\right), \qquad R_u = \left(\frac{v^2}{r} - u\frac{\partial u}{\partial r}\right), \qquad R_\Phi = -\frac{1}{r}\frac{\partial}{\partial r}(ru\Phi)$$

In Section 6.6 we assumed an infinite domain. This time, assume the domain is bounded at $r = r_0$ with the following boundary conditions,

$$u,\ v,\ \psi',\ \chi',\ \Phi' = 0 \qquad \text{at} \qquad r = 0,\ r_0 \tag{10.3.4}$$

and $'$ indicates differentiation with respect to r. These boundary conditions correspond to no-slip conditions.

Assume ψ, χ, and Φ have the following spatial dependence,

$$\begin{vmatrix}\psi(r,t)\\ \chi(r,t)\\ \Phi(r,t)\end{vmatrix} = \begin{vmatrix}\hat{\psi}\\ i\hat{\chi}\\ f_0\sqrt{K}\hat{\Phi}\end{vmatrix} J_0(kr/r_0) \tag{10.3.5}$$

where $K = k^2\tilde{\Phi}/r_0^2 f_0^2$, and the normalization is as in expansion (6.4.11). $J_0(r)$ is the Bessel function of the first kind and order zero plotted in Figure G1 (Appendix G); k ($\simeq 3.8317$) is the first root of $J_0(r) = 0$, and $-k^2/r_0^2$ is the eigenvalue of the Laplacian operator. This functional form (10.3.5) satisfies the boundary conditions (10.3.4) but cannot be used to find the general solution of (10.3.1–3) because only a single value of k has been chosen.

Insert (10.3.5) into (10.3.1–3), multiply by $rJ_0(kr/r_0)$, and integrate from $r=0$ to $r=r_0$. The result is

$$\frac{d\hat{\psi}}{dt}+if_0\hat{\chi}=if_0C_\psi\hat{\chi}\hat{\psi}$$

$$i\frac{d\hat{\chi}}{dt}-f_0\hat{\psi}+f_0\sqrt{K}\hat{\Phi}=-f_0(C_\chi^\psi\hat{\psi}\hat{\psi}+C_\chi^\chi\hat{\chi}\hat{\chi}) \tag{10.3.6}$$

$$\frac{d\hat{\Phi}}{dt}-if_0\sqrt{K}\,\hat{\chi}=if_0C_\Phi\hat{\Phi}\hat{\chi}$$

where $C_\psi=C_0[\bar{F}_1+\bar{F}_2]$, $C_\chi^\psi=C_0\bar{F}_2$, $C_\chi^\chi=C_0\bar{F}_1$, $C_\Phi=-\bar{F}_3/f_0I_1$, $\bar{F}_n=\int_0^{r_0}J_0(kr)F_n'(r)\,dr$, $I_1=\int_0^{r_0}J_0(kr)J_0(kr)r\,dr$, $C_0=r_0^2/(f_0k^2I_1)$, $F_1(r)=rJ_0'(kr)J_0''(kr)$, $F_2(r)=J_0'(kr)J_0'(kr)$, $F_3(r)=rJ_0(kr)J_0'(kr)$. C_ψ, C_χ^ψ, C_ψ^χ, C_Φ, and I_1 are real numbers, known technically as interaction coefficients.

Consider the linear case, $R_u=R_v=R_\Phi=0$, or $C_\psi=C_\chi^\psi=C_\chi^\chi=C_\Phi=0$. Then (10.3.6) is formally identical to (6.4.13), and all the results of Section 6.4 follow identically. Define Rossby and inertia–gravity mode amplitudes, y, z_1, and z_2 as in (6.4.21). Then, linear initialization consists in setting z_1 and z_2 initially to zero, which gives (6.5.1).

The real goal here is the analysis of the nonlinear case (C_ψ, C_χ^ψ, C_χ^χ, and C_Φ not equal to zero). The Machenhauer condition (6.5.1) implies that $\dot{z}_1=\dot{z}_2=0$. This condition can be applied to (10.3.6) by rewriting in terms of y, z_1, z_2, $\dot{y}$, $\dot{z}_1$, and $\dot{z}_2$ using (6.4.21) and setting $\dot{z}_1=\dot{z}_2=0$. This is rather tedious; a simpler way to obtain the same result is to note that (6.4.21) implies that

$$\frac{d\hat{\chi}}{dt}=\frac{\dot{z}_1-\dot{z}_2}{\sqrt{2}},\qquad \frac{d\hat{\psi}}{dt}-\sqrt{K}\frac{d\hat{\Phi}}{dt}=\sqrt{\frac{1+K}{2}}\,[\dot{z}_1+\dot{z}_2] \tag{10.3.7}$$

In other words, the Machenhauer balance condition is satisfied if

$$\frac{d\hat{\chi}}{dt}=0,\qquad \frac{d}{dt}[\hat{\psi}-\sqrt{K}\,\hat{\Phi}]=0 \tag{10.3.8}$$

Application of (10.3.8) to (10.3.6) yields

$$(1+K)\hat{\chi}=C_\psi\hat{\chi}\hat{\psi}-\sqrt{K}C_\Phi\hat{\chi}\hat{\psi},\qquad \hat{\psi}-\sqrt{K}\hat{\Phi}=C_\chi^\psi\hat{\psi}\hat{\psi}+C_\chi^\chi\hat{\chi}\hat{\chi} \tag{10.3.9}$$

The first equation of (10.3.9) implies that

$$\hat{\chi}=0,\qquad \hat{\psi}-\sqrt{K}\hat{\Phi}=C_\chi^\psi\hat{\psi}\hat{\psi} \tag{10.3.10}$$

It is left as Exercise 10.1 to show that (10.3.10) implies that the Rossby mode y is stationary. In real-space form, (10.3.10) is a form of the gradient wind equation (7.6.12):

$$u=\chi=0,\qquad \frac{1}{r}\frac{\partial}{\partial r}r\left(\frac{\partial\Phi}{\partial r}-f_0v-\frac{v^2}{r}\right)=0 \tag{10.3.11}$$

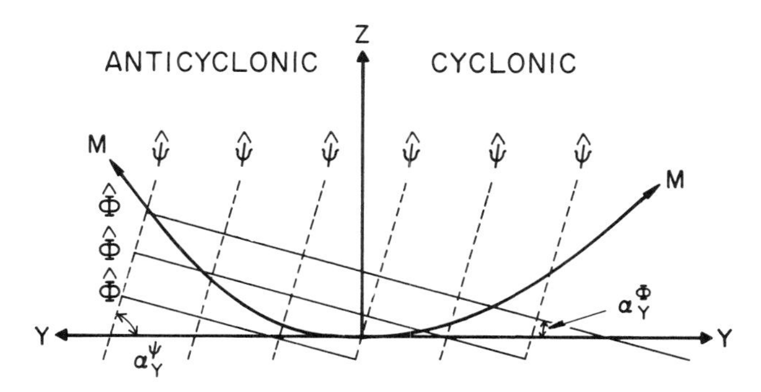

Figure 10.7 Slow manifold diagram for Tribbia vortex model: Z, fast manifold; Y, Rossby manifold; M, slow manifold; ψ and Φ, data manifolds. *Anticyclonic* and *cyclonic* denote flow direction in the vortex (cyclonic is anticlockwise in the northern hemisphere). (After Daley, *Mon. Wea. Rev.* **108**: 1719, 1980. The American Meteorological Society.)

The features of the Tribbia vortex model are illustrated graphically in Figure 10.7, in which the abscissa is the Rossby mode amplitude y and the ordinate an inertia–gravity mode amplitude z_1. (The diagram would not be changed if z_2 were used instead of z_1.) Also drawn are lines of constant streamfunction $\hat{\psi}$ and constant geopotential $\hat{\Phi}$, which are derived from (6.4.21) and so must be straight lines. Curve M corresponds to (10.3.10), which is quadratic in $\hat{\psi}$.

The form of Figure 10.7 depends only on K (10.3.5). Thus, this diagram is uniquely specified by $\bar{\Phi}$, r_0, and f_0, which were chosen as follows: $\bar{h} = g^{-1}\bar{\Phi} = 1$ km, $f_0 = f(45°)$, and $r_0 = 1000$ km. Changing the normalization assumed in (10.3.5) or replacing the streamfunction with the rotational wind or vorticity would only cause a relabeling of the isopleths of streamfunction or geopotential. The angles with which $\hat{\psi}$ and $\hat{\Phi}$ intersect y or z_1 and the shape of curve M would remain unchanged.

The angles with which ψ and Φ intersect y are the Rossby projection angles α_y^ψ and α_y^Φ, respectively. The values can be derived by noting that in (6.4.21), $\hat{\psi} = $ constant, $\hat{\chi} = $ constant, and $\hat{\Phi} = $ constant represent planes in a three-dimensional phase space whose coordinates are y, z_1, and z_2. The direction of a line normal to the $\hat{\psi} = $ constant plane with respect to the y (Rossby) axis is given by

$$\cos\left(\frac{\pi}{2} - \alpha_y^\psi\right) = \frac{R_\psi}{[(R_\psi)^2 + (G_\psi^1)^2 + (G_\psi^2)^2]^{1/2}}$$

where $\cos(\pi/2 - \alpha_y^\psi)$ is the direction cosine and α_y^ψ the Rossby projection angle. R_ψ, G_ψ^1, and G_ψ^2 are defined in (6.4.17–18). Subscript y represents the Rossby axis, and superscript ψ denotes the $\hat{\psi} = $ constant plane. Thus,

$$\sin^2 \alpha_y^\psi = \frac{K}{1+K}, \qquad \sin^2 \alpha_y^\Phi = \frac{1}{1+K} \tag{10.3.12}$$

The Rossby projection angle is related to the geostrophic adjustment theory discussed in Section 6.6. If $K \gg 1$, then $\alpha_y^{\psi} \to \pi/2$ and $\alpha_y^{\Phi} \to 0$; this corresponds to the case in which the geopotential adjusts to the windfield. Conversely, if $K \ll 1$, then $\alpha_y^{\psi} \to 0$ and $\alpha_y^{\Phi} \to \pi/2$; $K \ll 1$ implies a small Rossby radius of deformation, which implies that the rotational windfield adjusts to the geopotential field. An extension of this concept to the spherical global case is given in Daley (1980b).

Figure 10.7 is called a *slow manifold diagram*, after Leith (1980). The Rossby mode axis y is the *Rossby manifold*, and the inertia–gravity mode axis z is the *gravity manifold*. The curve M is the *slow manifold*, and the isopleths of $\hat{\psi}$ and $\hat{\Phi}$ are *data manifolds*. The Tribbia vortex model is so simple that the slow manifold diagram can be completely determined analytically.

In more complicated models, it is not so easy to determine features of the slow manifold diagram; numerical methods must be used (Daley 1980b). For a general model, the slow manifold diagram represents the amplitude of the model normal modes in a multidimensional phase space. The Rossby and inertia–gravity modes represent subsets or manifolds of this phase space. (For the purpose of this section, the Rossby manifold could also include very low frequency inertia–gravity modes.) The slow manifold is defined to be the locus of all points evolving slowly in time in the model.

In a linear model, only the Rossby modes are evolving slowly in time, so the slow manifold and Rossby manifolds coincide. In a nonlinear model, the situation is different. From (9.5.19), we see that the linear term $2\Omega i\sigma_z z$ produces high-frequency oscillations whereas the nonlinear term R_z acts primarily as a low-frequency forcing term. Thus, the solution for z contains both low and high frequencies. The low-frequency part of the solution is the balanced solution sought by initialization processes. In a nonlinear model, the slow manifold corresponds to the balanced solution and differs from the Rossby manifold because the projection on the fast modes z is nonzero. The balanced part of the solution is denoted z_B for a single mode or $\underline{Z}_B$ for the ensemble of fast modes.

In principle, the slow manifold can be obtained from (9.6.12). In practice, a lower-order approximation such as the Machenhauer condition (9.6.5) is usually used. Thus, an approximate slow manifold can be considered to be the locus of all points where $\dot{\underline{Z}} = 0$. It is assumed that if a model state is on the slow manifold M, subsequent model states will remain on or close to the slow manifold. Thus, model integrations will contain a low level of high-frequency oscillations.

Consider again the slow manifold diagram of Figure 10.7 which corresponds to the Tribbia vortex model. The slow manifold represents (10.3.10), which is derived by using the Machenhauer condition $\dot{\underline{Z}} = 0$ in (10.3.6). However, this case is rather special because $\dot{\underline{Z}} = 0$ implies that $\hat{\chi} = 0$, which implies $\dot{\underline{Y}} = 0$ (Exercise 10.1). Thus, in the Tribbia vortex, the slow manifold describes a steady-state solution.

More realistic models have many degrees of freedom, and so the slow manifold cannot be represented as a single curve on a plane. In addition, because the Rossby mode amplitudes are not usually independent of time, the slow manifold does not represent a steady-state solution. A particular point in this multidimensional phase

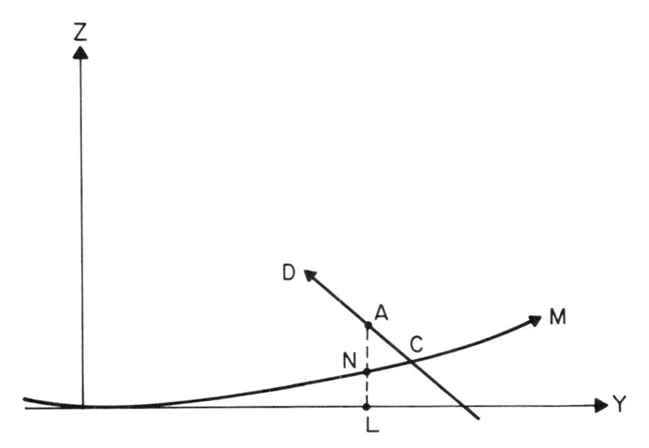

Figure 10.8 Schematic slow manifold diagram for a comprehensive model: Z and Y, fast and Rossby manifolds: M, slow manifold; D, data manifold. Points A, N, L, and C are discussed in the text.

space (i.e., a model state) can have a unique representation in terms of the modes $\underline{Z}$ and $\underline{Y}$. This point in the multidimensional space also corresponds to a single spatial configuration of the dependent variables of the model (geopotential, wind, etc.).

The slow manifold diagram is particularly useful for illustrating initialization procedures. A schematic slow manifold diagram is shown in Figure 10.8. Although the diagram is two dimensional, it is meant to represent a multidimensional phase space. Here Z refers to the gravity manifold, Y to the Rossby manifold, and M to the slow manifold. Suppose the spatial structures of one of the model variables (geopotential, say) were held fixed while the other variables were allowed to vary. Then, the locus of model states that satisfy these conditions is called a data manifold. A particular data manifold is denoted by the line D in this figure. The nonlinear normal mode initialization procedure discussed in Sections 9.6 and 9.7 is also shown schematically, by the dashed line joining the points A, N, and L:

Point A indicates the observed/analyzed state of the atmosphere. If this point were to be used as an initial condition for the forecast model, then inertia–gravity waves of amplitude proportional to the distance of A from M would be excited. All examples in Sections 9.6 and 9.7 considered the case in which $\underline{Y}$ is unchanged (i.e., slow mode constrained in the nomenclature of Chapter 6).

Point L indicates linear normal initialization (9.6.2), which consists of setting $\underline{Z}$ to zero without changing $\underline{Y}$, and is the intersection of a vertical line (independent of $\underline{Y}$) with the horizontal Rossby ($\underline{Y}$) axis.

Point N indicates the slow mode constrained nonlinear normal mode initialization (9.6.6–9) or (9.7.9–12). It is the intersection of the slow manifold M with a vertical line drawn from the observed/analyzed state A. Point N is on the slow manifold and thus has no high frequencies; it also has the same projection on the Rossby or slow modes as the observed/analyzed state.

The normal mode initialization procedure outlined in (9.6.6–9) consists of proceeding from point A to point L and then iterating toward point N. Some initialization procedures dispense with the linear step, and iteration proceeds directly from point A to point N.

The slow manifold is the locus of all points evolving slowly in time and represents a subspace in the multidimensional phase space of all possible model states. The slow manifold concept is exact for the Tribbia vortex and is helpful in understanding initialization procedures for complicated models, but does it really apply to the atmosphere? In the slow manifold concept, the inertia–gravity modes evolve slowly in time, with their amplitudes determined essentially from the Rossby modes by relationships such as (9.6.4) or (9.6.11). In other words, the inertia–gravity modes have a slave–master relationship with the Rossby modes. Moreover, a model or atmospheric state that was on the slow manifold would remain on the slow manifold indefinitely.

A number of studies (Errico 1982, 1984c; Kopell 1985; Vautard and Legras 1986; Warn and Menard 1986; Lorenz and Krishnamurthy 1987) have shown that this is not the case. Free (unbalanced) inertia–gravity modes do exist in models and in the atmosphere, propagating at their linear phase speeds. These freely propagating inertia–gravity waves can be excited by nonlinear interactions of the balanced flow, and the effect becomes more pronounced at higher Rossby numbers. Usually, the amplitude of these freely propagating inertia–gravity waves is not large, and is prevented from growing larger by dissipative processes. Thus, the slow manifold is really a region in multidimensional phase space where freely propagating inertia–gravity activity is a minimum but not zero.

The implications of these studies for initialization are that conditions such as (9.6.5) or (9.6.12) should not be applied exactly, but rather that $\dot{\underline{Z}}$, $\ddot{\underline{Z}}$, and so on, should be reduced to a tolerable level. In the terminology of Chapter 8, the initialization constraints should be weak rather than strong.

10.4 Normal mode and quasi-geostrophic initialization

Normal mode initialization applied to a linear model requires that $z = 0$ for all fast modes. From the Tribbia vortex model of the previous section, this implies (6.5.1) or

$$\chi = 0, \qquad f_0\psi - \Phi = 0 \tag{10.4.1}$$

in real-space form.

Application of the Machenhauer balance condition (9.6.5) requires that $\dot{z} = 0$ for all fast modes. For the Tribbia vortex model, the Machenhauer condition implies (10.3.8) or

$$\dot{\chi} = 0, \qquad f_0\dot{\psi} - \dot{\Phi} = 0 \tag{10.4.2}$$

An obvious correspondence exists between the condition (10.4.1) and the lowest-order quasi-geostrophic relation (7.4.5–6). Equation (10.4.2) suggests a connection between the Machenhauer balance condition and the next order quasi-geostrophic relations (7.4.18) and (7.4.21). This relationship was explored by Leith (1980) and Kasahara (1982a). In particular, they showed that the normal mode initialization procedure (9.6.6–7) applied to an f-plane baroclinic model was equivalent to the unscaled form of the quasi-geostrophic initialization constraints (7.4.18) and (7.4.21) for the case of slow mode/potential vorticity constrained initialization. Their proofs are rigorous but fairly lengthy and are not repeated here. It is, in fact, not necessary to run through these proofs to be convinced of the link between quasi-geostrophic and normal mode initialization.

Equation (10.4.2) showed that the application of the Machenhauer condition (9.6.5) to a shallow water f-plane model is equivalent to setting to zero the initial time tendencies of the geostrophic discrepancy and the velocity potential. We now apply conditions (10.4.2) to the baroclinic equations (7.3.1–4) on an f-plane.

Equations (7.3.1–4) can be written in streamfunction/velocity potential form by taking $\mathbf{k}\cdot\nabla\times$ and then $\nabla\cdot$ of (7.3.1) and making use of (6.4.7):

$$\frac{\partial}{\partial t}\nabla^2\psi + f_0\nabla^2\chi = -\mathbf{k}\cdot\nabla\times\left(\mathbf{v}\cdot\nabla\mathbf{v} + \omega\frac{\partial\mathbf{v}}{\partial P}\right) \tag{10.4.3}$$

$$\frac{\partial}{\partial t}\nabla^2\chi - f_0\nabla^2\psi + \nabla^2\Phi = -\nabla\cdot\left(\mathbf{v}\cdot\nabla\mathbf{v} + \omega\frac{\partial\mathbf{v}}{\partial P}\right) \tag{10.4.4}$$

$$\frac{\partial}{\partial t}\frac{\partial\Phi}{\partial P} + \tilde{\Gamma}\omega = -\mathbf{v}\cdot\nabla\frac{\partial\Phi}{\partial P} - \omega\Gamma' \tag{10.4.5}$$

where

$$\Gamma = \frac{1}{P}\frac{\partial}{\partial P}\left(P\frac{\partial\Phi}{\partial P} - \kappa\Phi\right)$$

The ∇^2 is the Laplacian operator in cartesian coordinates, $\tilde{}$ indicates the horizontal average, and $'$ indicates deviation from the horizontal average. Here we interpret Φ as the deviation from the horizontally averaged geopotential $\tilde{\Phi}$, as in Section 7.3. The frictional forcing and diabatic terms $\mathbf{F}$ and Q have been neglected.

Now apply the initial conditions (10.4.2). Setting the time derivative term in (10.4.4) equal to zero yields

$$\nabla^2\Phi - f_0\nabla^2\psi = -\nabla\cdot\left(\mathbf{v}\cdot\nabla\mathbf{v} + \omega\frac{\partial\mathbf{v}}{\partial P}\right) \tag{10.4.6}$$

Taking ∇^2 of (10.4.5), subtracting $f_0(\partial/\partial P)$ of (10.4.3), and requiring that the time tendency of the geostrophic discrepancy be equal to zero yields

$$\tilde{\Gamma}\nabla^2\omega + f_0^2\frac{\partial^2\omega}{\partial P^2} = -\nabla^2\left(\mathbf{v}\cdot\nabla\frac{\partial\Phi}{\partial P} + \omega\Gamma'\right) + f_0\mathbf{k}\cdot\nabla\times\frac{\partial}{\partial P}\left(\mathbf{v}\cdot\nabla\mathbf{v} + \omega\frac{\partial\mathbf{v}}{\partial P}\right) \tag{10.4.7}$$

Equations (10.4.6) and (10.4.7) can be compared with (7.4.18) and (7.4.21), respectively. Note that (7.4.18) and (7.4.21) contain β terms but do not contain nonlinear terms in $\mathbf{v}_\chi$ or ω (which are of $O(\varepsilon)$ or $O(\varepsilon^2)$ and were previously dropped).

Leith (1980) showed that the first two iterations of the scheme (9.6.6–7) on an f-plane are exactly equivalent to the f-plane versions of the quasi-geostrophic conditions (7.4.18) and (7.4.21). This can be verified by simply examining the real-space forms of (9.6.6–7) on an f-plane.

Denote the observed/analyzed fields as ψ_A, χ_A, and Φ_A and the values after the kth iteration as ψ_k, χ_k, and Φ_k. Slow mode constrained initialization requires that the Rossby mode amplitudes be conserved. On the f-plane, this requires the conservation of potential vorticity:

$$\nabla^2\psi_k + \frac{\partial}{\partial P}\frac{f_0}{\tilde{\Gamma}}\frac{\partial \Phi_k}{\partial P} = \nabla^2\psi_A + \frac{\partial}{\partial P}\frac{f_0}{\tilde{\Gamma}}\frac{\partial \Phi_A}{\partial P}, \qquad \text{for all } k \tag{10.4.8}$$

Linear initialization (9.6.6) gives

$$f_0\psi_0 - \Phi_0 = 0, \qquad \chi_0, \omega_0 = 0 \tag{10.4.9}$$

which must be solved simultaneously with (10.4.8) for $k=0$. Once ψ_0, χ_0, and Φ_0 have been determined, they are inserted into (10.4.6–7) to yield

$$\nabla^2\Phi_1 - f_0\nabla^2\psi_1 = -\nabla\cdot(\mathbf{v}_\psi^0\cdot\nabla\mathbf{v}_\psi^0) \tag{10.4.10}$$

$$\tilde{\Gamma}\,\nabla^2\omega_1 + f_0^2\frac{\partial^2\omega_1}{\partial P^2} = -\nabla^2\left(\mathbf{v}_\psi^0\cdot\nabla\frac{\partial\Phi_0}{\partial P}\right) + f_0\frac{\partial}{\partial P}(\mathbf{v}_\psi^0\cdot\nabla\,\nabla^2\psi_0) \tag{10.4.11}$$

which is solved together with (10.4.8) for $k=1$. Here the subscript 1 indicates the first nonlinear iteration, as in (9.6.7), and $\mathbf{v}_\psi^0$ is the nondivergent wind corresponding to ψ_0. Because $\chi_0=0$, then $\omega_0=0$, $\mathbf{v}_\chi^0=0$, and (10.4.6–7) simplify to (10.4.10–11). The latter two equations are identical to the unscaled versions of (7.4.18) and (7.4.21). Equation (10.4.8) simply indicates that the slow mode amplitudes y are unchanged during the initialization.

Thus, on the f-plane, Equations (9.6.6–7) set the divergent wind to zero on the first iteration and calculate it from the quasi-geostrophic omega equation on the second. On the first iteration, the geostrophic relationship and the conservation of potential vorticity are solved simultaneously for the geopotential and rotational wind. On the second iteration, a nonlinear balance equation and the conservation of potential vorticity are solved simultaneously.

10.5 Variational normal mode initialization

Slow mode constrained normal mode initialization, as outlined in Sections 9.6 and 9.7, provides a means of proceeding from the observed/analyzed state to the slow manifold. Once on the slow manifold, the model state is essentially free of high-frequency oscillations. However, the slow mode constrained procedure is not the only way to get to the slow manifold. It is clear from Figure 10.8 that any line drawn between the observed/analyzed state A and the slow manifold M is a possible initialization procedure. In particular, it is possible to consider proceeding from point A to M via a data manifold. This is indicated schematically in the figure by the line D, which intersects M at point C.

To fix ideas, we again consider the slow manifold diagram of the simple Tribbia vortex model (Figure 10.7). This diagram shows the data manifolds for fixed rotational wind ($\psi = \text{constant}$) and fixed geopotential ($\Phi = \text{constant}$). Slow mode constrained initialization in Figure 10.8 would consist of moving vertically from the observed/analyzed state to the slow manifold. Geopotential constrained initialization would consist of moving from the observed/analyzed state along a line of constant Φ until it intersects M. Similarly, rotational wind constrained initialization would consist of moving from the observed/analyzed state along a line of constant ψ until it intersects M.

Suppose that D (in Figure 10.8) is, for example, a geopotential manifold and that the spatial structure of the geopotential field is invariant along D. Then, the point C is the state that is on the slow manifold and yet has the same geopotential field as the observed/analyzed state A. Clearly, the rotational and divergent windfield is then different in states A and C. Moreover, the slow mode amplitudes $\underline{Y}$ have changed during the operation. However, if D is a rotational wind manifold, the spatial structure of the rotational wind is invariant between A and C but the other dependent variables (and $\underline{Y}$) change.

If we have accurate geopotential information but no wind information in the observed/analyzed state A, then the appropriate procedure is geopotential constrained initialization. In the case of accurate wind information but no geopotential information in the observed/analyzed state, then wind constrained initialization is appropriate. In practice, both wind and mass information of varying quality and data density are available, and neither geopotential constrained nor wind constrained initialization is appropriate. The variational methods of Chapter 8 can be used in this case. The following discussion is based on Daley (1978), Phillips (1981), Puri (1983), Temperton (1984), and especially Tribbia (1982).

Assume that observations or analyzed grid values of the wind components u_A and v_A and the geopotential Φ_A are available. Then, u_A, v_A, and Φ_A define the observed/analyzed state A in Figure 10.8. The objective is to produce initialized values of these variables, u_I, v_I, and Φ_I, that are on the slow manifold. Here, u_A, v_A, Φ_A, u_I, v_I, and Φ_I are functions of the spatial coordinates of the domain. Suppose also that the forecast model to be initialized is the shallow water model with equivalent depth $\tilde{h}$. The variational formulation attempts to minimize the integral

$$I = \int_S \{\tilde{\Phi}(u_I - u_A)^2 w_v + \tilde{\Phi}(v_I - v_A)^2 w_v + (\Phi_I - \Phi_A)^2 w_\Phi\}\, dS \qquad (10.5.1)$$

where $\tilde{h} = g^{-1}\tilde{\Phi}$ is the equivalent depth and S the domain, subject to the side conditions that u_I, v_I, and Φ_I are on the slow manifold. We consider a strong constraint formulation here; that is, u_I, v_I, and Φ_I are on the slow manifold and some condition such as $\dot{\underline{Z}} = 0$ is satisfied identically. Weak constraint formulations in which the $\dot{\underline{Z}}$ are minimized are also possible in principle. w_v and w_Φ are prespecified weight functions that indicate confidence in the observed/analyzed values of the variables. The weights can be spatially variable.

The variational initialization procedure is indicated schematically in Figure 10.9(a). Point A indicates the observed/analyzed state and the Rossby $\underline{Y}$, gravity $\underline{Z}$ and slow manifolds are indicated as in Figure 10.8. Data manifolds of geopotential Φ and wind $\mathbf{v}$ are also shown. The ellipses are isopleths of constant I (10.5.1), with the value of I increasing away from 0. Thus, the geopotential and wind manifolds are the major and minor axes of the ellipses. The aspect ratio of the ellipses is determined by the ratio of w_v to w_Φ. If $w_\Phi \gg w_v$, then the ellipse becomes elongated along the line of constant Φ; and it becomes elongated along the line of constant $\mathbf{v}$ if $w_v \gg w_\Phi$. The solution to the variational problem is the point C, which is on the slow manifold and yet minimizes the value of I. At the point C, the curves M and I have the same tangent lines.

In normal mode space, the observed/analyzed state is defined to be $(\underline{Y}_A, \underline{Z}_A)$. Suppose the final state (C) is to satisfy the Machenhauer balance condition (9.6.5). Then,

$$\underline{Z}_I = (2\Omega i)^{-1} \underline{\underline{\Delta}}_z^{-1} \underline{R}_z(\underline{Y}_I, \underline{Z}_I) \qquad (10.5.2)$$

and

$$\underline{Y}_I \neq \underline{Y}_A, \qquad \underline{Z}_I \neq \underline{Z}_A$$

A method of approximately reaching point C is demonstrated in Figure 10.9(b), which is a blow-up of a portion of (a). Starting at point A, we use a slow mode constrained iteration to reach the point 1, which is on the slow manifold:

$$\underline{Z}_1 = (2\Omega i)^{-1} \underline{\underline{\Delta}}_z^{-1} \underline{R}_z(\underline{Y}_A, \underline{Z}_A), \qquad \underline{Y}_1 = \underline{Y}_A \qquad (10.5.3)$$

The next step is to move horizontally from point 1 to point 2, as in (b). Moving horizontally changes the slow mode projection $\underline{Y}$ but does not change the fast mode projection $\underline{Z}$. Point 2 defines a minimum in I; that is, it is the point on the ellipse where a horizontal line is also a tangent line of the ellipse. Mathematically, the point 2 is obtained by minimizing the integral

$$I_1 = \int_S \{\tilde{\Phi}(u_2 - u_A)^2 w_v + \tilde{\Phi}(v_2 - v_A)^2 w_v + (\Phi_2 - \Phi_A)^2 \omega_\Phi\} \, dS \qquad (10.5.4)$$

subject to the side conditions $\underline{Z}_2 = \underline{Z}_1$. Thus, I_1 is minimized with respect to the change in $\underline{Y}$ (i.e., $\Delta\underline{Y}$).

Define u_2, v_2, and Φ_2 as follows:

$$\begin{bmatrix} u_2 \\ v_2 \\ \Phi_2 \end{bmatrix} = \begin{bmatrix} u_1 \\ v_1 \\ \Phi_1 \end{bmatrix} + \sum_i \Delta y_i \begin{bmatrix} u_y^i \\ v_y^i \\ \Phi_y^i \end{bmatrix} \qquad (10.5.5)$$

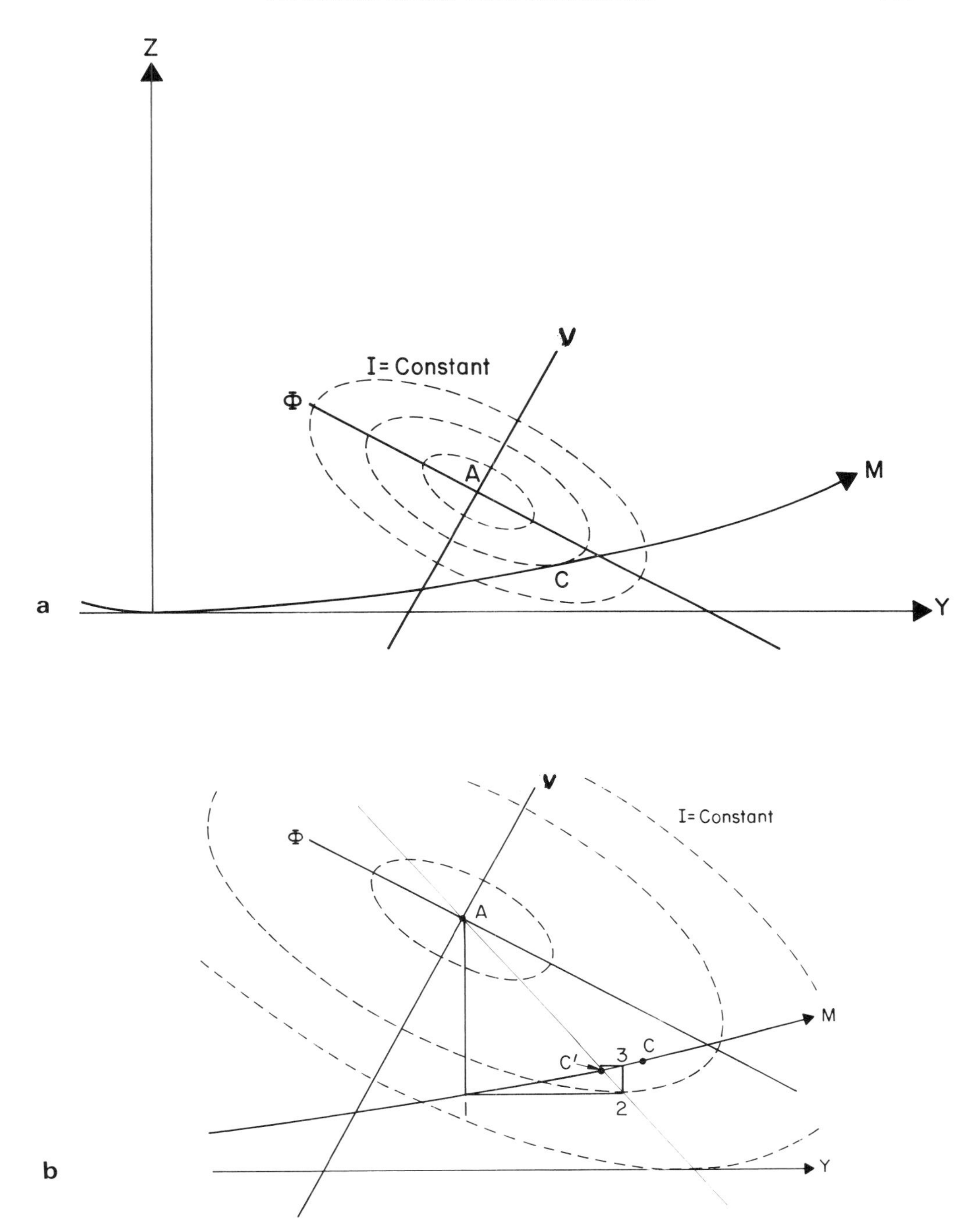

Figure 10.9 Schematic slow manifold diagram illustrating variational normal mode initialization. (After Daley, "The Application of Variational Methods of Initialization on the Sphere," in Saksaki (ed): *Variational Methods in Geoscience*, New York: Elsevier Science Publishing Co.)

where Δy_i are the changes in the slow mode amplitudes moving from point 1 to point 2. The u_y^i, v_y^i, and Φ_y^i are the normal mode structure functions corresponding to the slow mode y_i. In (10.5.5), $u_1, v_1, \Phi_1, u_y^i, v_y^i$, and Φ_y^i are known; the remaining variables are unknown at this point.

Now insert (10.5.5) into (10.5.4), differentiate with respect to an arbitrary slow increment Δy_j, and set the result to zero. The result (using 10.5.5) is

$$\int_S \{w_v \tilde{\Phi}(u_2 - u_A)u_y^j + w_v \tilde{\Phi}(v_2 - v_A)v_y^j + w_\Phi(\Phi_2 - \Phi_A)\Phi_y^j\}\, dS = 0 \tag{10.5.6}$$

Substituting into (10.5.6) from (10.5.5) for u_2, v_2, and Φ_2 gives

$$\sum_i \alpha_j^i \, \Delta y_i = f_j \tag{10.5.7}$$

where

$$\alpha_j^i = \int_S \{\tilde{\Phi}(u_y^i u_y^j + v_y^i v_y^j)w_v + \Phi_y^i \Phi_y^j w_\Phi\}\, dS$$

$$f_j = \int_S \{\tilde{\Phi}(u_A - u_1)u_y^j w_v + \tilde{\Phi}(v_A - v_1)v_y^j w_v + (\Phi_A - \Phi_1)\Phi_y^j w_\Phi\}\, dS$$

The u_A, v_A, Φ_A, u_1, v_1, and Φ_1 are already known; so f_j is already known for all j. The α_j^i depend only on the structure functions of the slow modes (u_y^i, v_y^i, and Φ_y^i) and the weights w_v and w_Φ, so they are known as well. Equation (10.5.7) describes a matrix relation that can be inverted to obtain the unknowns Δy_i. The Δy_i can then be inserted back into (10.5.5) to give the variables u_2, v_2, and Φ_2.

The thin line passing through A and point 2 in Figure 10.9(b) joins all points where horizontal lines are tangent to the family of ellipses. After we reach point 2, the next step is a slow mode constrained initialization that moves vertically to point 3 (which is on the slow manifold). Following that step, a repeat of the variational step moves horizontally leftward from point 3 until it reaches the thin line. The process is repeated until convergence. The iteration converges to point C′, which is the intersection of the thin line and the slow manifold. Points C and C′ are not coincident because the slow manifold makes a nonzero angle with horizontal lines. Thus, integrals such as (10.5.4) minimize the distance of point 2 from A, with respect to a horizontal line, but the total iterative process is not exactly equivalent to the minimization of (10.5.1). In practice, C and C′ are not far apart because the angle between the slow manifold and the Rossby manifold is not great.

The special case $w_v = 1$, $w_\Phi = 0$ is wind constrained initialization whereas the case $w_\Phi = 1$, $w_v = 0$ is geopotential constrained. A third special case is $w_\Phi = 1$, $w_v = 1$. Then, from (9.3.15), $\alpha_j^i = 0$ for $i \neq j$. Also, $f_j = 0$ for all j because the change from point A to point 1 is only in the fast modes $\underline{Z}$. Thus, $\Delta y_i = 0$ for all y_i and this case corresponds to slow mode constrained initialization.

Figure 10.10 illustrates the application of variational initialization to a shallow

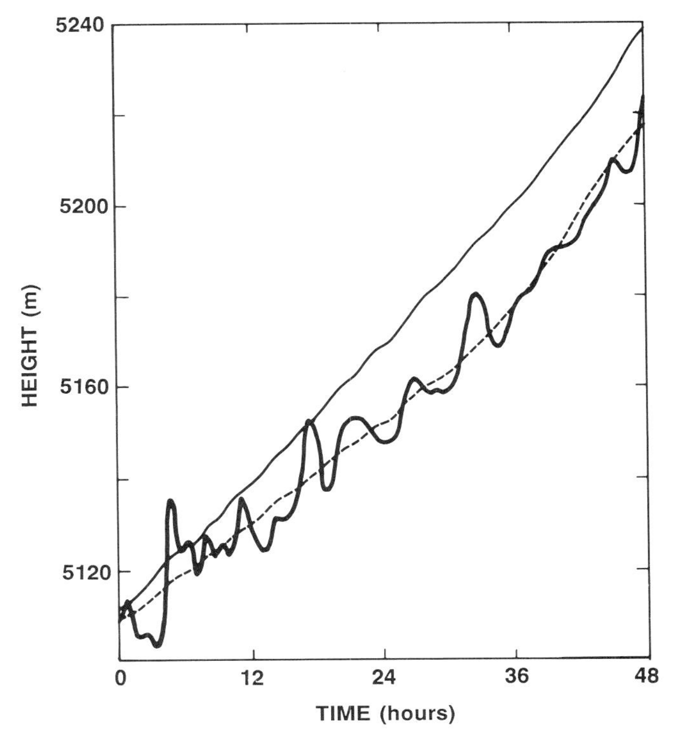

Figure 10.10 Time trace of 500 mb height at a fixed gridpoint during model integration. Heavy line, no initialization; dashed line, slow mode constrained initialization; light line, variational initialization. (After Fillion and Temperton, *Mon. Wea. Rev.* **117**: 2219, 1989. The American Meteorological Society.)

water model. Plotted is the 500 mb height at a gridpoint in northern Canada during a 48 hour integration of the model. It is clear that the slow mode constrained initialization eliminates most higher frequencies from the integration but does not really alter the slowly varying component of the solution. Variational initialization, as expected, not only removes the high-frequency components but also alters the time evolution of the low-frequency solution.

Returning to Figure 10.7, we see that some data manifolds do not intersect the slow manifold M. In particular, certain geopotential manifolds at the lower left of Figure 10.7 never intersect the slow manifold; these correspond to intense anticyclones. If the observed/analyzed state has a geopotential corresponding to one of these nonintersecting geopotential manifolds, then geopotential constrained initialization will not converge to the slow manifold. As shown by Tribbia (1981), such geopotential fields violate the ellipticity condition (7.6.7). This means that an initialization procedure that is geopotentially constrained in the tropics will not converge, implying that w_v must not equal zero in the tropics.

The concept of variational normal mode initialization is attractive because it optimizes the fit to the observations while suppressing high-frequency inertia–gravity oscillations from the model integration. The principal difficulty is computational because the global variational fit requires the inversion of large matrices. Williamson and Daley (1983) have replaced each variational step in Figure 10.9 with statistical interpolation (Chapter 4). The descent methods of Section 13.1 can also be used for this problem.

10.6 Initialization for limited area models

Numerical models written in latitude/longitude coordinates on the sphere can be initialized by normal mode techniques because they have simple (periodic) lateral boundary conditions and because the linearized equations are easily separable. There is an important class of numerical models for which these conditions do not apply. In particular, mesoscale models are usually integrated over limited horizontal domains and can have complex lateral boundary conditions. Moreover, these models often have coordinate systems (polar stereographic, Lambert conformal) for which the Coriolis terms are nonseparable (Temperton 1988).

As might be expected, normal mode initialization cannot be applied directly to limited area problems. However, the conceptual framework of the normal mode method is so powerful that it has led to some very effective solutions to the limited area initialization problem. This work is described in papers by Briere (1982), Bourke and McGregor (1983), Juvanon du Vachat (1986), Temperton (1988), Fillion and Temperton (1989), and others.

The difficulties with the lateral boundary conditions and the nonseparable Coriolis terms are essentially horizontal; the vertical decomposition is not affected. Thus, for limited area models, the determination of the vertical structure functions and the projection of the dependent variables onto these structures proceed exactly as in the global models. Consider the projection of the primitive equations onto the jth vertical mode, given by (9.5.7–10). Rewrite these equations in vector form as

$$\dot{\mathbf{v}}_j + \mathbf{k} \times f\mathbf{v}_j + \nabla\Phi_j = \mathbf{R}^j_{\mathbf{v}} \tag{10.6.1}$$

$$\dot{\Phi}_j + \tilde{\Phi}_j \nabla \cdot \mathbf{v}_j = R^j_{\Phi} \tag{10.6.2}$$

where $\mathbf{v}_j$, $\mathbf{R}^j_{\mathbf{v}}$, and R^j_{Φ} are the projections of the horizontal velocity vector and $\mathbf{v}$ of the nonlinear and forcing terms of the equations of motion and thermodynamic equation onto the jth vertical mode.

We simplify the notation by dropping the vertical mode designator j and in the remainder of the section will be concerned only with the horizontal problem. Temperton (1988) has developed a particularly powerful technique called *implicit normal mode initialization*. The following abbreviated treatment is based on this technique, with the primary emphasis on development of the appropriate balance conditions. A detailed practical implementation is contained in Temperton's paper.

Operate with the horizontal curl and divergence. Then (10.6.1–2) can be written as

$$\nabla^2\dot{\psi} + F(\chi) = R_\psi \tag{10.6.3}$$

$$\nabla^2\dot{\chi} - F(\psi) - B(\chi) + \nabla^2\Phi = R_\chi \tag{10.6.4}$$

$$\dot{\Phi} + \tilde{\Phi}\,\nabla^2\chi = R_\Phi \tag{10.6.5}$$

where $R_\psi = \mathbf{k}\cdot\nabla\times\mathbf{R}_v + B(\psi)$, $R_\chi = \nabla\cdot\mathbf{R}_v$, $F(\alpha) = \nabla\cdot f\nabla\alpha$ and $B(\alpha) = \mathbf{k}\cdot\nabla\times f\nabla\alpha$. Here, $\nabla^2\psi = \mathbf{k}\cdot\nabla\times\mathbf{v}$ and $\nabla^2\chi = \nabla\cdot\mathbf{v}$ are the vorticity and divergence.

Consider the linear case and set $R_\psi = R_\chi = R_\Phi = 0$ in (10.6.3–5). Then these equations become equivalent to (9.3.17–20), except that the term $2\Omega a^{-2}\psi_\lambda$ in (9.3.17) has not been included. Thus, apart from the beta term in (10.6.3), Equations (10.6.3–5) are the same as the equations used for nonlinear normal mode initialization on the sphere.

Continuing with the linear case ($R_\psi = R_\chi = R_\Phi = 0$), we note that there are solutions to (10.6.3–5) that are stationary ($\dot{\psi} = \dot{\chi} = \dot{\Phi} = 0$). These are the Rossby modes, which have a form similar to their f-plane counterparts (6.4.17), satisfying

$$\chi = 0 \qquad \text{and} \qquad \nabla^2\Phi = F(\psi) = \nabla\cdot f\,\nabla\psi \tag{10.6.6}$$

That is, these Rossby mode solutions are nondivergent and satisfy the linear balance equation (7.5.3). It is clear (in the absence of a basic state flow) that the Rossby modes become nonstationary only when the $B(\psi)$ term in (10.6.3) is included in the linearized equations.

Following (6.5.7), define a potential vorticity for (10.6.3–5) as follows:

$$Q = \nabla^2\psi + f - \tilde{\Phi}^{-1}F(\nabla^{-2}\Phi) \tag{10.6.7}$$

where ∇^{-2} is the inverse Laplacian operator. The definition (10.6.7) is purely formal because the operator ∇^{-2} requires the solution of a Poisson equation subject to appropriate lateral boundary conditions (see Section 7.5).

When $R_\psi = R_\chi = R_\Phi = 0$ in (10.6.3–5), we can see that $\dot{Q} = 0$ and that this must be true for any normal mode of these equations. Following (9.3.4), we can write the potential vorticity of any normal mode of the linear version of (10.6.3–5) in the form $Q(\mathbf{r}, t) = Q_0(\mathbf{r})\exp(i\omega t)$. Here, $\mathbf{r}$ is the horizontal coordinate, $Q_0(\mathbf{r})$ is a complex function and ω is the modal frequency. Using the methods of Section 9.5, we can show that ω is real. Then,

$$\dot{Q}(\mathbf{r}, t) = i\omega Q_0(\mathbf{r})\exp(i\omega t) = 0 \tag{10.6.8}$$

which requires that either $Q_0(\mathbf{r})$ or ω be equal to zero. Rossby modes are stationary ($\omega = 0$) in this system. For inertia–gravity modes, $\omega \neq 0$, which requires that $Q(\mathbf{r}, t) = Q_0(\mathbf{r}) = 0$. In other words, the inertia–gravity modes have no potential vorticity. Thus, the potential vorticity for the linear equations is entirely in the Rossby modes, as was the case for the f-plane equations of Section 6.4.

Slow mode/potential vorticity constrained initialization for the linear version of (10.6.3–5) simply requires the zeroing of inertia–gravity mode amplitudes while conserving the potential vorticity during the adjustment (Section 6.5). Thus, the initialized state $[\psi_I, \chi_I, \Phi_I]^T$ satisfies

$$\chi_I = 0, \qquad \nabla^2\Phi_I - \nabla\cdot f\,\nabla\psi_I = 0, \qquad Q_I = Q_A \tag{10.6.9}$$

where subscript A indicates the uninitialized state.

The Machenhauer balance condition (9.6.5) for the system (10.6.3–5), with R_ψ, R_χ, and R_Φ nonzero, can be obtained following the arguments of Sections 10.3 and 10.4. These conditions are simply

$$\dot{\chi}_I = 0, \qquad \nabla^2\dot{\Phi}_I - \nabla\cdot f\,\nabla\dot{\psi}_I = 0, \qquad Q_I = Q_A \tag{10.6.10}$$

Higher-order initialization is also possible by applying second or higher time derivative conditions, as in Section 9.6. In baroclinic models, the balancing condition (10.6.10) would usually be applied by vertically projecting the dependent variables on *only* the vertical modes with the largest equivalent depths before performing the horizontal operations.

Temperton (1988) successfully applied the balance conditions (10.6.10) to a variable resolution gridpoint shallow water model in polar stereographic coordinates. Because the technique requires no horizontal transformation into normal mode space, the nonseparable Coriolis terms and the lateral boundary conditions could be handled without difficulty. The fact that there is no transformation into normal mode space (apart from the vertical) also makes the technique attractive for variational initialization (Section 10.5). Fillion and Temperton (1989) have successfully applied balance conditions (10.6.10) in a variational context.

These limited area initialization techniques clearly have advantages over pure normal mode methods. However, there are one or two minor disadvantages. The neglect of the $B(\psi)$ term of (10.6.3) in the linear equations is not really serious because though the Rossby modes are affected, the inertia–gravity modes are only slightly modified (Temperton 1989). However, (10.6.10) is a coarser instrument for initialization than the normal mode methods. In effect, *all* the inertia–gravity modes for a given vertical mode are initialized by this technique. In normal mode initialization, the modes to be initialized can be individually selected, based on appropriate criteria (Section 10.2).

The techniques of this section also bear a strong resemblance to the quasi-geostrophic methods of Chapter 7. As noted in Section 7.9, the quasi-geostrophic methods have difficulties for small vertical scales. This problem is avoided here because condition (10.6.10) is applied only to the largest vertical scales. The ellipticity problem (Section 7.6) is also avoided because condition (10.6.10) is applied only in slow mode constrained or variational formulations.

10.7 Convergence properties and diabatic initialization

The normal mode initialization procedure is a powerful and precise initialization tool. Selection of model normal modes for initialization can be based on linear frequency, horizontal or vertical scale, or kind of mode (Rossby, Kelvin, etc.). During the early 1980s, this procedure was implemented in global forecast/analysis cycles in many operational centers. These schemes were very successful in suppressing high-frequency oscillations in global baroclinic primitive equation models and thus achieved the major objective of initialization discussed in Section 6.3.

It quickly became apparent, however, that the normal mode initialization schemes of that period had some major failings. First, the Picard iteration procedure (9.6.6–9) failed to converge for inertia–gravity modes of small equivalent depths or when R included parametrizations of diabatic processes (convection, radiation, hydrological processes, boundary layer processes, etc.). Second, the tropical circulation was adversely affected by initialization. In particular, the Hadley circulation (the zonally averaged upward and poleward transport of heat and moisture in the intertropical convergence zone) was artificially suppressed by the initialization process.

The convergence of the nonlinear iteration schemes was examined by Ballish (1981), Errico (1983), and Rasch (1985). Some understanding of the convergence properties can be gained from the following simple analysis.

The fast mode equation (9.5.19) can be written as

$$\dot{z} = -i\omega z + R \tag{10.7.1}$$

where $\omega = 2\Omega\sigma_z$, and $R = R_z(\underline{Y}, \underline{Z})$. The kth iteration step of the Picard procedure (9.6.6–9) can be written as

$$z_{k+1} = \frac{R_k}{i\omega}, \qquad \underline{Z}_{k+1} = -i\underline{\underline{\omega}}^{-1}\underline{R}_k \tag{10.7.2}$$

where $\underline{Z}$ is the column vector of fast mode amplitudes and $\underline{\underline{\omega}}$ the diagonal matrix of all fast mode frequencies. Insertion of (10.7.1) into (10.7.2) leads to the more convenient form (see Williamson and Temperton 1981) of

$$z_{k+1} = z_k + \frac{\dot{z}_k}{i\omega} \qquad \text{or} \qquad \underline{Z}_{k+1} = \underline{Z}_k - \mathrm{i}\underline{\underline{\omega}}^{-1}\dot{\underline{Z}}_k \tag{10.7.3}$$

The advantage of (10.7.3) over (10.7.2) is that the numerical model itself produces time tendencies of the dependent variables u, v, T, ψ, χ, and so on, and that these can be projected directly into normal mode space to calculate $\dot{\underline{Z}}$. This means that no special codes have to be written to calculate separately the linear and nonlinear terms of the model.

Consider the following simple form for R in (10.7.1):

$$R = -i\omega_r z + F_0 \tag{10.7.4}$$

where F_0 and ω_r are time-independent. Here the term $-i\omega_r z$ is intended to represent the slowly varying (low frequency) nonlinear advective terms, and F_0 represents the diabatic forcing terms (physical parametrizations of the model). Then (10.7.1) has the following solution:

$$z(t) = \left[z(0) - \frac{F_0}{i(\omega_r + \omega)}\right]\exp(-i(\omega_r + \omega)t) + \frac{F_0}{i(\omega_r + \omega)}, \qquad \omega \neq -\omega_r \tag{10.7.5}$$

and

$$z(t) = z(0) + F_0 t, \qquad \text{when } \omega_r + \omega = 0$$

The Machenhauer balance condition (9.6.5) applied to (10.7.1) with R given by (10.7.4) yields

$$z = \frac{F_0}{i(\omega + \omega_r)} \tag{10.7.6}$$

Now apply the Picard iteration scheme (10.7.2) to (10.7.1) for R given by (10.7.4),

$$z_0 = 0, \qquad z_1 = \frac{F_0}{i\omega}, \qquad z_2 = \frac{F_0}{i\omega}\left(1 - \frac{\omega_r}{\omega}\right)$$

$$z_k = \frac{F_0}{i\omega} \sum_{j=0}^{k-1} \left(-\frac{\omega_r}{\omega}\right)^j$$

Then

$$\lim_{k\to\infty} z_k = \frac{F_0}{i\omega}\left(1 + \frac{\omega_r}{\omega}\right)^{-1} = \frac{F_0}{i(\omega + \omega_r)}, \qquad \text{provided } \left|\frac{\omega_r}{\omega}\right| < 1 \tag{10.7.7}$$

For $|\omega_r/\omega| < 1$, the Machenhauer condition gives solution (10.7.6), and the Picard procedure (10.7.7) converges to it. For $\omega_r/\omega < -1$ or $\omega_r/\omega \geq 1$, the Machenhauer condition gives solution (10.7.6), but the Picard procedure (10.7.7) does not converge to it. For $\omega_r/\omega = -1$, the Machenhauer condition (9.6.5) cannot be applied.

If $|\omega|$ is comparable to the frequencies of the slowly varying advective and forcing terms, then the convergence of the Picard iteration is much less likely. Thus, it is not surprising that convergence was not obtained for the relatively low frequency inertia–gravity waves of small equivalent depth (see Figure 9.2).

Kitade (1983) proposed a modified form of Picard iteration. Replace (10.7.3) with

$$z_{k+1} = z_k + \frac{\lambda \dot{z}_k}{i\omega} \tag{10.7.8}$$

where $0 < \lambda \leq 1$ is a coefficient to be specified. Procedure (10.7.8) converges for R given by (10.7.4) if

$$-\omega < \omega_r < \left(\frac{2}{\lambda} - 1\right)\omega \tag{10.7.9}$$

Thus, if $\lambda = \frac{1}{2}$, convergence occurs for $-\omega < \omega_r < 3\omega$, which is less restrictive than (10.7.7).

Rasch (1985) proposed using a variant of Newton's method. Recall that if $f(x_k)$ and its derivative with respect to x, $f'(x_k)$, are known, then the kth iteration of Newton's method for finding a zero of $f(x)$ is given by

$$x_{k+1} = x_k - \frac{f(x_k)}{f'(x_k)} \tag{10.7.10}$$

Now consider a vector version of (10.7.10) with x replaced by the L elements z of the vector $\underline{Z}$ and the time-derivative operator replacing the function f. The elements of $\underline{Z}$ can be thought of as the L coordinates of an L-dimensional space. The generalization of (10.7.10) for this space is

$$\underline{Z}_{k+1} = \underline{Z}_k - \underline{\underline{N}}_k^{-1}\,\dot{\underline{Z}}_k \qquad \text{with} \qquad \underline{\underline{N}}_k = \frac{\partial \dot{\underline{Z}}_k}{\partial \underline{Z}_k} \tag{10.7.11}$$

$\underline{\underline{N}}_k$ is the derivative of a vector with respect to another vector, which is known as a *Jacobian* matrix. Ignore for the moment the iteration number k and consider the elements of $\underline{Z}$. Denote an individual fast mode as z_l with dimensional frequency ω_l. For the lth mode, (10.7.1) can be written as

$$\dot{z}_l = -i\omega_l z_l + R_l$$

Then, the element n_{lm} of $\underline{\underline{N}}$ is given by $n_{lm} = -i\omega_l \delta_{lm} + \partial R_l/\partial z_m$. If $|\partial R_l/\partial z_m|$, $l \neq m$, is small compared with $|-i\omega_l + \partial R_l/\partial z_l|$, then we can approximate (10.7.11) by

$$\underline{Z}_{k+1} = \underline{Z}_k - [\mathrm{diag}(\underline{\underline{N}}_k)]^{-1}\,\dot{\underline{Z}}_k \tag{10.7.12}$$

where $\mathrm{diag}(\underline{\underline{N}}_k)$ is a matrix whose only nonzero elements are the main diagonal elements of $\underline{\underline{N}}_k$. Now any diagonal element of $\underline{\underline{N}}_k$ has the form $\partial \dot{z}_k/\partial z_k$. Approximate this element by the simple finite difference approximation

$$\frac{\partial \dot{z}_k}{\partial z_k} = \frac{\dot{z}_k - \dot{z}_{k-1}}{z_k - z_{k-1}}$$

Insertion into (10.7.12) gives

$$z_{k+1} = z_k - \left[\frac{z_k - z_{k-1}}{\dot{z}_k - \dot{z}_{k-1}}\right]\dot{z}_k \tag{10.7.13}$$

This requires two starting values z_0 and z_1. Application of (10.7.13) to (10.7.1) with R given by (10.7.4) and the first two steps taken from (10.7.7) yields

$$z_0 = 0, \qquad z_1 = \frac{F_0}{i\omega}, \qquad z_2 = \frac{F_0}{i(\omega + \omega_r)} \tag{10.7.14}$$

The Newton iteration procedure converges here in only three steps. Moreover, it converges for all values of ω except $\omega = -\omega_r$. Rasch (1985) has demonstrated that the application of (10.7.13) to the initialization of global baroclinic models often leads to convergence when (10.7.2) does not.

At the end of Section 10.3, we noted that reduction of $\dot{z}$ to zero may not be either necessary or desirable. However, the Newton procedure (10.7.13) can be used to reduce $\dot{z}$ to a tolerably small value.

The deleterious effect of initialization on the tropical circulation has been examined by a number of authors, including Puri and Bourke (1982), Wergen (1983), Errico (1984), Puri (1985), Puri (1987), Errico and Rasch (1988), and Errico (1989). The Hadley cell and large-scale nonzonal divergent components of the tropical circulation are primarily driven by convective processes. Because these circulations are largely divergent, they project substantially onto the large-scale tropically trapped inertia–gravity and Kelvin waves with smaller equivalent depths.

Diabatic normal mode initialization, which includes physical parametrization, was not found to converge very well using the Picard iteration scheme (9.6.6–9), as noted earlier. Consequently, adiabatic normal mode initialization (which ignores the physical parametrizations of the model) was widely used in the early 1980s. The Hadley cell, however, is maintained by convection, which is generally parametrized in numerical models. Thus, leaving convective parametrization out of the initialization procedure resulted in a suppression of the Hadley cell.

The Hadley cell problem was tackled both by modifying the adiabatic initialization procedures and by introducing more practical diabatic initialization schemes. Puri and Bourke (1982) demonstrated that if the large-scale inertia–gravity modes on which the Hadley cell projected were not initialized, then adiabatic initialization would not suppress the Hadley cell. Wergen (1983) used a form of diabatic initialization that successfully retained the Hadley cell. He first generated a time series of diabatic heating fields (including convective heating) from a short time integration of the uninitialized model. The diabatic heating field was time and space filtered and introduced as part of the R_z term in the Picard iteration process (9.6.6–9). This diabatic heating field was held constant during the iteration process, and convergence was readily attained. Errico and Rasch (1988) used diabatic normal mode initialization and the Newton iteration scheme and were also able to retain the Hadley circulation.

Despite the apparent success of diabatic initialization procedures in overcoming the shortcomings of the adiabatic schemes, a number of problems remain. Errico and Rasch (1988) demonstrated that if the diabatic terms are dominant, then the Machenhauer balance condition (9.6.5) is inappropriate. This can be illustrated by (10.7.4). When $F_0 \gg i(\omega + \omega_r)z$, then the Machenhauer balance condition gives essentially a singular solution. In this case, $\dot{z}$ is the same order as F_0 and cannot be neglected. This can also be seen on the left-hand side of Figure 10.6, where the time tendency terms are not small compared with the other terms. Unfortunately, the diabatic terms are often dominant, particularly in the tropics.

The large-scale divergent tropical circulation can be largely determined from observations; and initialization processes such as that of Puri and Bourke (1982), Wergen (1983), or Errico and Rasch (1988) essentially reproduce this circulation. However, the smaller-scale divergent tropical circulation is very important in the hydrological cycle and is not observed well (see Section 13.6). As noted in Section 6.3, an initialization procedure should provide diagnostic constraints that can, in principle, be used to generate approximate but model-consistent information that is not available from the observation network. Thus, one might hope that information on the tropical divergent flow could be generated from the rotational windfield or mass field.

Unfortunately, this does not appear to be the case. The tropical divergent flow has a large projection on inertia–gravity modes of small equivalent depth (tropically trapped). These modes have small linear frequencies, and the time tendency term $\dot{z}$ is often as large as some of the other terms in (9.5.19) and cannot be neglected. Thus, a diagnostic relationship among the various dependent variables may not exist. Moreover, even if it is appropriate to neglect $\dot{z}$, the resulting diagnostic relation may be so complex as to be unusable. The divergent tropical flow depends strongly on convection and boundary layer processes, which themselves depend strongly on the divergence. The resulting balance between linear terms, adiabatic (nonlinear) terms, and diabatic terms may be too highly implicit to be useful. Consequently, the tropical divergent windfield may not be easily derivable from other variables. Ultimately, there is no substitute for observations. The diagnostic relationships derivable from any initialization scheme can go only so far in generating missing information.

Exercises

10.1 Show that the Machenhauer balance condition (10.3.10) for the simple model (10.3.6) implies that the Rossby mode (y) is stationary.

10.2 Consider the fast mode equation (10.7.1) with $R = F_0 \exp(i\omega_r t)$, which is intended to simulate the low-frequency forcing and $|\omega_r/\omega| = 0.1$. Find the general solution to (10.7.1) in this case and show that there will be no high-frequency oscillations if $z(0) = F_0/[i(\omega + \omega_r)]$. Apply the Machenhauer balance condition $\dot{z} = 0$ and show that there will still be high-frequency oscillations in this case, but with an amplitude only one-tenth that of the low-frequency oscillations.

10.3 Consider again the example of the previous exercise. Find the initial conditions when the higher-order balance conditions $\ddot{z} = 0$ or $\dddot{z} = 0$ are applied at $t = 0$. Show that when

$$\left.\frac{d^p z}{\partial t^p}\right|_{t=0} = 0, \qquad \text{then} \qquad z(0) = \frac{F_0}{i\omega}\sum_{n=0}^{p-1}(-1)^n\left(\frac{\omega_r}{\omega}\right)^n$$

Show that in the limit as $p \to \infty$, $z(t)$ will have no high-frequency oscillations.

10.4 Consider a fast mode equation of the form $\dot{z} = -i\omega z + R(t)$ and approximate R by the first two terms in a power series in t. $R(t) = R_0 + R_1 t$, where R_n is the nth time derivative of R evaluated at $t = 0$. Show that the Laplace transform of z is

$$L(z) = \frac{z(0)s^2 + R_0 s + R_1}{s^2(s + i\omega)}$$

where s is a complex frequency. Use the method of residues (Section 13.4) to find the general solution for $z(t)$. Show that there will be no high-frequency oscillations when $z(0) = -iR_0/\omega + R_1/\omega^2$ and that this is equivalent to setting $\ddot{z}(0) = 0$. Repeat the calculations for $R(t) = R_0 + R_1 t + R_2 t^2/2$ and show that there are no high frequencies when $\dddot{z}(0) = 0$. Note the three-way correspondence between (a) adding an extra term to the power series for $R(t)$, (b) setting the next-highest derivative of z equal to zero at $t = 0$, and (c) adding another pole at $s = 0$ in the complex frequency plane (see Sections 7.1 and 13.4).

11

Dynamic initialization

The initialization procedures discussed in Chapters 7–10 (with the exception of Section 8.6) can be characterized as *static* in nature. That is, *at a fixed time*, these procedures adjust the data so that they conform exactly or approximately to externally imposed dynamical constraints. Usually, we derive these constraints from quasi-geostrophic theory or by demanding that the initialized state be on or near the slow manifold of the model for which the data are being initialized.

Dynamic initialization (also known as dynamic balancing) was introduced by Miyakoda and Moyer (1968) and Nitta and Hovermale (1969). It takes a very different approach. No dynamical constraints are explicitly applied; instead, the forecast model itself is used to produce the initialized state. This is done by integrating the forecast model forward and backward in time from the observed/analyzed state using time integration procedures that tend to damp high-frequency oscillations. After many forward and backward applications of this damping procedure, an initialized state results. Hopefully, subsequent model integrations run from this initialized state will contain little or no high-frequency variance. To be successful, the damping time integration schemes must be strongly frequency selective. That is, they must rapidly damp high-frequency gravity modes without seriously affecting the low-frequency Rossby modes. To understand dynamic initialization, it is first necessary to understand damping time integration procedures.

11.1 Damping time integration procedures

Damping time integration procedures are used in both dynamic initialization and continuous data assimilation (discussed in Chapter 12). A knowledge of damping time integration schemes is prerequisite to an understanding of dynamic initialization. Numerical time integration schemes are discussed more generally in Haltiner and Williams (1980) and Richtmeyer and Morton (1967).

Consider the following differential equation,

$$\frac{du}{dt} - F(u) = 0 \tag{11.1.1}$$

where t is the independent variable, u the dependent variable, and $F(u)$ some linear or nonlinear function of u and/or t. Many of the equations considered in Chapters 6–10 can be written in a form similar to (11.1.1). For example, (6.4.1), with Φ, $v = 0$ is of the form (11.1.1). Equation (11.1.1) can be discretized in time by a number of different methods. Two different time discretization procedures, which are both specifically designed to damp high-frequency oscillations, are discussed here.

Scheme 1 (due to Nitta and Hovermale 1969)

$$\begin{array}{ll}
\text{step 1} & \left.\begin{array}{l} u_a = u_{n-1} + F(u_{n-1})\,\Delta t \\ u_b = u_{n-1} + F(u_a)\,\Delta t \end{array}\right\} \text{forward} \\
\text{step 2} & \\
\text{step 3} & \left.\begin{array}{l} u_c = u_b - F(u_b)\,\Delta t \\ u_n = u_b - F(u_c)\,\Delta t \end{array}\right\} \text{backward} \\
\text{step 4} &
\end{array} \tag{11.1.2}$$

where Δt is some finite time increment (time-step) as yet unspecified; u_{n-1} is the value of u after $n-1$ iterations of scheme 1; u_n is the value of u after n iterations; and u_a, u_b, and u_c are intermediate values of u within an iteration. The first two steps correspond to forward integration in time. Thus, in step 1 the differential equation (11.1.1) is approximated by

$$\frac{u_a - u_{n-1}}{\Delta t} = F(u_{n-1})$$

The third and fourth steps correspond to backward integration in time. Thus, step 4 approximates (11.1.1) by

$$\frac{u_n - u_b}{(-\Delta t)} = F(u_c)$$

The final value u_n, though not in general equal to u_{n-1}, is defined at the same time.

Scheme 2 (due to Okamura; see Temperton 1976)

$$\begin{array}{ll}
\text{step 1} & u_a = u_{n-1} + F(u_{n-1})\,\Delta t \\
\text{step 2} & u_b = u_a - F(u_a)\,\Delta t \\
\text{step 3} & u_n = 3u_{n-1} - 2u_b
\end{array} \tag{11.1.3}$$

where u_n and u_{n-1} are understood to be the values of u after n and $n-1$ iterations, and u_a and u_b are intermediate values of u. For each iteration, scheme 2 is a three-step process.

We now examine the damping properties of schemes 1 and 2, using a very simple form of $F(u)$ in (11.1.1). Consider the ordinary differential equation

$$\frac{du}{dt} = -i\omega u \tag{11.1.4}$$

where $i = \sqrt{-1}$ and ω is a dimensional frequency. Equation (11.1.4) describes a simple harmonic oscillator and has the solution

$$u(t) = u(0) \exp(-i\omega t)$$

The application of scheme 1 (11.1.2) to Equation (11.1.4) yields

$$\begin{aligned} &\text{step 1} \qquad u_a = u_{n-1} - i\omega\,\Delta t\, u_{n-1} \\ &\text{step 2} \qquad u_b = u_{n-1} - i\omega\,\Delta t(u_{n-1} - i\omega\,\Delta t\, u_{n-1}) \\ &\qquad\qquad\;\; = (1 - i\omega\,\Delta t - \omega^2\,\Delta t^2)u_{n-1} \end{aligned} \tag{11.1.5}$$

Similarly, steps 3 and 4 of (11.1.2) give

$$u_n = (1 + i\omega\,\Delta t - \omega^2\,\Delta t^2)u_b$$

Substitution into (11.1.5) yields

$$u_n = (1 - \omega^2\,\Delta t^2 + \omega^4\,\Delta t^4)u_{n-1} \tag{11.1.6}$$

A response function $R_1(\omega\,\Delta t)$ for scheme 1 can be defined as follows:

$$R_1 = \frac{u_n}{u_{n-1}} = 1 - \omega^2\,\Delta t^2 + \omega^4\,\Delta t^4 \tag{11.1.7}$$

where R_1 indicates the change in u during iteration n. If $|R_1| < 1$, then the scheme is damping u. If $|R_1| > 1$, then the scheme is amplifying u. If $|R_1| = 1$, then u is not changed and the scheme is neutral. R_1, then, is the damping or amplification that occurs during a single iteration of scheme 1. After n iterations, the total damping or amplification is

$$[R_1]^n = \frac{u_n}{u_0} = (1 - \omega^2\,\Delta t^2 + \omega^4\,\Delta t^4)^n \tag{11.1.8}$$

The application of scheme 2 (11.1.3) to Equation (11.1.4) yields the following response function:

$$R_2 = \frac{u_n}{u_{n-1}} = 1 - 2\omega^2\,\Delta t^2 \tag{11.1.9}$$

The response functions R_1 and R_2 corresponding to schemes 1 and 2, respectively, are plotted as functions of $\omega\,\Delta t$ in Figure 11.1. On this diagram, damping occurs if $|R| < 1$ whereas amplification occurs if $|R| > 1$, and $|R| = 1$ is neutral. If $R < 0$, then the solution u changes sign after every iteration. R is symmetric about $\omega\Delta t = 0$.

The goal of dynamic initialization is to damp modes with high frequency (ω large)

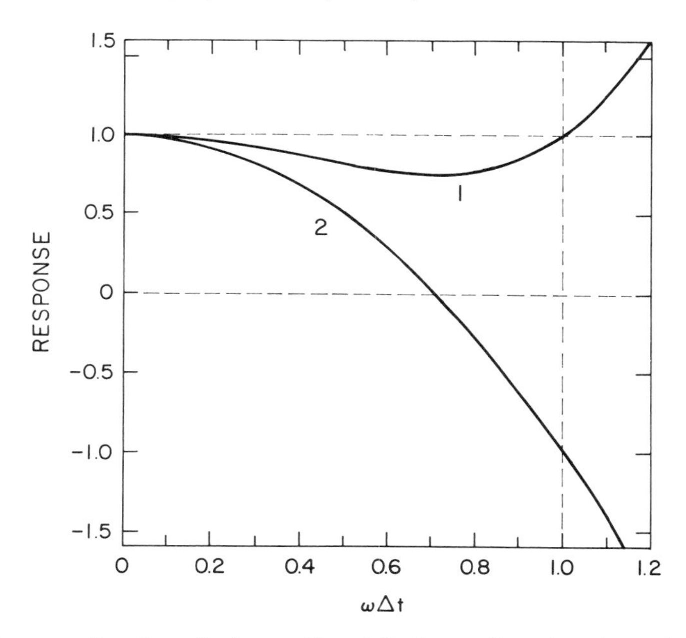

Figure 11.1 Response functions R_1 (curve 1) and R_2 (curve 2) as functions of $\omega\,\Delta t$. Dashed lines denote responses $R=0$ and $R=1$

and leave untouched modes with low frequency (ω small). At this point, Δt has not been specified, so we can choose it in such a way as to strongly damp high-frequency modes while minimally damping low-frequency modes.

Consider scheme 1 first. For $\omega\,\Delta t$ small (low-frequency modes), there is little damping. The maximum damping – found by differentiating (11.1.7) with respect to $\omega\,\Delta t$ and setting the result to zero – occurs at $\omega\,\Delta t = 1/\sqrt{2}$. If $\omega\,\Delta t > 1$, then the iteration scheme amplifies instead of damping. Denote the modulus of the highest model frequency as ω_{m}. Then, clearly, Δt must be chosen to be less than $1/\omega_{\mathrm{m}}$ to avoid amplifying solutions. Another choice of Δt might be

$$\Delta t = \frac{1}{\sqrt{2}\omega_m} \tag{11.1.10}$$

Then, the highest-frequency mode corresponding to ω_{m} would be maximally damped. The damping in scheme 1 for each iteration stop is not very large even at the maximum.

Scheme 2 is clearly a more effective damping scheme. Again, the damping is slight for ω small. The maximum damping ($R_2 = 0$) corresponds to $\omega\,\Delta t = 1/\sqrt{2}$, and amplification occurs for $\omega\,\Delta t > 1$. For $\omega\,\Delta t > 1/\sqrt{2}$, there is a change of sign at every iteration. In this scheme, Δt must be less than $1/\omega_{\mathrm{m}}$, but $\Delta t = 1/(\sqrt{2}\omega_{\mathrm{m}})$ might be a better choice.

It is not necessary to know anything about the model normal modes or their frequencies in either scheme 1 or scheme 2. It is sufficient to have a rough estimate of ω_{m} and then specify Δt accordingly.

11.2 Application to simple models

We now apply the two damping time integration schemes just discussed to two simple models: the linearized shallow water model (Section 6.4) and the Hinkelmann–Phillips model (Section 7.1). These simple examples reveal many of the properties of dynamic initialization and serve to illuminate subsequent results from experiments with sophisticated models.

Consider first the linearized shallow water f-plane model described by (6.4.8–10). The frequencies of the normal modes of this model were determined in Section 6.4 and are given by (6.4.17–18) for the Rossby and inertia–gravity modes, respectively. To analyze the effect of dynamic initialization on this model, it suffices to insert the value of frequency for each type of mode (from 6.4.17–18) into the response functions for scheme 1 (11.1.7) or scheme 2 (11.1.9). The dimensional Rossby mode frequency ω_R and the inertia–gravity mode frequency ω_G for this model (given by 6.4.17–18) can be written as

$$\omega_R = 0, \qquad \omega_G = \pm\sqrt{f_0^2 + \frac{4\pi^2 g\tilde{h}}{L_H^2}} \tag{11.2.1}$$

Here, $\tilde{h} = g^{-1}\tilde{\Phi}$, g is the gravitational constant, $\tilde{h}$ the equivalent depth, and f_0 the Coriolis parameter. The horizontal wavelength L_H is defined as

$$L_H = \frac{2\pi a}{\sqrt{n^2 + m^2}}$$

The Rossby mode frequency $\omega_R = 0$, so the responses R_1 (11.1.7) and R_2 (11.1.9) for this mode both equal 1. Thus, the Rossby modes of this simple model are undamped by either scheme 1 or scheme 2.

Consider now the response R_1 for the inertia–gravity modes of this model. To get numerical values, we must specify f_0, L_H, $\tilde{h}$, and Δt. We choose f_0 to have the midlatitude value of $10^{-4}\ \mathrm{s}^{-1}$ and Δt to be 600 s. The values of L_H and $\tilde{h}$ are chosen as follows: $L_H = 10{,}000$ km and $L_H = 2000$ km; $\tilde{h} = 10$ km and $\tilde{h} = 100$ m. For the case $L_H = 10{,}000$ km, $\tilde{h} = 10$ km, substitution into (11.2.1) and (11.1.7) gives $R_1 = 0.983$. The damping after n iterations of scheme 1 is $[R_1]^n$. The total damping for the inertia–gravity modes after 12, 36, and 144 iterations is given in Table 11.1 for the four cases. It is clear from Table 11.1 that scheme 1 is most effective in damping inertia–gravity modes with large vertical scales and small horizontal scales and is least effective for small vertical scales and large horizontal scales. In the nomenclature of Section 6.6, dynamic initialization for this model is most effective when $L_H \ll L_R$, where L_R is the Rossby radius of deformation equal to $2\pi\sqrt{g\tilde{h}}/f_0$.

The damping after 12, 36, and 144 iterations of scheme 2 are given by Table 11.2 in the same format as Table 11.1. As expected, scheme 2 is more effective in damping inertia–gravity modes than scheme 1.

Slow mode constrained initialization (defined in Section 6.5) adjusted the inertia–gravity modes but left the Rossby modes unchanged. For this simple model, the

Table 11.1 *Total damping for inertia–gravity modes, scheme 1*

No. of iterations	$\tilde{h} = 10$ km		$\tilde{h} = 100$ m	
	$L_H = 10{,}000$ km	$L_H = 2000$ km	$L_H = 10{,}000$ km	$L_H = 2000$ km
12	0.812	0.044	0.956	0.918
36	0.535	0.000	0.874	0.775
144	0.082	0.000	0.584	0.361

Table 11.2 *Total damping for inertia–gravity modes, scheme 2*

No. of iterations	$\tilde{h} = 10$ km		$\tilde{h} = 100$ m	
	$L_H = 10{,}000$ km	$L_H = 2000$ km	$L_H = 10{,}000$ km	$L_H = 2000$ km
12	0.651	0.000	0.913	0.842
36	0.276	0.000	0.763	0.598
144	0.006	0.000	0.339	0.128

application of dynamic initialization has the same effect, damping the inertia–gravity modes without altering the Rossby modes. However, the linearized f-plane shallow water model is very special in one sense. Because there is no beta effect, the Rossby modes of this model are stationary, and $\omega_R = 0$. Consequently, for the Rossby modes, $R_1 = R_2 = 1$, and there is no damping. For more general models, there is a beta effect or a mean flow effect, and ω_R becomes small but nonzero. In more general models, the Rossby modes are likely to be slightly damped by the application of schemes 1 or 2. Thus, the application of these dynamic initialization schemes to more general models is only approximately slow mode constrained because the Rossby modes are not unaltered.

If model gridpoints are spaced a distance Δr apart, then the analysis of Appendix H suggests that the smallest horizontal-scale wave that can be resolved by the model has a wavelength of $2\,\Delta r$. Equation (11.2.1) shows that the highest frequency ω_m in the model will correspond to the external inertia–gravity mode with this horizontal scale. The time-step Δt can be chosen accordingly.

Dynamic initialization of the Hinkelmann–Phillips model (Sections 7.1 and 7.2) has been discussed by Temperton (1976). The model equations for the complex streamfunction $\hat{\psi}$, velocity potential $\hat{\chi}$, and geopotential $\hat{\Phi}$ are given by (7.1.9–11). Schemes 1 (11.1.2) and 2 (11.1.3) are applied to the dynamic initialization of this model using the same representative midlatitude values of the parameters given in (7.2.9). Two values of the equivalent depth are chosen, $\tilde{h} = 10$ km and $\tilde{h} = 100$ m, and Δt has been chosen to equal 600 s.

Observed/analyzed values of $\hat{\psi}$, $\hat{\chi}$, and $\hat{\Phi}$ are chosen exactly as in Section 7.2: $\hat{\psi}_0 = (1, 1)$, $\hat{\Phi}_0 = f_0\hat{\psi}_0 = (10^{-4}, 10^{-4})$, and $\hat{\chi} = (0, 0)$. Thus, $\hat{\psi}_0$ and $\hat{\Phi}_0$ are assumed to be related geostrophically, and the velocity potential is set to zero.

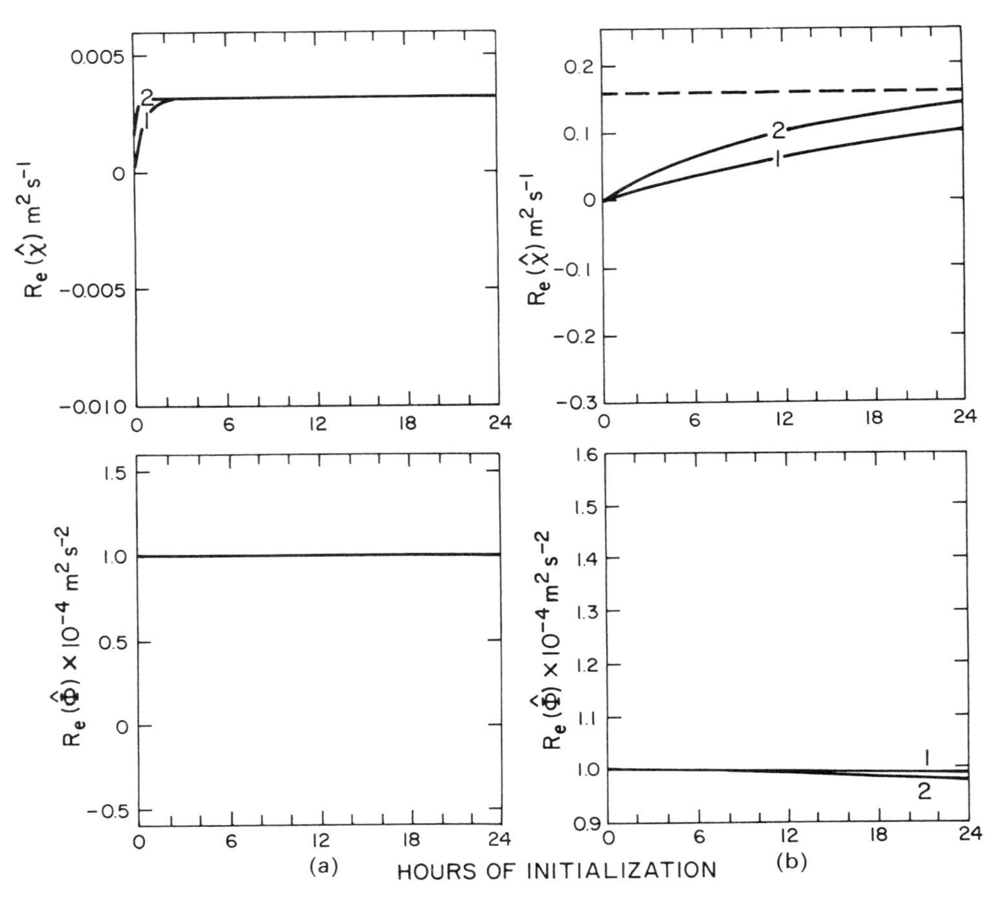

Figure 11.2 Evolution of velocity potential $\hat{\chi}$ and geopotential $\hat{\Phi}$ during the application of dynamic initialization: (a) $\bar{h} = 10$ km; (b) $\bar{h} = 100$ m. The abscissa is the number of hours of initialization, and curves 1 and 2 corresponds to schemes 1 and 2. The quasi-geostrophic solutions are shown by dashed lines.

As noted in earlier sections, one of the roles of any initialization process is to generate realistic, consistent fields of variables that are difficult to observe. One such variable is the divergent wind. Quasi-geostrophic initialization of the Hinkelmann–Phillips model generated a divergent wind given by (7.2.2). A good test of dynamic initialization is to see if it is capable of generating an equivalent divergent windfield.

Figure 11.2, in the same format as Figure 7.1, shows the evolution of $\hat{\chi}$ and $\hat{\Phi}$ during the application of dynamic initialization for (a) $\bar{h} = 10$ km and (b) $\bar{h} = 100$ m. ($\hat{\psi}$ is not shown because its evolution is essentially the same as that of $\hat{\Phi}$.) The amplitude of the real parts of $\hat{\chi}$ and $\hat{\Phi}$ are plotted against the iteration number for 24 hours (144 iterations). The units of the ordinate are exactly the same as in Figure 7.1.

On the plots of $\hat{\chi}$, the quasi-geostrophic values of $\hat{\chi}$ given by (7.2.2) are plotted as horizontal dashed lines. These values are $\hat{\chi} = 0.003213$ for the case $\bar{h} = 10$ km and

$\hat{\chi} = 0.1596$ for the case $\tilde{h} = 100$ m. It is clear that both schemes 1 and 2 are effective in generating a consistent divergent windfield for the large equivalent depth case ($\tilde{h} = 10$ km). For the small equivalent depth case ($\tilde{h} = 100$ m), neither scheme is very effective though scheme 2 is better than scheme 1.

Turning now to the geopotential field $\hat{\Phi}$, we see that the dynamic initialization procedure scarcely alters this field in the case $\tilde{h} = 10$ km and slightly damps it for $\tilde{h} = 100$ m. The damping of $\hat{\Phi}$ is slightly greater for scheme 2.

The results with the Hinkelmann–Phillips model are consistent with those of the linearized f-plane shallow water model. Dynamic initialization is effective for large vertical scales but less effective at small vertical scales. For simple models, dynamic initialization is capable of generating a consistent divergent windfield.

Note that these dynamic initialization schemes are expensive. The most costly part in any model or initialization scheme (with nonlinear and/or forcing terms) is the calculation of the term F in (11.1.1). Scheme 1 requires four calculations of F per iteration, and scheme 2 requires two. During the integration of the model itself, generally only one calculation of F occurs per time-step. Consequently, for scheme 1, 144 iterations with $\Delta t = 600$ s is equivalent to four days of integration with the model itself. By way of contrast, normal mode initialization requires one calculation of F per time-step and converges in less than four iterations.

Sections 11.1 and 11.2 have concentrated on a form of dynamic initialization that is almost analogous to slow mode constrained initialization. It is also possible to employ dynamic initialization in an approximately geopotential constrained or rotational wind constrained form. For geopotential constrained initialization, simply substitute the observed/analyzed geopotential for the initial geopotential after each iteration. This will continuously force the initialized geopotential toward the observed/analyzed geopotential. An analogous procedure can be carried out with respect to the rotational windfield. Geopotential or rotational wind constrained dynamic initialization may take longer to converge than slow mode constrained initialization. This type of initialization is often called forced dynamic initialization (see also Section 12.8).

11.3 Experiments with sophisticated models

Dynamic initialization experiments with simple models were sufficiently successful to encourage experiments with more sophisticated models. Miyakoda and Moyer (1968), Nitta and Hovermale (1969), Temperton (1976), and Williamson and Temperton (1981) all performed dynamic initialization experiments with baroclinic models of varying degrees of sophistication. The experiments of Williamson and Temperton (1981) were particularly illuminating for three reasons: They used a very complete global baroclinic primitive equations model; they did a careful comparison between dynamic and normal mode initialization; and they used some of the powerful tools developed in Chapter 10 to examine the balance achieved in the initialized state.

The forecast model and observed/analyzed state in the dynamic initialization experiments of Williamson and Temperton (1981) were identical to those discussed in Section 10.1, which produced Figures 10.1 and 10.2. The experiments were all of the (approximately) slow mode constrained type. An integration was performed from an initialized state that was obtained by dynamic initialization using 36 iterations of scheme 2 (11.1.3). The results of this experiment showed the clear superiority of nonlinear normal mode initialization with respect to dynamic initialization using (11.1.3). Calculation of the BAL statistic, defined by 10.1.1–2, showed that even after 36 iterations of dynamic initialization, the remaining imbalance in the inertia–gravity modes was much larger than after only two iterations of nonlinear normal mode initialization. Because each iteration of dynamic initialization required two complete calculations of the nonlinear and forcing terms, dynamic initialization was computationally inefficient as well as being relatively ineffective.

Dynamic initialization had two other problems. First, some physical processes such as precipitation and dissipation are inherently irreversible, and though a model containing such processes can be run backward in time, the physical meaning of such an operation is unclear (Winninghoff 1973; Miyakoda, Strickler, and Chludzinski 1978). Second, dynamic initialization is not a very precise initialization tool. Because it damps modes on the basis of frequency alone, there is no possibility of selection on the basis of vertical or horizontal scale, type of model, or other factor. Thus, Rossby modes are damped (albeit weakly) by dynamic initialization. This puts an upper limit on the number of dynamic initialization iterations that are advisable.

These problems with dynamic initialization as well as the results of Williamson and Temperton (1976) indicated the clear superiority of normal mode initialization. Thus, dynamic initialization algorithms fell rapidly into disfavor. However, this was not the end of the road for the dynamic initialization concept.

Bratseth (1982) proposed and Sugi (1986) implemented a new dynamic initialization algorithm that overcame most of the problems. The development of this algorithm was strongly influenced by the theory of nonlinear normal mode initialization. The key idea was to hold the nonlinear and forcing terms fixed while performing dynamic initialization on the linear terms only. For example, in (9.1.1–3), the terms R_u, R_v, and R_Φ would be held fixed during the iteration process. (See also Exercise 11.3.) Sugi (1986) held the nonlinear and forcing terms fixed while performing 20 iterations of dynamic initialization on the linear terms only. After he had recalculated the nonlinear and forcing terms using the new values of the dependent variables, he fixed the nonlinear and forcing terms again while 20 more iterations of dynamic initialization were performed on the linear terms. He repeated this procedure five times, thus performing a total of 100 iterations but with only five recalculations of the nonlinear and forcing terms. Because the calculation of the linear terms is much less costly than the calculation of the nonlinear and forcing terms, the Sugi (1986) procedure was as efficient as nonlinear normal mode initialization. Calculation of the BAL statistic (10.1.2) indicated that the balance obtained by the procedure was comparable to that obtained from nonlinear normal mode initialization.

Figure 11.3 shows the BAL statistic for five vertical modes calculated in Sugi's

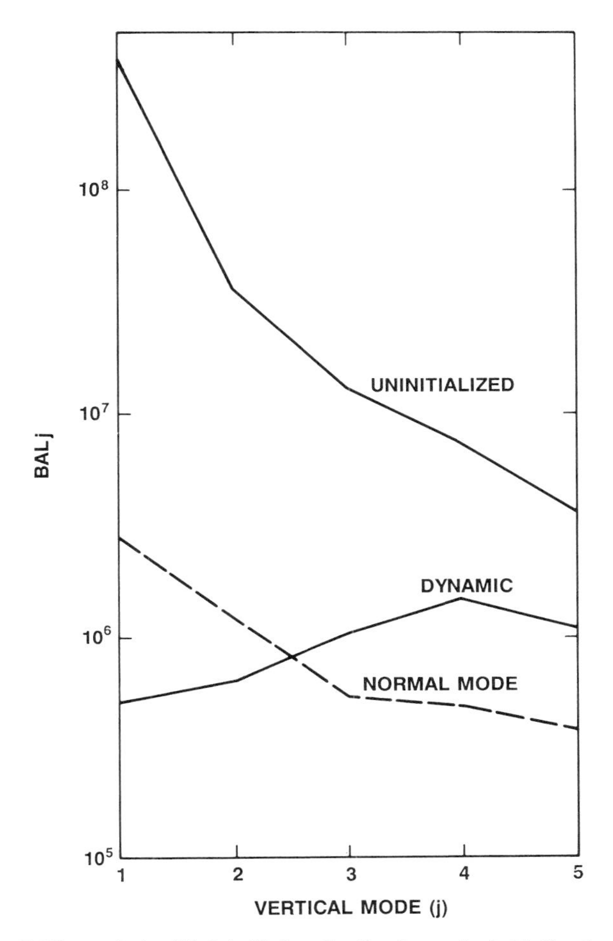

Figure 11.3 The BAL statistic (10.1.1–2) for Sugi's dynamic initialization experiment as a function of vertical model number j. (After Sugi 1986)

experiment. (Note that the BAL statistic is normalized differently in Figures 10.2 and 11.3.) The uninitialized case, nonlinear normal mode initialized case, and dynamic balanced case are shown. The balance obtained by Sugi's procedure is comparable to that obtained from nonlinear normal mode initialization. Moreover, because the nonlinear and forcing terms were not integrated forward and backward in time, there was no problem with irreversible physical parametrizations.

Dynamic initialization in the form suggested by Sugi (1986) has a number of attractive properties. It is efficient, produces a good balance, and can be used successfully with the full physical parametrizations of the model. It can be implemented very easily, with no need for a normal mode analysis. Krishnamurti et al. (1988) have used this technique very successfully in their initialization of the tropical flow (see Section 13.6). Satomura (1988) has applied the technique successfully to limited area models. A significant improvement in convergence rate was noted with the use of an implicit iteration scheme (Exercise 11.2).

Dynamic initialization treats the forecast model as a "black box" which is to be integrated with damping time integration schemes that are frequency selective. This philosophy is the basis for the combined analysis initialization technique called *continuous data assimilation*, to be discussed in the next chapter.

Exercises

11.1 Consider the equation $\dot{u} = -i\omega u + F_0$, where F_0 is time independent. A more general form of the dynamic initialization scheme (11.1.3) can be developed by replacing the third step by $u_n = u_{n-1} - \gamma(u_b - u_{n-1})$, where subscript n indicates iteration number. Apply this scheme to the initialization of this equation and derive a relationship between u_n, u_0, and F_0. Under what conditions will this iteration converge? Show that in the limit as $n \to \infty$, the solution is the same as that obtained by applying Machenhauer's condition $[\dot{u}(t=0)=0]$.

11.2 Repeat Exercise 11.1, except for the implicit dynamic initialization scheme of Satomura (1988),

$$u_a = u_{n-1} + F(u_a)\,\Delta t$$

$$u_n = u_a - F(u_n)\,\Delta t$$

Under what conditions would this scheme converge and to what limit? In what sense is this scheme implicit?

11.3 Consider the equation $\dot{u} = -i\omega u + F_0 \exp(i\omega_r t)$, where ω is a high (inertia–gravity) frequency and ω_r a low frequency. The term $F_0 \exp(i\omega_r t)$ is intended to simulate the slowly varying nonlinear and forcing terms. Apply the scheme (11.1.3) and derive a relationship between u_0, u_n, and F_0. Find the condition for convergence and the limit as $n \to \infty$. Show that as $\Delta t \to 0$, this limiting solution is equivalent to assuming $\ddot{u}(t=0)=0$. Now apply Sugi's (1986) idea: That is, hold the term $F_0 \exp(i\omega_r t)$ invariant during the entire iteration procedure. Compare the two solutions as $n \to \infty$. Repeat the exercise using the scheme of Exercise 11.2.

12

Continuous data assimilation

The weather satellite NIMBUS III, launched in the late 1960s, contained an instrument that was to have an enormous impact on the theory and practice of objective analysis and initialization. This instrument, a vertical sounder, measured emitted radiance from the atmosphere in several different wavelength channels. The radiance data were then transformed (using complicated inversion algorithms) into estimates of the temperature structure in individual atmospheric columns (see Houghton, Taylor, and Rodgers 1985). Unlike most previous meteorological instruments, the satellite-borne radiometer produced observations that were not tied to a spatially fixed observation network or to a particular observing interval.

The observations from satellite-borne systems were asynoptic, and their spatial distributions were governed by the orbital parameters of the satellites. An example of two orbits over North America from the later satellite TIROS-N is shown in Figure 12.1. The first orbit is over eastern North America and the subsequent orbit over western North America. The dots on Figure 12.1 indicate the position of individual temperature profiles (retrievals). As was noted in Chapter 4, there have been many problems with the quality of this type of data, but there is no denying that its horizontal spatial coverage is impressive.

The existence of this new type of observational data was a challenge to those interested in analyzing the state of the atmosphere. To meet this challenge, a totally new form of objective analysis/initialization was conceived: continuous data assimilation.

12.1 The basic philosophy

Numerical model development proceeded steadily through the 1950s, and by the late 1960s early baroclinic primitive equations models were being successfully integrated. Charney, Halem, and Jastrow (1969) proposed that the numerical models themselves be used to *assimilate* this newly available asynoptic data. The idea went as follows:

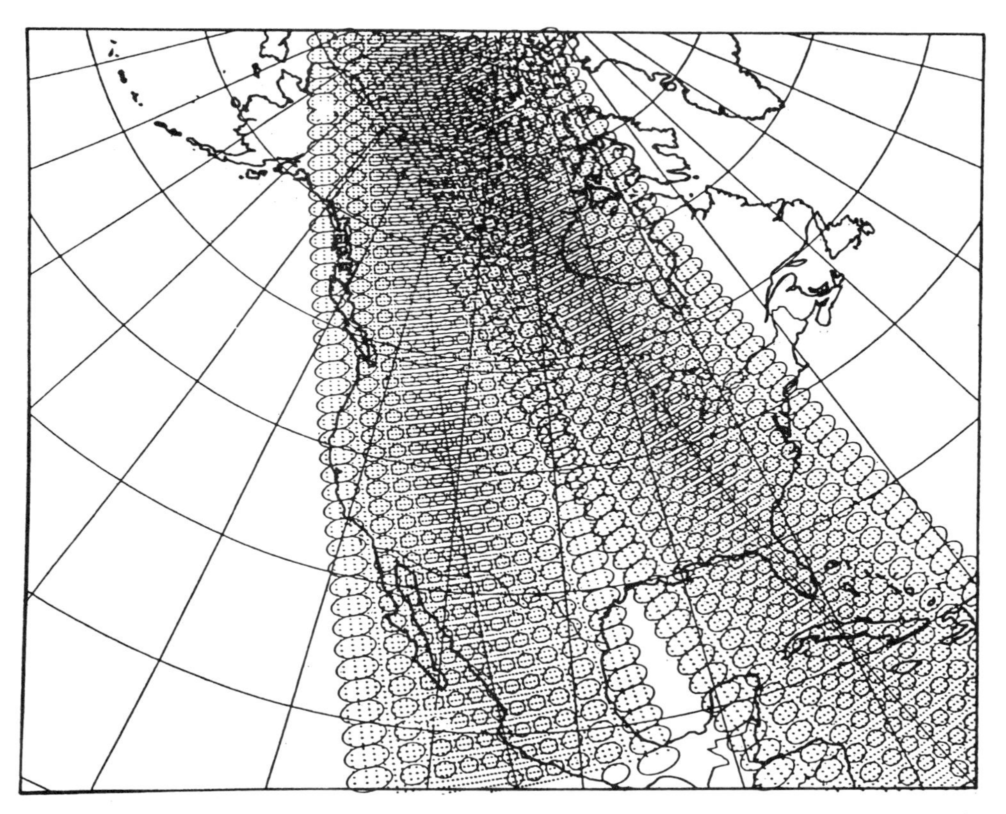

Figure 12.1 Scanning pattern for two successive orbits of a polar orbiting satellite. (After Smith et al., *Bull. Amer. Met. Soc.* **60**: 1177, 1979. The American Meteorological Society.)

There are always insufficient data to completely specify the state of the atmosphere; a model integrated from such imperfect initial conditions will eventually diverge from the true state of the atmosphere because of the imperfections in the initial state and inherent model faults. Now, over a period of a few days or less, satellite-borne radiometers provide temperature information over the whole globe. If asynoptic temperature information is *inserted* into the model at its true (asynoptic) time, then the model integrations will gradually be forced closer to the true atmospheric state. The idea is illustrated in Figure 12.2. Data are presumed to be asynoptically available before time $t = 0$ and are assimilated into the model. An improved estimate of the atmospheric state at $t = 0$ is thus obtained and serves as initial conditions for a subsequent model prediction.

This process has come to be known as *continuous data assimilation* or four-dimensional data assimilation; we depicted it schematically in Figure 1.13. The term *continuous data assimilation* is used in this chapter because the phrase *four-dimensional data assimilation* (three space dimensions and one time dimension) is often used to refer to the entire field covered by this book. In the present context, the word

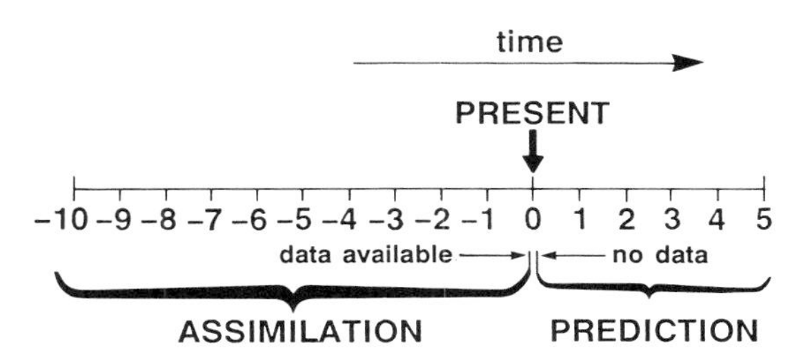

Figure 12.2 Schematic illustration of continuous data assimilation. (After Miyakoda et al. 1978)

continuous is not to be taken in its strict mathematical sense but rather as a synonym for *frequent*. Thus, continuous data assimilation should be assumed to refer specifically to the frequent insertion of asynoptic data. What distinguishes continuous data assimilation from other methods described so far (except in Section 8.6) is its rigid requirement for temporal consistency.

In many continuous data assimilation schemes, the model is integrated forward as observations are inserted. This is referred to as *continuous forward data assimilation*. In this case, the analyzed values at time t_A depend on all observations for times $t_O \leq t_A$. This process was depicted schematically in Figure 1.13 and is the main subject of this chapter. However, it is also possible, in principle, to insert observations into the assimilation model while it is being integrated forward and backward over the interval between t_1 and t_2. In this case, the analyzed values at time $t_1 \leq t_A \leq t_2$ depend implicitly on all observations $t_O \leq t_1$ and explicitly on all observations $t_1 \leq t_O \leq t_2$. This subject will be pursued further in Chapter 13.

Charney et al. (1969) performed an experiment with a general circulation model of the time that suggested that continuous data assimilation might work in practice. A similar idea was put forward by Tadjbakhsh (1969) at the same time, but it was really Charney's proposal that stimulated further development. As you will see in later sections, the road from Charney's proposal to successful operational implementation of continuous data assimilation was long and difficult. Before we proceed along that road, it is instructive to examine Charney's proposal with two simple examples.

Consider the simple one-dimensional advection equation:

$$\frac{\partial T}{\partial t} + U \frac{\partial T}{\partial x} = 0 \tag{12.1.1}$$

where T is the temperature, x the spatial coordinate, t the temporal coordinate, and U an advecting velocity that is independent of x and t. Assume periodic boundary conditions in x, $T(2\pi, t) = T(0, t)$.

Equation (12.1.1) is an example of a first-order partial differential equation:

$$a \frac{\partial T}{\partial t} + b \frac{\partial T}{\partial x} = c \tag{12.1.2}$$

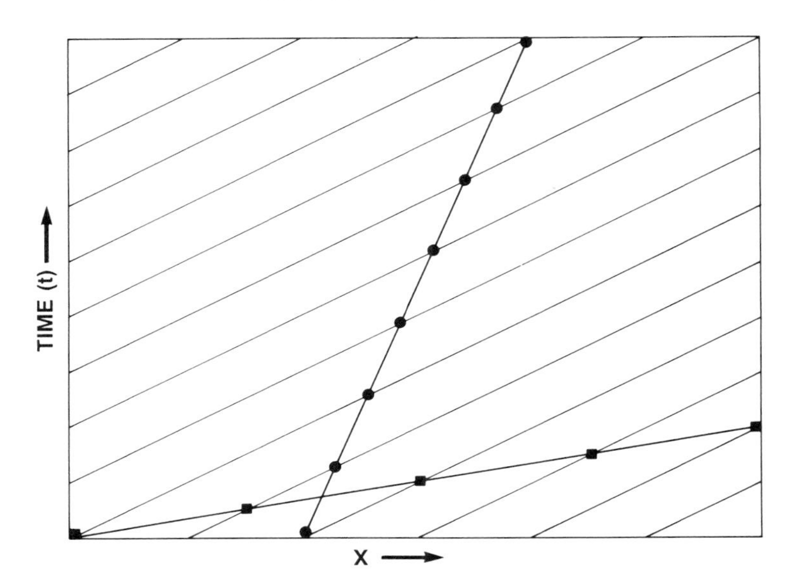

Figure 12.3 The characteristics of (12.1.2); also, two asynoptic orbits (solid dots and solid squares).

where $a = 1$, $b = U$, and $c = 0$. The solution to (12.1.2) can be defined in terms of its *characteristics* (see Garbedian 1964), which are the curves $x = x(t)$ in the x-t plane satisfying $dx/dt = b/a$. Equation (12.1.2) simply says that T satisfies the ordinary differential equation $dT/dt = c/a$ along the characteristics. Because $b/a = U = \text{constant}$ in our example, the characteristics of (12.1.1) are straight lines,

$$x - Ut = \text{constant} \tag{12.1.3}$$

The temperature T is constant along a given characteristic because $c/a = 0$. Therefore, the general solution of (12.1.1) must be a function of $x - Ut$ alone. The solution arising from an initial condition $T(x, 0) = T_0(x)$ is then

$$T(x, t) = T_0(x - Ut) \tag{12.1.4}$$

where the argument of T_0 is taken modulo 2π to reflect the periodic boundary condition.

This solution is illustrated in Figure 12.3, where time t is plotted against x. The characteristics are the lightweight slanted lines (U is presumed to be positive, so our example describes a wave traveling from left to right). Along these characteristics, the solution is invariant. The normal situation for integrating a model is to start at $t = 0$ and integrate forward in time. In this simple model, if $T(x, 0)$ is known at some point $x = x_0$, then the temperature is known for all subsequent times along the characteristic that passes through the point $(x_0, 0)$.

The situation of asynoptic data assimilation corresponds to one of the two arbitrary orbits (solid squares or circles) in Figure 12.3. In these cases, the information is available along the orbit at many different times and places. However, every time the orbit intersects a characteristic, the solution is known subsequently for all time along that characteristic. If an orbit in space-time intersects each characteristic once, the error-free observation of T along that orbit will specify the solution for all time. However, if the orbit coincides with a characteristic, the value of T along the orbit will not specify the entire solution but only the value of T along that particular characteristic.

Real models, of course are not as simple as (12.1.1). Primitive equation models have three spatial variables instead of one, and nonlinearity can lead to very complicated characteristics. A legitimate question to ask about continuous data assimilation in primitive equation models is the following. Is it possible to reconstruct the windfield from asynoptic temperature information alone?

Ghil (1980) considered this question in the context of the linear (6.4.4–6) shallow water equations. Suppose a complete history of the geopotential field everywhere is available. Then, differentiating (6.4.4) with respect to x, (6.4.5) with respect to y, and (6.4.6) with respect to t yields

$$u_{tx} + \Phi_{xx} - f_0 v_x = 0 \tag{12.1.5}$$

$$v_{ty} + \Phi_{yy} + f_0 u_y = 0 \tag{12.1.6}$$

$$\Phi_{tt} + \bar{\Phi}(u_{xt} + v_{yt}) = 0 \tag{12.1.7}$$

where $u_{tx} = \partial^2 u/(\partial t\, \partial x)$, and so on.

Now substitute into (12.1.7) for u_{tx} and for v_{ty} from (12.1.6). The resulting equation, together with (6.4.6), gives two equations for u and v in terms of Φ, Φ_t, and Φ_{tt}:

$$\nabla^2 \chi = u_x + v_y = -\frac{\Phi_t}{\bar{\Phi}} \tag{12.1.8}$$

$$\nabla^2 \psi = v_x - u_y = f_0^{-1}\left(\nabla^2 \Phi - \frac{\Phi_{tt}}{\bar{\Phi}}\right) \tag{12.1.9}$$

If Φ is known everywhere for all time, then Φ_t and Φ_{tt} are known; so the streamfunction ψ and velocity potential χ are known as well. Consequently, for the linearized shallow water equations, if the complete time history of the geopotential is known, then the windfield is also known (see also Bubé and Ghil 1980).

Ghil (1980) also examined the nonlinear shallow water equations (6.4.1–3) and showed that the equations corresponding to (12.1.8–9) in the nonlinear case were similar in form to the nonlinear balance equation (7.6.1), even to the point of having an ellipticity condition. Ghil's work suggests the validity of Charney's proposal for the primitive equations. It does not, however, demonstrate that it would work in practice. For example, if the geopotential height had to be known everywhere for all previous time to the nearest millimeter to construct windfields that were accurate to 5 ms^{-1}, then the whole idea would not be very attractive in practice. Proof of concept could be established only by practical experimentation with realistic numerical models. The early experimental results of Charney et al. (1969), Williamson and

Kasahara (1971), and Gordon, Umscheid, and Miyakoda (1972) were sufficiently encouraging that work on continuous data assimilation was pursued vigorously.

12.2 Damping time integration schemes

Before discussing the problems and practice of continuous data assimilation, we must digress briefly to reconsider damping time integration schemes. Observations that are inserted into primitive equation models often excite spurious inertia–gravity oscillations in the model. Damping time integration schemes are used in continuous data assimilation to damp these high-frequency oscillations. The use of damping integration schemes in continuous data assimilations is similar in purpose but different in application to procedures used in dynamic initialization.

In dynamic initialization, the model equations are integrated forward and backward in time with a damping initialization scheme until the initial state contains an acceptable minimum of inertia–gravity wave energy. At the end of the dynamic initialization process, the model state has not advanced in time. In continuous data assimilation, the model is integrated *forward* in time while observations are inserted. Thus, the damping procedures used in continuous data assimilation are marching procedures that actually advance the solution in time as well as damping the high-frequency components.

The most commonly used damping scheme in continuous data assimilation was due to Matsuno (1966). When applied to (11.1.1), the Matsuno scheme takes the following form:

$$\tilde{u}(t + \Delta t) - u(t) = F[u(t)]\,\Delta t \tag{12.2.1}$$

$$u(t + \Delta t) - u(t) = F[\tilde{u}(t + \Delta t)]\,\Delta t \tag{12.2.2}$$

where Δt is the time-step, u the dependent variable, and $\tilde{u}$ an intermediate value of u. Note that at the end of the second step, the solution has advanced from t to $t + \Delta t$. Equations (12.2.1–2) are equivalent to the forward component (steps 1 and 2) of (11.1.2).

Consider the application of (12.2.1–2) to the simple differential equation that was examined in Section 11.1:

$$\frac{du}{dt} = -i\omega u \tag{12.2.3}$$

where $i = \sqrt{-1}$ and ω is the frequency. Following the same arguments as those in Section 11.1, we find that the response for the Matsuno scheme is

$$R = \left|\frac{u(t + \Delta t)}{u(t)}\right| = \sqrt{1 - \omega^2\,\Delta t^2 + \omega^4\,\Delta t^4} \tag{12.2.4}$$

which is similar to the response of scheme 1 in Section 11.1, except for the square root. The response curve of the Matsuno scheme would be like curve 1 in Figure

11.1 except that the minimum response (at $\omega\,\Delta t = \pm 1/\sqrt{2}$) is only $\sqrt{3}/2$. Thus, in damping inertia–gravity waves, two time-steps of the Matsuno scheme are equivalent to one iteration of scheme 1.

The damping in the Matsuno scheme is controlled by the length of the time-step. Thus, the time-step could be chosen such that $\Delta t = 1/\sqrt{2}\omega_m$, where ω_m is the modulus of the highest model frequency. The scheme damps high frequencies most effectively but also damps lower frequencies (including Rossby modes) slightly. Because of the similarity of its response to that of (11.1.2), the effective damping of inertia–gravity modes by the Matsuno scheme can be determined directly from Table 11.1. Thus, for $\Delta t = 600$ s, the 12, 36, and 144 iterations of scheme 1 are equivalent to 4, 12, and 48 hours of integration with the Matsuno scheme.

In another scheme, called divergence damping (see Morel and Talagrand 1974), the divergence equation (6.4.9) in the shallow water equations is modified as follows:

$$\frac{\partial}{\partial t} D - f_0 \zeta + \nabla^2 \Phi = c\,\nabla^2 D \tag{12.2.5}$$

where $D = \nabla^2 \chi$ is the divergence, $\zeta = \nabla^2 \psi$ is the vorticity, and c is a damping coefficient that is specified by the user. The extra term in (12.2.5) has the effect of damping the smaller-scale divergence. The rationale behind this procedure is that the high-frequency inertia–gravity oscillations have a large divergence and a small vorticity. Thus, damping the divergence field tends to suppress high-frequency oscillations. Unfortunately, this procedure can also damp the low-frequency balanced component of the divergence.

Both of the procedures just described are frequency selective to some extent. However, neither could be described as precision instruments for damping inertia–gravity oscillations (see Exercises 12.1–3).

12.3 Identical twin configurations: convergence of solutions

The mechanics of continuous data assimilations are, in principle, straightforward: Data are inserted directly into a numerical model as it is being integrated in time. What actually happens to the model integration each time data are inserted is not straightforward at all. It is certainly possible to examine experimentally the process of data insertion in a complicated baroclinic primitive equation model. However, if real understanding of the process is the goal, both the model and the data have to be considerably simplified.

Models can be simplified by linearization and/or using the shallow water equations. The temporal and spatial characteristics of the data insertion can be made more regular than they are in practice. Use of synthetic data eliminates the inherent problems of accuracy, representativeness, and so forth, that are found in real data. This section and the following two sections (12.4 and 12.5) are concerned with synthetic data experiments. In these experiments, the observational data are actually generated by the assimilating model itself in a previous integration. This type of experiment is called an *identical twin* experiment because the simulated "atmosphere" from which the data were derived was created by the same model that is used for assimilating the data. The identical twin configuration is really not very realistic because it is equivalent to assuming that the assimilating model has no error. In the real world, of course, no models are perfect, and the results of identical twin experiments necessarily err on the optimistic side.

Having made the identical twin assumption, we still can make the experiment realistic in other respects. The model can be as complicated as desired, and the data, though synthetic, can be inserted in a realistic manner. It is even possible to simulate partially the effect of observation error by adding noise to the synthetic observations.

This section considers identical twin experiments with the linearized shallow water equations using error-free data. Section 12.4 will consider the effects of nonlinearities and Section 12.5 the effect of observational errors (see also Appendix B). The approach here is inspired by the work of Williamson and Dickinson (1972) and Talagrand (1981a).

We consider both geopotential and wind insertion. Geopotential insertion in the shallow water equations is a surrogate for temperature (from 7.3.2) insertion in the baroclinic primitive equations. Our emphasis is on the temporal rather than the spatial aspects of the problem. Thus, we assume that when geopotential is inserted, it is inserted everywhere simultaneously. Similarly, when wind is inserted, it is inserted everywhere simultaneously. This makes the analysis more tractable but obviously does not simulate the insertion pattern from an orbiting satellite very faithfully.

The linearized shallow water equations (6.4.8–10) can be written in normal mode form from (6.4.26). Assume that the inertia–gravity mode equations have been modified to include a damping term:

$$\dot{y} = 0 \tag{12.3.1}$$

$$\dot{z}_1 = -i\sigma_G^1 f_0 z_1 - \sigma_D z_1 \tag{12.3.2}$$

$$\dot{z}_2 = -i\sigma_G^2 f_0 z_2 - \sigma_D z_2 \tag{12.3.3}$$

where σ_D is an as yet unspecified frequency-dependent damping coefficient. Equations (12.3.1–3) have the solutions

$$y(t) = y(0), \qquad z_1(t) = z_1(0)\exp(-i\alpha t - \sigma_D t), \qquad z_2(t) = z_2(0)\exp(i\alpha t - \sigma_D t) \tag{12.3.4}$$

where $\alpha = f_0\sqrt{1+K}$ is as in (6.4.28). In this case, the solutions for $\hat{\psi}(t)$, $\hat{\chi}(t)$, and $\hat{\Phi}(t)$ given in (6.4.29–31) are modified slightly in that the terms $\cos(\alpha t)$ and $\sin(\alpha t)$ are both to be multiplied by $\exp(-\sigma_D t)$.

The simple form of damping adopted in (12.3.1–3) is very idealized and could be realized in practice only if the model equations were written in normal mode form as in Chapter 9. Operational damping schemes such as those discussed in the previous section often have undesirable side effects, such as modification of the eigenfrequencies (see Section 12.8) or damping of the Rossby modes. However, the most important property of operational damping schemes, their frequency dependence, can be examined using (12.3.1–3).

Equations (12.3.1–3) will be the governing equations for the identical twin assimilation experiment. All solutions to these equations are given by the damped form of (6.4.29–31). In the first run of the model, we create "synthetic" data or a "true" solution. To mimic the properties of the atmosphere as closely as possible, we assume the "true" solution to be in balance. Thus,

$$\hat{\psi}_{\mathrm{T}}(t) = \hat{\psi}_{\mathrm{T}}(0) = \sqrt{K}\hat{\Phi}_{\mathrm{T}}(0), \qquad \hat{\chi}_{\mathrm{T}}(t) = 0, \qquad \hat{\Phi}_{\mathrm{T}}(t) = \hat{\Phi}_{\mathrm{T}}(0)$$

which satisfies the damped form of (6.4.29–31).

Suppose the assimilating model is integrated a second time, from arbitrarily chosen initial conditions $\hat{\psi}_{\mathrm{A}}(0)$, $\hat{\chi}_{\mathrm{A}}(0)$, and $\hat{\Phi}_{\mathrm{A}}(0)$, where the subscript A stands for actual. Then the actual solution $\hat{\psi}_{\mathrm{A}}(t)$, $\hat{\chi}_{\mathrm{A}}(t)$, and $\hat{\Phi}_{\mathrm{A}}(t)$ will, in general, differ from the true solution. Define

$$\varepsilon_{\psi}(t) = \hat{\psi}_{\mathrm{A}}(t) - \hat{\psi}_{\mathrm{T}}(t), \qquad \varepsilon_{\chi}(t) = i(\hat{\chi}_{\mathrm{A}}(t) - \hat{\chi}_{\mathrm{T}}(t)), \qquad \varepsilon_{\Phi}(t) = \hat{\Phi}_{\mathrm{A}}(t) - \hat{\Phi}_{\mathrm{T}}(t)$$

to be the "errors" in the actual solution. Because the governing equations are linear, the errors ε_{ψ}, ε_{χ}, and ε_{Φ} also satisfy the damped version of (6.4.29–31).

We first consider geopotential insertion. For the purpose of analyzing the effect of observation error in Section 12.5, we assume that the inserted geopotential field contains random observation error. Data insertion is assumed to take place at times

$$t_0, \ldots, \qquad t_0 + j\,\Delta_{\mathrm{I}}t, \ldots, \qquad t_0 + J\,\Delta_{\mathrm{I}}t \tag{12.3.5}$$

$\Delta_{\mathrm{I}}t$ is the data insertion interval, as the subscript I denotes; it may differ from the time-step of the model. Then the geopotential error at any insertion time $t_0 + j\,\Delta_{\mathrm{I}}t$ is $\varepsilon_{\Phi}(t_0 + j\,\Delta_{\mathrm{I}}t)$ and is actually a complex number. Multiplying by the complex conjugate and taking expectation values gives

$$\langle \varepsilon_{\Phi}(t_0 + j\,\Delta_{\mathrm{I}}t)\varepsilon_{\Phi}^{*}(t_0 + j\,\Delta_{\mathrm{I}}t)\rangle = \langle \varepsilon_{\Phi}^2 \rangle = \mathrm{E}_{\Phi}^2 \tag{12.3.6}$$

which is the expected observation error variance for geopotential and is assumed to be independent of j. What we want to determine is the error in the expectation values of the windfield, $\langle \varepsilon_{\psi}^2(t)\rangle$ and $\langle \varepsilon_{\chi}^2(t)\rangle$.

Expressions for $\varepsilon_{\psi}(t + \Delta_{\mathrm{I}}t)$ and $\varepsilon_{\chi}(t + \Delta_{\mathrm{I}}t)$ can be written in terms of $\varepsilon_{\psi}(t)$, $\varepsilon_{\chi}(t)$, and $\varepsilon_{\Phi}(t)$, using the damped versions of (6.4.29–30):

$$\varepsilon_{\psi}(t + \Delta_{\mathrm{I}}t) = a_1\varepsilon_{\psi}(t) + a_2\varepsilon_{\chi}(t) + b_1\varepsilon_{\Phi}(t) \tag{12.3.7}$$

$$\varepsilon_{\chi}(t + \Delta_{\mathrm{I}}t) = a_3\varepsilon_{\psi}(t) + a_4\varepsilon_{\chi}(t) + b_2\varepsilon_{\Phi}(t) \tag{12.3.8}$$

where

$$a_1 = \frac{K + D\cos\beta}{1 + K}, \qquad a_2 = -\frac{D\sin\beta}{\sqrt{1 + K}}, \qquad a_3 = \frac{D\sin\beta}{\sqrt{1 + K}},$$

$$a_4 = D\cos\beta, \qquad b_1 = \sqrt{K}\,\frac{(1 - D\cos\beta)}{1+K}, \qquad b_2 = -\frac{D\sqrt{K}\sin\beta}{\sqrt{1+K}}$$

and where $D = \exp(-\sigma_{\mathrm{D}}\,\Delta_{\mathrm{I}}t)$ and $\beta = f_0\sqrt{1+K}\,\Delta_{\mathrm{I}}t$.

Now ε_ψ, ε_χ, and ε_Φ are complex numbers. The complex conjugate forms of (12.3.7–8) also exist (because the original streamfunction, velocity potential and geopotential fields are real), and they are exactly the same except that they involve the complex conjugates ε_ψ^*, ε_χ^*, and ε_Φ^*. A prediction equation for $\varepsilon_\Phi(t + \Delta_{\mathrm{I}}t)$ also exists, but it is not required because the predicted value of Φ is always directly replaced by the observed value. $\varepsilon_\psi(t)$ and $\varepsilon_\chi(t)$ can be thought of as forecast errors.

Taking (12.3.7–8), multiplying by the complex conjugates, and applying the expectation operator gives

$$\langle \underline{\varepsilon}_{\mathrm{w}}^2(t + \Delta_{\mathrm{I}}t)\rangle = \underline{\underline{S}}\langle \underline{\varepsilon}_{\mathrm{w}}^2(t)\rangle + \langle \varepsilon_\Phi^2\rangle \underline{b} \tag{12.3.9}$$

where

$$\underline{\underline{S}} = \begin{pmatrix} a_1^2 & a_2^2 & 2a_1a_2 \\ a_3^2 & a_4^2 & 2a_3a_4 \\ a_1a_3 & a_4a_2 & a_1a_4 + a_2a_3 \end{pmatrix}$$

$$\underline{b} = [b_1^2 \quad b_2^2 \quad b_1b_2]^{\mathrm{T}}$$

$$\langle \underline{\varepsilon}_{\mathrm{w}}^2\rangle = [\langle \varepsilon_\psi^2\rangle \quad \langle \varepsilon_\chi^2\rangle \quad 0.5\langle \varepsilon_\psi\varepsilon_\chi^* + \varepsilon_\psi^*\varepsilon_\chi\rangle]^{\mathrm{T}}$$

It has been assumed here that

$$\langle \varepsilon_\psi\varepsilon_\Phi^* + \varepsilon_\psi^*\varepsilon_\Phi\rangle = \langle \varepsilon_\chi\varepsilon_\Phi^* + \varepsilon_\chi^*\varepsilon_\Phi\rangle = 0$$

because $\varepsilon_\Phi(t)$ is a purely random observation error and is therefore uncorrelated with the forecast errors $\varepsilon_\psi(t)$ and $\varepsilon_\chi(t)$.

Now $\langle \underline{\varepsilon}_{\mathrm{w}}^2(t + \Delta_{\mathrm{I}}t)\rangle$ is a measure of the wind error after one geopotential insertion. After J geopotential insertions, the wind error at time $t_0 + J\,\Delta_{\mathrm{I}}t$ is

$$\langle \underline{\varepsilon}_{\mathrm{w}}^2(t_0 + J\,\Delta_{\mathrm{I}}t)\rangle = \underline{\underline{S}}^J\langle \underline{\varepsilon}_{\mathrm{w}}^2(t_0)\rangle + \langle \varepsilon_\Phi^2\rangle\left[\underline{\underline{I}} + \sum_{j=1}^{J} \underline{\underline{S}}^j\right]\underline{b} \tag{12.3.10}$$

where $\underline{\underline{I}}$ is the identity matrix.

At this point, assume that the geopotential observations are perfect, that is, $\langle \varepsilon_\Phi^2\rangle = 0$. Then clearly the matrix $\underline{\underline{S}}$ defines a response matrix, that is, it relates the values of $\langle \varepsilon_\psi^2(t + j\,\Delta_{\mathrm{I}}t)\rangle$ and $\langle \varepsilon_\chi^2(t + j\,\Delta_{\mathrm{I}}t)\rangle$ to the values at the previous insertion time. If the procedure is to converge, then the values of $\langle \varepsilon_\psi^2\rangle$ and $\langle \varepsilon_\chi^2\rangle$ must decrease after every insertion. This can be investigated by examining the spectral radius of matrix $\underline{\underline{S}}$. The spectral radius of $\underline{\underline{S}}$, denoted $\rho(\underline{\underline{S}})$, is the modulus of the eigenvalue of $\underline{\underline{S}}$ having the largest modulus. The procedure is said to converge if

$$\langle \underline{\varepsilon}_{\mathrm{w}}^2(t_0 + J\,\Delta_{\mathrm{I}}t)\rangle \to 0 \qquad \text{as} \qquad J \to \infty$$

and it can be shown that this happens for all initial errors $\langle \underline{\varepsilon}_{\mathrm{w}}^2(t_0)\rangle$ if and only if

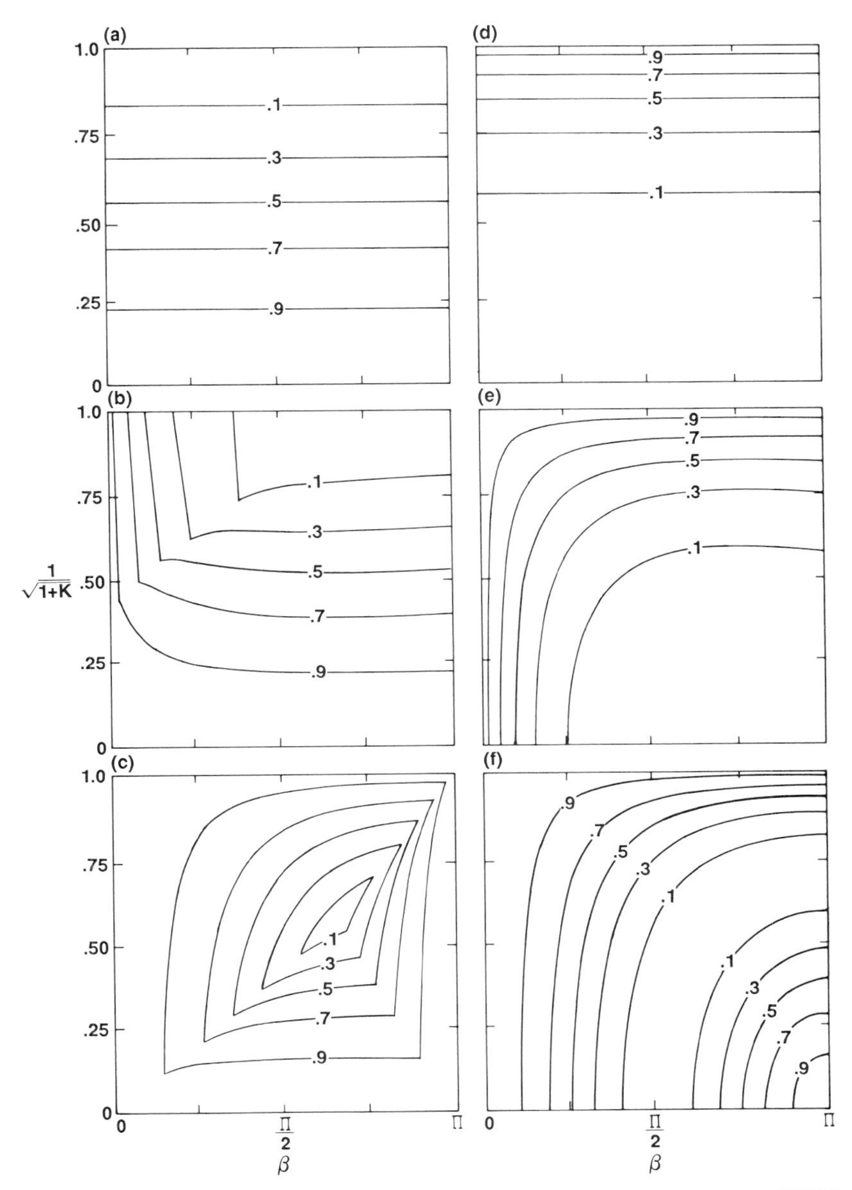

Figure 12.4 Spectral radii of matrix $\underline{\underline{S}}$ as a function of $\beta = \alpha\,\Delta_I t$ and $1/\sqrt{K+1}$: (a–c), geopotential insertion; (d–f), wind insertion. (a,d), $\sigma_D = \infty$; (b,e), $\sigma_D = \alpha$; (e,f) $\sigma_D = 0$.

$\rho(\underline{\underline{S}}) < 1$. Note, however, that $\rho(\underline{\underline{S}}) < 1$ does not guarantee that each component of $\langle \underline{\varepsilon}_w^2 \rangle$ decreases after each iteration.

The spectral radii $\rho(\underline{\underline{S}})$ for several cases are plotted in Figures 12.4(a–c). Case (a) corresponds to $D = 0$ or $\sigma_D = \infty$, and case (c) to $D = 1$ or $\sigma_D = 0$. Case (b) refers to

$\sigma_D = \alpha = f_0\sqrt{1+K}$. In these diagrams, the abscissa is $\beta = \alpha\,\Delta_I t$ and the ordinate is $1/\sqrt{K+1}$. Thus, $K = 0$ is at the top of each diagram and $K = \infty$ at the bottom.

In (a), $D = 0$ and the matrix $\underline{\underline{S}}$ is reduced to the one-by-one matrix a_1^2, which has a spectral radius $K^2/(1+K)^2$ that is independent of $\alpha\,\Delta_I t$. This case corresponds to complete damping; convergence is most rapid for small K and does not take place at all for $K = \infty$. In other words, for large horizontal scale, small vertical scale, or high latitudes, geopotential insertion converges rapidly; but for small horizontal scales, large vertical scales, and low latitudes, convergence is slow. This result is consistent with geostrophic adjustment theory discussed in Section 6.6. In this case, at a given time-step, all the error is in the geostrophic mode.

In (c), where there is no damping at all, we see there is generally convergence, albeit slow under most conditions. It can be shown that $\rho(\underline{\underline{S}}) \le 1$ for this model (see Exercise 12.4). However, convergence does not occur – $\rho(\underline{\underline{S}}) = 1$ – on the boundaries $K = 0$, $K = \infty$, $\beta = 0$, and $\beta = \pi$. The case $K = \infty$ is as in (a). $\beta = 0$ or $\beta = 2n\pi$, where n is an integer, occurs when exactly the same data are being inserted at every insertion time and is analogous to inserting along a characteristic in Figure 12.3. The case $\beta = \pi$ is similar, except that data 180 degrees out of phase are being inserted every other insertion, and there still are not enough new data for convergence.

In (b), we see the result of error-free geopotential insertion with damping that is directly proportional to the frequency. Note that for $\alpha = 0$ there is no damping, which is consistent with not damping the Rossby mode. In this case, convergence occurs most rapidly for K small and $\beta \gg 0$. Convergence slows for very large K or when α or $\Delta_I t$ approach zero. Incidentally, the discontinuities in the slopes of the isolines in Figure 12.4 are caused by the fact that different eigenvalues of $\underline{\underline{S}}$ are predominant in different regions of parameter space.

We now consider the case of wind insertion. Assume that the wind data are inserted at an interval of $\Delta_I t$. The inserted winds contain random observation errors with time-independent expectation values of $\langle\varepsilon_\psi^2\rangle$ and $\langle\varepsilon_\chi^2\rangle$. What is sought is the expectation value of the geopotential error $\langle\varepsilon_\Phi^2(t)\rangle$. From the damped version of (6.4.31), we can write an expression analogous to (12.3.7):

$$\varepsilon_\Phi(t+\Delta_I t) = c_1\varepsilon_\Phi(t) + d_1\varepsilon_\psi(t) + d_2\varepsilon_\chi(t) \tag{12.3.11}$$

where

$$c_1 = \frac{1+KD\cos\beta}{1+K}, \qquad d_1 = \frac{\sqrt{K}(1-D\cos\beta)}{1+K}, \qquad d_2 = \frac{\sqrt{K}D\sin\beta}{\sqrt{1+K}}$$

Multiplying by the complex conjugate and taking expectation values gives

$$\langle\varepsilon_\Phi^2(t+\Delta_I t)\rangle = c_1^2\langle\varepsilon_\Phi^2(t)\rangle + d_1^2\langle\varepsilon_\psi^2\rangle + d_2^2\langle\varepsilon_\chi^2\rangle \tag{12.3.12}$$

where $\langle\varepsilon_\psi^2\rangle$ and $\langle\varepsilon_\chi^2\rangle$ are the time-independent expectation values of the wind errors.

It has been assumed that ε_ψ and ε_χ are uncorrelated with each other or with ε_Φ. The analogous expression to (12.3.10) for the wind insertion case is

$$\langle\varepsilon_\Phi^2(t_0 + J\,\Delta_\mathrm{I}t)\rangle = c_1^{2J}\langle\varepsilon_\Phi^2(t_0)\rangle + (d_1^2\langle\varepsilon_\psi^2\rangle + d_2^2\langle\varepsilon_\chi^2\rangle)\left(1 + \sum_{j=1}^{J-1} c_1^{2j}\right) \quad (12.3.13)$$

Now assume perfect wind observations, so that $\langle\varepsilon_\psi^2\rangle$ and $\langle\varepsilon_\chi^2\rangle$ are equal to zero. Following the earlier arguments, we find that wind insertion converges if $c_1^2 \leq 1$.

The values of c_1^2 are plotted in Figures 12.4(d–f) for the wind insertion case, in the same format as for geopotential insertion. Case (d) corresponds to $D = 0$ or $\sigma_\mathrm{D} = \infty$ and case (f) to $D = 1$ or $\sigma_\mathrm{D} = 0$. Case (e) corresponds to $\sigma_\mathrm{D} = f_0\sqrt{1 + K}$.

In (d), where there is complete damping, $c_1^2 = [1/(1 + K)]^2$. Convergence is most rapid for large K, and there is no convergence for $K = 0$.

In (f), convergence generally occurs when there is no damping, except at small K and small $\Delta_\mathrm{I}t$ for reasons that are to be expected. The maximum at the lower right-hand corner corresponds to $\varepsilon_\Phi(t)$ changing sign after each insertion. In (e), the maximum at the lower right is removed when there is damping.

The results of the analysis with an identical twin experiment using a linearized shallow water model and error-free insertion suggest that both wind and geopotential insertion almost always converge. Convergence is enhanced by damping, whether complete or not. Geopotential insertion is most effective at large horizontal scales, small vertical scales, and high latitudes. Wind insertion is most effective at small horizontal scales, large vertical scales, and low latitudes. If $\Delta_\mathrm{I}t$ is too small, convergence is not as rapid. (It should be noted, though, that Figure 12.4 indicates the response for a single insertion. More rapid insertion gives more insertions in a given period of time, and the matrix $\underline{\underline{S}}$ or the factor c_1^2 is taken to a higher power.) The results of the present analysis have been largely confirmed by identical twin experiments using nonlinear baroclinic primitive equation global models, performed by Williamson and Kasahara (1971) and Gordon et al. (1972).

This analysis has been concerned only with temporal aspects of the problem. The simple spectral model used cannot properly account for inertia–gravity wave dispersion discussed in Section 6.6. In practice, the observations are inserted irregularly in space. The insertion of a few observations in an isolated area excites inertia–gravity waves, which propagate inertia–gravity wave energy away from the region of excitation as well as being damped by the model dissipation mechanisms.

12.4 The effect of nonlinearities

The previous section considered error-free geopotential and wind insertion in identical twin experiments with a linear shallow water model. The results of that analysis demonstrate that wind and geopotential insertion almost always converges though the convergence rate is a strong function of the insertion interval $\Delta_\mathrm{I}t$, the damping rate, the latitudes, and the horizontal and vertical scales. In particular, convergence generally occurs when $\Delta_I t$ gets very large. In fact, convergence of wind or geopotential insertion for large $\Delta_I t$ is true only in linear models. Nonlinear models have a characteristic, called *predictability error growth*, that is not present in most linear models and that frustrates data insertion with large $\Delta_I t$.

Predictability error occurs in the atmosphere and in nonlinear atmospheric models. It has been studied for many years and is discussed in detail in Holloway and West (1984). For the purpose of analyzing data assimilation, only a few results from predictability theory are needed; we state these rather than prove them.

We discuss the problem in the context of the nonlinear shallow water equations (6.4.1–3). Consider two randomly chosen atmospheric states defined by u_a, v_a, Φ_a and u_b, v_b, Φ_b. Define the following difference energy measure:

$$\varepsilon^2(a, b) = \int_S [(\Phi_a - \Phi_b)^2 + \tilde{\Phi}(u_a - u_b)^2 + \tilde{\Phi}(v_a - v_b)^2]\, dS \qquad (12.4.1)$$

where $\int dS$ is an integral over the domain; u, v, and Φ are defined as in Section 6.4, and Φ is assumed to be the deviation from the domain average $\tilde{\Phi}$.

Next, set $u_b = v_b = \Phi_b = 0$ and take the expectation value of (12.4.1). Then $\langle \varepsilon^2(a, 0) \rangle = V$ is the variance for any atmospheric state. If u_a, v_a, Φ_a define a state that is statistically independent of the state u_b, v_b, Φ_b, then $\langle \varepsilon^2(a, b) \rangle = 2V$. That is, the expected square difference of two independent atmospheric states is twice the variance of any individual state. Normally, atmospheric states are independent if they are far apart in time.

Now, define two model integrations starting at time $t = 0$ from slightly different initial conditions, using the same model (identical twin assumption). Also assume there is no data insertion. As time progresses, two series of simulated atmospheric states evolve: $u_a(t)$, $v_a(t)$, $\Phi_a(t)$ and $u_b(t)$, $v_b(t)$, $\Phi_b(t)$. Define the expected normalized error (really the difference) between two such states as a function of time:

$$E^2(t) = \frac{\langle \varepsilon^2(a, b, t) \rangle}{V} \qquad (12.4.2)$$

where the initial error is $E^2(0)$.

The results of an experiment of this type are shown schematically in Figure 12.5. For an undamped linear model (such as the linearized shallow water model), the normalized error variance $E^2(t) = E^2(0)$. For a damped linear model, $E^2(t) \leq E^2(0)$ and tends to zero. Thus, the asymptotic normalized error depends only on the initial error. For a nonlinear model (such as the nonlinear shallow water model), $E^2(\infty) = 2$. That is, the asymptotic error is independent of the initial error. In fact, the asymptotic error is equal to the expected difference between two completely independent states.

A similar result applies for the baroclinic primitive equations. The error curve for the nonlinear model is known as a predictability error growth curve. Results taken from turbulence theory and identical twin experiments with general circulation models suggest that for small values of time, the predictability curve can be approximated

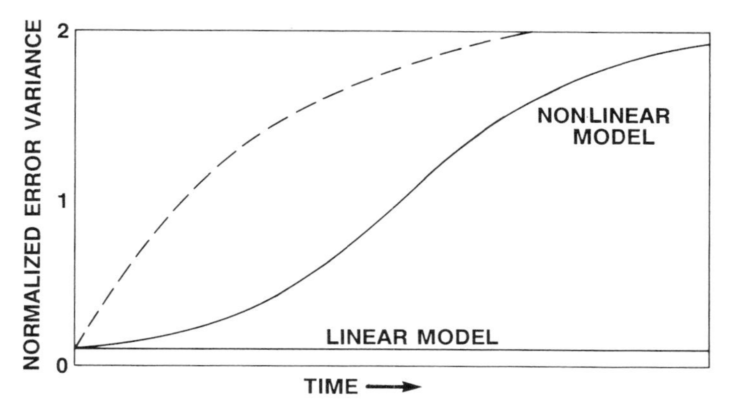

Figure 12.5 Schematic illustration of the expected value of differences between model integrations with slightly different initial conditions as a function of time. Linear and nonlinear models indicate predictability results whereas the dashed curve gives the difference between model forecast and actual atmosphere.

by an exponential with an *e*-folding time of about two days. In these experiments, two model states that initially differed by an amount equal to the observation error, eventually became as different as two randomly chosen states. The time period for this to occur is called the *limit of deterministic predictability* and is normally about 10–14 days for synoptic scales. This limit is indicated as the dark solid line at 10^6 s in Figure 1.1. (The dashed curve in Figure 12.5 will be discussed in Section 12.6.)

One other aspect of predictability error growth should be mentioned here. Predictability experiments performed by Daley (1980b, 1981) with nonlinear barotropic and baroclinic models have shown that the evolution of the low-frequency component of the model integration is altered when the initial projection onto the slow modes is altered. Altering the initial projection onto the fast modes does not substantially alter the low-frequency solution. This result is illustrated by Figure 10.10: Comparison of the heavy and dashed curves shows that the initial differences were projected entirely onto the fast modes and that the low-frequency evolutions of the two integrations were essentially the same; the light curve shows that the initial slow mode projection was altered and that the difference between this case and the other two cases actually grew with time, consistent with predictability theory.

What does predictability error growth have to do with continuous data assimilation? Consider perfect geopotential or wind insertion (no observation error) in a nonlinear shallow water model. Again make the identical twin assumption and consider what happens to the error after each data insertion.

In the terminology of Section 12.3, each data insertion brings the actual state closer to the true state. In a linear model, the error decreases or remains constant after each insertion. As Figure 12.5 suggests, in a nonlinear model the data insertion itself would decrease the error, but after the insertion the error would start to grow again. The implications of this are that in a linear model, $\Delta_I t$ could be very large

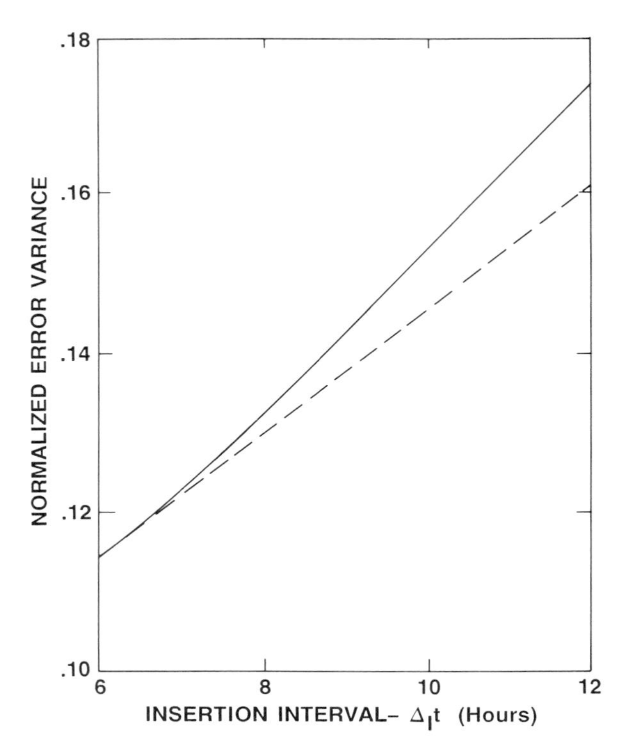

Figure 12.6 Normalized asymptotic error variance as a function of insertion interval. Solid line, an analytic result; dashed line, experimental result. (After Blumen, *J. Atmos. Sci.* **33**: 170, 1976. The American Meteorological Society.)

without harming the convergence. In a nonlinear model, if $\Delta_I t$ is too large, the predictability error growth may be larger than the error reduction due to the data insertion, so there may be no convergence. Clearly, in nonlinear models $\Delta_I t$ must not be too large. It is worth emphasizing that predictability error occurs even when the model is perfect (identical twin assumption).

Stern (1974) and Blumen (1976) examined the effect of predictability error on data assimilation using simple (low-order) nonlinear models. Both analyses examined error-free data insertion in identical twin configurations. Blumen (1976) also compared his analytic results with equivalent experimental results obtained by Williamson and Kasahara (1971) using a general circulation model. The results are shown in Figure 12.6. In nonlinear models, the asymptotic error variance is nonzero and increases as $\Delta_I t$ is increased.

12.5 The effect of observation error

The effect of observation error in identical twin experiments can be examined by using the simple model developed in Section 12.3. The model is based on the linearized shallow water equations, and there is no predictability error growth to affect the results at large $\Delta_{\mathrm{I}}t$.

In the case of geopotential insertion, the expected error in the windfield $\langle \underline{\varepsilon}_{\mathrm{w}}^2 \rangle$ after J insertions is given by (12.3.10). $\langle \varepsilon_{\Phi}^2 \rangle$ is the geopotential observation error variance after each insertion. The asymptotic error can be obtained by letting $J \to \infty$. For all cases in which the spectral radius of $\underline{\underline{S}}$ is less than one, the first term on the right-hand side of (12.3.10) goes to zero as $J \to \infty$. The second term is a geometric series in $\underline{\underline{S}}$, and providing that $\rho(\underline{\underline{S}}) < 1$, it converges to give the following asymptotic values:

$$\langle \underline{\varepsilon}_{\mathrm{w}}^2(\infty) \rangle = \langle \varepsilon_{\Phi}^2 \rangle [\underline{\underline{I}} - \underline{\underline{S}}]^{-1} \underline{b} \tag{12.5.1}$$

Two elements of the vector $\langle \underline{\varepsilon}_{\mathrm{w}}^2(\infty) \rangle$ are of primary interest: namely, $\langle \varepsilon_{\psi}^2(\infty) \rangle$ and $\langle \varepsilon_{\psi}^2(\infty) \rangle$, the asymptotic rotational and divergent wind errors, respectively.

In the case of wind insertion, the expected error in the geopotential field$\langle \varepsilon_{\Phi}^2 \rangle$ after J insertions is given by (12.3.13). Letting $J \to \infty$ gives the asymptotic expected error in the geopotential:

$$\langle \varepsilon_{\Phi}^2(\infty) \rangle = \frac{d_1^2 \langle \varepsilon_{\psi}^2 \rangle + d_2^2 \langle \varepsilon_{\chi}^2 \rangle}{1 - c_1^2} \tag{12.5.2}$$

where the geometric series in (12.3.13) can be summed provided $c_1^2 < 1$. If we make the reasonable assumption that streamfunction and velocity potential observation error variances are identical, then the asymptotic geopotential error variance for wind insertion is given by

$$\langle \varepsilon_{\Phi}^2(\infty) \rangle = \langle \varepsilon_{\psi}^2 \rangle \frac{d_1^2 + d_2^2}{1 - c_1^2} \tag{12.5.3}$$

As in Section 12.3, we calculate $\langle \varepsilon_{\psi}^2(\infty) \rangle$ and $\langle \varepsilon_{\chi}^2(\infty) \rangle$ for geopotential insertion and $\langle \varepsilon_{\Phi}^2(\infty) \rangle$ for wind insertion for various values of the dissipation parameter σ_{D}, $\alpha\,\Delta_{\mathrm{I}}t$, and K. The first case corresponds to $\sigma_{\mathrm{D}} = 0$ or $D = 1$. In this case, $\langle \varepsilon_{\psi}^2(\infty) \rangle / \langle \varepsilon_{\Phi}^2 \rangle = \langle \varepsilon_{\chi}^2(\infty) \rangle / \langle \varepsilon_{\Phi}^2 \rangle = 1$ in (12.5.1) and $\langle \varepsilon_{\Phi}^2(\infty) \rangle / \langle \varepsilon_{\psi}^2 \rangle = 1$ in (12.5.3); see Exercise 12.5. The case $\sigma_{\mathrm{D}} = 0$ corresponds to no damping of inertia–gravity modes after each insertion and constitutes a standard of comparison for the remaining cases. Remember that $\langle \varepsilon_{\psi}^2 \rangle$, $\langle \varepsilon_{\chi}^2 \rangle$, and $\langle \varepsilon_{\Phi}^2 \rangle$ are the scaled errors corresponding to the normalization assumed in (6.4.11). Multiplication by $(m^2 + n^2)/a^2$ and application of (6.4.11) gives the unscaled errors $\langle E_{\psi}\, \nabla^2 E_{\psi} \rangle$, $\langle E_{\chi}\, \nabla^2 E_{\chi} \rangle$, and $\tilde{\Phi}^{-1} \langle E_{\Phi}^2 \rangle$, which have units of meters^2 per seconds^2 ($\mathrm{m}^2\,\mathrm{s}^{-2}$) and are energy measures. Here E_{ψ}, E_{χ}, and E_{Φ} are the real space counterparts of ε_{ψ}, ε_{χ}, and ε_{Φ}, and ∇^2 is the Laplacian operator.

Figures 12.7 and 12.8 show the asymptotic errors of geopotential insertion and wind insertion corresponding to various values of the dissipation parameter σ_{D}. $\beta = \alpha\,\Delta_{\mathrm{I}}t$ and $(K+1)^{-1/2}$ are the abscissa and ordinate, as in Figure 12.4. In Figure 12.7, (b) and (d) show $\langle \varepsilon_{\psi}^2(\infty) \rangle / \langle \varepsilon_{\Phi}^2 \rangle$ and (c) and (e) show $\langle \varepsilon_{\chi}^2(\infty) \rangle / \langle \varepsilon_{\Phi}^2 \rangle$. In (a), only $\langle \varepsilon_{\psi}^2(\infty) \rangle / \langle \varepsilon_{\Phi}^2 \rangle$ is shown because $\langle \varepsilon_{\chi}^2(\infty) \rangle$ is identically zero in this case; Figure 12.8 shows $\langle \varepsilon_{\Phi}^2(\infty) / \langle \varepsilon_{\psi}^2 \rangle$ for three cases of wind insertion. All values in Figures 12.7 and 12.8 are between zero and one. When the normalized error equals one, the damping

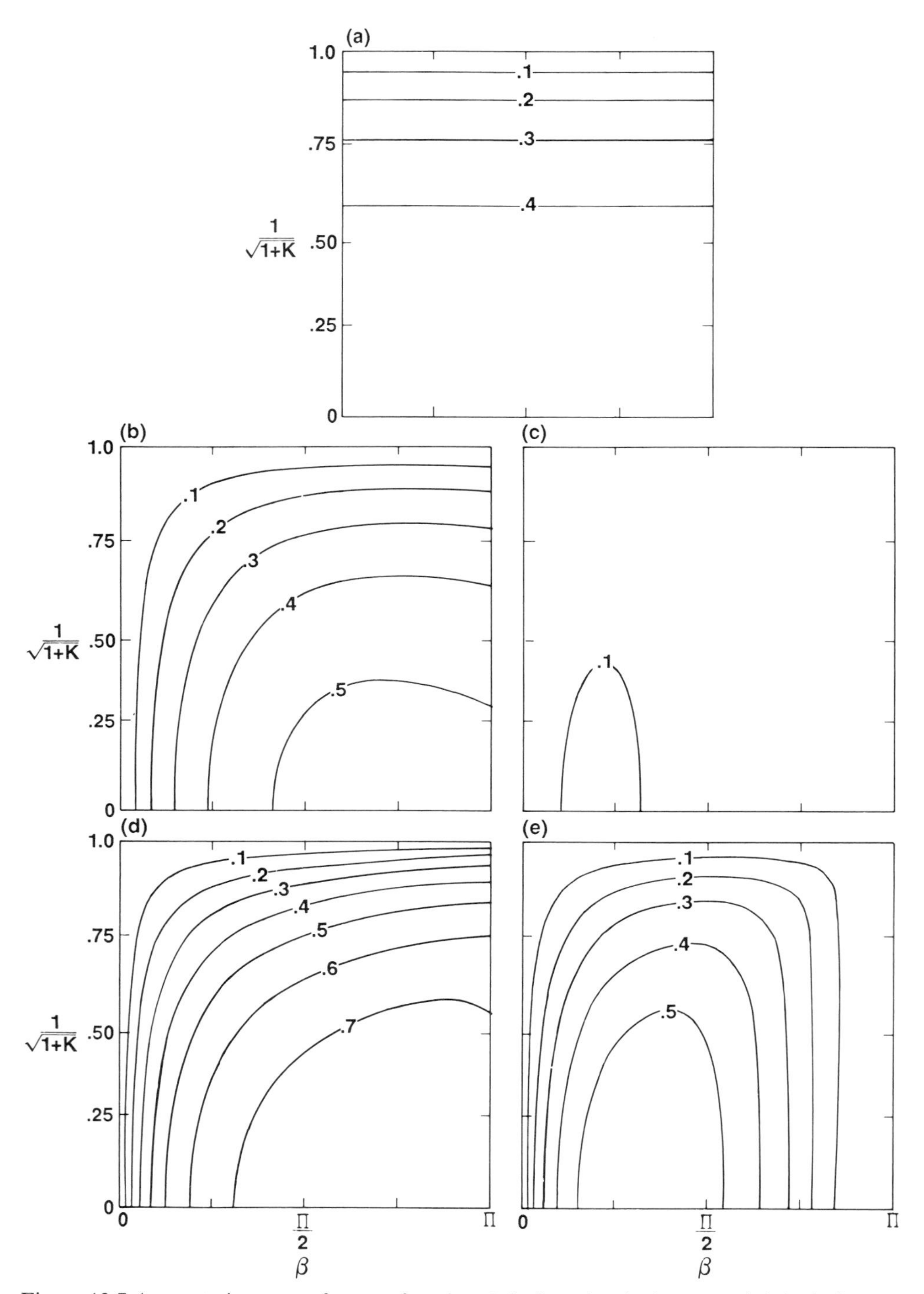

Figure 12.7 Asymptotic error of streamfunction (a,b,d) and velocity potential (c,e) due to geopotential insertion in same format as Figure 12.4. $\sigma_D = \infty$, α, and 0.2α in (a), (b,c), and (d,e), respectively.

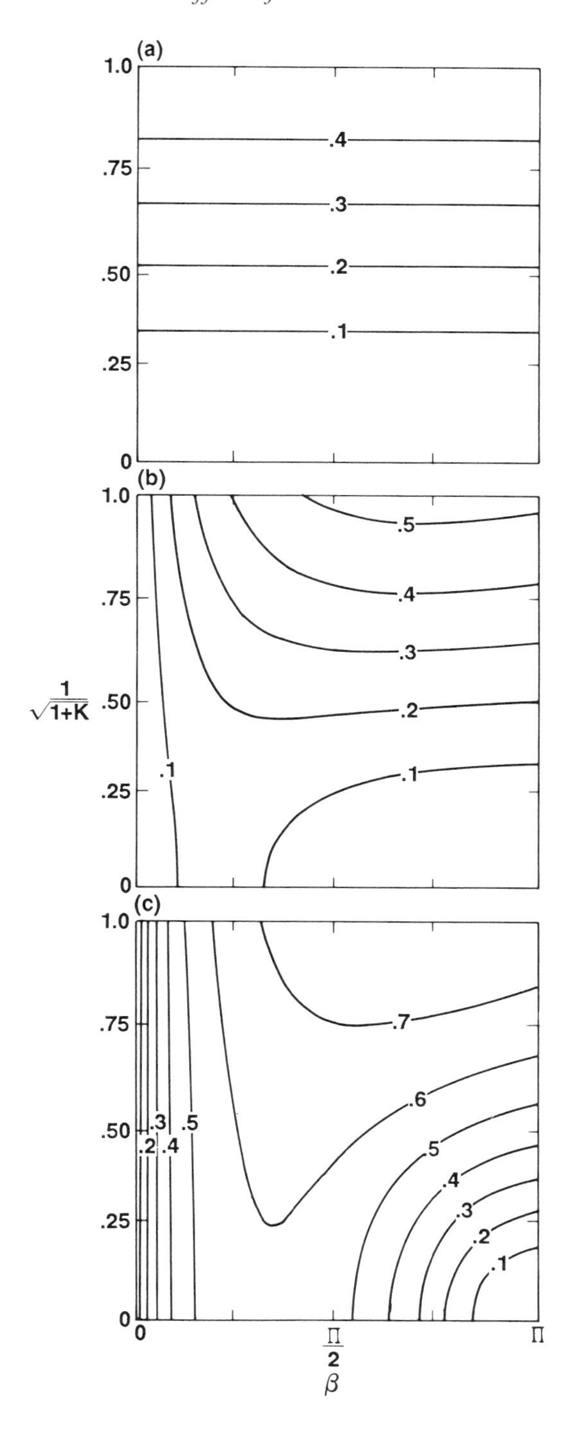

Figure 12.8 Same as Figure 12.7, except for effect of wind insertion on geopotential.

has had no effect; when it equals zero, the damping has reduced the asymptotic error to zero.

The first case, Figures 12.7(a) and 12.8(a), corresponds to complete damping of the inertia–gravity modes excited after each data insertion; that is, $\sigma_D = \infty$, $D = 0$. These errors are completely independent of $\Delta_I t$ (consistent with Section 12.3), and there is no asymptotic divergent wind error. It is straightforward to show that, for this case,

$$\frac{\langle \varepsilon_\psi^2(\infty) \rangle}{\langle \varepsilon_\Phi^2 \rangle} = \frac{K}{1+2K} \tag{12.5.4}$$

$$\frac{\langle \varepsilon_\Phi^2(\infty) \rangle}{\langle \varepsilon_\psi^2 \rangle} = \frac{1}{K+2} \tag{12.5.5}$$

For geopotential insertion, the asymptotic wind error approaches zero when K is small and 0.5 when K is large. Conversely, for wind insertion, the asymptotic error approaches zero when K is large and 0.5 when K is small.

Figures 12.7(b,c) and 12.8(b) correspond to the case $\sigma_D = \alpha = f_0\sqrt{1+K}$ (moderate damping) and Figures 12.7(d,e) and 12.8(c) to the case $\sigma_D = 0.2\alpha$ (very weak damping). It is evident that as the damping decreases, the asymptotic errors generally increase. $\langle \varepsilon_\chi^2(\infty) \rangle$ in the geopotential insertion case and $\langle \varepsilon_\Phi^2(\infty) \rangle$ in the wind insertion case are relatively sensitive to the values of $\alpha\, \Delta_I t$. At large values of β, all asymptotic errors eventually approach the completely damped limit given by $\sigma_D = \infty$, $D = 0$. An interesting and perhaps unexpected result is that, as $\beta \to 0$, all asymptotic errors approach zero when σ_D is finite. This can be confirmed by examining the case of small $\Delta_I t$ in (12.5.1) and (12.5.3); see Exercise 12.5.

Consider wind insertion and $\langle \varepsilon_\Phi^2(\infty) \rangle / \langle \varepsilon_\psi^2 \rangle$ as a function of $\Delta_I t$, for fixed K (Figure 12.8). At insertion time, the observation error is entirely in the windfield. Because this error is unbalanced, geostrophic adjustment occurs, affecting the geopotential field. At this point, the inertia–gravity wave error (but not the Rossby wave error) can be damped by the frequency selective dissipation mechanism. The maximum asymptotic error (for given K) occurs when the insertion interval $\Delta_I t$ is roughly equal to the geostrophic adjustment time-scale (Section 6.6). For larger $\Delta_I t$, the asymptotic error is smaller because the inertia–gravity wave error is damped; for smaller $\Delta_I t$, the asymptotic error is smaller because the geostrophic adjustment process has not been completed before the next insertion of random wind error.

The optimum choice of insertion interval clearly depends on the spectrum of inertia–gravity waves excited by the data insertion. However, if α_m is the maximum inertia–gravity wave frequency in the assimilating model, then the asymptotic errors are very small if $\alpha_m\, \Delta_I t$ is very small. From Section 11.2, α_m will correspond to the frequency of an external inertia–gravity wave with the minimum resolvable scale.

In one sense, the limit $\Delta_I t \to 0$ is unrealistic in Figures 12.7 and 12.8. It is not reasonable to assume that the observation errors ε_Φ in (12.5.1) and ε_ψ and ε_χ in (12.5.2) are temporally uncorrelated when the insertion interval gets very small.

Indeed, they are correlated in time, that is,

$$\langle \varepsilon_{\Phi}(t + j\,\Delta_I t)\varepsilon^*_{\Phi}(t + (j+1)\,\Delta_I t)\rangle \neq 0$$

In this case, the assumptions in Section 12.3 about ε_{Φ} being uncorrelated with ε_{ψ} and ε_{χ} are no longer true, and the analysis of this section is not valid. The results of Appendix B suggest that temporally correlated observation errors are not likely to affect the convergence rate. They may, however, increase the asymptotic error level for extremely rapid insertion.

12.6 Real data experiments: the rejection problem

The results of identical twin experiments (discussed in the previous three sections) suggest that continuous data assimilation is a viable procedure. Convergence usually occurs (sometimes not very rapidly), and the asymptotic error level is tolerable provided that suitable damping procedures are employed.

Early results with continuous data assimilation using real observations were not encouraging at all. Convergence was slow, and the asymptotic error levels were intolerably high. The basic procedure of four-dimensional data assimilation as outlined in the previous sections did not seem to work very well when applied to real observations – for two basic reasons. First, real data contain many types of error. The random noise model of observation error assumed in the previous section does not faithfully simulate many of the characteristics of true observation error. Real observations are often inconsistent with each other, contain biases, sometimes have spatially and temporally correlated errors, and occasionally contain gross errors due to transmission or human problems.

Second, and perhaps more important, the assimilating model does not faithfully simulate the atmosphere. In identical twin experiments, the model that created the simulated "atmosphere" from which the synthetic data were derived is identical to the assimilating model. In other words, in identical twin experiments there is no model error. In real data experiments, the assimilating model has many deficiencies in its simulation of the true atmosphere, so model errors are an important factor. To put the effect of model error in perspective, consider again Figure 12.5. The nonlinear model curve shows the error (really difference) between two integrations with the same model from slightly different initial conditions.

When we use numerical models to produce forecasts, these forecasts can be verified against the true (really perceived) atmospheric state. Difference between forecast and true atmospheric state can be normalized and plotted as a function of time in the format of Figure 12.5. A schematic rendering of this forecast error is shown by the dashed curve in Figure 12.5. We assume that the normalized initial error is the same as for the other curves. It is apparent that the forecast error grows more rapidly than the predictability error. In other words, ignoring the effect of model errors, as is done in identical twin experiments of continuous data assimilation, is a gross over-simplification of reality. Despite the advantages of identical twin experiments in producing comprehension and even analytic results, these experiments always err on the optimistic side.

It was found that real observational data inserted into assimilating models introduced an *inertia–gravity wave shock* into the model. Of course, we might have anticipated this from the identical twin experiments discussed previously. The problem was that the inertia–gravity wave shock was much worse than before because of the incompatibility between the model and the data; so this shock caused the inserted data to be *rejected* by the model. (Note the use of medical terminology derived from organ transplant experiments of the 1970s.) If a temperature observation was inserted in the model, examination of the model temperature fields shortly after the insertion would show little trace of the inserted temperature observation. In other words, the model integration tended to revert to what it would have been without data insertion. Again, this phenomenon might have been anticipated from identical twin experiments, but rejection in real data experiments was catastrophic because of incompatibility between model and data.

Inertia–gravity wave shock can be studied using geostrophic adjustment theory (Section 6.6). The transient inertia–gravity waves excited by the impulsive insertion of geopotential and rotational wind data were illustrated in Figures 6.3 and 6.4. Barwell and Bromley (1988) examined inertia–gravity wave shock using simple analytic models and comprehensive numerical models.

The rejection phenomenon can be graphically illustrated by Figure 12.9, derived from Daley and Puri (1980). This experiment was performed with a nonlinear global shallow water model using real observations. Figure 12.9(a) shows the positions of geopotential observations that were inserted at the given insertion time. Partial satellite tracks are clearly visible. Figure 12.9(b) shows the difference (in meters) between the geopotential observations and the model geopotential field at the insertion time. The contour interval is 20 m. (This field could be likened to the observation increment used in statistical interpolation.) Figure 12.9(d) shows the projection of the geopotential difference field, in (b), onto the slow manifold of the model, and Figure 12.9(c) shows the projection onto the high-frequency inertia–gravity waves. As discussed in Section 12.4, projection onto the slow modes affects the subsequent evolution of the model integration. Conversely, projection on the inertia–gravity modes causes high-frequency transients but does not seriously affect the evolution of the model on the advective time-scale. Consequently, the geopotential information shown in Figure 12.9(b) causes inertia–gravity wave shock and is essentially rejected by the model.

When inserted temperatures, say, are rejected by the model, the model windfield is not substantially altered from what it would have been in the absence of data insertion. Thus, the model temperatures and windfields after a period of data insertion do not strongly reflect the inserted data. Consequently, atmospheric analyses obtained after a period of data assimilation tend to reflect too strongly the model state that would have been obtained without data insertion.

A number of remedies exist for these shortcomings of continuous data assimilation

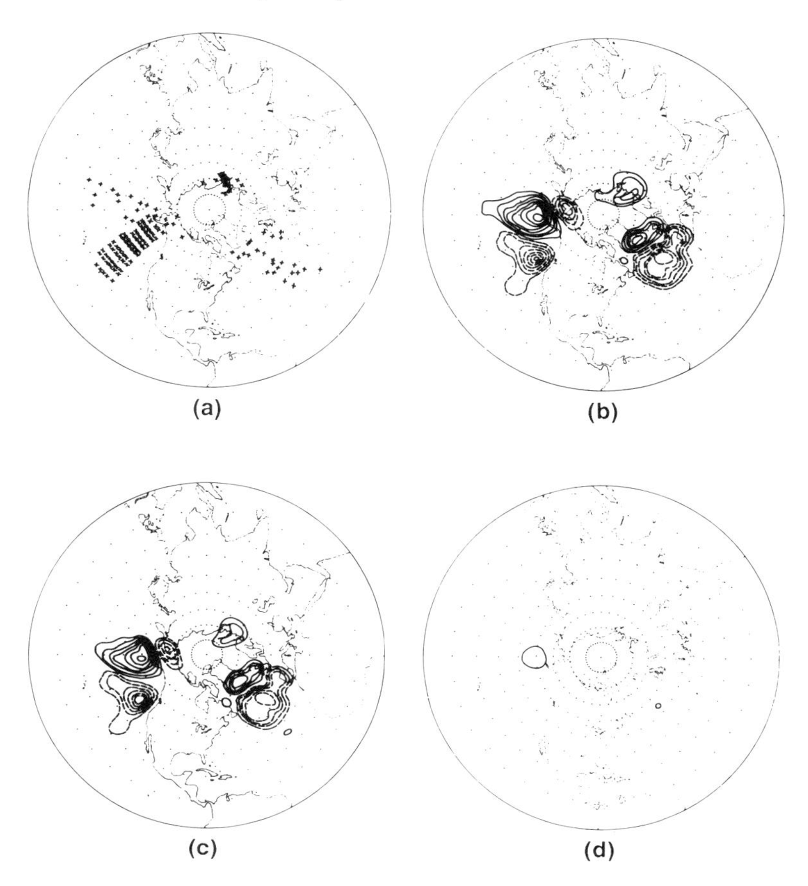

Figure 12.9 Illustration of the rejection phenomenon: (a) locations of geopotential insertions, (b) difference between geopotential observations and model forecasts (meters), (c) projection of difference on fast manifold, (d) projection on slow manifold. (After Daley and Puri, *Mon. Wea. Rev.* **108**: 85, 1980. The American Meteorological Society.)

with real data. One of them is simply to reduce the insertion interval to the time-step of the model (Miyakoda et al. 1976), that is, to keep reinserting the same observation every time-step during a window of several hours centered around the actual observation time. We examined this approach in a simple identical twin context in Sections 12.3 and 12.5 and Appendix B. We look at other approaches in the following two sections.

12.7 The geostrophic wind correction

The characteristics of the rejection phenomenon discussed in the previous section can be examined by using the analysis of Section 12.3. Rejection can be thought of as nonconvergence or excessively slow convergence. Figures 12.4(a,d) demonstrate the conditions under which this is likely to occur for completely damped inertia–gravity waves.

Insertion of a single observation of temperature or wind will excite a full spectrum of spatial scales (horizontal and vertical). In the horizontal, primarily the shorter scales will be excited, however. For fixed vertical scale and latitude, this implies a large value of K. Figure 12.4 suggests that under these conditions convergence will be rapid for wind insertion and slow for geopotential insertion. In other words, individual inserted geopotential or temperature observations will tend to be rejected whereas wind observations will be more readily accepted by the assimilating model. This effect is increased at lower latitudes but is somewhat ameliorated by the dependence on vertical scale.

The original motivation behind continuous data assimilation was to find a way objectively to analyze asynoptic temperature observations from satellite-borne radiometers. The rejection of these temperature observations by the assimilating model would be embarrassing to say the least. A simple technique was devised by Hayden (1973) and Kistler and McPherson (1975) that substantially inhibits temperature and geopotential rejection, at least in the extratropics. Suppose Φ_p is the predicted geopotential and u_p and v_p are the forecast wind components at some insertion time. Suppose that Φ_i is the geopotential that is to be inserted. The normal procedure would be simply to replace Φ_p by Φ_i, that is, directly insert Φ_i into the model. Defining Φ_a to be the final value of Φ after the insertion, this process can be written as

$$\Phi_a = \Phi_p + (\Phi_i - \Phi_p) \tag{12.7.1}$$

The investigators made the following modification. They presumed that Φ_i was known not only at a particular observation location but also in a small horizontal neighborhood around the observation point. As can be seen in Figure 12.1, this is a reasonable assumption for a satellite track. They then calculated

$$u_a = u_p - \frac{1}{f}\left(\frac{\partial \Phi_i}{\partial y} - \frac{\partial \Phi_p}{\partial y}\right) \tag{12.7.2}$$

$$v_a = v_p + \frac{1}{f}\left(\frac{\partial \Phi_i}{\partial x} - \frac{\partial \Phi_p}{\partial x}\right) \tag{12.7.3}$$

where f is the local Coriolis parameter. In other words, they inserted not only the geopotential observation but also a synthetic wind observation derived geostrophically from the observed-minus-forecast geopotential field.

In the simple model described in Section 12.4, this *geostrophic wind correction* would cause instantaneous convergence if the inserted geopotential were error free. In sophisticated assimilating models, this procedure substantially inhibits the rejection phenomenon.

The effect of the geostrophic wind correction can be seen in Figure 12.10. This diagram, in the same format as Figure 12.9, shows (b) the projection on the slow

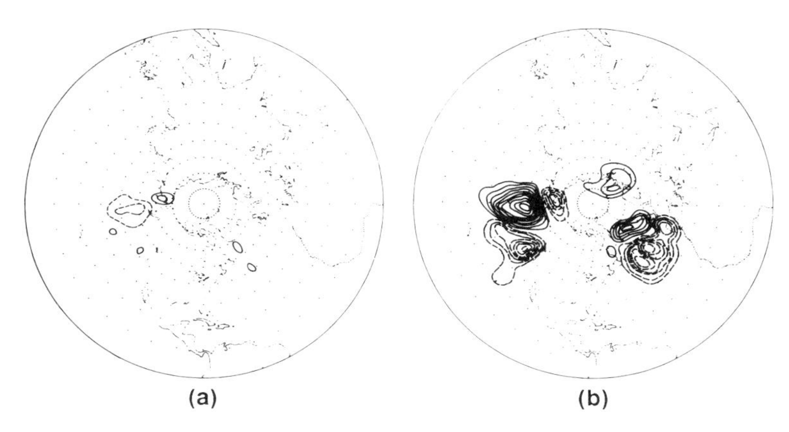

Figure 12.10 Effect of geostrophic wind correction in same format as Figure 12.9. (a) fast manifold projection, (b) slow manifold projection. (After Daley and Puri, *Mon. Wea. Rev.* **108**: 85, 1980. The American Meteorological Society.)

manifold and (a) the inertia–gravity waves for the observed-minus-forecast geopotential field shown in Figure 12.9(b). In Figure 12.10, a procedure analogous to the geostrophic wind correction has been used (Puri 1981). Clearly, most of the geopotential information has been forced onto the slow manifold of the assimilating model, and there is little energy in the inertia–gravity waves. In other words, the geostrophic wind correction procedure has prevented the geopotential information shown in Figure 12.9(b) from being rejected by the model. The results of Figure 12.9 are consistent with the discussion of Section 12.4.

Now, clearly, this procedure has validity only to the extent that the geostrophic wind law relates the observed-minus-forecast geopotentials and winds. It is unlikely to be of much value in the tropics or where $\Phi_i - \Phi_p$ is very large (for then the neglect of nonlinear terms in the geostrophic wind law would be serious). Under some conditions, the geostrophic wind correction could be potentially damaging because it might cause the model to accept a very inaccurate synthetic wind as well as an observed geopotential.

12.8 Dynamic relaxation

Another technique that has been applied to continuous data assimilation with real data is *dynamic relaxation*. Variations on this technique are known as *nudging* or *Newtonian relaxation*. It can also be used for forced dynamic initialization as discussed in Section 11.2. Dynamic relaxation has been investigated theoretically by Hoke and Anthes (1976) and Davies and Turner (1977) and used operationally by Lyne, Swinbank, and Birch (1982). Another application of this technique is in the initialization of the hydrological cycle (see Section 13.6).

Consider geopotential data assimilation with the nonlinear shallow water model

(6.4.1–3). Equations (6.4.1–2) remain unchanged, but Equation (6.4.3) is modified as follows:

$$\frac{\partial h}{\partial t}+u\frac{\partial h}{\partial x}+v\frac{\partial h}{\partial y}+h\left(\frac{\partial u}{\partial x}+\frac{\partial v}{\partial y}\right)=-\gamma_{\mathrm{N}}(h-h_{\mathrm{O}})+\gamma_{\mathrm{D}}\nabla^{2}(h-h_{\mathrm{O}}) \qquad (12.8.1)$$

where h_O is the observed geopotential, γ_N a Newtonian relaxation coefficient, and γ_D a diffusive relaxation coefficient. Both γ_D and γ_N are assumed positive. Both additional terms in (12.8.1) nudge the model geopotential h toward the observed geopotential h_O. The Newtonian term will be most effective at large horizontal scales and the diffusive term at small horizontal scales.

This technique can be analyzed by using the linearized shallow water model of (6.4.8–10). Consider error-free geopotential insertion. Make the identical twin assumption that the observed geopotential $h_O = g^{-1}\Phi_O$ is the true geopotential and the true geopotential and winds satisfy (6.4.8–10). Denote the true streamfunction, velocity potential, and geopotential as ψ_T, χ_T, and Φ_T.

Now define an actual streamfunction, velocity potential, and geopotential ψ_A, χ_A, and Φ_A that satisfy the set of dynamic relaxation equations. In this case, these equations consist of (6.4.8–9) whereas (6.4.10) is modified to

$$\frac{\partial}{\partial t}\Phi_{\mathrm{A}}+\tilde{\Phi}\nabla^{2}\chi_{\mathrm{A}}=-\gamma_{\mathrm{N}}(\Phi_{\mathrm{A}}-\Phi_{\mathrm{T}})+\gamma_{\mathrm{D}}\nabla^{2}(\Phi_{\mathrm{A}}-\Phi_{\mathrm{T}}) \qquad (12.8.2)$$

Subtract (6.4.8–10) for the true fields ψ_T, χ_T, and Φ_T from the dynamic relaxation equations. This gives

$$\frac{\partial}{\partial t}\nabla^{2}\psi'+f_{0}\nabla^{2}\chi'=0 \qquad (12.8.3)$$

$$\frac{\partial}{\partial t}\nabla^{2}\chi'-f_{0}\nabla^{2}\psi'+\nabla^{2}\Phi'=0 \qquad (12.8.4)$$

$$\frac{\partial}{\partial t}\Phi'+\tilde{\Phi}\nabla^{2}\chi'=-\gamma_{\mathrm{N}}\Phi'+\gamma_{\mathrm{D}}\nabla^{2}\Phi' \qquad (12.8.5)$$

where $\psi' = \psi_A - \psi_T$, $\chi' = \chi_A - \chi_T$, and $\Phi' = \Phi_A - \Phi_T$ indicate the errors.

By expanding ψ', χ', and Φ' using (6.4.11–14), we derive a matrix equation similar to (6.4.15). The matrix $\underline{\underline{L}}$ is unchanged, except that the lower right-hand corner element changes from zero to $-ig$, where

$$g=\frac{1}{f_{0}}\left(\gamma_{\mathrm{N}}+\gamma_{\mathrm{D}}\frac{n^{2}+m^{2}}{a^{2}}\right)=\frac{1}{f_{0}}\left(\gamma_{\mathrm{N}}+\gamma_{\mathrm{D}}\frac{Kf_{0}^{2}}{\tilde{\Phi}}\right) \qquad (12.8.6)$$

Proceeding as in Section 6.4 yields a cubic frequency equation,

$$\sigma^{3}+ig\sigma^{2}-(K+1)\sigma-ig=0 \qquad (12.8.7)$$

which collapses to (6.4.16) when $g = 0$. Equation (12.8.7) generally has complex roots, which can be written formally as

$$\sigma = \mathrm{Re}(\sigma) + i\,\mathrm{Im}(\sigma)$$

where $\mathrm{Re}(\sigma)$ and $\mathrm{Im}(\sigma)$ are the real and imaginary parts. $\mathrm{Re}(\sigma)$ is the frequency, and $\mathrm{Im}(\sigma) < 0$ indicates a damped mode whereas $\mathrm{Im}(\sigma) > 0$ indicates a growing mode.

For $K = 0$, the roots of (12.8.7) are

$$\begin{aligned} \sigma_{\mathrm{R}} &= -ig = -\frac{i\gamma_n}{f_0} \\ \sigma_{\mathrm{G}} &= \pm 1 \end{aligned} \tag{12.8.8}$$

Thus, the Rossby mode errors are damped and the inertia–gravity mode errors are not damped. The frequencies of all these modes are unaltered.

For $K > 0$, it can be shown (using Descartes' rule of signs) that each root has $\mathrm{Im}(\sigma) < 0$ provided that $g > 0$. In other words, dynamic relaxation damps all modes except the two inertia–gravity modes for $K = 0$. Also, $\mathrm{Re}(\sigma_{\mathrm{R}}) = 0$ for all $K \geq 0$ and $g \geq 0$, meaning that dynamic relaxation never alters the frequency of any Rossby mode.

To calculate approximately the roots of (12.8.7) for all K, assume that $g \ll K + 1$. According to (12.8.6), this is the case if the relaxation coefficients are small in the sense that

$$\gamma_{\mathrm{N}} \ll f_0 \qquad \text{and} \qquad \gamma_{\mathrm{D}} \ll \frac{\tilde{\Phi}}{f_0} \tag{12.8.9}$$

We can express σ_{R} as a convergent series in odd powers of $g/(K+1)$:

$$i\sigma_{\mathrm{R}} = \frac{g}{K+1} + \frac{K}{K+1}\left(\frac{g}{K+1}\right)^3 + \frac{K}{K+1}\left(\frac{2K-1}{K+1}\right)\left(\frac{g}{K+1}\right)^5 + \cdots \tag{12.8.10}$$

as can be verified by substituting the series into (12.8.7). Thus, the Rossby mode errors are damped by a factor that is approximately $g/(K+1)$. For large K, this factor asymptotes to

$$\frac{g}{K+1} \sim \frac{\gamma_{\mathrm{D}} f_0}{\tilde{\Phi}} \tag{12.8.11}$$

which is independent of K. Hence, each Rossby mode of small horizontal scale is about equally damped by a factor of approximately $\gamma_{\mathrm{D}} f_0/\tilde{\Phi}$.

Given one root of a cubic equation, the other two satisfy a quadratic equation. Thus,

$$i\sigma_{\mathrm{G}} = \frac{g - i\sigma_{\mathrm{R}}}{2} \pm \sqrt{\left(\frac{g - i\sigma_{\mathrm{R}}}{2}\right)^2 - \frac{g}{i\sigma_{\mathrm{R}}}} \tag{12.8.12}$$

and for $i\sigma_{\mathrm{R}} \cong g/(K+1)$, this results in

$$\sigma_{\mathrm{G}} \cong -i\frac{g}{K+1}\frac{K}{2} \pm \sqrt{K + 1 - \left(\frac{g}{K+1}\right)^2\left(\frac{K}{2}\right)^2} \tag{12.8.13}$$

The inertia–gravity mode errors are damped more than the Rossby mode errors for $K \gtrsim 2$. The inertia–gravity frequencies are reduced for all $K > 0$. For all sufficiently large K, the radicand in (12.8.13) is negative. Thus, at horizontal scales so small that

$$\frac{1}{4}\left(\frac{g}{K+1}\right)^2 K \gtrsim 1 \tag{12.8.14}$$

the inertia–gravity errors become stationary, and one mode is damped very rapidly whereas the other is damped very little.

There are important differences between the damping scheme employed in dynamic relaxation and those discussed in Section 12.2. In dynamic relaxation, it is only the error that is being damped. Consequently, it is desirable to damp both the Rossby mode error and the inertia–gravity mode error. The damping schemes discussed in Section 12.2 damp the total model solution. Thus, they are designed primarily to damp the inertia–gravity modes and to affect the Rossby modes as little as possible.

Further analysis of dynamic relaxation, including the effects of observational error, can be seen in Davies and Turner (1977). The case of wind insertion is left as Exercise 12.6.

12.9 Data assimilation on the slow manifold

Continuous data assimilation has been largely technology driven. That is, the potential availability of large quantities of asynoptic radiance data from earth orbiting satellites has forced the development of a suitable technique for assimilating these data. The data specifications were set by the producer (the satellite engineers) and not by the consumer (the meteorologists). As is often the case in technology-driven fields, the initial response of the consumers was entirely empirical.

Empirical experimentation commenced with the work of Charney et al. (1969) and continued through the early and mid-1970s. Even in retrospect, many of the experimental results are contradictory and confusing. Numerical models were not as well understood as they are today, and interpretation of the experimental results was sometimes misguided. However, out of these experiments gradually came an empirical understanding of the phenomena associated with data assimilation, leading ultimately to a sounder theoretical foundation. The key theoretical work was done by Williamson and Dickinson (1972), Blumen (1975a,b, 1976), Davies and Turner (1977), Ghil (1980), and Talagrand (1981a,b) and provides the intellectual basis for this chapter.

In operational practice, continuous data assimilation has borrowed techniques from other data assimilation methodologies. For example, in the continuous data assimilation experiments of Ploshay, White, and Miyakoda (1983), both normal mode initialization and statistical interpolation are employed to inhibit data rejection.

Continuous data assimilation, in principle, is a unified analysis/initialization procedure. At the time of writing, it is one of the two most common analysis/ initialization procedures used in operational large-scale forecast centers. (The other is a form of intermittent data assimilation based on statistical interpolation followed by normal mode initialization.) After years of refinement, it is now a very successful and competitive technique, particularly espoused by the Geophysical Fluid Dynamics Laboratory (GFDL) and the United Kingdom Meteorological Office.

The assimilating models used in continuous data assimilation are invariably based on the baroclinic primitive equations. These models provide the most realistic simulation of the atmosphere and therefore minimize the forecast error. Unfortunately, they admit inertia–gravity wave solutions and, when used in assimilation mode, suffer from inertia–gravity wave shock and data rejection. One attactive possibility for eliminating inertia–gravity wave shock is to assimilate data with a model that does not admit inertia–gravity wave solutions. Such a model is the quasi-geostrophic potential vorticity model discussed in Section 7.8. In such models, the quasi-geostrophic potential vorticity, defined by (7.8.7) for a baroclinic model and (6.5.7) for a barotropic model, is conserved following the flow. Data assimilation would proceed by adjusting the assimilating model quasi-geostrophic potential vorticity to the quasi-geostrophic potential vorticity of the observation. In this process, no inertia–gravity waves would be excited. Unfortunately, the quasi-geostrophic potential vorticity model has the usual quasi-geostrophic limitations in the tropics and for the planetary scales, so it is not very useful for global data assimilation.

Recently, however, several attempts have been made to create more general models that do not permit inertia–gravity waves. Such models, which we shall call slow manifold models, would be very attractive for global data assimilation. It is interesting that several of these slow manifold models have become possible due to advances in initialization theory. Slow manifold models can be constructed using the intermediate model approach of Gent and McWilliams (1982) or the bounded derivative approach of Browning and Kreiss (1986). Van Isacker and Struylaert (1985) and Lynch (1986, 1987) have constructed slow manifold models based on the Laplace transform methodology; see (13.4.15).

Another procedure for constructing a slow manifold model is based on the normal mode theory of Chapter 9. Daley (1980a) has successfully used this procedure for integrating the global baroclinic equations. The model normal modes are broken up into a fast set $\underline{Z}$ and a slow set $\underline{Y}$. A baroclinic primitive equations model is integrated in physical space in the usual fashion. Then, after each time-step, there is a vertical projection – using an appropriate form of the vertical structure operator (9.1.11) – for each equivalent depth. Following this, there is a horizontal projection for each of the fast modes z. The fast mode amplitudes are calculated (usually without iteration) using the Machenhauer condition (9.6.4). The results are projected back into physical space, and the amplitudes of the physical space variables (wind, temperature, etc.) are adjusted accordingly. The method is stable and efficient and does not allow high-frequency oscillations.

The philosophy of using a sophisticated numerical model to directly assimilate data was and remains very attractive. Even after the techniques of present-day continuous data assimilation have been abandoned, the philosophical basis will remain, and new techniques based on this approach will be introduced. This theme is developed further in the first three sections of the next chapter.

Exercises

12.1 Consider the equation $\dot{u} = -i\omega u + F_0 \exp(i\omega_r t)$, where ω is a fast frequency and ω_r is a slow frequency. A special analytic solution of this equation that contains no high frequencies is given by $u(t) = F_0 \exp(i\omega_r t)/[i(\omega + \omega_r)]$; see Exercise 10.2. Assume $u(0)$ is arbitrary and apply the Matsuno scheme (12.1.1–2) to this equation. Find the solution $u(n\,\Delta t)$. Now find the solution for large time ($n \to \infty$). Assume $\omega = 0.75$, $\omega_r = 0.50$, and $\Delta t = 1$ and calculate the Matsuno solution (for large time). Compare the amplitude and phase of this discrete solution with the special low-frequency solution just given. Show that in the limit as $\Delta t \to 0$, the discrete solution approaches the low-frequency analytic solution.

12.2 Consider the divergence damping scheme (12.2.5). Add the extra term $c\,\nabla^2 D = c\,\nabla^4 \chi$ to the right-hand side of (6.4.9). Find the modified matrix $\underline{\underline{L}}$ corresponding to (6.4.15) and the modified frequency equation. Show that the Rossby modes are not damped. Show that the gravity modes with positive and negative frequencies are unequally damped and can become stationary for very large values of c or very small horizontal scales.

12.3 The damping scheme in the previous exercise is a special case of the following scheme. Add a term of the form $c\,\nabla^4(\alpha_{\chi\psi}\psi + \alpha_{\chi\chi}\chi + \alpha_{\chi\Phi}\Phi)$ to the right-hand side of (6.4.9) and a term of the form $c\,\nabla^2(\alpha_{\Phi\psi}\psi + \alpha_{\Phi\chi}\chi + \alpha_{\Phi\Phi}\Phi)$ to the right-hand side of (6.4.10). c, $\alpha_{\chi\psi}$, $\alpha_{\Phi\Phi}$, and so on, are specified constants (with all $\alpha_{xy} = 0$ except $\alpha_{\chi\chi} = 1$ in Exercise 12.2). Define the eigenvalues of the system as σ_R, σ_G^1, and σ_G^2. Choose the constants $\alpha_{\chi\Phi}$, $\alpha_{\Phi\psi}$, and so on, such that $\text{Re}(\sigma_R)$, $\text{Re}(\sigma_G^1)$, and $\text{Re}(\sigma_G^2)$ are given by (6.4.16), $\text{Im}(\sigma_R) = 0$ and $\text{Im}(\sigma_G^1) = \text{Im}(\sigma_G^2) \leq 0$.

12.4 Define

$$\underline{\underline{G}} = \begin{pmatrix} a_1 & a_2 \\ a_3 & a_4 \end{pmatrix}$$

where a_1, a_2, a_3, and a_4 are defined in (12.3.8). Define the eigenvalues of $\underline{\underline{G}}$ and σ_+ and σ_-. Show that the eigenvalues of $\underline{\underline{S}}$, from (12.3.9), are σ_+^2, $\sigma_+\sigma_-$, and σ_-^2. Using this result, demonstrate that the spectral radius of $\underline{\underline{S}}$ (for $0 \leq \beta \leq \pi$) is always less than or equal to 1, with equality occurring only for $K = 0$, ∞ or $D \cos \beta = \pm 1$.

12.5 Consider (12.5.3). Show that $\langle \varepsilon_\Phi^2(\infty) \rangle / \langle \varepsilon_\psi^2 \rangle$ equals 1 for all K and $\Delta_I t$ if $\sigma_D = 0$ and equals $1/(K+2)$ if $\sigma_D = \infty$. Show that for all other positive values of σ_D, $\langle \varepsilon_\Phi^2(\infty) \rangle / \langle \varepsilon_\psi^2 \rangle \to 0$ as $\Delta_I t \to 0$.

12.6 Consider dynamic relaxation applied to wind observations rather than height observations. Add terms of the form $-\gamma(u - u_O)$ and $-\gamma(v - v_O)$ to the right-hand sides of (6.4.4) and (6.4.5), respectively. Here u_O and v_O are the observed wind components (with no observation errors) and γ is space and time independent. Perform the same type of mathematical analysis as in Section 12.8. Show that the gravity mode errors are always damped, whereas the Rossby mode errors are damped except when the horizontal wavenumber is zero. Show that the frequencies of the Rossby modes are unaltered by this form of dynamic relaxation.

13

Future directions

Atmospheric data assimilation, as practiced today, has progressed enormously since Panofsky's pioneering work in the late 1940s. But, what of the future?

Clearly, a number of difficult challenges lie ahead. Atmospheric simulation with numerical models is becoming increasingly important in a variety of contexts other than short-range numerical weather prediction. A host of climate and interdisciplinary environmental problems exist for which numerical simulation models are ideally suited. However, these models must be very sophisticated indeed. Future numerical models will have higher resolution (well into the mesoscale), will routinely simulate the stratosphere and lower mesosphere, will interact with coupled ocean and cryosphere models, will simulate many surface hydrological and biological processes, and will include chemical species equations.

In the future, objective analyses will be required as initial conditions for models and for diagnostic purposes, as they are at present; but in addition to the dynamic variables discussed in previous chapters, objective analyses of hydrological, biological, and chemical variables will be required. New observation systems (mostly space based) will be developed, and the new data bases will contain mostly asynoptic, nonconventional data.

Present data assimilation methods will not be capable of assimilating future observations into future models. Fortunately, a number of promising new techniques for objective analysis, initialization, and continuous data assimilation are under active investigation. These new techniques have their roots in methods introduced in the previous chapter. However, they are not simply extensions of the older techniques but are radical departures.

We discuss new analysis and initialization techniques in Sections 13.1–5. Then, in Sections 13.6–13.8, we briefly cover data assimilation for the hydrological cycle, the mesoscale, and the oceans. The final section discusses new challenge for data assimilation.

13.1 Descent methods

The variational formalism provides an elegant framework for problems of atmospheric data analysis. Unfortunately, variational formulations, though instructive, have not been very useful in practice because of the computationally intensive nature of the resulting algorithms. For certain more restricted problems involving limited amounts of data, the computational limitations can be overcome (see, e.g., Sasaki 1970c). However, for the global data assimilation problem, variational approaches have been out of the question until very recently.

One reason for renewed interest in variational formulations is the greatly increased computing capacity available in recent years. A more important reason is the application of procedures for directly minimizing cost functions. Sections 13.1 and 13.2 explore these new developments, examining the three-dimensional variational problem and the four-dimensional problem, respectively. Current development in this area is extremely rapid, and consequently we make no attempt here to develop useable algorithms. Instead, our emphasis is on providing suitable background material to make future operational algorithms comprehensible.

Before we discuss the direct minimization of a particular three-dimensional cost function, a generalized form of (5.6.11), we introduce a new notation, general enough for all the mathematical analysis of Sections 13.1–3.

Consider state variables defined at gridpoints $\mathbf{r}_j$, $1 \leq j \leq J$, at time $t_n = t_0 + n\,\Delta t$ on a regular time-independent three-dimensional grid. Here Δt is a constant time interval. Define $\underset{\sim}{s}_n$ to be the column vector of length J of *true* values of all the state variables of interest on the regular grid $\mathbf{r}_j$ at time t_n. Denote $\underset{\sim}{s}_n^{\mathrm{A}}$ as the analyzed values of $\underset{\sim}{s}_n$ on the grid. Assume there exist background values of the state variables on the same grid and these background values have been produced by a forecast model. Denote the background vector as $\underset{\sim}{s}_n^{\mathrm{F}}$. Then the forecast error at time t_n is

$$\underset{\sim}{\varepsilon}_n^{\mathrm{F}} = \underset{\sim}{s}_n^{\mathrm{F}} - \underset{\sim}{s}_n \tag{13.1.1}$$

Assume that $\langle \underset{\sim}{\varepsilon}_n^{\mathrm{F}} \rangle = 0$ and define $\underset{\approx}{P}_n^{\mathrm{F}} = \langle \underset{\sim}{\varepsilon}_n^{\mathrm{F}} (\underset{\sim}{\varepsilon}_n^{\mathrm{F}})^{\mathrm{T}} \rangle$ as the forecast (background) error covariance matrix.

Now assume that there exists an observation network $\mathbf{r}_k(t_n)$, $1 \leq k \leq K(n)$, where $K(n)$ is the number of observations available at time t_n. The observation network may be time dependent, so the observation locations can change at every time-step. Define a vector of observations $\underset{\sim}{d}_n$ of length $K(n)$ with elements $d_n(\mathbf{r}_k)$ at time t_n. Suppose H_n is a linear or nonlinear forward interpolation operator of variables on the time-independent grid $\mathbf{r}_j$ to the time-dependent observation network $\mathbf{r}_k(t_n)$,

$$\underset{\sim}{d}_n = H_n(\underset{\sim}{s}_n) + \underset{\sim}{\varepsilon}_n^{\mathrm{R}} \tag{13.1.2}$$

and $\underset{\sim}{\varepsilon}_n^{\mathrm{R}}$ is a column vector of length $K(n)$ of errors. H_n could be a simple linear univariate forward interpolation operator (like $\underset{\approx}{\Omega}$ in Section 5.7). Alternatively, it could be a complicated nonlinear operator relating one set of state variables to another set of state variables. For example, $\underset{\sim}{s}_n$ could represent the true values of temperature on a regular grid whereas $\underset{\sim}{d}_n$ could be observed radiances obtained from a satellite radiometer. H_n in this case would involve some form of radiative transfer equation. When H_n is a linear operator,

$$H_n(\underline{s}_n) = \underline{\underline{H}}_n \underline{s}_n \tag{13.1.3}$$

with $\underline{\underline{H}}_n$ being a $K(n) \times J$ rectangular matrix whose elements are independent of $\underline{s}$. The $\underline{\varepsilon}_n^R$ in (13.1.2) can be written as the sum of two different terms:

$$\underline{\varepsilon}_n^R = [\underline{d}_n - \underline{d}_n^*] + [\underline{d}_n^* - H_n(\underline{s}_n)]$$

where $\underline{d}_n^*$ is the column vector of length $K(n)$ of true values at the observation stations. The first term on the right-hand side is the measurement error, and the second term is the error in the forward interpolation. Assume that $\underline{\varepsilon}_n^R$ is unbiased and not temporally correlated:

$$\langle \underline{\varepsilon}_n^R \rangle = 0 \qquad \text{and} \qquad \langle \underline{\varepsilon}_n^R (\underline{\varepsilon}_\eta^R)^T \rangle = \delta_{n\eta} \underline{\underline{R}}_n \tag{13.1.4}$$

where $\underline{\underline{R}}_n$ is the $K(n) \times K(n)$ observation error covariance matrix and can be identified with the sum $\underline{\underline{O}} + \underline{\underline{F}}$ of (5.6.9).

Because this section considers the three-dimensional problem and there is no explicit time dependence, we drop the time-step index n for the remainder of this section. Thus, $\underline{d}$ is a vector of observations of length K; $\underline{s}_A$ and $\underline{s}_F$ are vectors of length J of analyzed and forecast values, respectively; $\underline{\underline{P}}_F$ is the $J \times J$ forecast error covariance matrix; and $\underline{\underline{R}}$ is the $K \times K$ observation error covariance matrix. Then, a generalized form of (5.6.11) is given by

$$I = 0.5\{[\underline{d} - H(\underline{s}_A)]^T \underline{\underline{R}}^{-1}[\underline{d} - H(\underline{s}_A)] + [\underline{s}_F - \underline{s}_A]^T \underline{\underline{P}}_F^{-1}[\underline{s}_F - \underline{s}_A]\} \tag{13.1.5}$$

where H is the nonlinear forward interpolation operator. If H is a linear operator (13.1.3), then from (5.6.9–10) and Exercise 5.5, the functional (13.1.5) is minimized if

$$\underline{s}_A = \underline{s}_F + [\underline{\underline{H}}^T \underline{\underline{R}}^{-1} \underline{\underline{H}} + \underline{\underline{P}}_F^{-1}]^{-1} \underline{\underline{H}}^T \underline{\underline{R}}^{-1}[\underline{d} - \underline{\underline{H}}\, \underline{s}_F] \tag{13.1.6}$$

or

$$\underline{s}_A = \underline{s}_F + \underline{\underline{P}}_F \underline{\underline{H}}^T[\underline{\underline{R}} + \underline{\underline{H}}\, \underline{\underline{P}}_F \underline{\underline{H}}^T]^{-1}[\underline{d} - \underline{\underline{H}}\, \underline{s}_F] \tag{13.1.7}$$

As discussed in Tarantola (1987), when H is linear and the forecast and observation errors are normally distributed, then so are the analysis errors; consequently, $\underline{s}_A$ is a maximum likelihood estimate. For the global data assimilation problem, Equations (13.1.6–7) are not easy to use because of the sizes of the matrices that must be inverted. One solution to this problem, which is also applicable when H is nonlinear, is to minimize (13.1.5) directly, This can be done, in principle, by using a *descent* algorithm.

All descent algorithms require calculation of the gradient of the functional I. Define the vector $\underline{q} = \underline{s}_A$ with elements $q_j = s_A(\mathbf{r}_j)$, $1 \leq j \leq J$, as the *control variable*. Then the J elements of $\underline{q}$ can be thought of as the coordinates of a J-dimensional space. It is possible to define a gradient operator in this space that is similar to the normal

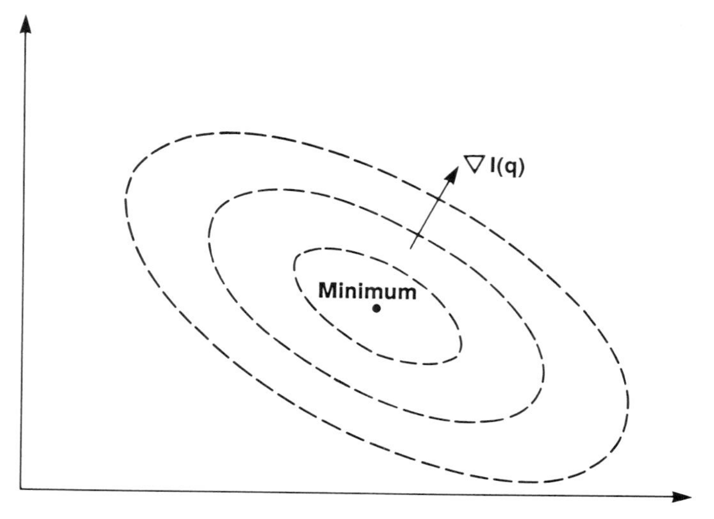

Figure 13.1 Illustration of the minimum of a functional and the gradient operator.

gradient operator in physical space. Define,

$$\nabla I(q) = \frac{\partial I}{\partial \underline{q}} \tag{13.1.8}$$

as a column vector of length J with elements $\partial I/\partial q_j$, $1 \le j \le J$.

Consider the application of (13.1.8) to (13.1.5) when H is a linear operator:

$$\nabla I(s_A) = \frac{\partial I}{\partial \underline{s}_A} = \underline{\underline{H}}^T \underline{\underline{R}}^{-1}[\underline{\underline{H}}\, \underline{s}_A - \underline{d}] + \underline{\underline{P}}_F^{-1}[\underline{s}_A - \underline{s}_F] \tag{13.1.9}$$

for $\underline{q} = \underline{s}_A$. When $I(s_A)$ is a minimum, then $\nabla I(s_A) = 0$, yielding (13.1.6–7).

Figure 13.1 is a two-dimensional schematic representation of the J-dimensional space with the axes indicating coordinates q_1 and q_2 for an arbitrary functional I. Here the dashed lines represent contours of constant values of I and the gradient $\nabla I(q)$ is indicated by the arrow normal to the dashed contours. $\nabla I(q)$ is the local gradient for a particular position in the space. A single minimum is indicated in this figure, but it is possible to have secondary minima in addition to the global minimum.

An expression for $\nabla I(s_A)$ can often be found when the forward interpolation operator $H(\underline{s}_A)$ is nonlinear (Lorenc 1986; Tarantola 1987). When the forecast vector $\underline{s}_F$ is reasonably accurate, then $H(\underline{s}_A)$ can be expanded in the first two terms of a Taylor series around $\underline{s}_F$:

$$H(\underline{s}_A) \approx H(\underline{s}_F) + \underline{\underline{H}}[\underline{s}_A - \underline{s}_F], \qquad \text{with } \underline{\underline{H}} = \left[\frac{\partial H(\underline{s})}{\partial \underline{s}}\right]_{\underline{s}=\underline{s}_F}$$

$\underline{\underline{H}}$ is the Jacobian matrix (see Section 10.7) whose elements are the partial derivatives of the forward interpolated values at the observation stations with respect to the elements of $\underline{s}_F$. Inserting this expression into (13.1.5) and differentiating with respect to $\underline{s}_A$ gives

$$\nabla I(\underline{s}_A) = \underline{\underline{H}}^T \underline{\underline{R}}^{-1}[H(\underline{s}_A) - \underline{d}] + \underline{\underline{P}}_F^{-1}[\underline{s}_A - \underline{s}_F] \tag{13.1.10}$$

which becomes equal to (13.1.9) if $H(\underline{s})$ is a linear function of $\underline{s}$. Although (13.1.9–10) are formally very similar, $\underline{\underline{H}}$ in (13.1.10) is no longer independent of $\underline{s}$ and must be reevaluated during the course of the minimization.

If $\nabla I(q)$ can be determined, it can be used in an iterative procedure to reduce the functional I to its minimum value. Algorithms that do this are called descent algorithms and have a large mathematical literature. For example see Gill, Murray, and Wright (1981) and Tarantola (1987); and for meteorological applications, see Navon and Legler (1987). The general subject of descent algorithms is beyond the scope of this book, but we present a few of the more basic ideas in the remainder of this section.

In Figure 13.1, $\nabla I(q)$ points in the direction of greatest (local) increase in I. Consequently, $-\nabla I(q)$ points in the direction of most rapid (local) decrease in I. This is called the direction of *steepest descent* of I. Examination of Figure 13.1 suggests a simple method for reducing the value of I and thus getting closer to the minimum. Define the following simple iteration:

$$\underline{q}_{m+1} - \underline{q}_m = -\gamma_m \nabla I(q_m) = -\gamma_m \frac{\partial I}{\partial \underline{q}_m} \tag{13.1.11}$$

where $\nabla I(q_m)$ is the value of the gradient of I with respect to $\underline{q}$ after the mth iteration and m is the iteration number. γ_m is a scalar constant called the *stepsize*, to be specified at each iteration. In Figure 13.1, it can be seen that $I(q_{m+1})$ should be smaller than $I(q_m)$; thus, $\underline{q}_{m+1}$ is closer to the desired solution (13.1.6) than $\underline{q}_m$. The distance moved toward the minimum depends on the stepsize γ_m, which thus should be chosen carefully. If the stepsize is too small, the descent toward the minimum does not progress. If it is too large, $I(q_{m+1})$ could be larger than $I(q_m)$. This is illustrated in Figure 13.2, which shows schematically $I(q_m)$, $I(q_{m+1})$ and $\nabla I(q_m)$, $\nabla I(q_{m+1})$ for an arbitrary functional $I(q)$. The direction $-\nabla I(q_m)$ is shown to be normal to the contour $I(q_m)$ in Figure 13.1. Note that if we descend along the direction $-\nabla I(q_m)$, eventually we reach a point at which $I(q)$ is a minimum *along that line*. Proceeding further in that direction would be counterproductive. This point is denoted $I(q_{m+1})$ in Figure 13.2. At that point, $-\nabla I(q_m)$ is parallel to the contours of constant I. Therefore,

$$\nabla I(q_{m+1}) \cdot \nabla I(q_m) = \left[\frac{\partial I}{\partial \underline{q}_{m+1}}\right]^T \left[\frac{\partial I}{\partial \underline{q}_m}\right] = 0 \tag{13.1.12}$$

where $(\cdot)$ indicates the dot product in the J dimensional space. Condition (13.1.12) can be used to find the optimum stepsize γ_m; this procedure is called a *line search* (see Exercise 13.1).

The algorithm (13.1.11–12) is known as a *steepest descent algorithm* because the

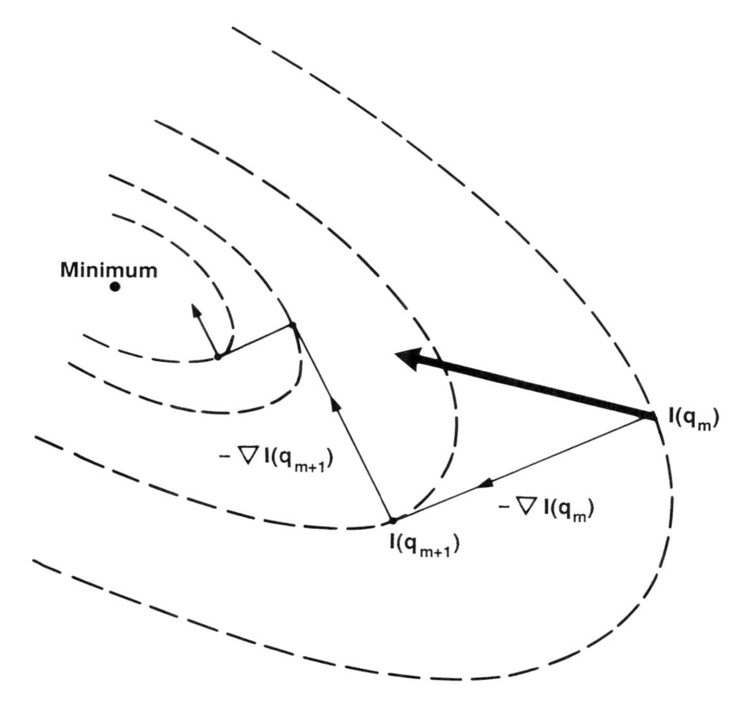

Figure 13.2 Illustration of the method of steepest descent and of obtaining the optimum stepsize γ_m.

descent always proceeds along the direction of the negative of the local gradient. Note how the descent curve changes direction after every iteration in Figure 13.2. Clearly, this method of descent is not very efficient, and many descent steps may be required to reach convergence. The problem is that the local negative gradient does not usually point toward the global minimum of I. In Figure 13.2, the solid vector at $I(q_m)$ points toward the absolute minimum of I. Clearly, descent would be much faster in that direction. An arbitrary descent direction in the J-dimensional space can be written, $-\underline{\underline{\theta}}(\partial I/\partial \underline{q})$, where $\underline{\underline{\theta}}$ is an arbitrary $J \times J$ matrix.

If $\underline{\underline{\theta}}$ is the identity matrix, then the descent direction is the steepest descent. One special choice of descent direction is the Newton descent. Define $\underline{\underline{\theta}}$ in the following way:

$$\underline{\underline{\theta}}^{-1} = \frac{\partial^2 I}{\partial \underline{q}^2} \tag{13.1.13}$$

An arbitrary element of $\underline{\underline{\theta}}^{-1}$ is $\partial^2 I/(\partial q_\mu \, \partial q_\eta)$, where μ and η indicate the row and column, respectively. The second derivative of a scalar I with respect to a vector $\underline{q}$ is known as a *Hessian matrix*. A descent algorithm similar to the Newton algorithm used for initialization in Section 10.7 can be written:

$$\underline{q}_{m+1} - \underline{q}_m = -\gamma_m \left(\frac{\partial^2 I}{\partial \underline{q}_m^2}\right)^{-1} \frac{\partial I}{\partial \underline{q}_m} \tag{13.1.14}$$

where γ_m is the stepsize as before.

It is straightforward to determine the Hessian matrix of (13.1.5) when $H(\underline{s})$ is linear:

$$\frac{\partial^2 I}{\partial \underline{q}^2} = \frac{\partial^2 I}{\partial \underline{s}_A^2} = \underline{\underline{H}}^T \underline{\underline{R}}^{-1} \underline{\underline{H}} + \underline{\underline{P}}_F^{-1} \tag{13.1.15}$$

Then, a Newton descent algorithm for the functional (13.1.5), with stepsize equal to 1, for all m, can be written:

$$\underline{s}_A^{m+1} = \underline{s}_A^m + \{\underline{\underline{H}}^T \underline{\underline{R}}^{-1} \underline{\underline{H}} + \underline{\underline{P}}_F^{-1}\}^{-1}\{\underline{\underline{H}}^T \underline{\underline{R}}^{-1}[\underline{d} - \underline{\underline{H}}\,\underline{s}_A^m] + \underline{\underline{P}}_F^{-1}[\underline{s}_F - \underline{s}_A^m]\} \tag{13.1.16}$$

where m is the iteration number and we have assumed that H is a linear operator (13.1.3). This expression can be rewritten in many ways, but a particularly useful form is the *preconditioned form*:

$$\underline{s}_A^{m+1} = \underline{s}_A^m + \{\underline{\underline{I}} + \underline{\underline{P}}_F \underline{\underline{H}}^T \underline{\underline{R}}^{-1} \underline{\underline{H}}\}^{-1}\{\underline{\underline{P}}_F \underline{\underline{H}}^T \underline{\underline{R}}^{-1}[\underline{d} - \underline{\underline{H}}\,\underline{s}_A^m] + \underline{s}_F - \underline{s}_A^m\} \tag{13.1.17}$$

where $\underline{\underline{I}}$ is the identity matrix.

We use (13.1.16) or (13.1.17) by choosing an initial guess $\underline{s}_A^0$ and then iterating to convergence. Note that if $\underline{s}_A^0 = \underline{s}_F$, then (13.1.16) converges to the minimum (13.1.6) in a single iteration. This occurs because (13.1.5) is a quadratic form and $H(\underline{s})$ is a linear operator. Unfortunately, the algorithm (13.1.16) is not directly useable for the global analysis problem because of the necessity of inverting the Hessian matrix. However, Lorenc (1986, 1988a,b) has discussed forms of (13.1.16–17) in which the Hessian matrix is diagonalized. For example, if the matrix $\underline{\underline{P}}_F\underline{\underline{H}}^T\underline{\underline{R}}^{-1}\underline{\underline{H}}$ is approximated by a diagonal matrix, then (13.1.17) becomes a more general version of the SCM algorithm (3.7.4).

A singular advantage of descent methods is that, in principle, they can be applied when $H(\underline{s})$ is nonlinear, or even when the cost function is not quadratic. A nonquadratic cost function is appropriate when the error statistics are not normally distributed (Section 2.2). Under these conditions the analyses will not be linear in the observations (Section 1.7). This subject is discussed further by Lorenc (1988b). Development of practical descent methods for large problems of this type is beyond the scope of this book, but the interested reader can find detailed descriptions in Gill et al. (1981) and Navon and Legler (1987). In general, conjugate gradient (Exercise 13.2) or quasi-Newton methods have the most applicability for these problems.

13.2 Four-dimensional variational analysis

The variational problem considered in Section 8.6 is, in principle, a very general formulation of the atmospheric data assimilation problem. It demands a close fit to the data, plus consistency with a dynamic model over an extended period of time. In a sense, this formulation can be thought of as a logical successor to the continuous data assimilation procedure discussed in Chapter 12. Unfortunately, application of the techniques of Section 8.6 to realistic problems can be very frustrating (Lewis and

Panetta 1983). The problem is that the Euler–Lagrange equations that result from this formulation are usually very difficult to solve.

In recent years, a number of techniques have been derived for handling variational problems with strong constraints. One example is the augmented Lagrange multiplier techniques described by Navon and deVilliers (1983), in which a penalty function is added to the functional. Another method that is particularly useful for time-dependent problems is the *adjoint method*. The name *adjoint* arises because of use of an operator that bears a precise relationship to the operator in the dynamical constraint. This operator arises in a natural fashion when the gradient of the functional is found. The terms in the expression for the gradient are precisely the same terms that arise in the Euler–Lagrange equations, and when the minimum of the functional is found by a descent process, the gradient vanishes and the Euler–Lagrange equations are satisfied.

Historically, the meteorological use of adjoint methods seems to have followed separate paths that originated in the USSR and France. In the Soviet Union, Marchuk (1974), Penenko and Bratsov (1976), and Sadokov and Schteinbok (1977) used these methods to test the sensitivity of climate models to parameters used in the formulation. The work by Penenko and Obratsov (1976) extended these ideas into analysis and initialization, but restricted them to linear constraints. Meanwhile, in France, the work by Lions (1971) in *optimal control theory* suggested a way to attack the assimilation problem in meteorology. The theory of optimum control addresses the dependence of output parameters in a model on the input parameters, or more specifically, how they can be controlled by the input. Francois LeDimet, who studied with Lions, read the work of Talagrand (1981b), discussed in Chapter 12, and interpreted the work in terms of optimal control. These scientists have stimulated other work in the West. There is now a growing literature on the subject, and the student is referred to papers by Lewis and Derber (1985), LeDimet and Talagrand (1986), Talagrand and Courtier (1987), Courtier and Talagrand (1987), Derber (1987), Lorenc (1988a,b), Thacker and Long (1988), and Derber (1989).

The essence of these variational techniques can be comprehended without recourse to optimal control theory or the mathematics of adjoint operators. In fact, it is possible to derive the equations of the adjoint formulation using Lagrange multipliers. This approach, based on the background material of Section 8.2, is taken here.

Before we consider the full-blown four-dimensional data assimilation problem, we demonstrate the technique for a simple, linear, continuous one-dimensional (time) variational problem. Consider the minimization of the following functional:

$$I = \frac{1}{2} \int_0^T w(t)[\tilde{u}(t) - u(t)]^2 \, dt \tag{13.2.1}$$

subject to the strong constraint

$$\dot{u} + i\sigma u = 0 \tag{13.2.2}$$

Here $w(t)$ is a specified weight function, $\tilde{u}(t)$ is observed, and $u(t)$ is to be determined; $\dot{u}$ is the derivative of u with respect to time, and σ is a constant frequency. The idea is to find the function $u(t)$ that minimizes the cost function (13.2.1) while exactly satisfying the forecast model (13.2.2). Following the ideas of the previous section, we do this by finding the gradient of I and then using a descent algorithm to find the minimum. The first step is to define a control variable. The solution to (13.2.2) can be written as

$$u(t) = u_0 \exp(-i\sigma t) \tag{13.2.3}$$

where $u_0 = u(t=0)$. A good choice of control variable is the initial value u_0. In this case, the control variable is a scalar, not a vector as in the previous section. Then the gradient I with respect to u_0, that is, $\nabla I(u_0)$, can be determined simply by substituting (13.2.3) into (13.2.1) and differentiating with respect to u_0:

$$\nabla I(u_0) = \frac{\partial I}{\partial u_0} = -\int_0^T \exp(-i\sigma t) w(t)[\tilde{u}(t) - u(t)]\, dt \tag{13.2.4}$$

Derivation of this expression for $\nabla I(u_0)$ requires knowledge of the closed-form solution for the model equations (13.2.3), which is not normally available for complex atmospheric models. However, $\nabla I(u_0)$ can also be obtained using the method of Lagrange multipliers; this approach results in a useable form for $\nabla I(u_0)$. Consider the new functional

$$I_1 = \frac{1}{2}\int_0^T w(t)[\tilde{u}(t) - u(t)]^2\, dt + \int_0^T \lambda(t)[\dot{u} + i\sigma u]\, dt \tag{13.2.5}$$

where $\lambda(t)$ is an unknown time-dependent Lagrange multiplier. Take the first variation of (13.2.5),

$$\delta I_1 = \int_0^T \{[-w(\tilde{u} - u) + i\sigma\lambda - \dot{\lambda}]\, \delta u + [\dot{u} + i\sigma u]\, \delta\lambda\}\, dt + \lambda\, \delta u|_0^T \tag{13.2.6}$$

The minimum value of I and I_1 occurs at $\delta I_1 = 0$, which implies that

$$\dot{u} + i\sigma u = 0 \tag{13.2.7}$$

and

$$\dot{\lambda} - i\sigma\lambda + w(\tilde{u} - u) = 0 \tag{13.2.8}$$

subject to the imposed boundary conditions $\delta u(0)$, $\delta u(T) = 0$ or the natural boundary conditions $\lambda(0)$, $\lambda(T) = 0$. In the present case, we want to determine the value of $u(t=0)$ that minimizes I, so the appropriate boundary conditions are the natural boundary conditions $\lambda(0) = \lambda(T) = 0$.

Equation (13.2.7), obtained by variation on the multiplier is, of course, the model equation (13.2.2). Equation (13.2.8) is the same equation that would arise in the adjoint theory. It is an inhomogeneous equation in λ forced by the weighted discrepancy between $\tilde{u}(t)$ and $u(t)$. Note that the homogeneous part of (13.2.8) for λ is very similar to (13.2.2) for u. In the adjoint theory, the homogeneous part of

(13.2.8) is referred to as the adjoint equation corresponding to the original model (13.2.2).

Our immediate object is not to minimize I but to find an expression for the gradient $\nabla I(u_0)$. This is achieved as follows. Choose some arbitrary value u_0 and integrate (13.2.7) forward to produce $u(t)$. Then integrate (13.2.8) backward in time from $t = T$ to $t = 0$, using the natural boundary condition $\lambda(T) = 0$. The inhomogeneous term in (13.2.8) is calculated by using the known values $w(t)$ and $\tilde{u}(t)$ and the values of $u(t)$ calculated from the forward integration of (13.2.7). The solution $\lambda(t)$ for (13.2.8) can be determined by defining a new dependent variable $\lambda(t)\exp(-i\sigma t)$ and integrating directly. The result is

$$\lambda(0) = \int_0^T \exp(-\mathrm{i}\sigma \mathrm{t})\mathrm{w}(t)[\tilde{u}(t) - u(t)]\, dt = -\nabla I(u_0) \qquad (13.2.9)$$

Thus, the forward integration of the model equation (13.2.7) starting from some choice of initial condition u_0, followed by the backward integration of (13.2.8) using the natural boundary condition $\lambda(T) = 0$, produces the negative of the desired gradient function. This gradient can then be used in a descent algorithm to produce a new value of u_0. The integration of (13.2.7–8) is then repeated with this new value of u_0, and the whole process is iterated until I is minimized. When the minimum of I is obtained, $\lambda(0) = -\nabla I(u_0) = 0$, which satisfies the natural boundary condition at $t = 0$. This method of minimizing I is useable if the adjoint of the original model can be determined, but it is not necessary to know the closed-form solution of the model.

Let us apply this technique to four-dimensional variational analysis. In Section 13.1, the observation, analysis, and forecast vectors at time t_n were denoted $\underline{d}_n$, $\underline{s}_n^A$, and $\underline{s}_n^F$ and the observation error covariance matrix $\underline{\underline{R}}_n$. Assume that the forward interpolation is linear and given by (13.1.3). In the four-dimensional algorithm, a forecast model is required. Assume that the forecast model is linear but time dependent. Then the model forecasts at time t_{n+1} and t_n on the grid $\mathbf{r}_j$, $1 \leq j \leq J$, are related by

$$\underline{s}_{n+1}^F = \underline{\underline{M}}_n \underline{s}_n^F, \qquad 0 \leq n \leq N-1 \qquad (13.2.10)$$

where $\underline{\underline{M}}_n$ is a $J \times J$ square matrix that depends on time t_n. The $\underline{\underline{M}}_n$ could represent a set of coupled, linear, discretized partial differential equations with time-dependent coefficients. The formulation is sufficiently powerful to handle nonlinear forecast equations (see illustration in Appendix C). We also assume that the forecast model is perfect and thus set $\underline{s}_n^A = \underline{s}_n^F$ for all n. Define the functional

$$I = \frac{1}{2}\sum_{n=0}^{N} [\underline{d}_n - \underline{\underline{H}}_n \underline{s}_n^F]^T\, \underline{\underline{R}}_n^{-1}[\underline{d}_n - \underline{\underline{H}}_n \underline{s}_n^F] \qquad (13.2.11)$$

which is a sum over time of the first term of (13.1.5). Equation (13.2.11) is a considerably modified form of (8.6.1). In particular, it is a discrete form in which the observation network and forecast model grid do not have to coincide, the forward interpolation is explicitly accounted for, it is not necessarily univariate, and the observation errors may be spatially (but not temporally) correlated. The problem is to minimize (13.2.11) subject to the constraints (13.2.10) for $0 \leq n \leq N-1$. As in the simple example at the beginning of this section, the immediate goal is to find the gradient $\nabla I(s_0^F)$: that is, the gradient of I with respect to the control variable $\underset{=}{s}_0^F$, the vector of forecast values at time t_0.

Following (13.2.3), we can write $\underset{=}{s}_{n+1}^F$ in terms of $\underset{=}{s}_0^F$ simply as

$$\underset{=}{s}_{n+1}^F = \underset{=}{M}_n \cdots \underset{=}{M}_0 \underset{=}{s}_0^F \tag{13.2.12}$$

Equation (13.2.11) can then be minimized with respect to $\underset{=}{s}_0^F$ by inserting (13.2.12) into (13.2.11) and differentiating with respect to $\underset{=}{s}_0^F$. The result is

$$\nabla I(s_0^F) = -\underset{=}{H}_0^T \underset{=}{R}_0^{-1}[\underset{-}{d}_0 - \underset{=}{H}_0 \underset{=}{s}_0^F] - \sum_{n=1}^{N} \underset{=}{M}_0^T \cdots \underset{=}{M}_{n-1}^T \underset{=}{H}_n^T \underset{=}{R}_n^{-1}[\underset{-}{d}_n - \underset{=}{H}_n \underset{=}{s}_n^F] \tag{13.2.13}$$

As noted earlier, this form is not easy to use because it requires knowledge of the model solution in closed form (13.2.12). Following (13.2.5), define a new functional

$$I_1 = I + \sum_{n=1}^{N} [\underset{=}{s}_n^F - \underset{=}{M}_{n-1} \underset{=}{s}_{n-1}^F]^T \underset{-}{\lambda}_n \tag{13.2.14}$$

where I is given by (13.2.11). Here $\underset{-}{\lambda}_n$ is a vector of length J of Lagrange multipliers for time t_n. Variation on $\underset{-}{\lambda}_n$ yields the N constraints (13.2.10) whereas variation on $\underset{=}{s}_n^F$ yields

$$-\underset{=}{H}_n^T \underset{=}{R}_n^{-1}[\underset{-}{d}_n - \underset{=}{H}_n \underset{=}{s}_n^F] + \underset{-}{\lambda}_n - \underset{=}{M}_n^T \underset{-}{\lambda}_{n+1} = 0 \tag{13.2.15}$$

Equation (13.2.15) is an inhomogeneous discrete equation, and the homogeneous part is the adjoint equation corresponding to the model equation (13.2.10). The model equation (13.2.10) is integrated forward from $t = t_0$ to t_N, using some arbitrary choice of $\underset{=}{s}_0^F$, to produce values of $\underset{=}{s}_n^F$ for all n. Following the procedure discussed earlier in the section, set $\underset{-}{\lambda}_{N+1} = 0$ and integrate (13.2.15) backward in time from $t = t_N$ to t_0. The result is

$$\begin{aligned}
\underset{-}{\lambda}_N &= \underset{=}{H}_N^T \underset{=}{R}_N^{-1}[\underset{-}{d}_N - \underset{=}{H}_N \underset{=}{s}_N^F] \\
\underset{-}{\lambda}_{N-1} &= \underset{=}{M}_{N-1}^T \underset{=}{H}_N^T \underset{=}{R}_N^{-1}[\underset{-}{d}_N - \underset{=}{H}_N \underset{=}{s}_N^F] + \underset{=}{H}_{N-1}^T \underset{=}{R}_{N-1}^{-1}[\underset{-}{d}_{N-1} - \underset{=}{H}_{N-1} \underset{=}{s}_{N-1}^F] \\
\underset{-}{\lambda}_0 &= -\nabla I(s_0^F)
\end{aligned} \tag{13.2.16}$$

The resulting gradient is then used iteratively in a descent algorithm to minimize (13.2.11). It might be noted that when $N = 0$, (13.2.16) reverts to the first term of (13.1.9), as it should. In applying (13.2.16), we must first obtain the adjoint equation $\underset{-}{\lambda}_n - \underset{=}{M}_n^T \underset{-}{\lambda}_{n+1} = 0$ corresponding to the model (13.2.10). This is generally possible, even when the model is nonlinear (see Appendix C).

Some simple properties of four-dimensional variational analysis schemes are illustrated in Figure 13.3. Rewrite (13.2.11) in the form

$$I = 0.5 \sum_{n=0}^{N} I_n \qquad \text{where} \qquad I_n = [\underset{-}{d}_n - \underset{=}{H}_n \underset{=}{s}_n^F]^T \underset{=}{R}_n^{-1}[\underset{-}{d}_n - \underset{=}{H}_n \underset{=}{s}_n^F] \tag{13.2.17}$$

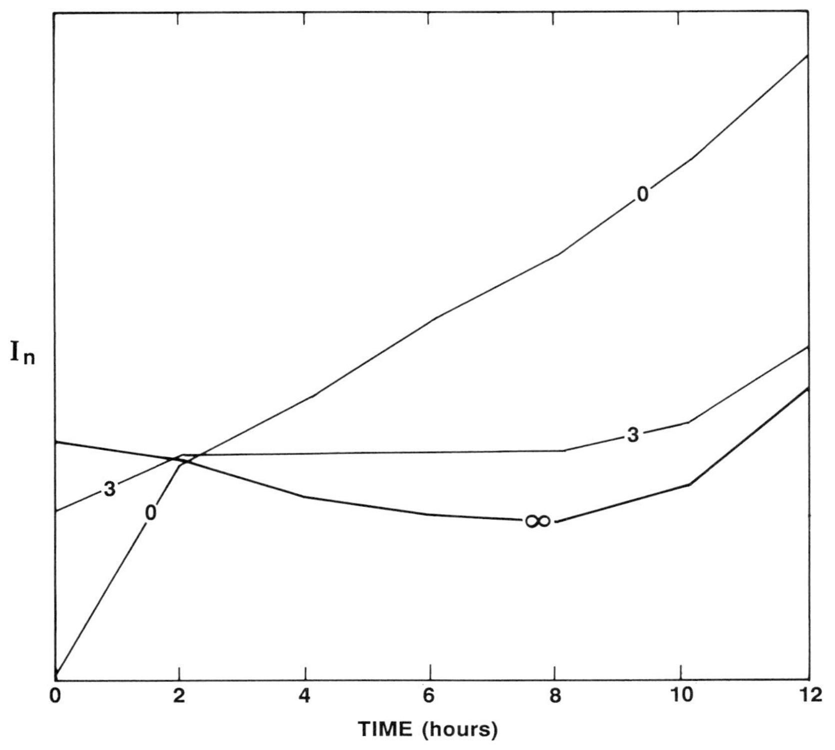

Figure 13.3 Plot of I_n from (13.2.17) as a function of timestep t_n. The iteration number is shown on each curve. The asymptotic curve is marked ∞. (After Derber, *Mon. Wea. Rev.* **117**: 2437, 1989. The American Meteorological Society.)

Derber (1989) minimized a form of (13.2.17) in which the observation network coincided with the forecast grid ($\underline{\underline{H}}_n = \underline{\underline{I}}$) and the observation errors were all equal and uncorrelated ($\underline{\underline{R}}_n = \underline{\underline{I}}$). The control variable was the initial forecast $\underline{s}_0^F$. Figure 13.3 shows the cost function I_n as a function of time during a 12 hour data assimilation experiment. Curves 0 and 3 show I_n after 0 and 3 iterations of the algorithm, and curve ∞ gives the asymptotic value. As can be seen, when convergence is reached, the cost function I_n has maxima at $t = t_0$ and $t = t_N$ and a relative minimum in the middle of the period. In other words, the forecasts or analyses (remember, $\underline{s}_n^A = \underline{s}_n^F$) produced by this algorithm most accurately fit the observations in the middle of the assimilating period. This is not surprising because the model equation is integrated forward in time and the adjoint equation is integrated backward; thus, at a given time t_n, information is propagated from both the future and the past.

There are two obvious problems. First, when a forecasting model is integrated to make a 1, 5, or 10 day forecast, it should be integrated from the latest available observations. However, the analyses produced by this algorithm seem to provide the most accurate fit to the data about four hours before the end of the data assimilating period. It would obviously be more desirable to have the best fit to the data at the end of the assimilating period.

Second, if a sequence of 12 hour variational assimilations of this type were performed, there would be a temporal discontinuity in the analyses at the beginning of each 12 hour period. Define $\tilde{\underline{s}}$ as the forecast produced at the end of the previous assimilation period and set $\underline{s}_0^F = \tilde{\underline{s}}$ before the first iteration of the descent algorithm. Then, after the variational assimilation had been completed, $\underline{s}_0^F$ would no longer be equal to $\tilde{\underline{s}}$, and there would be a temporal discontinuity at t_0. Furthermore, $\tilde{\underline{s}}$ was obtained by assimilating observations prior to t_0, and all this information has been essentially disregarded after the minimization has been completed.

Derber (1989) suggested a variant of the algorithm that addresses these problems. Suppose the model (13.2.10) is replaced by

$$\underline{s}_{n+1}^F = \underline{\underline{M}}_n \underline{s}_n^F + \alpha_{n+1} \underline{c} \tag{13.2.18}$$

where α_n is a specified time-dependent scalar and $\underline{c}$ is a time-independent vector of length J, which is to be determined. The idea is to calculate $\nabla I(c)$ rather than $\nabla I(s_0^F)$: that is, to find the vector $\underline{c}$ that minimizes (13.2.11). In the modified algorithm, the forecast model is no longer assumed to be perfect. $\underline{s}_0^F$ is specified to equal $\tilde{\underline{s}}$ and does not change during the descent procedure. This eliminates the temporal discontinuity at t_0. α_n is specified to increase as n increases, thus simulating the increase of forecast error during the period. Examination of (13.2.14) indicates that this will force the forecast more closely to the observations as t_n approaches t_N, as desired. Derivation of the actual algorithm is left as Exercise 13.4.

Another approach (Lorenc 1986) is to replace (13.2.11) by

$$I = \frac{1}{2} \sum_{n=0}^{N} [\underline{d}_n - \underline{\underline{H}}_n \underline{s}_n^F]^T \underline{\underline{R}}_n^{-1} [\underline{d}_n - \underline{\underline{H}}_n \underline{s}_n^F] w_n + \tfrac{1}{2} [\tilde{\underline{s}} - \underline{s}_0^F]^T \underline{\underline{P}}^{-1} [\tilde{\underline{s}} - \underline{s}_0^F] \tag{13.2.19}$$

where $\tilde{\underline{s}}$ is the forecast (at t_0) produced by data assimilation for $t < t_0$, and the w_n are specified time dependent but spatially independent weights. $\underline{\underline{P}}$ is the forecast error covariance matrix at t_0. The extra cost function in (13.2.19) forces the forecast $\underline{s}_0^F$ toward the previous forecast $\tilde{\underline{s}}$, thus reducing the temporal discontinuity. The weights w_n are designed to increase with increasing n, thus forcing the forecast closer to the observations, late in the assimilating period.

The procedure (13.2.10–16) can, in principle, be applied when the forecast model ($\underline{\underline{M}}_n$) or forward interpolation ($\underline{\underline{H}}_n$) are nonlinear, or even when the minimization is with respect to norms other than the l_2-norm (Lorenc 1988b).

The four-dimensional variational algorithms discussed in this section are very powerful; but they require that the forecast model be introduced as a strong constraint, which is appropriate only if the model is perfect. The four-dimensional algorithm to be discussed in the next section does not require a perfect forecast model.

13.3 The Kalman–Bucy filter

A mathematic framework well-suited to the four-dimensional data assimilation problem is the *state-space* approach. It deals with estimation of stochastic processes that are generated by randomly perturbed differential or difference equations. This approach was first formulated by Kalman (1960) and Kalman and Bucy (1961) for linear systems of ordinary differential equations. This formulation is usually referred to as the Kalman or Kalman–Bucy filter and has been widely applied to signal processing, optimal control, and aerospace problems. It is ironic that almost 200 years after Gauss invented least squares estimation to study the orbits of comets, we now use Kalman–Bucy filtering to determine the orbits of artificial satellites. The general theory is discussed rigorously in Jazwinski (1970) and on a more elementary level by Gelb (1974).

Early attempts to apply these ideas to meteorological problems were made by Jones (1965) and Petersen (1973), but they made little impact on the operational practices of the time. More recently, a group at Courant Institute of New York University has made a concerted attempt to apply the Kalman–Bucy methodology to the meteorological data assimilation problem (Ghil et al. 1981; Dee et al. 1985; Parrish and Cohn 1985).

We have already introduced many of the elements of the Kalman–Bucy (or KB) filter in the previous two sections. In the Kalman–Bucy formulation, the analyzed and forecast values are not the same. Thus, (13.2.10) is replaced by

$$\underset{\sim}{s}^{F}_{n+1} = \underline{\underline{M}}_n \underset{\sim}{s}^{A}_n \tag{13.3.1}$$

That is, the model uses the analyzed values on the grid at time t_n to produce forecast values on the same grid at time t_{n+1}. This model is assumed to be imperfect and

$$\underset{\sim}{s}_{n+1} = \underline{\underline{M}}_n \underset{\sim}{s}_n + \underset{\sim}{\varepsilon}^{M}_n \tag{13.3.2}$$

where $\underset{\sim}{s}_n$ is the vector of true values on the grid (13.1.1) and $\underset{\sim}{\varepsilon}^{M}_n$ is the column vector of length J of errors due to the imperfections in the model. Assume that $\underset{\sim}{\varepsilon}^{M}_n$ is unbiased and uncorrelated in time. Thus,

$$\langle \underset{\sim}{\varepsilon}^{M}_n \rangle = 0 \qquad \text{and} \qquad \langle \underset{\sim}{\varepsilon}^{M}_n (\underset{\sim}{\varepsilon}^{M}_\eta)^{T} \rangle = \delta_{n\eta} \underline{\underline{Q}}_n \tag{13.3.3}$$

where T indicates matrix transpose, $\delta_{n\eta}$ is the Kronecker delta, and $\underline{\underline{Q}}_n$ is the $J \times J$ positive definite covariance matrix of the model error at time t_n. $\underline{\underline{Q}}_n$ is called the *system* or *plant* error covariance matrix; it may be spatially correlated and include correlations between model errors in different state variables. Now define $\underset{\sim}{\varepsilon}^{F}_n$ (13.1.1) to be the forecast error and

$$\underset{\sim}{\varepsilon}^{A}_n = \underset{\sim}{s}^{A}_n - \underset{\sim}{s}_n \tag{13.3.4}$$

to be the analysis error at time t_n. Subtraction of (13.3.2) from (13.3.1) gives

$$\underset{\sim}{\varepsilon}^{F}_{n+1} = \underline{\underline{M}}_n \underset{\sim}{\varepsilon}^{A}_n - \underset{\sim}{\varepsilon}^{M}_n \tag{13.3.5}$$

Equation (13.3.3) implies that

$$\langle \underline{\varepsilon}_{n+1}^{\mathrm{F}} \rangle = \underline{\underline{\mathrm{M}}}_n \langle \underline{\varepsilon}_n^{\mathrm{A}} \rangle \tag{13.3.6}$$

Thus, if the analysis at time t_n is unbiased, the forecast at time t_{n+1} is also unbiased. In other words, the assumption $\langle \underline{\varepsilon}_n^{\mathrm{M}} \rangle = 0$ implies that the model has no climate drift (Section 4.2). If the model does have a (known) climate drift, the algorithm can be modified to account for it. Right multiply (13.3.5) by $(\underline{\varepsilon}_{n+1}^{\mathrm{F}})^{\mathrm{T}}$ and apply the expectation operator, giving

$$\underline{\underline{\mathrm{P}}}_{n+1}^{\mathrm{F}} = \underline{\underline{\mathrm{M}}}_n \underline{\underline{\mathrm{P}}}_n^{\mathrm{A}} \underline{\underline{\mathrm{M}}}_n^{\mathrm{T}} + \underline{\underline{\mathrm{Q}}}_n \tag{13.3.7}$$

where $\underline{\underline{\mathrm{P}}}_n^{\mathrm{F}}$, the prediction error covariance matrix, is defined by (13.1.1), and $\underline{\underline{\mathrm{P}}}_n^{\mathrm{A}} = \langle \underline{\varepsilon}_n^{\mathrm{A}} (\underline{\varepsilon}_n^{\mathrm{A}})^{\mathrm{T}} \rangle$ is the analysis error covariance matrix at time t_n. In the derivation of (13.3.7), the terms $\langle \underline{\varepsilon}_n^{\mathrm{A}} (\underline{\varepsilon}_n^{\mathrm{M}})^{\mathrm{T}} \rangle = \langle \underline{\varepsilon}_n^{\mathrm{M}} (\underline{\varepsilon}_n^{\mathrm{A}})^{\mathrm{T}} \rangle = 0$ because of (13.3.3).

The prediction part of the Kalman–Bucy filter consists of the model equation (13.3.1) plus the equation for the prediction error covariance matrix (13.3.7). The forecast error covariance $\underline{\underline{\mathrm{P}}}_n^{\mathrm{F}}$ can be identified with the background error covariance $\underline{\underline{\mathrm{B}}}$ of Chapters 4 and 5. In the same way, $\underline{\underline{\mathrm{P}}}_n^{\mathrm{A}}$ can be identified with the analysis error covariance discussed in Sections 4.5 and 5.6. Three more equations are required to complete the algorithm, and they have already been introduced in Sections 5.6 and 13.1:

$$\underline{\mathrm{s}}_n^{\mathrm{A}} - \underline{\mathrm{s}}_n^{\mathrm{F}} = \underline{\underline{\mathrm{K}}}_n [\underline{\mathrm{d}}_n - \underline{\underline{\mathrm{H}}}_n \underline{\mathrm{s}}_n^{\mathrm{F}}] \tag{13.3.8}$$

$$\underline{\underline{\mathrm{K}}}_n = \underline{\underline{\mathrm{P}}}_n^{\mathrm{F}} \underline{\underline{\mathrm{H}}}_n^{\mathrm{T}} [\underline{\underline{\mathrm{R}}}_n + \mathrm{H}_n \underline{\underline{\mathrm{P}}}_n^{\mathrm{F}} \mathrm{H}_n^{\mathrm{T}}]^{-1} \tag{13.3.9}$$

$$\underline{\underline{\mathrm{P}}}_n^{\mathrm{A}} = [\underline{\underline{\mathrm{I}}} - \underline{\underline{\mathrm{K}}}_n \underline{\underline{\mathrm{H}}}_n] \underline{\underline{\mathrm{P}}}_n^{\mathrm{F}} \tag{13.3.10}$$

Here $[\underline{\mathrm{d}}_n - \underline{\underline{\mathrm{H}}}_n \underline{\mathrm{s}}_n^{\mathrm{F}}]$ is the observation increment (known as the *innovation vector*), and $[\underline{\mathrm{s}}_n^{\mathrm{A}} - \underline{\mathrm{s}}_n^{\mathrm{F}}]$ is the analysis increment (known as the *correction vector*). $\underline{\underline{\mathrm{K}}}_n$ is called the *gain matrix* and is the same as the rectangular a posteriori weight matrix $\underline{\underline{\mathrm{W}}}^{\mathrm{T}}$ of Section 5.6 (not to be confused with $K(n)$, the number of observations, defined in Section 13.1). We have assumed that the forward interpolation operator H is linear.

The equations (13.3.1) and (13.3.7) are the prediction portion of the algorithm whereas (13.3.8–10) are the analysis portion. If we assume that $\underline{\underline{\mathrm{R}}}_n$, $\underline{\underline{\mathrm{Q}}}_n$, and $\underline{\mathrm{d}}_n$ are known, the algorithm proceeds from an initial guess at time $t = t_0$ of the forecast state vector on the analysis grid $\underline{\mathrm{s}}_0^{\mathrm{F}}$ and the forecast error covariance $\underline{\underline{\mathrm{P}}}_0^{\mathrm{F}}$. $\underline{\underline{\mathrm{P}}}_0^{\mathrm{A}}$ is calculated from (13.3.9–10) and $\underline{\mathrm{s}}_0^{\mathrm{A}}$ from (13.3.8–9). $\underline{\mathrm{s}}_1^{\mathrm{F}}$ can be calculated from (13.3.1) and $\underline{\underline{\mathrm{P}}}_1^{\mathrm{F}}$ from (13.3.7). The sequence is then repeated. The algorithm is sequential or recursive in that $\underline{\mathrm{s}}_n^{\mathrm{A}}$ depends explicitly only on the observations at time t_n (though implicitly it depends on all previous observations).

The Kalman–Bucy formalism includes *direct insertion* (Chapter 12) and *statistical interpolation* (Chapters 4 and 5) as special cases. In direct insertion, only (13.3.1) is used, with direct replacement of elements of $\underline{\mathrm{s}}_n^{\mathrm{A}}$ by spatially interpolated observations if available. The more general form of statistical interpolation of Section 5.6 uses (13.3.1), (13.3.8), and (13.3.9), but with $\underline{\underline{\mathrm{P}}}_n^{\mathrm{F}}$ in (13.3.9) approximated in some way.

In statistical interpolation (13.3.10) is used only for diagnostic purposes (if at all), and (13.3.7) is not used.

Before discussing the properties and extensions of the KB filter, we apply it to a very simple example, one-dimensional multivariate data assimilation. The model is based on the simple linearized f-plane shallow water model discussed in Sections 6.4 and 12.3. The state variables are complex spectral coefficients of streamfunction $\hat{\psi}$, velocity potential $\hat{\chi}$, and geopotential $\hat{\Phi}$, with the analyzed, observed, and forecast values being denoted by superscripts A, O, and F, respectively, and the true values with no superscript. The model equation (13.3.1) in this case is written as

$$\underline{s}_{n+1}^{F} = [\hat{\psi}_{n+1}^{F} \quad \hat{\chi}_{n+1}^{F} \quad \hat{\Phi}_{n+1}^{F}]^{T} = \underline{\underline{M}}[\hat{\psi}_{n}^{A} \quad \hat{\chi}_{n}^{A} \quad \hat{\Phi}_{n}^{A}]^{T} = \underline{\underline{M}}\,\underline{s}_{n}^{A} \tag{13.3.11}$$

where

$$\underline{\underline{M}} = \begin{pmatrix} \dfrac{K + D\cos\beta}{1+K} & \dfrac{-D\sin\beta}{\sqrt{1+K}} & \dfrac{\sqrt{K}(1 - D\cos\beta)}{1+K} \\ \dfrac{D\sin\beta}{\sqrt{1+K}} & D\cos\beta & \dfrac{-D\sqrt{K}\sin\beta}{\sqrt{1+K}} \\ \dfrac{\sqrt{K}(1 - D\cos\beta)}{1+K} & \dfrac{D\sqrt{K}\sin\beta}{\sqrt{1+K}} & \dfrac{1 + KD\cos\beta}{1+K} \end{pmatrix}$$

is time invariant. Here $\beta = f_0\alpha\,\Delta t$, $\alpha = \sqrt{1+K}$, $D = \exp(-\sigma_D\,\Delta t)$, f_0 is the (constant) Coriolis parameter, and K (not to be confused with the gain matrix) is given by (6.4.12). σ_D is the frequency-dependent damping coefficient introduced in Section 12.3. The remaining equations corresponding to (13.3.7–10) are (assuming no correlation between system errors in spectral coefficients of different wavenumbers) as follows:

$$\underline{\underline{P}}_{n+1}^{F} = \underline{\underline{M}}\,\underline{\underline{P}}_{n}^{A}\,\underline{\underline{M}}^{T} + \underline{\underline{Q}} \tag{13.3.12}$$

$$\underline{s}_{n}^{A} = \underline{s}_{n}^{F} + \underline{\underline{K}}_{n}[\underline{d}_{n} - \underline{s}_{n}^{F}] \tag{13.3.13}$$

$$\underline{\underline{K}}_{n} = \underline{\underline{P}}_{n}^{F}[\underline{\underline{P}}_{n}^{F} + \underline{\underline{R}}]^{-1} \tag{13.3.14}$$

$$\underline{\underline{P}}_{n}^{A} = [\underline{\underline{I}} - \underline{\underline{K}}_{n}]\,\underline{\underline{P}}_{n}^{F} \tag{13.3.15}$$

where $\underline{d}_n = [\hat{\psi}_n^{O} \quad \hat{\chi}_n^{O} \quad \hat{\Phi}_n^{O}]^{T}$ and superscript O indicates observation.

There is no explicit spatial dependence in (13.3.11) and consequently no forward interpolation, so $\underline{\underline{H}}_n$ is the identity matrix, assuming that we observe directly all three spectral components. $\underline{\underline{Q}}$ and $\underline{\underline{R}}$ are assumed to be time independent. All matrices $\underline{\underline{P}}_n^{F}$, $\underline{\underline{P}}_n^{A}$, $\underline{\underline{K}}_n$, $\underline{\underline{Q}}$, and $\underline{\underline{R}}$ are symmetric 3×3 matrices. The elements of the error covariance matrix $\underline{\underline{P}}$ are denoted as

$$\underline{\underline{P}} = \begin{pmatrix} p_{\psi\psi} & p_{\psi\chi} & p_{\psi\Phi} \\ p_{\chi\psi} & p_{\chi\chi} & p_{\chi\Phi} \\ p_{\Phi\psi} & p_{\Phi\chi} & p_{\Phi\Phi} \end{pmatrix} \tag{13.3.16}$$

with $p_{xy} = 0.5\langle \varepsilon_x \varepsilon_y^{*} + \varepsilon_y \varepsilon_x^{*} \rangle$ as $\hat{\psi}$, $\hat{\chi}$, and $\hat{\Phi}$ are complex as in Section 12.3.

Define the observation error covariance as

$$\underline{\underline{R}} = \langle(\varepsilon_\Phi^R)^2\rangle \begin{pmatrix} \frac{K}{2} & & 0 \\ & \frac{K}{2} & \\ 0 & & 1 \end{pmatrix} \tag{13.3.17}$$

where $\langle(\varepsilon_\Phi^R)^2\rangle = \langle(\Phi_n^O - \Phi_n)^2\rangle$ is the expected geopotential observation error variance and is time invariant. The nonzero elements of $\underline{\underline{R}}$ are $r_{\psi\psi}$, $r_{\chi\chi}$, and $r_{\Phi\Phi}$, and in real space form they are given by

$$\langle(\hat{\psi}_n^O - \psi_n)\,\nabla^2(\psi_n^O - \psi_n)\rangle, \qquad \langle(\chi_n^O - \chi_n)\,\nabla^2(\chi_n^O - \chi_n)\rangle, \qquad \text{and} \qquad \tilde{\Phi}^{-1}\langle(\Phi_n^O - \Phi_n)^2\rangle$$

The trace of $\underline{\underline{R}}$ (in real-space form) has units of (meters per second)^2 and corresponds to an error energy. From (13.3.17), we can see that $r_{\psi\psi} = r_{\chi\chi} = Kr_{\Phi\Phi}/2$, and in real-space form (see Section 12.3) we have

$$\langle(\psi_n^0 - \psi_n)^2\rangle + \langle(\chi_n^0 - \chi_n)^2\rangle = f_0^{-2}\langle(\Phi_n^0 - \Phi_n)^2\rangle$$

which says that the total expected wind observation error *variance* is related geostrophically to the expected geopotential observation error *variance*. However, $r_{\Phi\psi} = r_{\Phi\chi} = 0$, and the observation errors themselves are *not* geostrophically related.

Define two system noise covariance matrices:

$$\underline{\underline{Q}}_U = c\begin{pmatrix} \frac{K}{2} & & 0 \\ & \frac{K}{2} & \\ 0 & & 1 \end{pmatrix} \qquad \text{and} \qquad \underline{\underline{Q}}_G = c\begin{pmatrix} K & 0 & \sqrt{K} \\ 0 & 0 & 0 \\ \sqrt{K} & 0 & 1 \end{pmatrix} \tag{13.3.18}$$

where $c = \Delta t\langle(\varepsilon_\Phi^R)^2\rangle/6$ hours. The system noise $\underline{\underline{Q}}$ really corresponds to the growth of model error between t and $t + \Delta t$. The choice of the constant c assumes that the increase in system geopotential error variance in 6 hours is equal to the expected geopotential observation error variance. The constant c has been specified arbitrarily here, but more realistically it would reflect the estimated error growth rate of the model. In $\underline{\underline{Q}}_U$, as in $\underline{\underline{R}}$, assume there is no correlation in the system noise between the state variables. In $\underline{\underline{Q}}_G$, by contrast, the system noise is assumed to be geostrophically correlated. Thus, the elements $q_{\psi\psi} = Kq_{\Phi\Phi}$ and $q_{\psi\Phi} = q_{\Phi\psi} = \sqrt{K}q_{\Phi\Phi}$, with remaining elements zero, which in real-space form is

$$\langle(E_\psi^M)^2\rangle = f_0^{-2}\langle(E_\Phi^M)^2\rangle \qquad \text{and} \qquad \langle E_\Phi^M E_\psi^M\rangle = |f_0|^{-1}\langle(E_\Phi^M)^2\rangle$$

where E_ψ^M and E_Φ^M are real-space model errors (system noise) for streamfunction and geopotential, respectively (see also Section 5.3). Note that $\text{Trace}(\underline{\underline{Q}}_U) = \text{Trace}(\underline{\underline{Q}}_G)$; thus in this sense the total expected system noise is the same for both choices of $\underline{\underline{Q}}$.

When compared with the statistical interpolation algorithm of Chapters 4 and 5, the Kalman–Bucy formulation has two primary attractions. First, it presents the data assimilation problem in a more complete and elegant form by explicitly including the prediction of the background error statistics. Second, because the background error covariance can actually be calculated, we can compare it with the specified background error covariances used in statistical interpolation. In this way, we can examine the validity of the assumptions (separability, homogeneity, isotropy, geostrophy, nondivergence) we made in deriving useful background error covariances in Chapters 4 and 5. Of course, in the simple model discussed here, the spatially dependent properties of separability, homogeneity, and isotropy are not accessible, but the geostrophic and nondivergence assumptions can certainly be examined. Consequently, we integrate the set consisting of (13.3.12), (13.3.14), and (13.3.15) to examine properties of the forecast error covariance $\underline{\underline{P}}_n^F$ and the analysis error covariance $\underline{\underline{P}}_n^A$. All that is required is an initial estimate of the forecast error covariance $\underline{\underline{P}}_0^F$ to initiate the integration. Assume that the initial expected forecast error covariance is uncorrelated and very large, $\underline{\underline{P}}_0^F = 3\underline{\underline{R}}$. We integrated four cases, corresponding to two choices of σ_D and three choices of $\underline{\underline{Q}}$.

In case 1, $\sigma_D = 0.2\alpha$, $\underline{\underline{Q}} = 0$, corresponding to a perfect model with weak frequency-dependent damping. In other words, this is an identical twin experiment and the choice of σ_D corresponds to cases examined in Sections 12.3 and 12.5.

In case 2, $\sigma_D = 0$, $\underline{\underline{Q}} = 0$, corresponding to a perfect model with no damping.

In case 3, $\sigma_D = 0.2\alpha$, and $\underline{\underline{Q}} = \underline{\underline{Q}}_U$ (an imperfect, weakly damped model with uncorrelated system noise).

In case 4, $\sigma_D = 0.2\alpha$, and $\underline{\underline{Q}} = \underline{\underline{Q}}_G$ (an imperfect, weakly damped model with geostrophically correlated system noise).

The remaining parameters to be specified are $\Delta t = 2$ hours, $\langle(\varepsilon_\Phi^R)^2\rangle = 1$, and $K = 6.04$, corresponding to wavenumber 5 in the north–south and east–west directions and an equivalent depth of 5 km.

The results for each of the four cases are shown in Figure 13.4. The plotted curves show $p_{\Phi\Phi}^F$ and $p_{\Phi\Phi}^A$ (the expected geopotential forecast and analysis error variances) at each time-step. As would be expected, $p_{\Phi\Phi}^A(t_n) \le p_{\Phi\Phi}^F(t_n)$, which accounts for the vertical lines in the sawtooth curves. The dashed line denotes $r_{\Phi\Phi}$ (the expected geopotential observation error variance). Also shown for each case, as elements of a 3×3 matrix, are $[p_{\Phi\Phi}^F(t_\infty)]^{-1}\underline{\underline{P}}_\infty^F$. These matrix elements correspond to the asymptotic values of the prediction error covariance matrix $\underline{\underline{P}}_\infty^F$ normalized by the asymptotic geopotential prediction error variance. Here asymptotic refers to values after five days of integration.

The algorithm is started by calculating $\underline{\underline{K}}_0$ from (13.3.14) and $\underline{\underline{P}}_0^A$ from (13.3.15) using $\underline{\underline{R}}$ and $\underline{\underline{P}}_0^F$. There is an immediate drop from $p_{\Phi\Phi}^F(t_0) = 3.0$ to $p_{\Phi\Phi}^A(t_0) = 0.75$. Because $\underline{\underline{P}}_0^F$ and $\underline{\underline{R}}$ are diagonal, it is easy to see that $p_{\Phi\Phi}^A(t_0) = (1^{-1} + 3^{-1})^{-1} = 0.75$. Clearly, $p_{\Phi\Phi}^A(t_0)$ is less than the expected observation error variance $r_{\Phi\Phi} = 1$. At that point, a forecast is made, and the forecast error covariance $\underline{\underline{P}}_1^F$ (i.e., for $t = t_1$) is

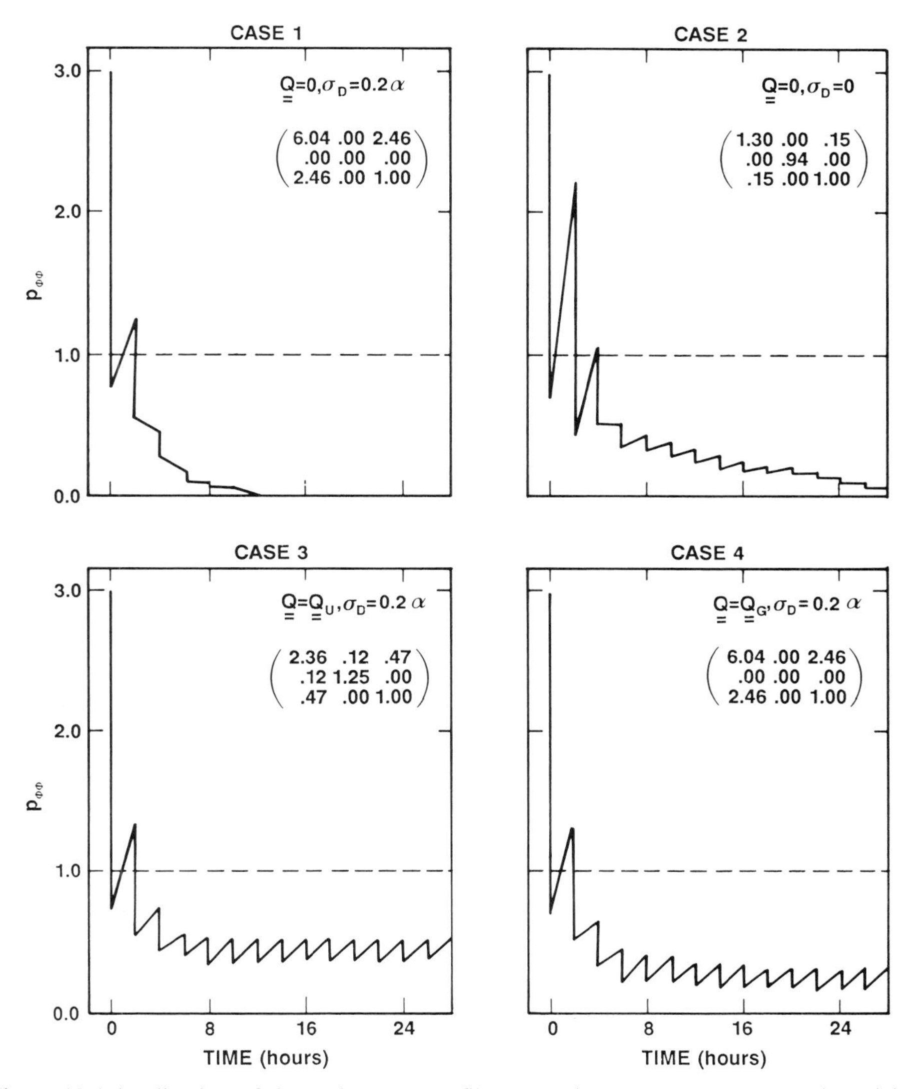

Figure 13.4 Application of the Kalman–Bucy filter to a three-component spectral model.

calculated from (13.3.12) using the known values of $\underline{\underline{P}}_0^A$ and $\underline{\underline{M}}$. The geopotential prediction error variance $p^F_{\Phi\Phi}(t_1)$ is larger than $p^A_{\Phi\Phi}(t_0)$, as might be expected. The process continues causing the sawtooth line sequence, which is particularly evident in cases 3 and 4.

Cases 1 and 2 have no system noise, and the forecast and analysis error variances gradually decline toward zero. This is to be expected in an identical twin experiment, but does not correspond to reality. Actual models are not perfect, and the analysis and prediction error variances cannot really asymptote to zero. The more realistic situation is captured in cases 3 and 4, where after a period of adjustment, an equilibrium is established in which the reduction of error after each data insertion is balanced by the growth of error during the prediction phase.

After the assimilation process has stabilized, we can establish the geostrophy or lack of geostrophy of the prediction error covariance $\underline{\underline{P}}^F_\infty$ by examining the 3×3 matrices shown in the figure. From the arguments given after (13.3.18), a geostrophically related covariance matrix should have a functional form similar to that of $\underline{\underline{Q}}_G$. That is,

$$p^F_{\psi\psi} = K p^F_{\Phi\Phi}, \qquad p^F_{\psi\Phi} = p^F_{\Phi\psi} = \sqrt{K} p^F_{\Phi\Phi}, \qquad \text{and} \qquad p^F_{\chi\chi} = p^F_{\chi\psi} = p^F_{\psi\chi} = 0$$

In the present instance, $K = 6.04$, and (apart from a common multiplicative constant) a perfectly geostrophically correlated covariance matrix would have all elements equal to zero, except 1 in the lower right-hand corner of the matrix, $K = 6.04$ in the upper left, and $\sqrt{K} = 2.46$ in the upper right and lower left. We see that $\underline{\underline{P}}^F_\infty$ is clearly geostrophic for cases 1 and 4 but not for cases 2 and 3, in which the wind errors are divergent and subgeostrophic in magnitude. In this model, the asymptotic prediction error will be geostrophic only if the system noise is zero or geostrophic and either the initial prediction error is geostrophic or there is frequency selective damping.

Let us now return to the more general Kalman–Bucy formulation (13.3.1) and (13.3.7–10). This formulation has been extensively studied in estimation theory and has some remarkable properties. The elements of $\underline{\underline{P}}^A_n$ are bounded from above and below. Under fairly general conditions (see Cohn and Dee 1988), the algorithm is asymptotically stable; thus, errors at $t = t_0$ in the analyzed values $\underline{s}^A_0$ and analysis error covariance $\underline{\underline{P}}^A_0$ have no lasting effect on the filter performance. This is consistent with Figure 13.4.

In Section 4.9, we showed that if the background and observation error covariance matrices $\underline{\underline{B}}$ and $\underline{\underline{O}}$ were incorrectly specified, then the expected analysis errors of the statistical interpolation algorithm would not be a minimum. In the same way, if the system noise $\underline{\underline{Q}}_n$ in (13.3.7) and the observation error covariance $\underline{\underline{R}}_n$ in (13.3.9) are *not* specified correctly, then the expected analysis error given by Trace($\underline{\underline{P}}^A_n$) of (13.3.10) is not a minimum. In this case, the KB filter is said to be suboptimal. Suppose a suboptimal gain matrix is denoted by $\tilde{\underline{\underline{K}}}_n$. Then the expected analysis error covariance is given by generalization of (4.9.7):

$$\underline{\underline{P}}^A_n = [\underline{\underline{I}} - \tilde{\underline{\underline{K}}}_n \underline{\underline{H}}_n] \underline{\underline{P}}^F_n [\underline{\underline{I}} - \tilde{\underline{\underline{K}}}_n \underline{\underline{H}}_n]^T + \tilde{\underline{\underline{K}}}_n \underline{\underline{R}}_n \tilde{\underline{\underline{K}}}^T_n \tag{13.3.19}$$

where $\underline{\underline{I}}$ is the identity matrix. If the optimal gain matrix $\underline{\underline{K}}_n$ of (13.3.9) is inserted into (13.3.19), it collapses to (13.3.10). Usually, $\underline{\underline{R}}_n$ can be estimated reliably, but estimation of $\underline{\underline{Q}}_n$ can be a problem, as was the estimation of $\underline{\underline{B}}$ in Chapter 4. However, if the elements of $\underline{\underline{Q}}_n$ and $\underline{\underline{R}}_n$ are *overestimated*, then the estimates of $\underline{\underline{P}}^A_n$ from (13.3.10) are known to be an upper bound on the true values of the analysis error covariance. This is consistent with the discussion of Section 4.9, in which we suggested that the background error covariance matrix $\underline{\underline{B}}$ should be overestimated rather than underestimated.

Clearly, if the Kalman–Bucy formulation is to be successfully employed, the system noise $\underline{\underline{Q}}_n$ must be reliably estimated. The performance of the KB filter is reflected in the statistical properties of the innovation sequence (time series of observation increments):

$$\underline{d}_0 - \underline{\underline{H}}_0\,\underline{s}_0^F,\ \underline{d}_1 - \underline{\underline{H}}_1\,\underline{s}_1^F,\ \ldots,\ \underline{d}_n - \underline{\underline{H}}_n\,\underline{s}_n^F$$

The innovation sequence is the only direct connection between the model forecasts and reality. Kailath (1968) has shown that if the gain matrix is optimal, then the observation increments are *not* temporally correlated. In other words, in the optimal case the innovation sequence should be a white noise sequence:

$$\langle(\underline{d}_n - \underline{\underline{H}}_n\,\underline{s}_n^F)(\underline{d}_\eta - \underline{\underline{H}}_\eta\,\underline{s}_\eta^F)^T\rangle = 0 \qquad \text{for} \qquad n \neq \eta \tag{13.3.20}$$

The innovation sequence (for $n = \eta$) was used in Section 4.3 to estimate the background error covariance for statistical interpolation. Equation (13.3.20) provides a direct check on the specification of $\underline{\underline{Q}}_n$ and $\underline{\underline{R}}_n$. If $\underline{\underline{Q}}_n$ and $\underline{\underline{R}}_n$ are incorrect, then the gain matrix $\tilde{\underline{\underline{K}}}_n$ is suboptimal, and the covariance of the innovation sequence (13.3.20) is directly proportional to the difference between $\tilde{\underline{\underline{K}}}_n$ and the optimal gain matrix $\underline{\underline{K}}_n$. A special case of (13.3.20) is left as Exercise 13.6. If $\tilde{\underline{\underline{K}}}_n = \underline{\underline{K}}_n$, then (13.3.20) is satisfied. This fact can be used to derive an *adaptive* filtering procedure for estimating $\underline{\underline{Q}}_n$ during the data assimilation.

The Kalman–Bucy formulation – (13.3.1), (13.3.7), and (13.3.8–10) – does not generally guarantee a slowly evolving (i.e., slow manifold) forecast. The experiments illustrated in Figure 13.4 suggest that the use of frequency-selective damping and geostrophic system noise tends to produce slowly evolving forecasts. Another technique is the *modified KB filter*; it defines the gain matrix as

$$\underline{\underline{K}}_n^\Pi = \underline{\underline{\Pi}}\,\underline{\underline{K}}_n \tag{13.3.21}$$

where $\underline{\underline{\Pi}}$ is a slow manifold projection matrix similar to that defined in Exercises 6.1 and 6.2. The gain matrix (13.3.21) is not necessarily optimal, however.

Experiments have been conducted in a meteorological context with models similar to the Hinkelmann–Phillips model of Section 7.1. Ghil et al. (1981) discuss two-dimensional (longitude and time) multivariate KB filtering using a midlatitude f-plane shallow water model linearized about a mean zonal flow. Parrish and Cohn (1985) discuss a three-dimensional (latitude, longitude, and time) multivariate KB filter with a similar model on a midlatitude beta plane.

In these experiments, the scientists introduced inhomogeneous data distributions and investigated the time evolution of the forecast and analysis error covariances in one or two spatial dimensions. They found that the KB filter advected information from data-rich into data-poor regions, as hoped. Because of the spatial dependence in these experiments, properties such as homogeneity, isotropy, and geostrophy could be examined. They found that the forecast error covariances are most homogeneous, isotropic (for geopotential autocovariances), and geostrophic in the center of data-rich regions. The forecast error covariances departed most noticeably from those specified in Chapters 4 and 5 in the regions of rapidly changing data density (i.e., on the edges of data voids). This result is not surprising because the forecast error covariances used in statistical interpolation are obtained from observation increments (innovations) in the center of data-rich areas (Section 4.3). The lack of agreement between the assumptions of Chapters 4 and 5 and the results of KB filter experiments is due both to the advection of information by the model (13.3.7) and the inhomogeneous data distribution.

The model (13.3.1) was assumed to be linear, which is not in accord with the usual situation in numerical weather prediction. If the model is nonlinear, then $\underline{\underline{M}}_n$ depends on $\underline{s}_n^A$ and in (13.3.7) $\underline{\underline{P}}_{n+1}^F$ depends on all higher moments of the analysis error and not just on the second moment $\underline{\underline{P}}_n^A$. For the nonlinear case, an *extended Kalman–Bucy filter* can be formulated in which the model is linearized about each successive analysis $\underline{s}_n^A$.

The Kalman–Bucy filter and the four-dimensional variational algorithm of the previous section are related. When the forecast model ($\underline{\underline{M}}_n$) is linear and perfect ($\underline{\underline{Q}}_n = 0$) and the forward interpolation ($\underline{\underline{H}}_n$) is linear, then the analyzed values produced by the Kalman–Bucy filter at $t = t_N$, can also be obtained by setting all $w_n = 1$ while minimizing (13.2.19). This relation is discussed further by Lorenc (1986), and a special case is left as Exercise 13.5.

The Kalman–Bucy formulation is an elegant and comprehensive mathematical description of the data assimilation problem. It suffers from two serious drawbacks, however. The first is the estimation of $\underline{\underline{Q}}_n$, to which we have already alluded. The second is the computational expense. The problem is not the matrix inversion required in (13.3.9), for which efficient numerical methods exist (see Section 5.6). The real difficulty is the integration of (13.3.7), which requires multiplication of $J \times J$ matrices. Because J is often greater than 10^6 in a four-dimensional global model, this operation is computationally expensive even when efficiently formulated.

13.4 Initialization by Laplace transform

Lynch (1985a,b) devised a method of initialization based on the properties of the Laplace transform. It does not require transformation of the equations into normal mode space and is thus useful for initializing limited-domain models that have complicated boundary conditions. At the time of writing, this technique has been used successfully for the initialization of both barotropic and baroclinic limited-area models.

Lynch (1985a,b; 1987) considers the application of the Laplace transform procedure to barotropic and baroclinic fully nonlinear models. The present discussion does not follow Lynch's work closely; instead, it applies the Laplace transform procedure to a simpler problem, which we have discussed extensively in earlier chapters. In this way, the relationship of the Laplace transform technique to other initialization techniques can be more clearly seen. Thus, the present analysis is based on the linearized shallow water equations on an f-plane (6.4.8–10). The case of periodic boundary conditions is considered first. This problem can be solved by classical Laplace transform theory using the method of residues. As a by-product of this analysis, the solution of (7.1.20) is obtained. (It was merely quoted in Chapter 7.)

Problems with periodic boundary conditions can be solved by a number of procedures including the normal mode techniques of Chapter 9. In limited-area models, the boundary conditions are usually much more complex and normal mode procedures are not applicable. It is for problems with complex boundary conditions, however, that the real power of the Laplace transform technique becomes apparent. The latter part of this section demonstrates how the Laplace transform technique can be applied when the boundary conditions are not simple.

The domain is assumed to be bounded $0 \le x \le D_x$, $0 \le y \le D_y$. Equations (6.4.8–10) can be Laplace transformed using relations (7.1.12–13). Thus,

$$s\,\nabla^2 L(\psi) + f_0\,\nabla^2 L(\chi) = \nabla^2\psi(0) \tag{13.4.1}$$

$$s\,\nabla^2 L(\chi) - f_0\,\nabla^2 L(\psi) + \nabla^2 L(\Phi) = \nabla^2\chi(0) \tag{13.4.2}$$

$$sL(\Phi) + \tilde{\Phi}\,\nabla^2 L(\chi) = \Phi(0) \tag{13.4.3}$$

where ψ, χ, and Φ are the streamfunction, velocity potential, and geopotential, which are functions of x, y, and t. $L(\psi)$, $L(\chi)$, and $L(\Phi)$ indicate the Laplace transform of the variables, defined by (7.1.12), and s can be thought of as a continuous frequency. The right-hand sides of (13.4.1–3) are evaluated at $t = 0$. The remaining symbols are defined as in Section 6.4.

$L(\psi)$ and $L(\chi)$ can be eliminated from (13.4.1–3) to produce the following equation for $L(\Phi)$:

$$[s^3 + sf_0^2 - s\tilde{\Phi}\,\nabla^2]L(\Phi) = s^2\Phi(0) + f_0^2\Phi(0) - s\tilde{\Phi}\,\nabla^2\chi(0) - \tilde{\Phi}f_0\,\nabla^2\psi(0) \tag{13.4.4}$$

In general, (13.4.1–3) have complicated boundary conditions at $x = 0$, D_x and $y = 0$, D_y. For the moment, however, we assume the boundary conditions are periodic. Thus, $\Phi(0, y) = \Phi(D_x, y)$ and $\Phi(x, 0) = \Phi(x, D_y)$; similarly for ψ and χ. In this case, ψ, χ, and Φ can be expressed in a double Fourier series:

$$\begin{bmatrix} \psi(x, y, t) \\ \chi(x, y, t) \\ \Phi(x, y, t) \end{bmatrix} = \sum_{m=-\infty}^{\infty}\sum_{n=-\infty}^{\infty} \begin{bmatrix} \hat{\psi}(t) \\ \hat{\chi}(t) \\ f_0\sqrt{K}\hat{\Phi}(t) \end{bmatrix} \exp\left\{\frac{2\pi imx}{D_x} + \frac{2\pi iny}{D_y}\right\} \tag{13.4.5}$$

where the indices (m, n) on $\hat{\psi}$, $\hat{\Phi}$, and $\hat{\psi}$ have been dropped. Note that $\hat{\chi}(t)$ is not multiplied by i as in (6.4.11).

Then

$$\nabla^2 = -4\pi^2\left(\frac{m^2}{D_x^2} + \frac{n^2}{D_y^2}\right)$$

and (13.4.4) becomes

$$L(\hat{\Phi}) = \hat{\Phi}(0)\,\frac{[s^2 + a_1 s + a_0]}{s[s^2 + a_2^2]} \tag{13.4.6}$$

where

$$a_0 = \frac{f_0^2}{\hat{\Phi}(0)}[\sqrt{K}\hat{\psi}(0) + \hat{\Phi}(0)], \qquad a_1 = f_0\sqrt{K}\frac{\hat{\chi}(0)}{\hat{\Phi}(0)}$$

$$a_2 = f_0\sqrt{1+K}, \qquad K = \frac{4\pi^2\tilde{\Phi}}{f_0^2}\left[\frac{m^2}{D_x^2} + \frac{n^2}{D_y^2}\right]$$

Note that the right-hand side of (13.4.6) has the same form as the right-hand side of (7.1.20).

The Laplace transform of (13.4.6) can be written (see Kuhfittig 1977) as

$$\hat{\Phi}(t) = L^{-1}[L(\hat{\Phi})] = \frac{1}{2\pi i}\int_{C_0} e^{st}L(\hat{\Phi})\,ds \tag{13.4.7}$$

where s is assumed to be a complex variable; C_0 is a line parallel to the imaginary axis on the complex s plane and lies to the right of all singularities of $L(\Phi)$ and extends to $\pm\infty$, as in Figure 13.5(a). The singularities of $L(\hat{\Phi})$, called poles, occur at $s=0$ and $s=\pm ia_2$. Thus, if we regard s as a frequency, the poles occur at the Rossby wave frequency ($s=0$ in the present case) and at the two inertia–gravity wave frequencies $s=\pm if_0\sqrt{1+K}$.

From the theory of complex variables, it can be shown that the line integral of $\exp(st)L(\hat{\Phi})$ along the curve C_0 is equivalent to a line integral along any curve C_1 that encloses all the poles of $L(\hat{\Phi})$. The curve C_1 is shown in Figure 13.5(a). Thus,

$$\hat{\Phi}(t) = \frac{1}{2\pi i}\int_{C_1} e^{st}L(\hat{\Phi})\,ds$$

Designate the poles of $L(\hat{\Phi})$ as α_j, $1 \leqq j \leqq J$. From complex variable theory (see Kuhfittig 1977), it can be shown that if $f(s)$ is analytic within and on a simple closed curve C_1 of finite length on the complex plane, except for a finite number of poles within C_1, then

$$\frac{1}{2\pi i}\int_{C_1} f(s)\,ds = S = \sum_j \lim_{s\to\alpha_j}(s-\alpha_j)f(s) \tag{13.4.8}$$

where $\lim_{s\to\alpha_j}(s-\alpha_j)f(s)$ is known as the residue at the pole α_j and S is the sum of the residues. Equation (13.4.8) is applicable if all the poles are distinct, as they are in (13.4.6). When there are multiple poles, we use the formula

$$\lim_{s\to\alpha_j}\frac{1}{(n-1)!}\frac{d^{n-1}}{ds^{n-1}}(s-\alpha_j)^n f(s)$$

where n is the pole multiplicity. Equation (13.4.8) makes it possible to evaluate

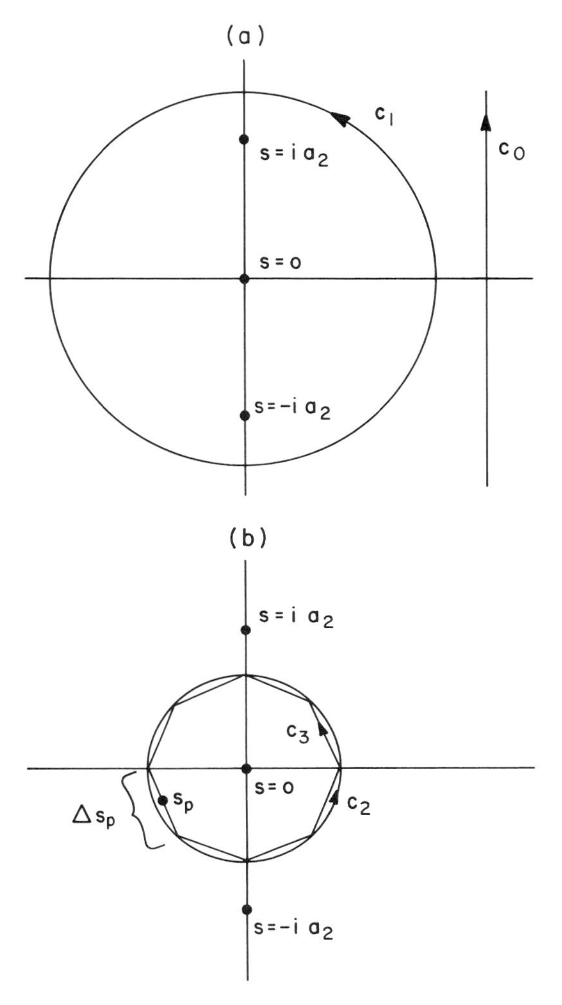

Figure 13.5 Complex s (frequency) plane with poles at $s = 0$ and $s = \pm ia_2$. The curves C_0, C_1, C_2, and C_3 are shown.

(13.4.6). Thus,

$$\hat{\Phi}(t) = \frac{1}{2\pi i}\int_{C_1} e^{st}L(\hat{\Phi})\,ds = \lim_{s\to 0}\left\{\frac{\hat{\Phi}(0)s(s^2 + a_1 s + a_0)e^{st}}{s(s^2 + a_2^2)}\right\}$$

$$+ \lim_{s\to -ia_2}\left\{\frac{\hat{\Phi}(0)(s + ia_2)(s^2 + a_1 s + a_0)e^{st}}{s(s^2 + a_2^2)}\right\}$$

$$+ \lim_{s\to ia_2}\left\{\frac{\hat{\Phi}(0)(s - ia_2)(s^2 + a_1 s + a_0)e^{st}}{s(s^2 + a_2^2)}\right\}$$

or

$$\hat{\Phi}(t) = \hat{\Phi}(0)\left[\frac{a_0}{a_2^2} + \left(1 - \frac{a_0}{a_2^2}\right)\cos a_2 t + \frac{a_1}{a_2}\sin a_2 t\right] \tag{13.4.9}$$

which has the same right-hand side as (7.1.21). Substituting in from (13.4.6) for a_0, a_1, and a_2 gives

$$\hat{\Phi}(t) = \frac{\sqrt{K}\hat{\psi}(0) + \hat{\Phi}(0)}{1+K} + \frac{K\hat{\Phi}(0) - \sqrt{K}\hat{\psi}(0)}{1+K}\cos(f_0\sqrt{1+K}\,t)$$

$$+ \frac{\sqrt{K}\hat{\chi}(0)}{\sqrt{1+K}}\sin(f_0\sqrt{1+K}\,t) \tag{13.4.10}$$

Equation (13.4.10) is identical to (6.4.31). The first term on the right-hand side of (13.4.10) corresponds to the residue at the pole $s = 0$ whereas the sine and cosine terms correspond to the residues at the poles $s = \pm if_0\sqrt{1+K}$.

Equation (13.4.10) is the complete time-dependent solution for the geopotential $\Phi(t)$ in the periodic boundary case and contains variation on both the slow and fast time scales. Initial conditions that eliminate high-frequency oscillations in this simple model were discussed in Section 6.5. From this equation, it is evident that if $\hat{\psi}(0) = \sqrt{K}\hat{\Phi}(0)$ and $\hat{\chi}(0) = 0$, the solution has no inertia–gravity waves. *This is equivalent to demanding that $\hat{\Phi}(t)$ be given by the residue at the pole $s = 0$ only*. Thus, we define $\hat{\Phi}_f$ to be the value of $\hat{\Phi}(t)$ obtained from the residue at the pole $s = 0$ only. Then, at time $t = 0$, for any closed curve C_2 that encloses only the pole at $s = 0$ and does not enclose the other poles $s = \pm if_0\sqrt{1+K}$,

$$\hat{\Phi}_f = \frac{1}{2\pi i}\int_{C_2} L(\hat{\Phi})\,ds = \frac{K\hat{\psi}(0) + \hat{\Phi}(0)}{1+K} \tag{13.4.11}$$

The curve C_2 is shown in Figure 13.5(b). This figure is the same as (a) except that it shows different closed curves. In particular, curve C_2 has a smaller radius than curve C_1 and encloses only the low-frequency pole at $s = 0$ and not the high-frequency poles at $s = \pm if_0\sqrt{1+K}$. Thus, for this simple model, the inverse Laplace transform calculated at $t = 0$ along any closed path on the s plane that encloses only the low-frequency poles produces initialized fields that do not excite the high frequencies.

In the present model with periodic boundary conditions, it is obviously simpler to perform this initialization by using (13.4.8) and summing only over the low-frequency poles. In more complicated models, (13.4.8) cannot be used because a simple algebraic form such as (13.4.6) cannot be used to derive the Laplace transform. However, the procedure (13.4.11) can still be used.

Consider again (13.4.4) and assume that the boundary conditions are not periodic. Then, a simple Fourier decomposition such as (13.4.5) is unavailable. Formally, an expression for $L(\Phi)$ can be written as

$$L[s, \Phi(x, y, t)] = [s^3 + sf_0^2 - s\tilde{\Phi}\,\nabla^2]^{-1}[s^2\Phi(0) + f_0^2\Phi(0) + s\tilde{\Phi}\,\nabla^2\chi(0) - \tilde{\Phi}f_0\,\nabla^2\psi(0)] \tag{13.4.12}$$

where $L[s, \Phi(x, y, t)]$ indicates the dependence of L on s, x, y, and t, $[s^3 + sf_0^2 - s\tilde{\Phi}\nabla^2]$ is a differential operator and the inverse is written formally. In general, the poles of (13.4.12) (that is, the frequencies of $[s^3 + sf_0^2 - s\tilde{\Phi}\nabla^2]$) are not known. But an equation analogous to (13.4.11) is still applicable. Thus,

$$\Phi_f(x, y) = \frac{1}{2\pi i} \int_{C_2} L[s, \Phi(x, y, 0)]\, ds \tag{13.4.13}$$

where $\Phi_f(x, y)$ is the filtered value of $\Phi(x, y, t = 0)$. The curve C_2 is a closed curve on the s-plane that has been chosen so as to enclose the low-frequency poles of $L[s, \Phi(x, y, t)]$ but exclude the high-frequency poles. It is not necessary to know the exact frequencies of the problem. Filtering of high frequencies occurs merely because the radius of the contour of integration has been reduced. This shrinking of the contour of integration is the essence of the Laplace transform filtering technique (see Lynch 1985a for further discussion on this point).

The line integral (13.4.13) around the curve C_2 can be approximated by the inscribed polygon C_3, shown in Figure 13.5(b). This particular curve is an octagon. Thus, (13.4.13) is approximated by

$$\Phi_f(x, y) \approx \frac{1}{2\pi i} \sum_{p=1}^{P} L[s_p, \Phi(x, y, 0)]\, \Delta s_p \tag{13.4.14}$$

where p indicates each side of the polygon and P is the number of sides of the polygon (eight in this figure). s_p is the value of s at the center point of the pth side of the polygon, and Δs_p is the length of that side. We approximate (13.4.13) by a discretization procedure known as numerical quadrature, in which a definite integral is approximated by a finite sum (see Lynch 1985a,b).

For each value s_p, $L[s_p, \Phi(x, y, 0)]$ must be calculated, which means inverting the operator $[s_p^3 + s_p f_0^2 - s_p\tilde{\Phi}\nabla^2]$ for $1 \le p \le P$. This operator is slightly more complicated than a Poisson operator (see Section 7.5) and must be inverted subject to whatever boundary conditions have been imposed. Corresponding filtered fields χ_f and ψ_f can be obtained by back substitution. Alternatively, we can initialize all three variables Φ_f, χ_f, and ψ_f together using the original coupled set (13.4.1–3). In the present case, the high-frequency inertia–gravity poles are well separated from the low-frequency Rossby poles, and the results are not sensitive to the exact choice of curve C_2 or C_3. For the beta plane or sphere, the Rossby mode poles are no longer at $s = 0$. In the baroclinic case, the frequency separation between fast and slow poles is more subtle (see Chapters 9 and 10), and the curves C_2 and C_3 have to be chosen with care.

The extension to the nonlinear case will not be considered here; it is discussed in detail in Lynch (1985a,b). The nonlinear extension follows very closely the methodology used in the Machenhauer initialization procedure (Section 9.6). Like dynamic initialization, the Laplace transform procedure discriminates between fast and slow modes on the basis of frequency alone.

Equation (13.4.11) is a special case of the more general equation

$$\hat{\Phi}_{\mathrm{f}}(t) = \frac{1}{2\pi i} \int_{C_2} e^{st} L(\hat{\Phi})\, ds \tag{13.4.15}$$

If (13.4.15) is applied at $t = 0$, it reduces to (13.4.11) and defines an initialization procedure. However, if (13.4.15) is applied to the model for all time, it defines the slow manifold of the model. We discussed the potential uses of this idea in continuous data assimilation in Section 12.9.

13.5 The bounded derivative initialization method

The bounded derivative method was introduced in Section 7.4 for the purpose of deriving the quasi-geostrophic constraints. It was also briefly encountered in Section 9.7 in connection with the Baer–Tribbia method of normal mode initialization. Bounded derivative initialization is a procedure for deriving an ordered sequence of increasingly effective initialization constraints. The general theory is due to Kreiss (1980) and was used in atmospheric applications by Browning, Kasahara, and Kreiss (1980), Browning and Kreiss (1982), Kasahara (1982a), and Browning and Kreiss (1986).

Section 7.4 considered a horizontal flow on a midlatitude f-plane. In this case, the application of the bounded derivative method produced the quasi-geostrophic constraints. As shown in Section 10.4, the application of normal mode initialization in this case also produced the quasi-geostrophic constraints. Thus, on the midlatitude f-plane, a three-way connection exists among normal mode initialization, bounded derivative method, and quasi-geostrophic initialization (discussed more fully in Kasahara 1982a).

The normal mode initialization procedure is particularly useful when the lateral boundary conditions are periodic, the spherical case being a good example. However, in problems with more complicated boundary conditions, it is not easy to find the normal modes. Consequently, a procedure such as the bounded derivative method, which does not require the construction of the normal modes, can be applied to these problems.

There is a whole class of numerical models that are integrated over limited domains. Some domains are as large as continents whereas others are only a few kilometers on a side. The lateral boundary conditions can be very complex. When this information is externally specified and time independent, these boundary conditions are called closed. In other numerical models, information is allowed to propagate through the lateral boundaries in either direction depending on the local flow at the boundary. These boundary conditions are called *open*. In either case, the boundary conditions can be included in the bounded derivative formulation by performing exactly the same scaling on the boundary conditions as was done to the interior equations. The case of closed boundaries was considered by Browning et al. (1980) and open boundaries by Browning and Kreiss (1982).

In principle, it is also possible to apply the bounded derivative technique to initialization of tropical models. Naturally, the scaling is different from that on a midlatitude beta plane, so the initialization constraints are also different. This case has been discussed by Kasahara (1982a). The bounded derivative and normal mode initialization methods have been compared using real data by Semazzi and Navon (1986).

Perhaps the most unsatisfactory aspect of the bounded derivative procedure is the requirement to use the scaled equations. The appropriate choice of scaling parameters is not always clear and can be controversial. This is not the case with the midlatitude beta plane though it is with tropical scaling.

13.6 Initialization of the hydrological cycle

Research in atmospheric data assimilation has been primarily concerned with the objective analysis and initialization of the dynamic variables such as temperature, wind, and geopotential. The previous chapters include little discussion of the moisture variables and the hydrological cycle. The water vapor in the atmosphere, if condensed, would form a layer of only 0.025 meters of liquid water on the surface of the earth. Nevertheless, water vapor plays a central role in the general circulation. Next to direct radiational heating, latent heat release is the most important source of heat in the atmosphere, and a substantial fraction of the poleward flux of energy takes place in the form of latent heat. Latent heat release plays an important role in the evolution of extratropical cyclones and is absolutely fundamental in tropical hurricane and monsoon dynamics. The solar and terrestrial radiation fluxes are affected both by the water vapor absorption bands and by liquid water in the form of cloud droplets.

Evaporation from the oceans (particularly the subtropical oceans), lakes, and land surface supplies the atmospheric branch of the hydrological cycle. Atmospheric water vapor is transported horizontally and vertically and is converted into cloud water droplets and precipitation (solid and liquid). Precipitation is the sink for the atmospheric branch of the hydrological cycle and is largest in the confluence zones of the midlatitudes and equatorial belts. The surface branch of the cycle involves hydrological processes such as ground water storage, river runoff, and storage in solid form, as well as biological processes such as evapotranspiration.

The atmospheric branch of the hydrological cycle is governed by the following conservation equation in pressure coordinates:

$$\frac{\partial \rho q}{\partial t} = -\nabla \cdot \rho q \mathbf{v} - \frac{\partial}{\partial P} \rho q \omega + E - C \tag{13.6.1}$$

where ρ is the density, $\mathbf{v}$ the horizontal velocity, ω the vertical velocity (in pressure coordinates), E the evaporation, and C the condensation. q is the specific humidity, defined by

$$q = \frac{\rho_{\mathrm{v}}}{\rho + \rho_{\mathrm{v}}} \tag{13.6.2}$$

where ρ_v is the density of water vapor. E includes evaporation from the earth's surface and from liquid water (cloud droplets and precipitation) in the atmosphere. C includes condensation into cloud droplets, which are suspended in the atmosphere, and into rain R, which falls to the earth's surface. The specific humidity q is a maximum in the tropical troposphere and decreases poleward and upward.

Equation (13.6.1) suggests that an adequate description of the hydrological cycle (at least the atmospheric branch) requires knowledge of the distribution of rainfall R, the horizontal wind $\mathbf{v}$ (particularly the divergent wind $\mathbf{v}_\chi$), the vertical motion ω, the specific humidity q, and the evaporation.

Let us begin with the rainfall R. Precipitation (particularly in the tropics) tends to be largely associated with mesoscale convective processes. Thus, it is highly variable and cannot be adequately resolved by conventional rain gauge networks (when they exist). Moreover, the rainfall distribution over the oceans has been a complete mystery until recently. Fortunately, space-based observation systems offer some hope. Precipitation data from rain gauges can be augmented using proxy estimates derived from satellite radiance information (Krishnamurty et al. 1984). In the 1990s, we hope that dedicated instruments for precipitation measurements will be in orbit.

Accurate objective analyses of the windfield are important in calculating the horizontal and vertical transport terms in (13.6.1). In particular, the determination of the winds in the tropics and subtropics is vital because of the major sources and sinks and because q itself is large there. Both the rotational and the divergent wind components are important in the tropics. The arguments of Sections 6.6, 10.3, and 12.3 suggest the value of accurate observations of the tropical rotational wind. However, the ratio of divergent wind to rotational wind is somewhat larger in the tropics than in midlatitudes, and many of the most important tropical circulations are largely divergent.

Daley and Mayer (1986) examined the errors in objective analyses produced from the GWE (Section 1.3) observation network. They demonstrated that wind errors were relatively small in the extratopics; but in the tropics, the wind errors were of the same order of magnitude as the winds themselves. They also showed that only the very largest-scale components of the divergent wind could be analyzed accurately. Daley et al. (1986) showed that these problems were due largely to deficiencies of the observation network but also to difficulties associated with the operational data assimilation procedures of the time. In particular, the lack of a proper wind/mass relationship in the tropics made it difficult to design appropriate multivariate statistical covariance models. The essentially univariate nature of tropical data assimilation meant that other types of data (temperature, for example) were not used to improve the wind analysis. Another problem was that data assimilation schemes tended to alias Rossby and Kelvin modes (see Section 9.4). In the extratropics, the vertical motion and divergent windfields could be inferred through initialization, but there were serious obstacles to this in the tropics (see Section 10.7). The inaccuracies of the tropical wind analyses were thought to be a major cause of forecast error in the tropics.

A number of attempts have been made to improve tropical wind analyses. Daley

(1985) relaxed the constraint of nondivergence in statistical interpolation to produce an improved analysis of the tropical divergent windfield. Julian (1984), and Kasahara, Balgovind, and Katz (1988) showed that satellite-measured radiances could be used directly to infer the divergence (and hence the vertical motion). Krishnamurti et al. (1984) demonstrated that the vertical motion could be deduced from the observed rainfall rate. Improved wind measuring instruments, such as space-based lidar, should provide more accurate and homogeneous wind measurements in the future.

Analyses of moisture variables such as specific humidity q have not traditionally been very reliable (Illari 1986). Krishnamurti et al. (1984) and Donner (1988) have devised a procedure called *physical initialization* for augmenting limited humidity observations with proxy data derived from rainfall observations. A simple expression for the rainfall rate in pressure coordinates is given by

$$R = -\frac{1}{g}\int_{P_T}^{P_s} \omega \frac{\partial q}{\partial P}\, dP \qquad (13.6.3)$$

where g is the gravitational constant, P_s and P_T are the pressure at the bottom and top of the atmosphere, and ω is the vertical motion. Suppose that at time $t = 0$, the rainfall rate derived from observations and the vertical motion ω derived from the data assimilation process are both considered reliable. At a given location, suppose $R > 0$ and $\omega < 0$ (ascending). Then, with a few simple assumptions, a vertical profile of q can be obtained by inverting (13.6.3).

Comprehensive numerical models simulate explicitly or through parametrization many elements of the hydrological cycle. When such models are integrated from observed initial conditions, for data assimilation or forecasting, the hydrological variables (in addition to the dynamic variables) should be initialized. Lack of initialization for the hydrological variables results in the so-called *spin-up problem*, which is illustrated in Figure 13.6. In (a), the globally averaged precipitation (PRECIP) and evaporation (EVAP) are plotted for a 120 hour integration of a comprehensive numerical model from an objective analysis. In the atmosphere, the globally averaged precipitation and evaporation should approximately balance and are thought to be between 2.4 and 2.8 mm/day. What happens in this model integration is that the precipitation is initially too low and the evaporation too high, and it takes more than 48 hours for them to come into balance. Despite the fact that sophisticated objective analysis and initialization procedures have been used, the initial model state in Figure 13.6(a) does not have a proper hydrological balance.

This problem has been examined by Krishnamurti et al. (1988). The initialization of the hydrological cycle was accomplished by a dynamic relaxation procedure (Section 12.8). Figure 13.6(b) shows the effect of 48 hours of dynamic relaxation prior to the start of the model integration. This time, at $t = 0$ (note shift of abscissa), the globally averaged evaporation and precipitation are in balance, and the hydrological cycle has effectively been spun up.

In the future, the variational techniques discussed in Section 13.2 are expected to be very useful in tackling the model spin-up problem.

A global experiment to measure all aspects of the hydrological cycle will be mounted

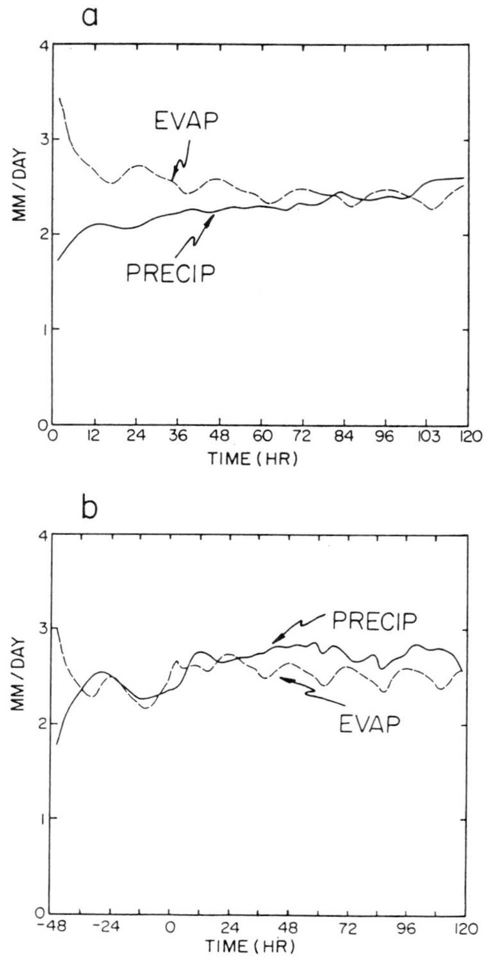

Figure 13.6 Illustration of the spin-up problem. Globally averaged precipitation as a function of time: (a) uninitialized; (b) after 48 hours of dynamic relaxation. (From Krishnamurti et al. *Mon. Wea. Rev.* **116**: 907, 1988. The American Meteorological Society.)

in the 1990s. This global energy budget and water cycle experiment (GEWEX) will have meteorological, hydrological, and oceanographic components. Dedicated orbiting platforms will measure rainfall, humidity, and windfields from space. Data assimilation will play a key role in the experiment.

13.7 Mesoscale data assimilation

Atmospheric phenomena can be characterized by their aspect ratio L_H/L_Z, where L_H is the characteristic horizontal scale and L_Z a characteristic vertical scale. In Section 7.3, the Rossby radius of deformation was defined by $L_R = N_0 L_z / 2\Omega$, where N_0 was the Brunt–Väisälä frequency and Ω the earth's rotation rate. Most of this book has been concerned with planetary and synoptic scales, for which $L_H \gtrsim L_R$ or the aspect ratio $\gtrsim N_0/2\Omega \approx 100$. For these large aspect ratios, the flow is hydrostatic, the earth's rotation is important, and there is a substantial degree of geostrophic balance (at least in the extratropics).

However, a number of interesting and important atmospheric phenomena are characterized by $L_H < L_R$, with correspondingly small aspect ratios. Emanuel (1983) groups these phenomena into two categories based on their aspect ratios. Mesoscale phenomena [aspect ratios of O(10)] have horizontal scales of 20–200 km, are largely hydrostatic, are affected by the earth's rotation, and have a substantial ageostrophic component. Included in this category are convective storm ensembles, frontal and jet stream phenomena, some orographic flows (lee cyclogenesis), polar lows, comma clouds, intensive extratropical storms (bombs), valley winds, sea breeze circulations, and the morning glory. Small-scale or convective phenomena [aspect ratios of O(1)] have horizontal scales of 0.2–2 km, are largely nonhydrostatic, and are not affected by the earth's rotation. In this category are cumulus clouds, thunderstorms, tornadoes, squall lines, urban circulations, Kelvin–Helmholtz waves, and some orographic flows. Flows with even smaller horizontal scales are referred to as microscale and will not be discussed here. The mesoscale and convective scales are discussed in detail by Orlanski (1975) and Atkinson (1981).

Lilly (1983) shows that the atmospheric kinetic energy spectrum obeys a k^{-3} power law for the synoptic scale and a $k^{-5/3}$ power law for the mesoscale and convective scale (see Figure 1.4). Here k is the horizontal wavenumber. Thus, the atmospheric power spectrum does not fall off as rapidly for decreasing scale for the mesoscale as for the synoptic scale. Mesoscale and convective scale phenomena are intermittent. The normally quiescent flow in these scales is occasionally interrupted by short-lived but very intense phenomena. We will discuss separately the mesoscale and convective scale data assimilation problems because they differ substantially. The extensive literature on these subjects can only be summarized briefly here.

The mesoscale data assimilation problem resembles in many ways the synoptic scale data assimilation problem, but there are substantial differences. The mesoscale power spectrum is flatter than that of the synoptic scale, so the aliasing problem (Appendix H) is likely to be worse in the mesoscale. The intermittency of mesoscale phenomena makes it more difficult to derive meaningful covariances for use in statistical objective analysis procedures. The same governing equations (the primitive equations of Section 7.3) apply in both cases. However, mesoscale models are usually integrated over only a portion of the globe, so correctly formulated lateral boundary conditions are very important. The data sources are similar and include surface stations, radiosondes, satellite radiances, and cloud track winds. But the mesoscale, with its smaller time and space scales, has an enormous requirement for additional data, from ground-based lidars, and wind profilers, for example. In fact, at one time, mesoscale phenomena were defined as those phenomena that could not be adequately resolved by the conventional surface and upper air network.

Mesoscale objective analysis techniques are similar to those used in synoptic data objective analysis. For example, Kuo and Anthes (1984) and Achtemeier (1987) have used a variant of the method of successive corrections (Chapter 3). Chang, Perkey, and Kreitzberg (1986) used a simplified statistical interpolation scheme in isentropic coordinates due to Bleck (1975). Many of the variational techniques discussed in Chapter 8 were originally introduced for mesoscale and convective scale data assimilation.

Mesoscale initialization is difficult because the implicit scaling assumed in the procedures of Chapters 6–11 may be less appropriate for the mesoscale than for the global and synoptic scales. The lateral boundaries of the limited domain introduce additional complications. Normal mode initialization is difficult (but not impossible; see Briere 1982) to apply here because the normal modes cannot be determined readily for complicated boundary conditions. A second difficulty is that many mesoscale models do not have a latitude–longitude projection and the Coriolis terms are nonseparable (Temperton 1988). Methods that do not require a normal mode decomposition, such as dynamic initialization (Section 11.3), bounded derivative (Section 13.5), Laplace transform procedures (Section 13.4), and implicit normal mode methods (Section 10.6), are clearly attractive for the mesoscale problem.

Continuous data assimilation is also possible for mesoscale models. The geostrophic adjustment arguments of Section 6.6 suggest that the geopotential adjusts to the rotational winds and that transient inertia–gravity waves excited by data insertion disperse rapidly (providing they pass through the lateral boundaries of the model without reflection). However, classical geostrophic adjustment arguments do not really take into account the highly divergent nature of mesoscale flows.

Gal-Chen (1983) has proposed a method for inserting satellite data into a mesoscale model. Because only the larger vertical scales of a satellite-derived temperature profile are considered reliable, Gal-Chen inserted only the vertically averaged temperature. The deviation (from the vertically averaged temperature) generated by the model and the observed vertically averaged temperature were combined variationally using the dynamic equations of the model as weak constraints. Kuo, Donall, and Shapiro (1987) have carried out mesoscale observing system simulation experiments to investigate the feasibility of short-range numerical weather prediction using a network of wind profilers to observe model initial conditions. Output from a mesoscale model was used in place of real profiler observations to produce a static initialization for further integration of the model. Sensitivity of model predictions to various network arrangements and errors in the data was determined. It was demonstrated that wind profiler observations have a positive impact on model forecasts and, further, that initial conditions retrieved from the profiler winds produced better forecasts than those initial states based on temperature observations.

Convective scale flows with aspect ratios of O(1) are nonhydrostatic. An appropriate set of governing equations on these scales are the quasi-Boussinesq or anelastic equations (Ogura and Phillips 1962; Miller and Moncrieff 1983). In cartesian coordinates, the inviscid, unforced versions of these equations can be written as shown in the following

$$\frac{du}{dt} + \frac{1}{\bar{\rho}} \frac{\partial P'}{\partial x} = 0 \tag{13.7.1}$$

$$\frac{dv}{dt} + \frac{1}{\bar{\rho}} \frac{\partial P'}{\partial y} = 0 \tag{13.7.2}$$

$$\frac{dw}{dt} - g \frac{\theta'}{\bar{\theta}} + \frac{\partial}{\partial z}\left(\frac{P'}{\bar{\rho}}\right) = 0 \tag{13.7.3}$$

$$\frac{\partial \bar{\rho} u}{\partial x} + \frac{\partial \bar{\rho} v}{\partial y} + \frac{\partial \bar{\rho} w}{\partial z} = 0 \tag{13.7.4}$$

$$\frac{d\theta'}{dt} + w \frac{\partial \bar{\theta}}{\partial z} = 0 \tag{13.7.5}$$

where u and v are the horizontal velocity components, w the vertical velocity component, and ρ, P, and θ are the density, pressure, and potential temperature, respectively; d/dt defines the total derivative, and an overbar denotes a horizontal average. Here, $\rho(x, y, z, t) = \bar{\rho}(z) + \rho'(x, y, z, t)$, and similarly for θ and P. The basic state ρ, θ, P is hydrostatic and may be adiabatic, $(\partial\bar{\theta}/\partial z = 0)$. Note that the vertical equation (13.7.3) is nonhydrostatic and that the Coriolis terms have been dropped. These equations permit internal gravity-wave solutions but filter sound waves because of the three-dimensional incompressibility implied by the equation of continuity (13.7.4).

The main observational tool for phenomena on this scale has been radar, supplemented on occasion by high-resolution local surface and radiosonde networks and aircraft measurements. The intensity of the reflected radar beam is a measure of the rainwater in a particular volume of the atmosphere. These radar reflectivities can be objectively analyzed to produce fields of rainwater distribution. Barnes (1978) used a variant of the method of successive corrections (Section 3.6) for this purpose.

The Doppler radar produces measurements of the radial component of velocity along the radar beam by measuring the Doppler shift of the radar echo. With two Doppler radars, it is possible to measure the u and v wind components, but the w component must be obtained from the continuity equation (13.7.4). With three Doppler radars, all three wind components can be observed in principle (Ray et al. 1978; Testud 1983). In practice, the direct measurements of w are not very accurate because of the relatively low elevation angles of the radar beam. The three-dimensional windfield can be objectively analyzed using the method of successive corrections. Another technique (for three Doppler radars) is to minimize the change to the observed values of u, v, and w using (13.7.4) as a nonholonomic constraint (Ray et al. 1980).

When hydrological processes are included, (13.7.3) can be written as

$$\frac{dw}{dt} - gB' + \frac{\partial}{\partial z}\left(\frac{P'}{\bar{\rho}}\right) = R_z - gq_r' \qquad (13.7.6)$$

where $B' = (\theta'/\bar{\theta}) + 0.61q' - q_c'$ is the buoyancy, q the specific humidity, q_c the mixing ratio of liquid cloud water and q_r the mixing ratio of rainwater. R_z represents turbulence flux terms. (′) indicates deviation from a horizontal average. The q_c and q_r tend to decrease the buoyancy and hence retard upward vertical acceleration.

Doppler measurements u, v, and w and radar reflectivity measurements of q_r can be used to determine P' and B' following the procedure of Gal-Chen (1978) and Hane, Wilhelmson, and Gal-Chen (1981). The horizontal momentum equations (13.7.1–2) can be written as

$$\frac{\partial P'}{\partial x} = F, \qquad \frac{\partial P'}{\partial y} = G \qquad (13.7.7)$$

where F and G contain the time derivatives, horizontal and vertical advection terms, and turbulent flux terms (not included in 13.7.1–2). Minimization of

$$\int_S \left[\left(\frac{\partial P'}{\partial x} - F\right)^2 + \left(\frac{\partial P'}{\partial y} - G\right)^2\right] dS \qquad (13.7.8)$$

gives the Euler–Lagrange equation

$$\nabla^2 P' = \frac{\partial F}{\partial x} + \frac{\partial G}{\partial y}$$

which can be solved subject to Neumann boundary conditions. The known values of P', u, v, w, q_r, and R_z can then be used in (13.7.6) to calculate B'. Where radar observations are collected in the convective boundary layer in the absence of precipitation, the cloudwater and rainwater terms are simply omitted (Gal-Chen and Kropfli 1984).

This procedure has two obvious deficiencies. First, only the deviation values P' and B' are obtained, rather than P and B themselves. Second, θ', q', and q_c can be obtained from B' only with further assumptions. Roux (1985) has shown that in certain types of convective circulations, the thermodynamic equation (13.7.5) can be used to circumvent the first problem. Through a different approach, Ziegler (1985) and Hauser and Amayenc (1986) have used continuity equations for water substance to arrive at distributions of cloudwater, rainwater, and water vapor.

Wolfsberg (1987) has shown, within the framework of a simple numerical model, that the full four-dimensional wind and temperature fields can be determined from a time history of one velocity component using the variational methods of Sections 13.1 and 13.2.

13.8 Data assimilation in the oceans

Seventy-one percent of the earth's surface is covered by oceans, which transport heat, salt, and momentum and thus play an important role in the global climate system. The coupling between oceans and atmosphere at the air/sea interface significantly affects both fluids.

In the 1980s, it became increasingly apparent that atmospheric models were inadequate for simulating phenomena with time scales longer than two weeks. In particular, simulation of the El Niño Southern Oscillation (ENSO) and of increased anthropogenic CO_2 concentrations required coupled atmosphere–ocean models (see Washington and Parkinson 1986). The international programs that evolved from the GWE (Section 1.3), such as Tropical Ocean Global Atmosphere (TOGA) and World Ocean Circulation Experiment (WOCE), had a predominant oceanographic component. Consequently, interest in oceanic data assimilation increased substantially during this period.

The governing equations of the atmosphere and ocean are similar, but the properties of the two fluids are not, resulting in substantial differences in their respective circulations. The density of water is approximately 800 times that of air. Water also has greater viscosity and heat capacity, which implies lower velocities $O(0.1 \text{ ms}^{-1})$, longer time scales, and less variation in temperature for the oceans. The equation of state for the ocean is complex and empirical, but phase changes are less important than in the atmosphere. The ocean is confined in interconnected basins and thus the lateral boundaries constrain the circulation.

Motion in the ocean is caused by buoyancy contrasts that are due to temperature, as in the atmosphere, and to salinity differences. The resulting circulation is called thermohaline (see Gill 1982). The ocean is stably stratified with warmer (lighter) water on the surface and colder (heavier) water at the ocean bottom. A surface mixed layer extends several hundred meters down to the thermocline, which can vary in depth seasonally. The ocean is forced thermally through direct insolation, through evaporation and precipitation, through sensible heat transfer from the overlying atmosphere, and through the surface wind stress.

The ocean has external modes with a Rossby radius of deformation of 2×10^6 m and internal modes with Rossby radii of about 3×10^4 m. The internal Rossby radius is much smaller than in the atmosphere, and thus the arguments of Section 6.6 suggest that the bulk of the kinetic energy should be at very small scales. The ocean admits a number of internal and external wave solutions: Poincaré (inertia–gravity) waves, coastal Kelvin waves, continental shelf waves, and a number of trapped equatorial waves, including Rossby, Kelvin, Yanai (mixed Rossby–gravity), and Poincaré modes.

The extratropical oceans are characterized by huge gyres, which are forced by the surface wind stress. In the gyre interior, the flow is largely governed by the Sverdrup balance between the surface wind stress and the beta term (see Chapter 7). Because of the westward propagation of planetary Rossby modes, the gyres have a marked asymmetry, with strong unstable western boundary currents such as the Gulf Stream.

Much of the small-scale eddy activity is in the vicinity of these western boundary currents.

The oceanic circulation is not as well understood as that of the atmosphere because of the relative paucity of observations. There has been little need for oceanic forecasting and hence no real-time network. Most oceanic observations have been obtained by research oceanographic vessels on a nonroutine basis; these vessels often take measurements in long, straight lines called sections. Because of the longer time scales in the oceans, the value of a given observation does not deteriorate as rapidly with time as in the atmosphere. Hydrographic observations (temperature, salinity, and pressure) are taken down to many hundreds of meters using bathythermographs. Currents are infrequently measured because of the cost of ocean moorings. Lagrangian tracer techniques that followed the dispersal of tritium from 1950s bomb tests have proved useful. New observing techniques such as expendable bathythermographs (XBT), satellite scatterometer measurements of surface winds, satellite infrared imagery from sea surface temperatures, satellite-tracked drifting buoys, and acoustic tomography techniques are also being tested. Advanced oceanographic observation systems are discussed by Munk and Wunsch (1982).

Analysis of the ocean circulation is somewhat different from that of the atmosphere:

The observational data base is much poorer though older data is much more useful.

Until recently (see Robinson et al. 1984) there has been little need for real-time prediction.

The kinetic energy-containing eddies are very small scale in the ocean. Eddy-resolving general circulation models (EGCM) have been developed to simulate the eddy processes, but they require very high resolution and are too expensive to be run globally in an assimilation cycle. Fortunately, the ocean is divided into basins, and it is much more logical than in the atmosphere to consider data assimilation over a limited region.

Many of the interesting phenomena can be modeled with quasi-geostrophic and/or linear dynamics; thus analysis methods that use simple models in a sophisticated way are very attractive. However, modeling of Gulf Stream meanders requires primitive equation models.

To a greater extent than in the atmosphere, the ocean circulation can be regarded as a response to an external forcing (usually the surface wind stress).

Because of the small internal Rossby radius, the currents adjust to the density fields except for very small scales and very close to the equator. Thus, the currents can usually be readily obtained from the density field.

Oceanographic data have been objectively analyzed directly, without the aid of an assimilating model by using statistical interpolation (often referred to in the oceanographic literature as the Gauss–Markov method). Bretherton, Davis, and Fandry (1976) and McWilliams (1976) analyzed hydrographic and current observations of midocean mesoscale eddies. Many of the assumptions discussed in Chapters 4 and 5, such as isotropy, homogeneity, nondivergence, and geostrophy, were made in deriving multivariate covariance functions. The results were found to be very sensitive to the covariances and means assumed, and unfortunately the data base was inadequate to estimate these with much certainty. Direct analysis of ocean data using statistical interpolation has been used in a variety of contexts since, including the analysis of temperature, geopotential, and even chlorophyll over the continental shelf, where many of the assumptions of homogeneity and isotropy cannot be applied (Denman and Freeland 1985).

Investigators also have shown increasing interest in oceanic data assimilation in which a model is implicitly or explicitly employed. There have been two approaches: inverse methods based on the ideas of Backus and Gilbert (1967) and continuous data assimilation (Chapter 12). Inverse methods (often called optimization methods or optimal control) are related to the variational methods of Sections 13.1 and 13.2. These methods have found application in two areas, determination of the steady state ocean circulation and tidal problems.

Wunsch (1978) applied the inverse technique to the determination of the time-invariant circulation of the northwest Atlantic. Hydrographic information was assumed to be available on sections that bound this region, and the circulation within this domain was to be determined subject to imposed simple linear constraints. Smaller-scale eddies were treated as observation noise (errors of representativeness in the terminology of Section 1.3). Later work demonstrated considerable refinement in both technique and application. In Schroter and Wunsch (1986), for example, the problem was the determination of the wind-driven circulation in an idealized basin. A simple, forced nonlinear QG model was used, and it was assumed that there were uncertainties in both the data and the forcing but not the model. Uncertainties in the data were expressed using inequalities, which required the use of slack variables (Section 8.5). An augmented Lagrangian procedure was used, and maps of the Lagrange multiplier indicated the sensitivity of the resulting analysis to the model and the observation network. Provost and Salmon (1986) studied the steady circulation of the Labrador Sea using a method not unlike that of Section 2.7, in which a penalty function (in addition to dynamical constraints) was added to ensure smoothness.

A quite different problem, the tides in a narrow strait, was studied by Bennett and McIntosh (1982) and McIntosh and Bennett (1984). This was a time-dependent problem in a limited domain. Linear shallow water dynamics was assumed with possible Poincaré and coastal Kelvin wave solutions. A variational approach using an adjoint formulation included the model equations, the coastal boundary conditions, the tidal gauge observations within the strait, and the forcing at the open ocean boundaries. All aspects (dynamics, boundaries, data, and forcing) were assumed to be uncertain, and the variational integral was completely weak in the sense of Section 8.4.

As mentioned earlier, continuous data assimilation (Chapter 12) has been attempted for oceanographic problems. Malanotte-Rizzola and Holland (1986) have used a nonlinear baroclinic quasi-geostrophic eddy resolving model in an idealized

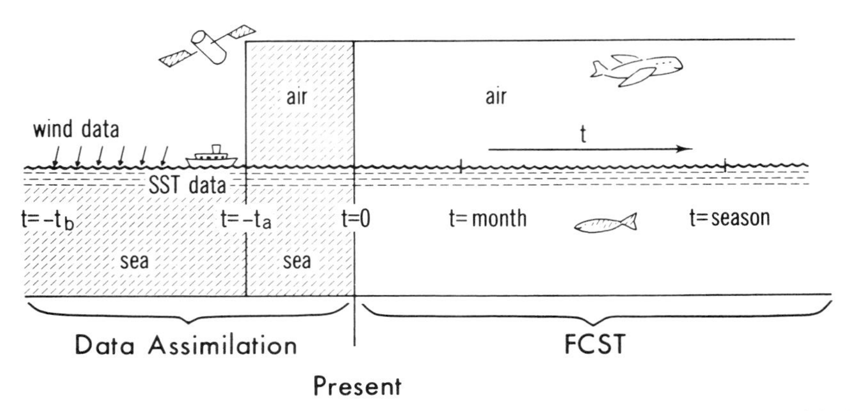

Figure 13.7 Illustration of data assimilation in a coupled atmosphere–ocean model. (After Miyakoda, "Assessment of results from different analysis schemes." In *International conference on the results of the Global Weather Experiment and their implication for the World Weather Watch*, Global Atmospheric Research Program (GARP) Publications Series No. 26. Geneva: World Meteorological Organization p. 217.)

ocean basin. An interesting result from this study was that the region of influence from a section was much greater to the west than to the east because of the westward propagation of Rossby modes. Data assimilation in the open ocean using statistical interpolation and a barolinic quasi-geostrophic model has been discussed by Robinson and Leslie (1985). Tropical ocean data assimilation has been investigated by Moore, Cooper, and Leslie (1987), and normal mode initialization has been used by Moore (1990). An operational real-time data assimilation scheme for the upper ocean has been developed by Clancy and Pollack (1983).

Other approaches to continuous data assimilation have also been considered. Thacker and Long (1988) investigated the adjoint equation method (see Section 13.2) using Lagrange multipliers and assuming that both observations and forcing were uncertain. The idea of minimizing a functional over a time period is particularly attractive in the ocean because the oceanic observations are not taken simultaneously. A Kalman filter approach has been discussed by Miller (1986).

A schematic diagram for global data assimilation for coupled ocean–atmospheric models is shown in Figure 13.7. Results for the global ocean have been described by Derber and Rosati (1989).

13.9 New challenges

Until the middle of this century, the state of the atmosphere was entirely diagnosed by subjective, manual techniques. The objective analysis procedures of the 1950s and

1960s were simple and expedient because of the limited computer resources of the time. With expanding computer power, true data assimilation became possible, and the techniques of analysis and initialization became more general and more elegant. At the time of writing, statistical interpolation and normal mode initialization define the state of the art in operational practice. But, these techniques will themselves be outmoded by some of the new ideas, such as variational procedures and Laplace transform initialization, discussed earlier in this chapter.

Despite the rapid progress over the past three decades, modern data assimilation techniques are not without flaws. One of these flaws was discussed recently by Ross Hoffman (1984):

> The variational analysis method used in this study performs well, but displays a disappointing lack of what might be called "common sense." This fault is basic and shared by all current analysis systems. The underlying statistical premise of these systems is that the data may be adequately represented by an idealized model plus "small" random errors. Therefore, a conservative (usually least squares) loss function will yield a good fit. The resulting objective analyses blend the different inputs and average out the errors. The theory is attractive, but in practice, the statistical premise of objective analysis is often violated.

Objective techniques have replaced subjective techniques because they are cheaper (not labor intensive), more consistent (same algorithm every day), and more accurate in an overall sense. As noted, they are also more conservative, which is sometimes very unfortunate. It has been demonstrated a number of times that the incipient stages of rare, extreme events in regions with marginal data coverage can often be analyzed more convincingly by skilled subjective analysts (with access to all the data and all the time they need to complete the analysis). The human brain, using a mixture of rules of thumb, pattern recognition, experience, simplified physics, and intuition can generate important features of the atmospheric circulation that escape the more conservative objective schemes. Can this human capability be objectivized? Perhaps advanced artificial intelligence (AI) techniques – expert systems and pattern recognition – may one day play a role here.

This book bears testimony to the progress in data assimilation methodology, but the scope of the data assimilation problem is also growing. In the late nineteenth century, analysis of surface parameters over continental land masses and adjacent waters was considered sufficient. The early twentieth century brought upper-air analyses of the traditional meteorological variables such as temperature, wind, and pressure. During the Global Weather Experiment (Section 1.3), the data assimilation spatial domain was extended to the whole globe. Data assimilation is now required on smaller scales (Section 13.7) and for the oceans (Section 13.8). It is not hard to imagine that there will soon be a requirement for regular analysis of the stratosphere and mesosphere.

Analysis of the traditional meteorological variables is no longer sufficient. Section 13.6 discussed data assimilation of the hydrological variables. In the future, data assimilation will be used for analyzing a host of new variables. Many surface properties are required in climate simulation models, such as soil moisture, surface albedo, snow cover, sea ice, surface temperature, and surface fluxes. Issacs, Hoffman, and Kaplan (1986) discuss the remote sensing of these parameters from satellite observing platforms. The global environment is threatened by dangerous new anthropogenic hazards such as increasing CO_2, thinning of the stratospheric ozone, and acid rain. New international programs such as the International Geosphere Biosphere Program (IGBP) have been mounted to study these threats. It is reasonable to expect that data assimilation techniques will be used to analyze chemical and biological variables to monitor the state of the global environment and to serve as initial conditions for simulation models.

Exercises

13.1 Define $\underset{\sim}{q} = [u, v]$ and $I(q) = I(u, v) = 0.5[au^2 + bv^2]$, with $a, b > 0$.

(a) Find $\nabla I(u, v)$, and for the steepest descent algorithm (13.1.11), solve (13.1.12) to find the stepsize γ_m.

(b) Show that this result can be obtained directly by minimizing $I(q_{m+1})$ with respect to γ_m.

(c) Choose initial values $[u_0, v_0]$ such that $a^2u_0^2 = b^2v_0^2$. Show that (13.1.11–12) imply that

$$|u_{m+1} \quad v_{m+1}|^{\mathrm{T}} = \frac{b-a}{b+a} |u_m \quad -v_m|^{\mathrm{T}}$$

Discuss the convergence of the algorithm as a function of the aspect ratio a/b of the ellipses of constant I.

(d) Find the Hessian matrix. Show that the Newton scheme (13.1.13), with $\gamma_0 = 1$, converges to the minimum $[u, v] = [0, 0]$ in a single iteration.

13.2 Define an arbitrary descent direction $\underset{\sim}{p}_m$ after the mth iteration. Then, a conjugate gradient descent algorithm is defined by

$$\underset{\sim}{p}_0 = -\nabla I(q_0)$$

$$\underset{\sim}{p}_{m+1} = -\nabla I(q_{m+1}) + \beta_m^2 \underset{\sim}{p}_m, \qquad \beta_m = \frac{\nabla I(q_{m+1}) \cdot \nabla I(q_{m+1})}{\nabla I(q_m) \cdot \nabla I(q_m)}$$

$$\underset{\sim}{q}_{m+1} = \underset{\sim}{q}_m + \gamma_m \underset{\sim}{p}_m$$

β_m is a scalar, and γ_m is the stepsize, obtained by minimizing $I(q_{m+1})$ with respect to γ_m as in Exercise 13.1. For the example of Exercise 13.1, show that $[u_2, v_2] = [0, 0]$ and that convergence is reached in two iterations.

13.3 Consider the one-dimensional linear advection equation (12.1.1). $\partial u/\partial t + c\, \partial u/\partial x = 0$, with c being a positive, constant advecting velocity. Discretize this equation as follows.

$$\frac{u_n^j - u_{n-1}^j}{\Delta t} + c\,\frac{u_{n-1}^j - u_{n-1}^{j-1}}{\Delta x} = 0$$

Δx and Δt are the constant space and time increments, and $x_j = j\,\Delta x$, $0 \le j \le J$ and $t_n = t_0 + n\,\Delta t$, $0 \le n \le N$. Here $u_n^j = u(j\,\Delta t, t_0 + n\,\Delta t)$, and choose Δx, Δt such that $c\,\Delta t/\Delta x = 1$. Following Section 12.1, show that along a characteristic of the preceding discrete equation $u_n^{j+n} = u_0^j$, that is, $u(x_{j+n}, t_n) = u(x_j, t_0)$.

Following Appendix C, define observed values $\tilde{u}_n^j = \tilde{u}(x_j, t_n)$. Minimize a functional I of the form given by (C2) in Appendix C, subject to the preceding $J \times N$ discrete constraints, using the Lagrange multiplier formulation of Appendix C. Show that the equation corresponding to (C10) in this case is given by

$$\lambda_n^j = \lambda_{n+1}^{j+1} + \tilde{u}_n^j - u_n^j$$

Set $\lambda_{N+1}^j = 0$ and find λ_0^j. Show that when I is minimized,

$$u_0^j = u(j\,\Delta x, t_0) = \frac{1}{N+1} \sum_{n=0}^{N} \tilde{u}_n^{j+n}$$

Discuss this result in terms of the characteristics of the equation.

13.4 Minimize the cost function (13.2.11) subject to the constraint (13.2.18). Specify $\underset{\sim}{c}$ as the control variable. Following (13.2.14), define Lagrange multipliers $\underset{\sim}{\lambda}_n$. Show that

$$\nabla I(c) = -\sum_{n=1}^{N} \alpha_n \underset{\sim}{\lambda}_n$$

13.5 Consider the minimization of the quadratic form (13.2.19) with respect to the control variable $\underset{\sim}{s}_0^F$, for the case $N = 1$. Assume all $w_n = 1$, the forecast model is linear (13.2.10), and the analysis grid and observation network coincide ($\underset{\approx}{H}_n = \underset{\approx}{I}$). Compare the vectors $\underset{\sim}{s}_0^F$, $\underset{\sim}{s}_1^F$ obtained by minimizing (13.2.19) with the vectors $\underset{\sim}{s}_0^A$, $\underset{\sim}{s}_1^A$ obtained from the Kalman–Bucy algorithm (13.3.1, 13.3.7–10) with $\underset{\approx}{Q}_n = 0$. Show that the two solutions are equivalent at t_1, but they are only equal at t_0 if there are no observations at t_1.

13.6 Consider the special case where the analysis grid and observation network coincide ($\underset{\approx}{H}_n = \underset{\approx}{I}$). Show that when $\eta = n + 1$, (13.3.20) is satisfied when $\tilde{\underset{\approx}{K}}_n = \underset{\approx}{K}_n$.

Appendix A
The normal modes of the spherical horizontal structure equations

The most widely used method for obtaining the normal modes of the spherical horizontal structure equations (9.3.17–19) is through expansion in spherical harmonic functions. This method was pioneered by Hough (1898) and extended by Longuet–Higgins (1968) and Kasahara (1976). The ψ, χ, and Φ in (9.3.17–19) are expanded as follows:

$$\begin{bmatrix} \psi \\ \chi \\ \Phi \end{bmatrix} = \sum_{m=-\infty}^{\infty} \sum_{l=|m|}^{\infty} \begin{bmatrix} \psi_l^m \\ i\chi_l^m \\ C_l \Phi_l^m \end{bmatrix} P_l^m(\sin\phi) \exp[im\lambda - 2\Omega i \sigma_m t] \tag{A1}$$

where

$$C_l = [2\Omega\sqrt{K}], \qquad K = \frac{l(l+1)\tilde{\Phi}}{4\Omega^2 a^2}$$

$P_l^m(\sin\phi)$ is an associated Legendre polynomial of the first kind with degree m and order l, $P_l^m(\sin\phi)\exp(im\lambda)$ a spherical harmonic, and σ_m the nondimensional frequency.

In (A1), the longitude is expressed trigonometrically, and the normalization is exactly as in the shallow water f-plane case (6.4.11). The only unfamiliar element is the latitudinal expansion in terms of associated Legendre polynomials.

Associated Legendre polynomials are widely used in forecast and climate models (see Bourke et al. 1977). The following properties of associated Legendre polynomials (and spherical harmonics) will be useful in the expansion procedure:

$$P_l^m(\mu) = \left\{\frac{(2l+1)(l-m)!}{(l+m)!}\right\}^{1/2} \frac{(1-\mu^2)^{m/2}}{2^l l!} \frac{d^{l+m}(\mu^2-1)^l}{d\mu^{l+m}} \tag{A2}$$

where $\mu = \sin\phi$ and (!) indicates factorial. Equation (A2) is a useful definition of the associated Legendre polynomial. The $P_l^m(\mu)$ are orthogonal. Thus,

$$\frac{1}{2}\int_{-1}^{1} P_l^m(\mu)P_k^m(\mu)\,d\mu = \delta_k^l \tag{A3}$$

The spherical harmonics are the natural eigenfunctions of the spherical Laplacian operator (9.3.2),

$$\nabla^2 P_l^m(\mu)e^{im\lambda} = -\frac{l(l+1)}{a^2}P_l^m(\mu)e^{im\lambda} \tag{A4}$$

Note that $l(l+1)/a^2$ is the eigenvalue of the spherical Laplacian operator and plays the same role as $(n^2+m^2)/a^2$ on the f-plane (6.4.12). Thus, there is an analogy between the definitions of K in (6.4.12) and (A1). Three other identities are useful:

$$\sin\phi\, P_l^m(\sin\phi) = \mu P_l^m = \varepsilon_{l+1}^m P_{l+1}^m + \varepsilon_l^m P_{l-1}^m \tag{A5}$$

where

$$\varepsilon_l^m = \left[\frac{l^2-m^2}{4l^2-1}\right]^{1/2}$$

$$\frac{\partial}{\partial\lambda}P_l^m(\mu)e^{im\lambda} = imP_l^m(\mu)e^{im\lambda} \tag{A6}$$

and

$$-\cos\phi\,\frac{dP_l^m}{d\phi} = (\mu^2-1)\frac{dP_l^m}{d\mu} = l\varepsilon_{l+1}^m P_{l+1}^m - (l+1)\varepsilon_l^m P_{l-1}^m \tag{A7}$$

Inserting (A1) into (9.3.17–19) and making use of (A2–A7) leads to the following result after considerable algebraic manipulation (see Errico 1984a):

$$[\sigma_m + \alpha_l^m]\psi_l^m - [\beta_l^m\chi_{l-1}^m + \beta_{l+1}^m\chi_{l+1}^m] = 0 \tag{A8}$$

$$[\sigma_m + \alpha_l^m]\chi_l^m - [\beta_l^m\psi_{l-1}^m + \beta_{l+1}^m\psi_{l+1}^m] + \sqrt{K}\Phi_l^m = 0 \tag{A9}$$

$$\sigma_m\Phi_l^m + \sqrt{K}\chi_l^m = 0 \tag{A10}$$

where

$$\alpha_l^m = \frac{m}{l(l+1)}, \qquad \beta_l^m = \sqrt{\frac{l^2-1}{l^2}}\,\varepsilon_l^m$$

$\psi_l^m = \chi_l^m = \Phi_l^m = 0$ for $l \le |m|$, by definition in (A1). The relationship in (A8) and (A9) are nonterminating, that is, they always involve the next highest values of l. To make system (A8–A10) tractable, it is necessary to set $\psi_l^m = \chi_l^m = \Phi_l^m = 0$ for $l \ge |m| + L + 1$, where L is some suitably large number [i.e., to terminate the sum in (A1) at $|m| + L + 1$ rather than ∞]. This turns (A8–A10) into a simple algebraic eigenvalue problem of the form

$$\underline{\underline{S}}\underline{X}_m = \sigma_m\underline{X}_m \tag{A11}$$

where $\underset{=}{S}$ is the matrix composed of the elements α_l^m, β_l^m, and so on, and

$$\underset{-}{X}_m = [\psi_{|m|}^m, \chi_{|m|}^m, \Phi_{|m|}^m, \ldots, \psi_{|m|+L}^m, \chi_{|m|+L}^m, \Phi_{|m|+L}^m]^{\mathrm{T}}$$

Matrix $\underset{=}{S}$ has degree $3L$; thus, for each value of m, there are $3L$ eigenvectors and $3L$ eigenvalues. Each eigenvector has a ψ component, a χ component, and a Φ component. The real-space form of the eigenvector can be constructed using (A1) in the same manner as in Chapter 6. Consistent with the analysis of Section 9.3, there are symmetric eigenvectors (χ, Φ symmetric, ψ antisymmetric) and antisymmetric eigenvectors (χ, Φ antisymmetric, ψ symmetric). The eigenvectors are orthogonal, which is to be expected from the analysis of Section 9.3. Standard algebraic eigenvalue techniques (Wilkinson 1965) can be used to find the eigenvalues and eigenvectors of $\underset{=}{S}$. Hough functions are defined to be the eigenvectors of (A11) in which $L \to \infty$. In practice, it suffices to take L large enough so that the structure of a particular eigenvector of interest does not change as L is increased.

If we want to initialize a global forecast model based on a spherical harmonic expansion, then (A11) is the appropriate eigenvalue problem to solve to obtain the normal modes of the model. In this case, however, the value of L is specified by the particular truncation of the forecast model to be initialized. In other words, the value of L in the initialization procedure must be the same as that used in the forecast model so that the initialization procedure and forecast model are discretized in exactly the same manner. In this case, L may not be particularly large, depending on the truncation assumed in the forecast model to be initialized. Eigenvectors obtained in this way will be called discrete Hough functions because they are only approximations to the true Hough modes. The lowest-order (largest horizontal scale) discrete Hough functions are indistinguishable from true Hough functions; but the higher-order discrete Hough modes may differ substantially from their Hough function counterparts.

The u, v, Φ form of the Hough functions can be obtained from the ψ, χ, Φ form simply by use of the Helmholtz equation (9.3.16). Examples of the u, v, Φ form of the Hough functions and their associated eigenfrequencies are shown in Section 9.4.

The normal modes of (9.3.1–3) or (9.3.17–19) can also be found if these equations are discretized by finite difference methods, as would be the case in a finite difference model. The procedure in this case is more complicated than that outlined in (A1–A11), but it has been applied successfully by a number of authors, notably Williamson and Dickinson (1976), Temperton (1977), and Temperton and Williamson (1981). If the normal mode problem is solved by finite differences, there may be additional computational modes that do not correspond to any of the Hough modes.

Appendix B
The effect of temporally correlated observation error on data assimilation

The white noise model of observation error discussed in Section 12.5 becomes unrealistic as the insertion interval $\Delta_I t$ approaches zero. For extremely frequent insertion, the observation error at a given insertion time is correlated with the observation error at previous insertions. Consider the case of geopotential insertion and model the observation error as follows:

$$\varepsilon_\Phi(t + \Delta_I t) = \mu \varepsilon_\Phi(t) + \sqrt{1 - \mu^2} r_\Phi(t + \Delta_I t) \tag{B1}$$

where $0 \leq \mu < 1$ and r_Φ is a random noise component. The observation error model of (B1) is known as a first-order Markov process. If r_Φ is normally distributed, then (B1) describes a Gauss–Markov process. It assumes that errors at time t and time $t + \Delta_I t$ are correlated and also contain a random (white noise) component. There is also an equation of the form (B1) for the complex conjugates ε_Φ^* and r_Φ^*. Assume that the error at $t = t_0$ is random, $\varepsilon_\Phi(t_0) = r_\Phi(t_0)$, and $\varepsilon_\Phi^*(t_0) = r_\Phi^*(t_0)$. Define E_Φ^2 as the expected variance of the random component and assume it is time independent:

$$E_\Phi^2 = \langle r_\Phi(t) r_\Phi^*(t) \rangle = \text{constant}$$

Because r_Φ and r_Φ^* are random,

$$\langle r_\Phi(t) r_\Phi^*(t + j\,\Delta_I t) \rangle = 0 \qquad \text{for } j \neq 0$$

and

$$\langle \varepsilon_\Phi(t) r_\Phi^*(t + j\,\Delta_I t) \rangle = 0 \qquad \text{for } j > 0$$

Then from (B1) it can easily be shown by induction that

$$\langle \varepsilon_\Phi(t) \varepsilon_\Phi^*(t) \rangle = \langle \varepsilon_\Phi^2 \rangle = E_\Phi^2 \tag{B2}$$

and

$$\langle \varepsilon_\Phi(t) \varepsilon_\Phi^*(t + j\Delta_I t) \rangle = \langle \varepsilon_\Phi^*(t) \varepsilon_\Phi(t + j\,\Delta_I t) \rangle = \mu^{|j|} E_\Phi^2 \tag{B3}$$

Thus, (B1) defines a stationary process (Appendix D) in which the correlation between the error at two different times depends only on the absolute time difference. Note that when $\mu = 0$, the first-order Markov process (B1) collapses to the white noise model of Section 12.5.

The Markov model can be used in an analysis similar to that of Section 12.3. The only difference is that the correlations

$$\langle \varepsilon_{\psi}^{*} \varepsilon_{\Phi} + \varepsilon_{\psi} \varepsilon_{\Phi}^{*} \rangle \qquad \text{and} \qquad \langle \varepsilon_{\chi}^{*} \varepsilon_{\Phi} + \varepsilon_{\chi} \varepsilon_{\Phi}^{*} \rangle$$

cannot be ignored. Proceeding as in Section 12.5 produces an analogue to (12.3.9):

$$\langle \underline{\varepsilon}_{\mathrm{w}}^{2}(t_0 + \Delta_{\mathrm{I}} t) \rangle = \tilde{\underline{\underline{S}}} \langle \underline{\varepsilon}_{\mathrm{w}}^{2}(t_0) \rangle + E_{\Phi}^{2} \tilde{\underline{b}} \tag{B4}$$

where

$$\tilde{\underline{\underline{S}}} = \begin{pmatrix} & & & 2a_1 b_1 & 2a_2 b_2 \\ & \underline{\underline{S}} & & 2a_3 b_2 & 2a_4 b_2 \\ & & & a_1 b_1 + b_1 a_3 & a_2 b_2 + a_4 b_1 \\ 0 & 0 & 0 & \mu a_1 & \mu a_2 \\ 0 & 0 & 0 & \mu a_3 & \mu a_4 \end{pmatrix}$$

$$\tilde{\underline{b}} = [b_1^2 \quad b_2^2 \quad b_1 b_2 \quad \mu b_1 \quad \mu b_2]^{\mathrm{T}}$$

$$\langle \underline{\varepsilon}_{\mathrm{w}}^{2} \rangle = [\langle \varepsilon_{\psi}^{2} \rangle \quad \langle \varepsilon_{\chi}^{2} \rangle \quad \langle \varepsilon_{\chi\psi} \rangle \quad \langle \varepsilon_{\Phi\psi} \rangle \quad \langle \varepsilon_{\Phi\chi} \rangle]^{\mathrm{T}}$$

$\langle \varepsilon_{pq} \rangle = \frac{1}{2} \langle \varepsilon_p \varepsilon_q^{*} + \varepsilon_p^{*} \varepsilon_q \rangle$, and $\underline{\underline{S}}$, a_1, b_1, and so forth, are defined in Section 12.3.

Convergence of (B4) is assured if the spectral radius of $\tilde{\underline{\underline{S}}}$ is less than one. $\tilde{\underline{\underline{S}}}$ is a block-upper-triangular matrix, and its eigenvalues are simply the eigenvalues of the blocks along the main diagonal. Therefore,

$$\sigma(\tilde{\underline{\underline{S}}}) = \text{maximum}[\rho(\underline{\underline{S}}), \mu \rho(\underline{\underline{G}})] \tag{B5}$$

where $\underline{\underline{G}}$ is the 2×2 matrix defined in Exercise 12.4. From the results of Exercise 12.4, $\rho(\underline{\underline{G}}) = \sqrt{\rho(\underline{\underline{S}})}$, so (B5) becomes simply

$$\rho(\tilde{\underline{\underline{S}}}) = \text{maximum}[\rho(\underline{\underline{S}}), \mu \sqrt{\rho(\underline{\underline{S}})}] \tag{B6}$$

Thus, (B4) converges (for $\mu < 1$) whenever (12.3.9) converges, though the convergence rate may differ. The asymptotic error level can be found in exactly the same way as in Section 12.5. Thus,

$$\langle \underline{\varepsilon}_{\mathrm{w}}^{2}(\infty) \rangle = \langle \varepsilon_{\Phi}^{2} \rangle [\underline{\underline{I}} - \tilde{\underline{\underline{S}}}]^{-1} \tilde{\underline{b}} \tag{B7}$$

At this point, μ has not yet been specified. Now, if $\mu = 1$, there is no convergence of the procedure. The forced, damped linear shallow water equations used in this analysis are similar to those of a forced, damped harmonic oscillator. It can easily be shown that a forced harmonic oscillator resonates (i.e., has a singular response) when the forcing frequency approaches the natural frequency of the oscillator. A

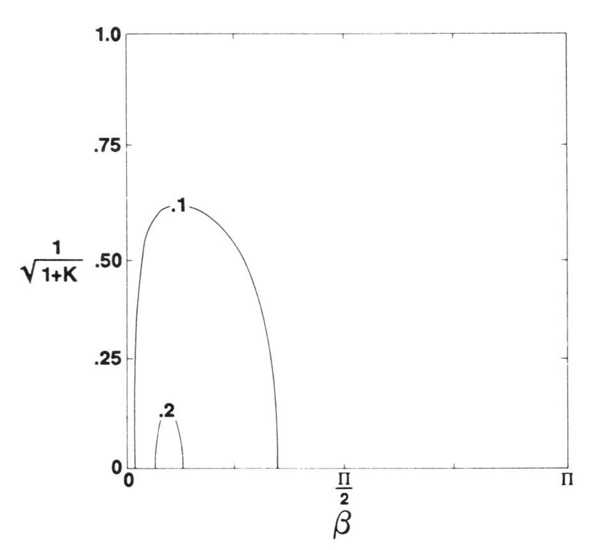

Figure B1 Asymptotic error of geopotential insertion on velocity potential with temporal correlation parameter $\mu_0 = 0.7$, for the same case as Figure 12.7(c).

resonance-like phenomenon occurs in the present analysis if $\mu = 1$. Clearly, then, μ must be less than one.

The correlation μ is specified as follows:

$$\mu = \mu_0 \exp(-\alpha\, \Delta_\text{I} t) \tag{B8}$$

where $0 \leq \mu_0 < 1$ is a constant and α and $\Delta_\text{I} t$ are defined as in Section 12.3. The form of μ in (B8) is such that the correlation increases as the insertion interval decreases, which is reasonable. The dependence on frequency α is merely for convenience, to allow the results to be plotted in the same format as in Section 12.5. There is, of course, no reason why the observation noise should depend on the natural frequencies of the assimilating model.

The asymptotic error $\langle \varepsilon_\chi^2(\infty)\rangle / \langle \varepsilon_\Phi^2 \rangle$ is plotted in Figure B1 for the same case shown in Figure 12.7(c). This case is for $\sigma_\text{D} = \alpha = f_0\sqrt{1 + K}$. The only difference is that $\mu_0 = 0.7$ instead of zero (as it did in Section 12.5). Figure B1 demonstrates that the major effect of temporally correlated error is to increase the asymptotic error level for frequent insertion. The convergence rate (not shown) is not affected because $\rho(\underline{\underline{S}})$ is always greater than $\mu\sqrt{\rho(\underline{\underline{S}})}$ in this case.

Temporally correlated error is likely to be particularly troublesome when the *same* observation is frequently inserted over a time interval.

Appendix C
Four-dimensional variational analysis with a nonlinear model

The forecast model $\underline{\underline{M}}_n$ of (13.2.10) was assumed to be linear. As noted in Section 13.2, four-dimensional variational analysis is also possible when the forecast model is nonlinear. The following simple example shows how the algorithm is developed in the nonlinear case.

Consider the simple one-dimensional nonlinear advection equation $\dot{u} + uu_x = 0$ in the domain $0 \leq x \leq 2\pi$. Here $u_x = \partial u/\partial x$. Discretize this nonlinear version of (12.1.1) as follows:

$$\frac{u_n^j - u_{n-1}^j}{\Delta t} = -u_{n-1}^j \frac{u_{n-1}^{j+1} - u_{n-1}^{j-1}}{2\,\Delta x} \tag{C1}$$

where Δx is a constant spatial increment and Δt a constant temporal increment. The independent variables x and t are given by $x_j = j\Delta x$, $0 \leq j \leq J$, and $t_n = t_0 + n\Delta t$, $0 \leq n \leq N$. u_n^j is the notation for $u(x_j, t_n)$. Assume cyclic boundary conditions in x, that is, $u_n^J = u_n^0$. Approximation (C1) is centered in x and forward in t. Those familiar with numerical time-differencing schemes will recognize that the forward time step in (C1) is numerically unstable (Haltiner and Williams 1980). However, it is used in this example because of its simplicity; its instability need not concern us.

Equation (C1) can be written in the form

$$u_n^j = u_{n-1}^j - \alpha u_{n-1}^j (u_{n-1}^{j+1} - u_{n-1}^{j-1}), \qquad \alpha = \frac{\Delta t}{2\,\Delta x} \tag{C2}$$

Now define a functional I, which is a simplified form of (13.2.17),

$$I = \frac{1}{2} \sum_{j=0}^{J-1} \sum_{n=0}^{N} (u_n^j - \tilde{u}_n^j)^2 \tag{C3}$$

where $\tilde{u}_n^j$ is the observed value of u at (x_j, t_n). We have assumed that the observations are on the same regular space/time grid as the model values u, so that $\underline{\underline{H}} = \underline{\underline{R}}_n = \underline{\underline{I}}$.

The functional (C3) is to be minimized subject to the $J \times N$ constraints of the form (C2). The goal is to find $\nabla I(u_0)$, the gradient of I with respect to the initial conditions u_0^j, $0 \leq j \leq J-1$, and then use it in an iterative stepwise descent as described in Section 13.1. To do this, define a new functional,

$$I_1 = I + \sum_{j=0}^{J-1} \sum_{n=1}^{N} \lambda_n^j [u_n^j - u_{n-1}^j + \alpha u_{n-1}^j (u_{n-1}^{j+1} - u_{n-1}^{j-1})] \tag{C4}$$

where the λ_n^j are the undetermined Lagrange multipliers. Because the constraints in (C4) all equal zero, the minimum of I_1 is also a minimum of I. The minimum of I_1 can be found by differentiating I_1 with respect to each of the variables λ_n^j *and* u_n^j and setting the results to zero:

$$\frac{\partial I_1}{\partial \lambda_n^j} = 0, \qquad N \times J \text{ equations} \tag{C5}$$

$$\frac{\partial I_1}{\partial u_n^j} = 0, \qquad (N+1) \times J \text{ equations} \tag{C6}$$

Equation (C5) simply returns the model constraints (C2), that is,

$$\frac{\partial I_1}{\partial \lambda_n^j} = 0 = u_n^j - u_{n-1}^j + \alpha u_{n-1}^j (u_{n-1}^{j+1} - u_{n-1}^{j-1}) \tag{C7}$$

To determine the calculation of (C6), we first list all terms in I_1 that contain u_n^j:

$$\begin{aligned} &\tfrac{1}{2}[u_n^j - \tilde{u}_n^j]^2 + \\ &\lambda_n^j [u_n^j - u_{n-1}^j + \alpha u_{n-1}^j (u_{n-1}^{j+1} - u_{n-1}^{j-1})] + \\ &\lambda_{n+1}^j [u_{n+1}^j - u_n^j + \alpha u_n^j (u_n^{j+1} - u_n^{j-1})] + \\ &\lambda_{n+1}^{j+1} [u_{n+1}^{j+1} - u_n^{j+1} + \alpha u_n^{j+1} (u_n^{j+2} - u_n^j)] + \\ &\lambda_{n+1}^{j-1} [u_{n+1}^{j-1} - u_n^{j-1} + \alpha u_n^{j-1} (u_n^j - u_n^{j-2})] \end{aligned} \tag{C8}$$

Differentiation of (C8) with respect to u_n^j yields

$$\frac{\partial I_1}{\partial u_n^j} = u_n^j - \tilde{u}_n^j + \lambda_n^j + \lambda_{n+1}^j [-1 + \alpha (u_n^{j+1} - u_n^{j-1})] - \alpha \lambda_{n+1}^{j+1} u_n^{j+1} + \alpha \lambda_{n+1}^{j-1} u_n^{j-1} = 0 \tag{C9}$$

or

$$\lambda_n^j = \lambda_{n+1}^j + \alpha [u_n^{j+1} (\lambda_{n+1}^{j+1} - \lambda_{n+1}^j) + u_n^{j-1} (\lambda_{n+1}^j - \lambda_{n+1}^{j-1})] - [u_n^j - \tilde{u}_n^j] \tag{C10}$$

Equation (C10) can be written in the form

$$\underline{\lambda}_n - \underline{\underline{M}}_n^* \underline{\lambda}_{n+1} + \underline{u}_n - \tilde{\underline{u}}_n = 0 \tag{C11}$$

where $\underline{\lambda}_n$ is the column vector whose elements are λ_n^j, $0 \le j \le J-1$, $\underline{u}_n$ the column vector with elements u_n^j, and $\underline{\underline{M}}_n^*$ the tri-diagonal $J \times J$ matrix whose nonzero elements on the jth row are

$$-\alpha u_n^{j-1}, \qquad 1 - \alpha(u_n^{j+1} - u_n^{j-1}), \qquad \alpha u_n^{j+1} \tag{C12}$$

Equation (C11) is of the same form as (13.2.15) with $\underline{\underline{H}} = \underline{\underline{R}}_n = \underline{\underline{I}}$. Whereas the matrix $\underline{\underline{M}}_n^{\mathrm{T}}$ in (13.2.15) is independent of $\underline{s}_n$, in this nonlinear example, the elements of the matrix $\underline{\mathrm{M}}_n^*$ are functions of the forecast variable $\underline{\mathrm{u}}_n$. However, $\underline{\mathrm{u}}_n$ is known, so $\underline{\mathrm{M}}_n^*$ can be determined. Note that (C11) is *linear* in $\underline{\lambda}_n$.

Backward integration of the Lagrange multiplier equation (C11) starting with $\underline{\lambda}_{N+1} = 0$ yields $\underline{\lambda}_0 = -\nabla I(u_0)$ in the same way as (13.2.16). The gradient $\nabla I(u_0)$ can then be used in a descent process following the discussion of Section 13.1.

The most important feature of this discussion is that the adjoint equation (C11) is linear even though the forecast model may have been nonlinear. The advantages of this approach are that nonlinear and physical processes (provided they are time differentiable) can be handled, that the model spin-up process (see Section 13.6) is handled implicitly, and that asynoptic and continuous data insertion can be treated in an elegant fashion.

Appendix D
Some statistical concepts

The statistical concepts used in this book are very elementary and can be found in any book on probability theory or random variables (e.g., Papoulis 1965).

A random variable is a variable that takes on values at random and can be thought of as a function of the outcomes of some random experiment. Consider the state variable s as a random variable. The expectation or expected value of s is defined as the sum of all values the variable might take, each weighted by the probability with which the value is taken. Suppose s can take on all values between $-\infty$ and ∞. Then, the expectation value of s is given by

$$\langle s \rangle = \int_{-\infty}^{\infty} sp(s)\, ds \tag{D1}$$

where $p(s)$ is called the density or probability density function of s. $p(s)$ is the probability that the value of s lies in the infinitesimal interval between s and $s + ds$, and it satisfies

$$p(s) \geq 0 \qquad \text{and} \qquad \int_{-\infty}^{\infty} p(s)\, ds = 1 \tag{D2}$$

The function $dp(s)/ds$ is called the distribution function and is a monotonically increasing function of s; $\eta = \langle s \rangle$, the expectation of s, is also known as the mean value or first moment of s.

If $f(s)$ is an arbitrary function of s, then,

$$\langle f(s) \rangle = \int_{-\infty}^{\infty} f(s)p(s)\, ds \tag{D3}$$

Suppose $f(s) = (s - \eta)^j$, where j is an integer and $\eta = \langle s \rangle$ is the expected value of s. Then,

$$\langle (s - \eta)^j \rangle = \int_{-\infty}^{\infty} (s - \eta)^j p(s)\, ds \tag{D4}$$

is defined as the jth moment of s around its expected value η. The second moment ($j = 2$) defines the variance of s,

$$\sigma_s^2 = \langle (s - \eta)^2 \rangle \tag{D5}$$

σ_s is the standard deviation of s and is a measure of the dispersion of s about η.

An important specific form of the probability distribution is the normal distribution. The normal probability density function is given by

$$p(s) = \frac{1}{\sigma_s \sqrt{2\pi}} \exp\left[-\frac{(s - \eta)^2}{2\sigma_s^2} \right] \tag{D6}$$

where σ_s is the standard deviation defined in (D5) and η the mean value defined in (D1).

The expectation operator commutes with the addition operator. Thus, if s and q are two random variables,

$$\langle s + q \rangle = \langle s \rangle + \langle q \rangle \tag{D7}$$

as can be verified from the definition (D1). The expectation operator does *not*, in general, commute with the multiplication operator. For example, consider the covariance of the two variables s and q, defined by

$$\text{cov}(s, q) = \langle [s - \langle s \rangle][q - \langle q \rangle] \rangle = \langle sq \rangle - \langle s \rangle \langle q \rangle \tag{D8}$$

The variables s and q are said to be independent or uncorrelated if $\langle sq \rangle = \langle s \rangle \langle q \rangle$. In the uncorrelated case, the expectation and multiplication operators *do* commute. Note that

$$\text{cov}(s, s) = \langle s^2 \rangle - \langle s \rangle^2 = \sigma_s^2 \tag{D9}$$

is the variance of s.

The correlation coefficient ρ_{sq} can be defined as

$$\rho_{sq} = \rho_{qs} = \frac{\langle [s - \langle s \rangle][q - \langle q \rangle] \rangle}{\sigma_s \sigma_q} \tag{D10}$$

where σ_s and σ_q are defined by (D5). Applications of (D1) and Schwartz's inequality demonstrates that

$$\langle [s - \langle s \rangle][q - \langle q \rangle] \rangle^2 \leq \langle [s - \langle s \rangle]^2 \rangle \langle [q - \langle q \rangle]^2 \rangle \tag{D11}$$

and $-1 \leq \rho_{sq} \leq 1$. If s and q are independent, then $\rho_{sq} = 0$. If s is a linear function of q, $\rho_{sq} = \pm 1$.

Now consider the case when s and q are functions of $\mathbf{r}$, with $\mathbf{r} = (x, y, z)$ being a three-dimensional spatial coordinate. Suppose s and q are defined at the spatial locations $\mathbf{r}_i$, $1 \leq i \leq I$. Define $\underline{s}$ and $\underline{q}$ to be the column vectors of length I with elements $s(\mathbf{r}_i)$ and $q(\mathbf{r}_i)$, respectively. Now form the expectation value of the *outer product* of $\underline{s} - \langle \underline{s} \rangle$ and $\underline{q} - \langle \underline{q} \rangle$:

$$\underline{\underline{C}}_{sq} = \langle [\underline{s} - \langle \underline{s} \rangle][\underline{q} - \langle \underline{q} \rangle]^{\mathrm{T}} \rangle$$

where $\underline{\underline{C}}_{sq}$ is a real, square $I \times I$ *covariance matrix* and T indicates matrix transpose.

The double underlining for matrices and single underlining for column vectors is the usual convention adopted in this book. An arbitrary element of $\underline{\underline{C}}_{sq}$ is given by

$$C_{sq}(\mathbf{r}_i, \mathbf{r}_j) = \text{cov}[s(\mathbf{r}_i), q(\mathbf{r}_j)] = \langle [s(\mathbf{r}_i) - \langle s(\mathbf{r}_i) \rangle][q(\mathbf{r}_j) - \langle q(\mathbf{r}_j) \rangle] \rangle \tag{D12}$$

By analogy with (D10), it is possible to define a *correlation matrix* $\underline{\underline{\rho}}_{sq}$ with elements $\rho_{sq}(\mathbf{r}_i, \mathbf{r}_j)$ given by

$$\rho_{sq}(\mathbf{r}_i, \mathbf{r}_j) = \frac{\langle [s(\mathbf{r}_i) - \langle s(\mathbf{r}_i) \rangle][q(\mathbf{r}_j) - \langle q(\mathbf{r}_j) \rangle] \rangle}{\sigma_s(\mathbf{r}_i)\sigma_q(\mathbf{r}_j)} \tag{D13}$$

where

$$\sigma_s^2(\mathbf{r}_i) = \langle [s(\mathbf{r}_i) - \langle s(\mathbf{r}_i) \rangle]^2 \rangle \qquad \text{and} \qquad \sigma_q^2(\mathbf{r}_j) = \langle [q(\mathbf{r}_j) - \langle q(\mathbf{r}_j) \rangle]^2 \rangle$$

$\rho_{sq}(\mathbf{r}_i, \mathbf{r}_j)$ can be related to $C_{sq}(\mathbf{r}_i, \mathbf{r}_j)$ and $\underline{\underline{\rho}}_{sq}$ to $\underline{\underline{C}}_{sq}$ as follows:

$$\rho_{sq}(\mathbf{r}_i, \mathbf{r}_j) = C_{sq}(\mathbf{r}_i, \mathbf{r}_j)/\sigma_s(\mathbf{r}_i)\sigma_q(\mathbf{r}_j) \tag{D14}$$

and

$$\underline{\underline{C}}_{sq} = \underline{\underline{\sigma}}_s \underline{\underline{\rho}}_{sq}$$

where $\underline{\underline{\sigma}}_s$ is the diagonal matrix with elements $\sigma_s(\mathbf{r}_i)$ and $\underline{\underline{\sigma}}_q$ is the diagonal matrix with elements $\sigma_q(\mathbf{r}_i)$. From (D11), $-1 \leq \rho_{sq}(\mathbf{r}_i, \mathbf{r}_j) \leq 1$.

In the special case $q = s$, $\underline{\underline{C}}_{ss}$ is symmetric and known as an autocovariance matrix. $\underline{\underline{C}}_{ss}$ is also positive semidefinite (2.3.12). This can be seen as follows. Suppose $\underline{z}$ is an arbitrary column vector with elements z_i, $1 \leq i \leq I$ (not all $z_i = 0$). Then,

$$\begin{aligned}
\underline{z}^{\mathrm{T}} \underline{\underline{C}}_{ss} \underline{z} &= \sum_{i=1}^{I} \sum_{j=1}^{I} C_{ss}(\mathbf{r}_i, \mathbf{r}_j) z_i z_j \\
&= \sum_{i=1}^{I} \sum_{j=1}^{I} z_i z_j \langle [s(\mathbf{r}_i) - \langle s(\mathbf{r}_i) \rangle][s(\mathbf{r}_j) - \langle s(\mathbf{r}_j) \rangle] \rangle \\
&= \left\langle \left[\sum_{i=1}^{I} z_i \{ s(\mathbf{r}_i) - \langle s(\mathbf{r}_i) \rangle \} \right]^2 \right\rangle \geq 0
\end{aligned}$$

From (2.3.15), the eigenvalues of $\underline{\underline{C}}_{ss}$ must be real and nonnegative. A zero eigenvalue of $\underline{\underline{C}}_{ss}$ occurs when any row is a linear combination of other rows. This happens if two spatial locations $\mathbf{r}_i$ and $\mathbf{r}_j$ coincide. Thus, if all spatial locations $\mathbf{r}_i$ are distinct, then $\underline{\underline{C}}_{ss}$ is strictly positive definite and all its eigenvalues are strictly positive. Clearly, the same is true for the correlation matrix $\underline{\underline{\rho}}_{ss}$.

The trace (sum of diagonal elements) of $\underline{\underline{C}}_{ss}$ is equal to the sum of the expected variances at each location $\mathbf{r}_i$ and is also equal to the expected value of the *inner product* of $\underline{s} - \langle \underline{s} \rangle$ with itself:

$$\text{Trace}(\underline{\underline{C}}_{ss}) = \sum_{i=1}^{I} \langle [s(\mathbf{r}_i) - \langle s(\mathbf{r}_i) \rangle]^2 \rangle = \langle [\underline{s} - \langle \underline{s} \rangle]^{\mathrm{T}} [\underline{s} - \langle \underline{s} \rangle] \rangle \tag{D15}$$

A well-known theorem of linear algebra (Wilkinson 1965) states that the trace of a matrix is equal to the sum of its eigenvalues. Suppose the eigenvalues of $\underline{\underline{C}}_{ss}$ are denoted μ_j, $1 \le j \le I$. Then $\sum_{j=1}^{I} \mu_j$ is equal to the sum of the expected variances at each location $\mathbf{r}_i$. Consequently, μ_j, $1 \le j \le I$, can be thought of as the expected variance of the jth eigenvector of $\underline{\underline{C}}_{ss}$.

$\underline{\underline{\rho}}_{ss}$ is symmetric, and the diagonal elements of $\underline{\underline{\rho}}_{ss}$ are all equal to 1. Trace$(\underline{\underline{\rho}}_{ss}) = I$, and the sum of the eigenvalues of $\underline{\underline{\rho}}_{ss}$ is equal to I.

For the sake of simplicity, assume that $\langle s(\mathbf{r}_i)\rangle = \langle q(\mathbf{r}_i)\rangle = 0$ for $1 \le i \le I$; thus, $\text{cov}[s(\mathbf{r}_i), q(\mathbf{r}_j)] = \langle s(\mathbf{r}_i)q(\mathbf{r}_j)\rangle$. The statistical structure of s and q is said to be spatially *homogeneous* if $\langle s(\mathbf{r}_i)q(\mathbf{r}_j)\rangle$ depends only on the relative displacement $\tilde{\mathbf{r}} = \mathbf{r}_j - \mathbf{r}_i$, rather than on the absolute locations $\mathbf{r}_i$ and $\mathbf{r}_j$. In the homogeneous case,

$$C_{sq}(\mathbf{r}_i, \mathbf{r}_j) = \langle s(\mathbf{r}_i)q(\mathbf{r}_i + \tilde{\mathbf{r}})\rangle = C_{sq}(\tilde{\mathbf{r}}) \tag{D16}$$

and the covariance in (D16) is a function only of $\tilde{\mathbf{r}}$ and not of $\mathbf{r}_i$ or $\mathbf{r}_j$. In particular, the diagonal elements of $\underline{\underline{C}}_{sq}$ are all equal in the homogeneous case:

$$C_{sq}(\mathbf{r}_i, \mathbf{r}_i) = \langle s(\mathbf{r}_i)q(\mathbf{r}_i)\rangle = C_{sq}(0) \tag{D17}$$

is independent of position. Then, using (D16–17) gives

$$\begin{aligned} C_{sq}(\tilde{\mathbf{r}}) &= \langle s(\mathbf{r}_i)q(\mathbf{r}_i + \tilde{\mathbf{r}})\rangle = \langle s(\mathbf{r}_i - \tilde{\mathbf{r}})q(\mathbf{r}_i)\rangle \\ &= \langle q(\mathbf{r}_i)s(\mathbf{r}_i - \tilde{\mathbf{r}})\rangle = C_{qs}(-\tilde{\mathbf{r}}) \end{aligned} \tag{D18}$$

Now consider $\underline{\underline{C}}_{ss}$ under conditions of homogeneity, and continue to assume that $\langle s(\mathbf{r}_i)\rangle = 0$. From (D18), it is evident that

$$C_{ss}(\tilde{\mathbf{r}}) = C_{ss}(-\tilde{\mathbf{r}}) \tag{D19}$$

Moreover, because $C_{ss}(0) = \langle s^2(\mathbf{r}_i)\rangle = \langle s^2\rangle$ is independent of location,

$$C_{ss}(\tilde{\mathbf{r}}) = \langle s^2\rangle \rho_{ss}(\tilde{\mathbf{r}}) = C_{ss}(0)\rho_{ss}(\tilde{\mathbf{r}})$$

and

$$C_{ss}(0) = \langle s^2\rangle \ge \langle s^2\rangle |\rho_{ss}(\tilde{\mathbf{r}})| = |C_{ss}(\tilde{\mathbf{r}})| \tag{D20}$$

The temporal analogue of spatial homogeneity is called stationarity. Define s and q at times t_1 and t_2, respectively, and define $\tau = t_2 - t_1$. Under stationary conditions,

$$C_{sq}(t_1, t_2) = \langle s(t_1)q(t_2)\rangle = \langle s(t_1)q(t_1 + \tau)\rangle = C_{sq}(\tau) \tag{D21}$$

is only a function of the time difference τ and not of the absolute times t_1 and t_2.

Now suppose that $s(\mathbf{r}, t)$ is the true value of s (the signal) and define an estimate $s_e(\mathbf{r}, t)$ that is in error:

$$s_e(\mathbf{r}, t) = s(\mathbf{r}, t) + \varepsilon(\mathbf{r}, t) \tag{D22}$$

where $\varepsilon(\mathbf{r}, t)$ is the error. The estimate $s_e(\mathbf{r}, t)$ of $s(\mathbf{r}, t)$ could be obtained by measurement (observation), by a forecast using a numerical model (background), or by an objective analysis algorithm. The expectation operator (D1) can be applied to the error $\varepsilon(\mathbf{r}, t)$ as well:

$$\langle\varepsilon\rangle = \int_{-\infty}^{\infty} \varepsilon p(\varepsilon)\, d\varepsilon \qquad \text{and} \qquad \langle\varepsilon^2\rangle = \int_{-\infty}^{\infty} \varepsilon^2 p(\varepsilon)\, d\varepsilon \tag{D23}$$

where $p(\varepsilon)$ is the probability that the error lies between ε and $\varepsilon + d\varepsilon$. A number of simplifying assumptions can be made about the error:

If $\langle\varepsilon(\mathbf{r}, t)\rangle = 0$, there is no systematic error or bias. (D24)

If $\langle\varepsilon(\mathbf{r}_i, t)\varepsilon(\mathbf{r}_j, t)\rangle = 0$ for $\mathbf{r}_i \neq \mathbf{r}_j$, the error is not spatially correlated. (D25)

If $\langle\varepsilon(\mathbf{r}, t_1)\varepsilon(\mathbf{r}, t_2)\rangle = 0$ for $t_1 \neq t_2$, the error is not temporally correlated. (D26)

If $\langle s(\mathbf{r}, t)\varepsilon(\mathbf{r}, t)\rangle = 0$, the error is not correlated with the signal. (D27)

If the error is unbiased (D24), then $\langle s_e(\mathbf{r}, t)\rangle = \langle s(\mathbf{r}, t)\rangle$. (D28)

That is, the expected value of the estimate over many realizations is equal to the expected value of the signal. $s_e(\mathbf{r}, t)$ is then called an unbiased estimate of $s(\mathbf{r}, t)$.

A measure of the importance of the error is the signal-to-noise ratio, which is defined as

$$\frac{\langle[s(\mathbf{r}, t) - \langle s(\mathbf{r}, t)\rangle]^2\rangle}{\langle\varepsilon^2(\mathbf{r}, t)\rangle} \tag{D29}$$

Appendix E
Classical interpolation

Interpolation is used in the data assimilation cycle primarily to determine values of the background field at observation stations. This is known as the forward problem and is discussed in Sections 1.6, 3.1, 4.2, and 5.6. Usually, we know the background field at the gridpoints of a regular mesh, and we use interpolation formulas of various types to obtain estimates of the background field at the observation stations. The interpolation algorithms are usually two- or three-dimensional formulations, but we examine only the one-dimensional case here. The subject is a classical numerical analysis topic and is discussed in many books on numerical analysis, such as Hildebrand (1956).

Suppose the function $y=f(x)$ is known exactly at K regularly or irregularly spaced values of the independent variable in the domain $x_a \leq x \leq x_b$. Thus, the $f(x_1), f(x_2), \ldots, f(x_K)$ are assumed known. The process of determining an approximate (or sometimes exact) value of the function at an arbitrary point x is known as interpolation. For example, function tables tabulate functional values at constant increments in the independent variables. To determine values of the function at intermediate values of the independent variable requires interpolation.

Perhaps the best-known interpolation formula is that of Lagrange:

$$
\begin{aligned}
f_{\mathrm{A}}(x) = & \frac{(x-x_2)(x-x_3)\cdots(x-x_K)}{(x_1-x_2)(x_1-x_3)\cdots(x_1-x_K)} f(x_1) \\
& + \cdots \\
& + \frac{(x-x_1)(x-x_2)\cdots(x-x_{K-1})}{(x_K-x_1)(x_K-x_2)\cdots(x_K-x_{K-1})} f(x_K)
\end{aligned}
\tag{E1}
$$

where $f_{\mathrm{A}}(x)$ is an approximation of the true function $f(x)$. This formula requires that $f(x)$ be known exactly at K regularly or irregularly spaced values of x. $f_{\mathrm{A}}(x)=f(x)$ at the points $x_1, x_2, \ldots, x_K$.

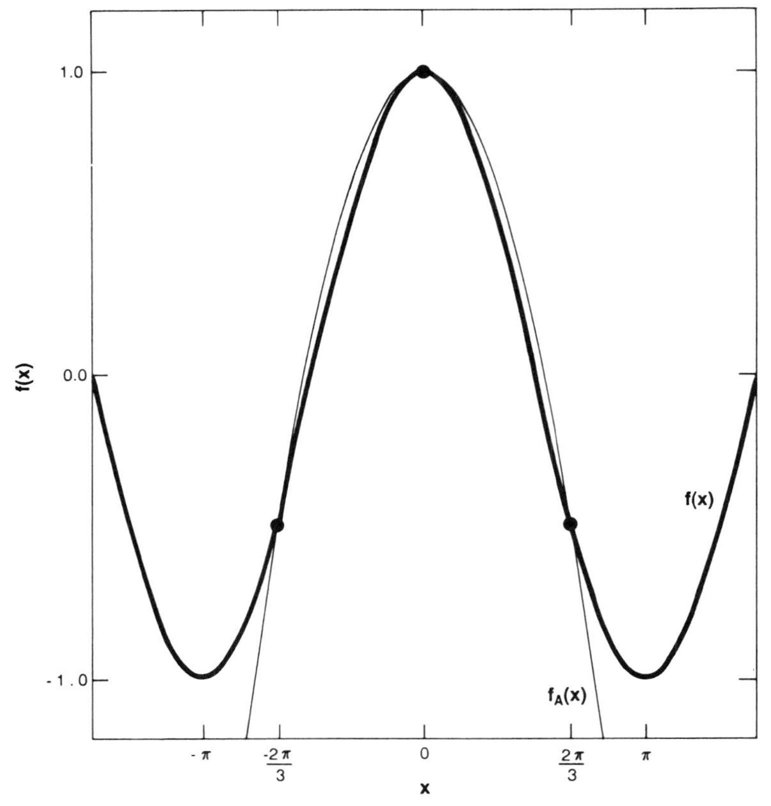

Figure E1 Plot of $f(x) = \cos(x)$ (heavy line) and the Lagrange interpolated approximation $f_A(x)$ (light line) as a function of x. The known values of $f(x)$ at $x = 0$, $\pm 2\pi/3$ are indicated by heavy dots.

If $f(x)$ is the polynomial

$$f(x) = a_0 + a_1 x + a_2 x^2 + \cdots + a_N x^N$$

then Lagrange's interpolation formula is exact for any value of x provided that $N + 1 \leq K$. If $f(x)$ is any other function, Lagrange's interpolation formula is not exact but only approximate. Suppose $f(x) = \cos(x)$ and $f(x)$ is assumed known at the points $x_1 = -2\pi/3$, $x_2 = 0$, and $x_3 = 2\pi/3$. Then the application of Lagrange's interpolation formula yields

$$f_A(x) = -\frac{27x^2}{8\pi^2} + 1 \tag{E2}$$

The functions $f(x)$ and $f_A(x)$ are plotted in Figure E1. It is apparent that for $-2\pi/3 \leq x \leq 2\pi/3$, $f_A(x)$ is a good approximation to $f(x)$; but outside this range, $f_A(x)$ is an increasingly poor approximation. For $x \leq -2\pi/3$ and $x \geq 2\pi/3$, Lagrange's formula is actually doing extrapolation, which is generally much less accurate than interpolation.

Another commonly used interpolation procedure is spline interpolation (Ahlberg, Nilson, and Walsh 1967). As above, assume $f(x_k)$, $1 \le k \le K$, is known. Find the function $f_A(x)$ that is continuous, has continuous first and second derivatives, and passes through the points $f(x_k)$. Let $S_k = f''_A(x_k)$ be the unknown values of the second derivative evaluated at the points x_k. Then, $f''_A(x)$ is linear between x_{k-1} and x_k:

$$f''_A(x) = S_{k-1}\left(\frac{x_k - x}{\Delta_k}\right) + S_k\left(\frac{x - x_{k-1}}{\Delta_k}\right) \tag{E3}$$

where $\Delta_k = x_k - x_{k-1}$.

Integrate (E3) twice and evaluate the constants of integration using the known values $f(x_k)$. Thus,

$$\begin{aligned} f_A(x) = S_{k-1}\frac{(x_k - x)^3}{6\Delta_k} + S_k\frac{(x - x_{k-1})^3}{6\Delta_k} + \left[f(x_{k-1}) - \frac{S_{k-1}\Delta_k^2}{6}\right]\left[\frac{x_k - x}{\Delta_k}\right] \\ + \left[f(x_k) - \frac{S_k\Delta_k^2}{6}\right]\left[\frac{x - x_{k-1}}{\Delta_k}\right] \end{aligned} \tag{E4}$$

$$f'_A(x) = -S_{k-1}\frac{(x_k - x)^2}{2\Delta_k} + \frac{S_k(x - x_{k-1})}{2\Delta_k} + \frac{f(x_k) - f(x_{k-1})}{\Delta_k} - \frac{(S_k - S_{k-1})\Delta_k}{6} \tag{E5}$$

Equation (E5) applies in the interval $x_{k-1} \le x \le x_k$. A similar equation applies in the interval $x_k \le x \le x_{k+1}$:

$$f'_A = -S_k\frac{(x_{k+1} - x)^2}{2\Delta_{k+1}} + S_{k+1}\frac{(x - x_k)^2}{2\Delta_{k+1}} + \frac{f(x_{k+1}) - f(x_k)}{\Delta_{k+1}} - \frac{(S_{k+1} - S_k)\Delta_{k+1}}{6} \tag{E6}$$

Because $f'_A(x)$ must be continuous, the limit of $f'_A(x)$ when $x \to x_k$ should be the same when approached from above or below. Setting $x = x_k$ in (E5) and (E6) and equating the two gives

$$\frac{\Delta_k}{6}S_{k-1} + \frac{(\Delta_k + \Delta_{k+1})}{3}S_k + \frac{\Delta_{k+1}}{6}S_{k+1} = \frac{f(x_{k+1}) - f(x_k)}{\Delta_{k+1}} - \frac{f(x_k) - f(x_{k-1})}{\Delta_k} \tag{E7}$$

Consider the periodic case, $S_{K+1} = S_1$, $f(x_{K+1}) = f(x_1)$, $\Delta_{K+1} = \Delta_1$. Then (E7) can be written in matrix form as

$$\underline{\underline{T}}\,\underline{S} = \underline{\underline{Q}}^{T}\underline{f} \tag{E8}$$

where $\underline{S}$ is the column vector of length K of unknown $S_k = f''_A(x_k)$ and $\underline{f}$ is the column vector of length K of known functional values of $f(x_k)$. $\underline{\underline{T}}$ is a symmetric positive definite, (almost) tri-diagonal $K \times K$ matrix with elements $t_{k,k} = (\Delta_k + \Delta_{k+1})/3$, $t_{k,k+1} = \Delta_{k+1}/6$, and $t_{k,k-1} = \Delta_k/6$. $\underline{\underline{Q}}$ is a $K \times K$ tri-diagonal diagonal matrix with elements $q_{k-1,k} = 1/\Delta_k$, $q_{k,k} = -1/\Delta_k - 1/\Delta_{k+1}$, $q_{k+1,k} = 1/\Delta_{k+1}$. In accord with usual matrix notation, the first index indicates the row and the second the column. $\underline{\underline{T}}$ is a simple sparse matrix and can be inverted simply and efficiently. Substitution of the S_k into (E4) yields the approximation to $f(x)$.

Spline functions have a minimum curvature property. It was shown by Holladay (see Ahlberg et al. 1967) that of all the functions that have continuous second derivatives in the domain and that exactly fit $f(x_k)$, $1 \leq k \leq K$, the spline function (E8) minimizes the integral

$$\int_{x_a}^{x_b} \{f''(x)\}^2 \, dx \tag{E9}$$

There are many types of spline functions; (E8) defines an interpolating cubic spline because (E4) is a cubic polynomial in the interval $x_{k-1} \leq x \leq x_k$.

Appendix F
An iterative approach to statistical interpolation

Bratseth (1986) has developed a version of the SCM algorithm that converges in the limit to the same a posteriori analysis weights as the statistical interpolation algorithm of Chapter 4. In the univariate case, the statistical interpolation algorithm (4.5.1) can be written in the form

$$f_A(\mathbf{r}_i) = f_B(\mathbf{r}_i) + \underline{B}_i^T \underline{M}^{-1}[(\underline{B} + \underline{O})\underline{M}^{-1}]^{-1}[\underline{f}_O - \underline{f}_B] \tag{F1}$$

where $f_A(\mathbf{r}_i)$ and $f_B(\mathbf{r}_i)$ are the analysis and background values at the analysis gridpoint; $\underline{f}_O$ and $\underline{f}_B$ are column vectors of length K of observation and background estimates at the observation stations; and $\underline{B}$, $\underline{B}_i^T$ and $\underline{O}$ are background and observation error covariance matrices as in (4.5.1). The matrix $\underline{M}$ is a diagonal matrix with elements $M_k = \sum_{l=1}^{K} |b_{kl} + o_{kl}|$, where b_{kl} and o_{kl} are elements of $\underline{B}$ and $\underline{O}$, respectively.

Introduce the following iteration cycle:

$$\underline{d}_{j+1} = [\underline{I} - (\underline{B} + \underline{O})\underline{M}^{-1}]\underline{d}_j + [\underline{f}_O - \underline{f}_B] \tag{F2}$$

where $\underline{d}_o = \underline{f}_O - \underline{f}_B$. Then,

$$\underline{d}_1 = [\underline{I} - (\underline{B} + \underline{O})\underline{M}^{-1}][\underline{f}_O - \underline{f}_B] + [\underline{f}_O - \underline{f}_B]$$

$$\underline{d}_j = \sum_{q=0}^{j} [\underline{I} - (\underline{B} + \underline{O})\underline{M}^{-1}]^q[\underline{f}_O - \underline{f}_B] \tag{F3}$$

Equation (F3) defines a geometric series in the matrix $[\underline{I} - (\underline{B} + \underline{O})\underline{M}^{-1}]$. In the limit as $j \to \infty$, this series converges to $[(\underline{B} + \underline{O})\underline{M}^{-1}]^{-1}$, if the spectral radius (Sections 12.3 and 12.5) of the matrix $[\underline{I} - (\underline{B} + \underline{O})\underline{M}^{-1}] < 1$. Following the arguments of Section 3.5, this occurs if all the eigenvalues of $(\underline{B} + \underline{O})\underline{M}^{-1}$ are <2 and ≥ 0. The condition (3.5.10) is automatically satisfied in this case because $\underline{B}$ and $\underline{O}$ are both covariance matrices and are thus positive definite. Moreover, the elements of $\underline{M}$ are all positive by definition. Consequently, the eigenvalues of $(\underline{B} + \underline{O})\mathrm{M}^{-1}$ all lie between

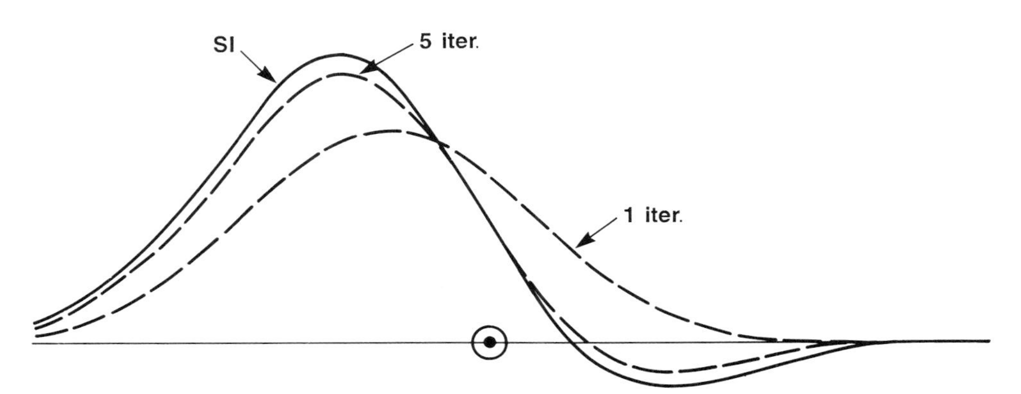

Figure F1 A simple example of the convergence of Bratseth's SCM algorithm to the statistical interpolation analysis. (After Bratseth 1986)

0 and 1 and the sequence (F2) is guaranteed to converge. The limit of the iteration cycle as $j \to \infty$ is

$$\underline{\mathrm{d}}_{\infty} = \lim_{j \to \infty} \underline{\mathrm{d}}_j = [(\underline{\underline{\mathrm{B}}} + \underline{\underline{\mathrm{O}}})\underline{\underline{\mathrm{M}}}^{-1}]^{-1}[\underline{\mathrm{f}}_{\mathrm{O}} - \underline{\mathrm{f}}_{\mathrm{B}}] \tag{F4}$$

Inserting (F4) into (F1) gives

$$f_{\mathrm{A}}(\mathbf{r}_i) - f_{\mathrm{B}}(\mathbf{r}_i) = \underline{\mathrm{B}}_i^{\mathrm{T}} \underline{\underline{\mathrm{M}}}^{-1} \underline{\mathrm{d}}_{\infty} = \underline{\mathrm{B}}_i^{\mathrm{T}} [\underline{\underline{\mathrm{B}}} + \underline{\underline{\mathrm{O}}}]^{-1} [\underline{\mathrm{f}}_{\mathrm{O}} - \underline{\mathrm{f}}_{\mathrm{B}}] \tag{F5}$$

which is the same solution as (4.5.1).

In effect, the iteration cycle (F2) is an indirect method of inverting the matrix $\underline{\underline{\mathrm{B}}} + \underline{\underline{\mathrm{O}}}$. Note that the iteration is all on the observation increments and that no forward interpolation is required, except initially to obtain $\underline{\mathrm{f}}_{\mathrm{B}}$. Like the SCM algorithm (3.7.4), the iteration (F4) does not converge to the observations at the observation stations; the asymptotic value is the minimum variance estimate of Chapter 4 and is thus an appropriately weighted linear combination of observations and background estimates.

Convergence occurs most rapidly (as in Section 3.5) for the eigenvectors of $(\underline{\underline{\mathrm{B}}} + \underline{\underline{\mathrm{O}}})\underline{\underline{\mathrm{M}}}^{-1}$ whose eigenvalues are just less than 1 and occurs most slowly for eigenvectors with eigenvalues just greater than 0. At least for the case of uncorrelated observation error ($\underline{\underline{\mathrm{O}}} = \varepsilon_{\mathrm{O}}^2 \underline{\underline{\mathrm{I}}}$), the largest/smallest eigenvalues of $(\underline{\underline{\mathrm{B}}} + \underline{\underline{\mathrm{O}}})$ occur for the eigenvectors of $\underline{\underline{\mathrm{B}}}$ with the largest/smallest spatial scales (Section 4.5). Thus, the largest spatial scales of the analysis increment converge in the early iterations of (F2), and the smallest scales converge in the later iterations. Incompletely iterated analysis increments will be smoother than completely iterated analysis increments. Bratseth (1986) and Franke (1988) have demonstrated the effectiveness of this SCM procedure.

Figure F1 shows a simple one-dimensional example of the univariate algorithm. The background field is assumed to be given by the constant straight line, and the two observations are indicated by the dotted circles. The background error is assumed to be spatially correlated whereas the observation error is not. The objective analysis obtained by statistical interpolation is indicated by the solid curve marked SI. The analyses obtained by terminating the iteration cycle after 1 and 5 iterations are shown by the dashed lines. The approach to the statistical interpolation analysis is clearly indicated, and the relative smoothness of the analysis increments in the earlier iterations is evident. The limiting analysis of the traditional SCM algorithm (3.2.4) would pass through the observations.

Appendix G
Bessel functions and Hankel transforms

Bessel functions and/or Hankel transforms are useful mathematical tools, which we employed for various purposes in Chapters 3, 6, and 10. Because many of the required identities, recursion formulas, series expansions, asymptotic limits, orthogonality properties, and so on, are not always instantly recognizable, the relevant ones have been gathered here for convenience. Most of these equations are discussed in more detail in the classical treatise on Bessel functions by Watson (1958), in mathematical tables such as those found in Abramovitz and Stegun (1965), or in standard mathematical physics textbooks such as those by Morse and Feshbach (1953).

The developments here are entirely concerned with Bessel functions of integer order (positive or negative) of real arguments. Bessel's equation,

$$\frac{1}{r}\frac{d}{dr}r\frac{dF}{dr}+\left(1-\frac{n^2}{r^2}\right)F=0, \qquad n \text{ integer} \tag{G1}$$

has solutions of the form

$$J_n(r)=\sum_{m=0}^{\infty}\frac{(-1)^m\left(\frac{r}{2}\right)^{n+2m}}{m!(n+m)!} \tag{G2}$$

where (!) indicates factorial. $J_n(r)$ is the Bessel function of the first kind of integer order n. The functions $J_0(r)$ and $J_1(r)$ are plotted in Figure G1. For Bessel functions of integer order,

$$J_{-n}(r)=(-1)^n J_n(r) \tag{G3}$$

Some useful identities concerning the $J_n(r)$ are

$$e^{ir\sin\phi}=\sum_{n=-\infty}^{\infty}e^{in\phi}J_n(r) \tag{G4}$$

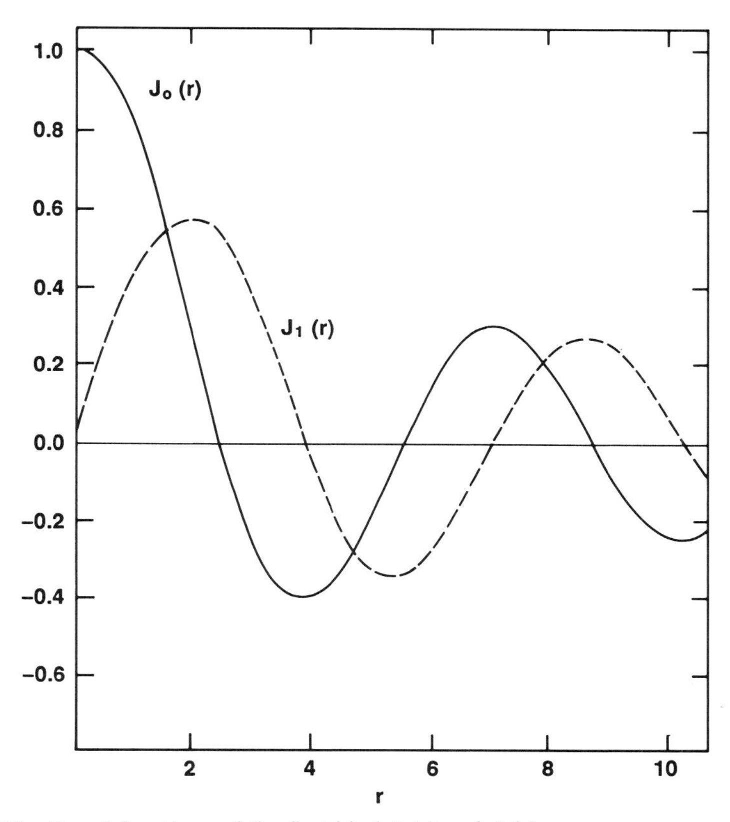

Figure G1 The Bessel functions of the first kind $J_0(r)$ and $J_1(r)$.

and

$$J_0(r) = \frac{2}{\pi} \int_0^\infty \sin[r \cosh(t)] \, dt \tag{G5}$$

The Bessel functions obey the recursion formulas

$$J_{n-1}(r) + J_{n+1}(r) = \frac{2n}{r} J_n(r) \tag{G6}$$

and

$$J_{n-1}(r) - J_{n+1}(r) = 2 \frac{dJ_n(r)}{dr} \tag{G7}$$

from which it can be shown that

$$\frac{1}{r} \frac{d}{dr} r J_1(r) = J_0(r) \qquad \text{and} \qquad \frac{d}{dr} J_0(r) = -J_1(r) \tag{G8}$$

Two useful asymptotic forms of the Bessel functions $J_n(r)$ are

$$\lim_{r\to 0} J_n(r) = \frac{r^n}{2^n n!} \tag{G9}$$

and

$$\lim_{r\to\infty} J_n(r) = \sqrt{\frac{2}{\pi r}} \cos\left[r - \frac{\pi}{2}(n + 0.5)\right] \tag{G10}$$

The Bessel functions satisfy orthogonality conditions on both finite and infinite domains. Suppose k_n^l is the lth root of $J_n(r) = 0$. Then, over the domain $0 \le r \le r_0$,

$$\int_0^{r_0} J_n\left(\frac{k_n^l r}{r_0}\right) J_n\left(\frac{k_n^j r}{r_0}\right) r\, dr = \delta_{lj} N_n^l \tag{G11}$$

where $N_n^l = -0.5 r_0^2 J_{n+1}(k_n^l) J_{n-1}(k_n^l)$ and δ_{lj} is the Kronecker delta function.

In the limit $r_0 \to \infty$, (G11) becomes

$$\sqrt{kk'} \int_0^\infty J_n(kr) J_n(k'r) r\, dr = \delta(k - k') \tag{G12}$$

where $\delta(k - k')$ is the Dirac delta function, which equals zero for $k' \ne k$ and is infinite for $k' = k$. $\delta(k - k')$ has the property

$$\int_0^\infty \delta(k - k') F(k')\, dk' = F(k) \tag{G13}$$

In plane polar coordinates, over the domain $0 \le r \le r_0$, $-\pi \le \phi \le \pi$, an arbitrary function $F(r, \phi)$ that satisfies $F(r_0, \phi) = 0$ can be expanded in a Fourier–Bessel expansion which is analogous to a double Fourier expansion in cartesian coordinates:

$$F(r, \phi) = \sum_{n=-\infty}^{\infty} \sum_{l=1}^{\infty} \hat{F}(k_n^l, n) e^{in\phi} J_n\left(\frac{k_n^l r}{r_0}\right)$$

$$\hat{F}(k_n^l, n) = \frac{1}{2\pi N_n^l} \int_{-\pi}^{\pi} \int_0^{r_0} F(r, \phi) e^{-in\phi} J_n\left(\frac{k_n^l r}{r_0}\right) r\, dr\, d\phi \tag{G14}$$

where k_n^l and N_n^l are defined in (G11). In the limit $r_0 \to \infty$, (G14) becomes

$$F(r, \phi) = \sum_{n=-\infty}^{\infty} \int_0^\infty \hat{F}(k, n) e^{in\phi} J_n(kr) k\, dk$$

$$\hat{F}(k, n) = \frac{1}{2\pi} \int_{-\pi}^{\pi} \int_0^\infty F(r, \phi) e^{-in\phi} J_n(kr) r\, dr\, d\phi \tag{G15}$$

The integrals over r and k in (G15) are known as Hankel transforms. Note that from Bessel's equation (G1), functions of the form $\exp(in\phi) J_n(kr)$ are eigenfunctions of the

Laplacian operator in plane polar coordinates. Thus,

$$\left[\frac{1}{r}\frac{\partial}{\partial r}r\frac{\partial}{\partial r}+\frac{1}{r^2}\frac{\partial^2}{\partial\phi^2}\right]e^{in\phi}J_n(kr)=-k^2e^{in\phi}J_n(kr) \tag{G16}$$

In cartesian coordinates over the infinite domain $-\infty\le x\le\infty$, $-\infty\le y\le\infty$, the equivalent to (G15) is given by the double Fourier integral:

$$\hat{F}(\mu,\nu)=\frac{1}{2\pi}\int_{-\infty}^{\infty}\int_{-\infty}^{\infty}F(x,y)e^{-i(\mu x+\nu y)}\,dx\,dy \tag{G17}$$

The relation between $\hat{F}(\mu,\nu)$ and $\hat{F}(k,n)$ can be obtained as follows. Define $x=r\cos\phi$, $y=r\sin\phi$, $x^2+y^2=r^2$, $\mu=-k\sin\lambda$, $\nu=+k\cos\lambda$, $k^2=\mu^2+\nu^2$. Then,

$$\hat{F}(\mu,\nu)=\frac{1}{2\pi}\int_{-\pi}^{\pi}\int_{0}^{\infty}F(r,\phi)e^{ikr(\cos\phi\sin\lambda-\sin\phi\cos\lambda)}r\,dr\,d\phi \tag{G18}$$

noting that the Jacobian of the transformation $J\left(\frac{x,y}{r,\phi}\right)=r$. But from $\cos\phi\sin\lambda-\sin\phi\cos\lambda=\sin(\lambda-\phi)$ and using (G4), we obtain

$$\hat{F}(\mu,\nu)=\sum_{n=-\infty}^{\infty}e^{in\lambda}\hat{F}(k,n)\qquad\text{where }\lambda=-\tan^{-1}\left[\frac{\mu}{\nu}\right] \tag{G19}$$

and $\hat{F}(k,n)$ is given by (G15). Inverting (G19) yields

$$\hat{F}(k,n)=\frac{1}{2\pi}\int_{-\pi}^{\pi}\hat{F}(\mu,\nu)e^{-in\lambda}\,d\lambda,\qquad k^2=\mu^2+\nu^2 \tag{G20}$$

If F is independent of ϕ, then,

$$\hat{F}(k,n)=0\quad\text{for}\quad n\neq 0\qquad\text{and}\qquad \hat{F}(k,0)=\int_0^{\infty}F(r)J_0(kr)r\,dr \tag{G21}$$

Appendix H
Aliasing, network design, and Observation System Simulation Experiments (OSSE)

In Section 3.4, we found that on a discrete observation network, small-scale waves were often misrepresented as much larger-scale waves. This phenomenon is called aliasing and is investigated further here.

In functional expansions such as (2.4.2), the coefficients are ordered by increasing wavenumber m. The wavenumber is inversely proportional to the scale or wavelength of the function. In the examples of Section 2.4, the signal was large scale (low wavenumber), and a reasonable approximation to the signal could be extracted, provided that the observational noise was not too large and that underfitting or overfitting was not excessive.

What happens when the signal is very small scale (high wavenumber)? We examine this case using a one-dimensional expansion in trigonometric basis functions on a finite domain. For the sake of simplicity, the real form of the expansion (2.4.1) is used. Suppose there are K observation stations x_k, $1 \leq k \leq K$, in the domain $-\pi \leq x \leq \pi$. Then, the maximum number of coefficients a_m, b_m in (2.4.1) that can be extracted in this case is $M < K/2$. For the sake of illustration, assume that K is even. Consider now a signal

$$f_{\mathrm{T}}(x) = a_m \cos(mx), \qquad \frac{K}{2} < m < K \tag{H1}$$

where a_m is the true amplitude of the signal. Assume that the observations are error free, and define $\tilde{m} = K - m$. Then,

$$f_{\mathrm{O}}(x_k) = a_m \cos(mx_k) = a_m \cos(K - \tilde{m})x_k, \qquad 1 \leq k \leq K \tag{H2}$$

Consider first the equally spaced observation network of (2.5.3). Then (H2) becomes

$$f_{\mathrm{O}}(x_k) = a_m \cos\left[(K - \tilde{m})\left(-\pi + \frac{2\pi k}{K}\right)\right]$$

$$= a_m \cos(-K\pi + 2\pi k) \cos[\tilde{m}(-\pi + k\,\Delta x)]$$
$$+ a_m \sin(-K\pi + 2\pi k) \sin[\tilde{m}(-\pi + k\,\Delta x)] \qquad \text{(H3)}$$

where $\Delta x = 2\pi/K$.

Because K is even, $\cos(-K\pi + 2\pi k) = 1$ and $\sin(-K\pi + 2\pi k) = 0$. So,

$$f_O(x_k) = a_m \cos(\tilde{m}x_k) = a_m \cos(K - m)x_k \qquad \text{(H4)}$$

Equation (H4) implies that at the observation point x_k, the function $a_m \cos(mx_k)$ is indistinguishable from $a_m \cos(K - m)x_k$. Note that $\tilde{m} = K - m$ represents a much larger-scale wave than m because $K - m < K/2$.

This case is illustrated in Figure H1(a), which has the same format as Figure 2.2 and plots the independent variable x against the dependent variable $f(x)$. There are eight equally spaced observation points x. The signal (heavy curve) is given by (H1) with $a_m = 1$ and $m = 6$. The objective analysis (light curve) was obtained by the function-fitting algorithm (2.4.4) followed by (2.4.2), with all a priori weights $w_k = 1$ and $M = 2$ (five degrees of freedom). As expected, the objective analysis is a perfect cosine with wavenumber $\tilde{m} = K - m = 2$.

The phenomenon illustrated by Figure H1(a) is called aliasing. In effect, the objective analysis process represents a small-scale signal incorrectly as a larger-scale wave. With $K = 8$ observation stations, only the first eight degrees of freedom in (2.4.1) can be extracted, and any signal of higher wavenumber is misrepresented. If K is odd, then all cosine and sine coefficients with $m \le (K - 1)/2$ can be resolved. If K is even, then all cosine coefficients with $m \le K/2$ and all sine coefficients with $m \le K/2 - 1$ can be resolved. The wavenumber $m = K/2$ is the Nyquist limit: Any higher wavenumber (with the slight differences between the even and odd cases just noted) is misrepresented or aliased. In other words, if Δx is the distance between equally spaced observation stations, the $2\,\Delta x$ wave is the shortest wavelength that can be resolved by the network.

Usually, the network is *not* regular. The case of an irregular network is illustrated in Figure H1(b). The signal (heavy curve) is identical to that in (a). There are again eight observation points and no observation error, but the distance between observations is variable. The function-fitting algorithm (2.4.4) followed by (2.4.2), with $M = 2$ and all $w_k = 1$, was applied to this irregular observation network. The resulting over-determined fit is shown (light curve); it is apparent that the objective analysis contains both sine and cosine components and has nonzero coefficients for wavenumbers 0, 1, and 2. The result is just as unsatisfactory as in (a).

For an irregular network, the Nyquist limit cannot be defined as precisely as on a regular network. Nonetheless, it is often useful to estimate the smallest-scale wave that can effectively be resolved by the network. If some measure of the network spacing L_N is defined, the network is said to be capable of resolving scales larger than $2L_N$ (see Section 3.6). Using a two-dimensional domain as an example, define the area of the domain as A and the number of observation stations as N. One estimate of L_N is simply $L_N = \sqrt{A/N}$. A second estimate of L_N is given by the average

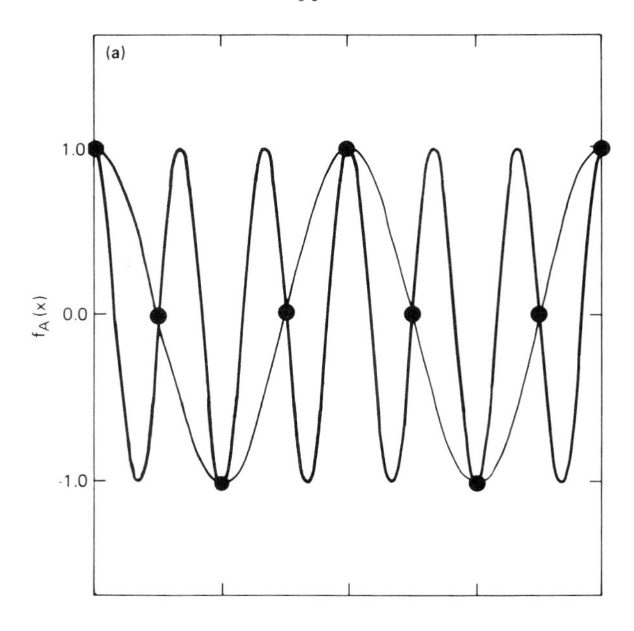

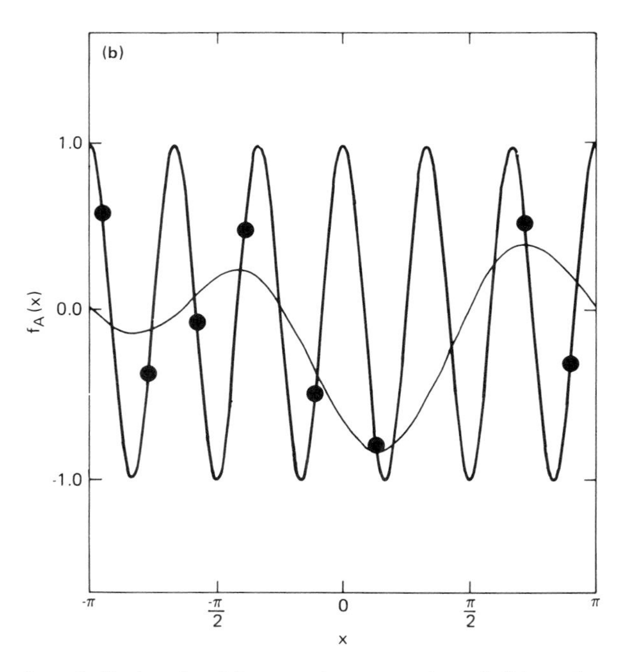

Figure H1 Examples of aliasing for (a) a regular network and (b) an irregular network. ● denotes an observation point; heavy curve, the signal; light curve, the objective analysis.

value of the distance between each observation station and its nearest neighbor. On an observation network with uniform observation density, both estimates give similar values for L_N; but on a network with highly variable data density, the second estimate of L_N is smaller than the first. This question is also discussed from the point of view of fractal theory by Lovejoy, Schertzer, and Ladoy (1986).

Aliasing can cause difficulties even when there is no observation error. Under what conditions is aliasing likely to cause serious deterioration in the quality of the objective analysis? Consider a signal

$$f(x) = \sum_{m=-\infty}^{\infty} c_m e^{imx} \tag{H5}$$

Now square $f(x)$, integrate over the domain $-\pi \leq x \leq \pi$, and apply the expectation operator; we obtain

$$\left\langle \frac{1}{2\pi} \int_{-\pi}^{\pi} f^2(x)\, dx \right\rangle = \sum_{m=-\infty}^{\infty} \langle c_m c_m^* \rangle = \sum_{m=-\infty}^{\infty} \langle |c_m|^2 \rangle \tag{H6}$$

where the identity (2.4.6) has been used. Equation (H6) defines the expected variance of the signal as a function of wavenumber and is the counterpart (for a finite domain) of the spectral density discussed in Section 4.3. A similar expression could be derived for the background error spectrum on a finite domain.

Consider two wavenumbers p, q with $q \gg p$. Then, the expected variances of wavenumbers p and q are $\langle |c_p|^2 \rangle$ and $\langle |c_q|^2 \rangle$, respectively. As noted in Sections 4.3 and 4.4, all realistic spectra must eventually asymptote to zero for very high wavenumber to have finite variance. However, over the range of interest, several types of spectra are possible. If $\langle |c_q|^2 \rangle = \langle |c_q|^2 \rangle$ for all p, q, the spectrum is said to be "white." That is, the expectation value of the variance is the same for all wavelengths. If $\langle |c_p|^2 \rangle \gg \langle |c_q|^2 \rangle$, the spectrum is said to be "red." That is, the expected variance is greater for the larger scales.

It is evident that for a red spectrum $\langle |c_p|^2 \rangle \gg \langle |c_q|^2 \rangle$, aliasing is not a serious problem. Even if wavenumber q is misrepresented in terms of much larger-scale waves, its relative amplitude is small and the aliasing errors also are small. For a white spectrum, or worse still for $\langle |c_q|^2 \rangle > \langle |c_p|^2 \rangle$, aliasing errors could be disastrous.

Fortunately, most spectra of interest are somewhat red. Figure 1.4 shows an atmospheric kinetic energy spectrum, which could be regarded as the spectrum of the signal. Adding the two components S and T gives the total signal (stationary and transient), which is clearly red with rapid fall off with increasing wavenumber. Generally speaking (see Section 5.4), geopotential spectra fall off more rapidly with increasing wavenumber than do wind spectra.

When observation increments are being analyzed, the spectrum of the background error is of primary interest. There are two cases – a climatological background and a forecast background – for which the background error spectra are quite different. When the background is a numerical forecast, Section 4.3 suggests that the background error spectrum (for the geopotential) is a monotonically decreasing function of increasing wavenumber and is therefore red; see (4.3.19), for example.

Climatological backgrounds have variance only in long time scales and large spatial scales similar to curve S in Figure 1.4; the background error spectrum in this case would contain relatively more of the higher-frequency, smaller-scale atmospheric components and would look more like curve T. Thus, the error spectrum for a climatological background is unlikely to have maximum variance in the largest scales (see Thiebaux 1981 and Section 4.3). In Figure 1.4, the T spectrum is clearly red only beyond global wavenumber 10. This argument, together with the results of Exercise 3.3, explain why the background error corrections for a climatological background (Figure 4.4) may be negative at large distance.

Throughout this book, we have assumed that the observation network is given and consequently that the goal of objective analysis should be to extract the optimum amount of information (signal) from the given network (Section 1.3). It was in this spirit that we discussed the minimum resolvable scale for a given network. However, network questions can be examined from a completely opposite viewpoint, in which one wants to obtain information about a given physical phenomenon and tries to determine the characteristics of an observation network that will optimally sample the phenomenon. This concept is particularly appropriate when the task is to develop an observation network to adequately sample a given phenomenon during a field experiment. *Network design*, as this concept is called, can be addressed in several ways.

A method that is commonly used when new observation systems are about to be introduced into the global observing network is the *Observation System Simulation Experiment* (*OSSE*). In this procedure, a simulated "truth" or "nature run" is created by a long integration of a sophisticated numerical model whose "model climate" approximates the true earth climate. This nature run is then sampled at observation locations that are identical to those in the actual global observing system. The pseudo-observations are modified by adding realistic observation errors and used as raw input to a data assimilation cycle. The analyzed values are then compared with the "true" values. Preferably, the model that created the nature run should not be the same as the assimilating model. Various experiments can be performed with and without a proposed new observing system, and the differences in the resulting objective analyses provide a measure of the effectiveness of the proposed system. A typical OSSE is described in Daley and Mayer (1986), and the OSSE literature is reviewed in Arnold and Dey (1986).

More general results of a mathematical nature can be obtained from sampling theory. The development of this theory has been extensively reported in the statistical and signal processing literature but has been discussed only briefly in the meteorological literature (Petersen and Middleton 1963; Stephens 1971).

Sampling theory starts with the concept of a *band-limited function*. Consider the one-dimensional case on an infinite domain and a signal $f(x)$. Define $\hat{f}(m)$ as the Fourier transform of $f(x)$ and assume that

$$\hat{f}(m) = 0, \qquad |m| > M \tag{H7}$$

Then the Fourier transform pairs for $f(m)$ and $f(x)$ are given by

$$\hat{f}(m) = \frac{1}{2\pi}\int_{-\infty}^{\infty} f(x)e^{-imx}\,dx, \qquad f(x) = \int_{-M}^{M} \hat{f}(m)e^{imx}\,dm \tag{H8}$$

The function $f(x)$ is said to be band-limited because it contains no spatial scales with wavenumber greater than M. In meteorological applications, this does not happen in reality; but if the spectrum of a function falls off sufficiently rapidly for $|m| > M$, then the band-limited approximation is useful.

Now consider

$$f\left(\frac{k\pi}{M}\right) = \int_{-M}^{M} \hat{f}(m)\exp\left(\frac{imk\pi}{M}\right)dm \tag{H9}$$

Then,

$$\hat{f}(m) = \frac{1}{2M}\sum_{l=-\infty}^{\infty} f\left(\frac{l\pi}{M}\right)\exp\left(-\frac{iml\pi}{M}\right) \tag{H10}$$

Equation (H10) can be proved by substitution of (H10) into (H9) and reversal of the order of integral and sum. Thus,

$$\begin{aligned} f\left(\frac{k\pi}{M}\right) &= \frac{1}{2M}\sum_{l=-\infty}^{\infty} f\left(\frac{l\pi}{M}\right)\int_{-M}^{M} \exp\left[\frac{im\pi(k-l)}{M}\right]dm \\ &= \sum_{l=-\infty}^{\infty} f\left(\frac{l\pi}{M}\right)\frac{\sin[\pi(k-l)]}{\pi(k-l)} = f\left(\frac{k\pi}{M}\right) \end{aligned} \tag{H11}$$

Now, substitute (H10) into the second equation of (H8) and reverse the order of integral and sum, giving

$$f(x) = \frac{1}{2M}\sum_{k=-\infty}^{\infty} f\left(\frac{k\pi}{M}\right)\int_{-M}^{M} \exp\left[im\left(x - \frac{k\pi}{M}\right)\right]dm$$

or

$$f(x) = \sum_{k=-\infty}^{\infty} f\left(\frac{k\pi}{M}\right)\frac{\sin(Mx - k\pi)}{(Mx - k\pi)} \tag{H12}$$

Equation (H12) is known as the sampling theorem for a band-limited signal on a one-dimensional infinite domain. This equation can be thought of as a spatial analysis algorithm of the form (1.7.1), where $f(x_k)$ is the observation at the observation station $x_k = k\pi/M$, and $\sin(Mx - k\pi)/(Mx - k\pi)$ represents the a posteriori analysis weight and has the same form as diffraction functions (3.3.12) plotted in Figure 3.3. The sampling theorem (H12) states that the band-limited signal can be *perfectly reconstituted anywhere in the domain* provided that it is sampled without error on the equally spaced discrete network $x_k = k\pi/M$, $-\infty \le k \le \infty$. Define $\Delta x = \pi/M$ as the distance between observations and $\lambda_M = 2\pi/M$ as the shortest wavelength in the band-limited signal. It follows that $\lambda_M = 2\,\Delta x$, which is the same result derived earlier in this appendix.

The weights $\sin(Mx - k\pi)/(Mx - k\pi)$ are the optimal weights for a perfectly sampled band-limited signal on an equally spaced network. Real life, unfortunately, is never so straightforward: Observation networks are irregular, the observations have error, and the signal is not band-limited. However, the sampling theorem (H12) does provide a basis for rational network design.

The extension of the sampling theorem (H12) to two dimensions is straightforward. It can also be extended to a space-limited (i.e., finite) domain (Goldman 1953). A sampling theorem has been derived for the case of a finite number of irregularly spaced observations by Yen (1956). Stephens (1971) has shown that if gradient observations are also available (i.e., geostrophic winds), then a coarser mesh is adequate to resolve the band-limited signal. Versions of the sampling theorem for stochastic (random) processes also exist (Papoulis 1965). Petersen and Middleton (1963) have discussed optimal lattices for meteorological sampling based on the multidimensional sampling theory of Petersen and Middleton (1962).

References

Abramovitz, M., and Stegun, I. (1965). *Handbook of mathematical functions* (New York: Dover).

Achtemeier, G. (1975). "On the initialization problem: a variational adjustment method," *Mon. Wea. Rev.* **103**: 1089–103.

Achtemeier, G. (1987). "On the concept of varying influence radii for a successive corrections objective analysis," *Mon. Wea. Rev.* **115**: 1760–71.

Ahlberg, J., Nilson, E., and Walsh, J. (1967). *The theory of splines and their applications* (New York: Academic Press).

Arnason, G. (1958). "A convergent method for solving the balance equation," *J. Meteor.* **15**: 220–5.

Arnold, C., and Dey, C. (1986). "Observing-system simulation experiments: past, present and future," *Bull. Amer. Meteor. Soc.* **67**: 687–95.

Atkinson, B. (1981). *Meso-scale atmospheric circulations* (London: Academic Press).

Atmospheric Environment Service of Canada (1954). Instrument Manual Number 10 (Toronto: Atmospheric Environment Service of Canada).

Backus, G., and Gilbert, J. (1967). "Numerical application of a formalism for geophysical inverse problems," *Geophys. J. Roy. Astron. Soc.* **13**: 247–76.

Baer, F. (1977a). "The spectral balance equation," *Tellus* **29**: 107–15.

Baer, F. (1977b). "Adjustments of initial conditions required to suppress gravity oscillations in non-linear flows," *Beitr. Phys. Atmosph.* **50**: 350–66.

Baer, F., and Tribbia, J. (1977). "On complete filtering of gravity modes through non-linear initialization," *Mon. Wea. Rev.* **105**: 1536–9.

Baker, W., Bloom, S., Woollen, J., Nestler, M., Brin, E., Schlatter, T., and Branstator, G. (1987). "Experiments with a three-dimensional statistical objective analysis scheme using FGGE data," *Mon. Wea. Rev.* **115**: 272–96.

Balgovind, R., Dalcher, A., Ghil, M., and Kalnay, E. (1983). "A stochastic-dynamic model for the spatial structure of forecast error statistics," *Mon. Wea. Rev.* **111**: 701–22.

Ballish, B. (1981). "A simple test of the initialization of gravity modes," *Mon. Wea. Rev.* **109**: 1318–21.

Bannon, J. (1948). "The estimation of large scale vertical currents from the rate of rainfall," *Quart. J. Roy. Meteor. Soc.* **74**: 57–66.

Barker, E., Haltiner, G., and Sasaki, Y. (1977). "Three dimensional assimilation using variational analysis." In: *Proceedings third conference on numerical weather prediction of the American Meteorological Society*, Ohama, Nebraska, May 1977 (Boston: American Meteorological Society), pp. 108–22.

Barnes, S. (1964). "A technique for maximizing details in numerical map analysis," *J. Appl. Meteor.* **3**: 395–409.

Barnes, S. (1978). "Oklahoma thunderstorms on 29–30 April 1970. Part I: Morphology of a tornadic storm," *Mon. Wea. Rev.* **106**: 673–84.

Barrodale, I. (1968). "L_1 approximations and the analysis of data," *Applied Stat.* **17**: 51–7.

Barwell, B., and Bromley, R. (1988). "The adjustment of numerical weather prediction models to local perturbations," *Quart. J. Roy. Meteor. Soc.* **114**: 665–84.

Barwell, B., and Lorenc, A. (1985). "A study of the impact of aircraft wind observations on a large-scale analysis and numerical weather prediction systems," *Quart. J. Roy. Meteor. Soc.* **111**: 103–29.

Batchelor, G. (1953). *The theory of homogeneous turbulence* (Cambridge: Cambridge University Press).

Bengtsson, L. (1975). *4-dimensional data assimilation of meteorological observations*, World Meteorological Organization GARP Publication No. 15 (Geneva: World Meteorological Organization).

Bengtsson, L., and Gustavsson, N. (1971). "An experiment in the assimilation of data in dynamic analysis," *Tellus* **23**: 328–36.

Bengtsson, L., and Shukla, J. (1988). "Integration of space and in situ observations to study global climate changes," *Bull. Amer. Meteor. Soc.* **69**: 1130–43.

Bennett, A., and McIntosh, P. (1982). "Open ocean modelling as an inverse problem: Tidal theory," *J. Phys. Oceanogr.* **12**: 1004–18.

Bergman, K. (1979). "Multivariate analysis of temperature and winds using optimum interpolation," *Mon. Wea. Rev.* **107**: 1423–44.

Bergman, K., and Bonner, W. (1976). "Analysis errors as a function of observation density for satellite temperature soundings with spatially correlated errors," *Mon. Wea. Rev.* **104**: 1308–16.

Bergthorsson, P., and Doos, B. (1955). "Numerical weather map analysis," *Tellus* **7**: 329–40.

Bjerknes, V. (1911). *Dynamic meteorology and hydrography. Part II. Kinematics* (New York: Carnegie Institute, Gibson Bros.).

Blackman, R., and Tukey, J. (1958). *The measurement of power spectra* (New York: Dover).

Bleck, R. (1975). "An economical approach to the use of wind data in the optimum interpolation of geo- and Montgomery potential fields," *Mon. Wea. Rev.* **103**: 807–16.

Blumen, W. (1972). "Geostrophic adjustment," *Rev. Geophys. Space Phys.* **10**: 485–528.

Blumen, W. (1975a). "An analytic view of updating meteorological variables. Part I. Phase errors," *J. Atmos. Sci.* **32**: 274–86.

Blumen, W. (1975b). "An analytic view of updating meteorological variables. Part II. Weighted assimilation," *J. Atmos. Sci.* **32**: 690–7.

Blumen, W. (1976). "Experiments in atmospheric predictability. Part II. Data assimilation," *J. Atmos. Sci.* **33**: 170–5.

Boer, G., and Shepherd, T. (1983). "Large-scale two-dimensional turbulence in the atmosphere," *J. Atmos. Sci.* **40**: 164–84.

Bolin, B. (1955). "Numerical forecasting with the barotropic model," *Tellus* **7**: 27–49.

Bolin, B. (1956). "An improved barotropic model and some aspects of using the balance equation for three-dimensional flow," *Tellus* **8**: 61–75.

Bourke, W., McAveney, B., Puri, K., and Thurling, R. (1977). "Global modelling of atmospheric flow by spectral methods." In: *Methods in computational physics, Vol. 17, General circulation models of the atmosphere*, Julius Chang, ed. (New York: Academic Press), pp. 267–324.

Bourke, W., and McGregor, J. (1983). "A nonlinear vertical mode initialization scheme for a limited area prediction model," *Mon. Wea. Rev.* **111**: 2285–97.

Bratseth, A. (1982). "A simple and efficient approach to the initialization of weather prediction models," *Tellus* **34**: 352–7.

Bratseth, A. (1986). "Statistical interpolation by means of successive corrections," *Tellus* **38A**: 439–47.

Bretherton, F., Davis, R., and Fandry, C. (1976). "A technique for objective analysis and design of oceanographic experiments applied to MODE-73," *Deep Sea Research* **23**: 559–92.

Briere, S. (1982). "Nonlinear normal mode initialization of a limited area model," *Mon. Wea. Rev.* **110**: 1166–86.

Brown, P., and Robinson, G. (1979). "The variance spectrum of tropospheric winds over Eastern Europe," *J. Atmos. Sci.* **36**: 270–86.

Browning, G., Kasahara, A., and Kreiss, H. (1980). "Initialization of the primitive equations by the bounded derivative method," *J. Atmos. Sci.* **37**: 1424–36.

Browning, G., and Kreiss, H. (1982). "Initialization of the shallow water equations with open boundaries by the bounded derivative method," *Tellus* **34**: 334–51.

Browning, G., and Kreiss, H. (1986). "Scaling and computation of smooth atmospheric motions," *Tellus* **38A**: 295–313.

Bube, K., and Ghil, M. (1980). "Assimilation of asynoptic data and the initialization problem." In: *Dynamic meteorology data assimilation methods*, L. Bengtsson, M. Ghil, and E. Kallen, eds. (New York: Springer-Verlag), pp. 111–38.

Buell, C. (1957). "An approximate relation between the variability of wind and the variability of pressure or height in the atmosphere," *Bull. Amer. Meteor. Soc.* **38**: 47–51.

Buell, C. (1960). "The structure of two-point wind correlations in the atmosphere," *J. Geophys. Res.* **65**: 3353–66

Buell, C. (1971). "Two-point wind correlations on an isobaric surface in a non-homogeneous non-isotropic atmosphere," *J. Appl. Meteor.* **10**: 1266–74.

Buell, C. (1972a). "Correlation functions for wind and geopotential on isobaric surfaces," *J. Appl. Meteor.* **11**: 51–9.

Buell, C. (1972b). "Variability of wind with distance and time on an isobaric surface," *J. Appl. Meteor.* **11**: 1085–91.

Buell, C., and Seaman, R. (1983). "The 'scissors effect': anisotropic and ageostrophic influences on wind correlation coefficients," *Aust. Met. Mag.* **31**: 77–83.

Bushby, F., and Huckle, V. (1956). "The use of a streamfunction in a two-parameter model of the atmosphere," *Quart. J. Roy. Meteor. Soc.* **82**: 409–18.

Chang, C., Perkey, D., and Kreitzberg, C. (1986). "Impact of missing wind observations on the simulation of a severe storm environment," *Mon. Wea. Rev.* **114**: 1278–87.

Chapman, S., and Lindzen, R. (1970). *Atmospheric tides* (Hingham, Mass.: D. Reidel).

Charney, J. (1955). "The use of the primitive equations of motion in numerical prediction," *Tellus* **7**: 22–6.

Charney, J., Fjortoft, R., and von Neumann, J. (1950). "Numerical integration of the barotropic vorticity equation," *Tellus* **2**: 237–54.

Charney, J., Halem, M., and Jastrow, R. (1969). "Use of incomplete historical data to infer the present state of the atmosphere," *J. Atmos. Sci.* **26**: 1160–3.

Charney, J., and Stern, T. (1962). "On the stability of internal baroclinic jets in a rotating atmosphere," *J. Atmos. Sci.* **19**: 159–72.

Claerbout, J., and Muir, F. (1973). "Robust modelling with erratic data," *Geophysics* **38**: 826–44.

Clancy, R., and Pollack, K. (1983). "A real-time synoptic ocean thermal analysis/forecast system," *Prog. Oceanogr.* **12**: 383–424.

Cohn, S., and Dee, D. (1988). "Observability of discretized partial differential equations," *SIAM J. Numer. Anal.* **25**: 586–617.

Cohn, S., and Dee, D. (1989). "An analysis of the vertical structure function for arbitrary thermal profiles." *Quart. J. Roy. Meteor. Soc.* **115**: 143–71.

Courant, R., and Hilbert, D. (1953). *Methods of mathematical physics. Volume 1* (New York: Interscience).

Courant, R., and Hilbert, D. (1962). *Partial differential equations. Methods of mathematical physics, Vol. II* (New York: Interscience).

Courtier, P., and Talagrand, O. (1987). "Variational assimilation of meteorological observations with the adjoint vorticity equations. Part II. Numerical results," *Quart. J. Roy. Meteor. Soc.* **113**: 1329–47.

Craven, P., and Wahba, G. (1979). "Smoothing noisy data with spline functions," *Numer. Math.* **31**: 377–403.

Cressman, G. (1959). "An operational objective analysis system," *Mon. Wea. Rev.* **87**: 367–74.

Cressman, G. (1960). "Improved terrain effects in barotropic forecasts," *Mon. Wea. Rev.* **88**: 327–42.

Daley, R. (1966). *Experimental large-scale rainfall predictions.* MSc. thesis (Montreal: McGill University).

Daley, R. (1978). "Variational non-linear normal mode initialization," *Tellus* **30**: 201–18.

Daley, R. (1979). "The application of normal mode initialization to an operational forecast model," *Atmosphere Ocean* **17**: 97–124.

Daley, R. (1980a). "The development of efficient time integration schemes using model normal modes," *Mon. Wea. Rev.* **108**: 100–10.

Daley, R. (1980b). "On the optimal specification of the initial state for deterministic forecasting," *Mon. Wea. Rev.* **108**: 1719–35.

Daley, R. (1981). "Predictability experiments with a baroclinic model," *Atmosphere Ocean* **19**: 77–89.

Daley, R. (1983a). "Linear non-divergent mass-wind laws on the sphere," *Tellus* **35A**: 17–27.

Daley, R. (1983b). *Spectral characteristics of the ECMWF objective analysis system*, ECMWF Technical Report No. 40 (Reading, England: European Centre for Medium Range Forecasts).

Daley, R. (1985). "The analysis of synoptic scale divergence by a statistical interpolation procedure," *Mon. Wea. Rev.* **113**: 1066–79.

Daley, R. (1986). "The application of variational methods of initialization on the sphere." In: *Variational methods in geosciences*, Y. Sasaki, ed. (Amsterdam: Elsevier), pp. 3–11.

Daley, R., and Mayer, T. (1986). "Estimates of global analysis error from the Global Weather Experiment observational network," *Mon. Wea. Rev.* **114**: 1642–53.

Daley, R., and Puri, K. (1980). "Four dimensional data assimilation and the slow manifold," *Mon. Wea. Rev.* **108**: 85–99.

Daley, R., Wergen, W., and Cats, G. (1986). "The objective analysis of planetary flow," *Mon. Wea. Rev.* **114**: 1892–908.

Davies, H., and Turner, R. (1977). "Updating prediction models by dynamic relaxation: An examination of the technique," *Quart. J. Roy. Meteor. Soc.* **103**: 225–45.

Dee, D., Cohn, S., Dalcher, A., and Ghil, M. (1985). "An efficient algorithm for estimating noise covariances in distributed systems," *IEEE Trans. Automatic Control* **30**(11); 1057–65.

Deepak, A. (1977). *Inversion methods in atmospheric remote sensing* (New York: Academic Press).

Denman, K., and Freeland, H. (1985). "Correlation scales, objective mapping and a statistical test of geostrophy over the continental shelf," *J. Mar. Res.* **43**: 517–39.

Derber, J. (1987a). "Variational four-dimensional analysis using quasi-geostrophic constraints," *Mon. Wea. Rev.* **115**: 998–1008.

Derber, J. (1989). "A variational continuous assimilation technique," *Mon. Wea. Rev.* **117**: 2437–46.

Derber, J., and Rosati, A. (1989). "A global oceanic data assimilation system," *J. Phys. Oceanogr.* **19**: 1333–47.

Dey, C., and Morone, L. (1985). "Evolution of the National Meteorological Center global data assimilation system: January 1982–December 1983," *Mon. Wea. Rev.* **113**: 304–18.

Dickinson, R., and Williamson, D. (1972). "Free oscillations of a discrete stratified fluid with application to numerical weather prediction," *J. Atmos. Sci.* **29**: 623–40.

Dixon, R., Spackman, E., Jones, I., and Francis, A. (1972). "The global analysis of meteorological data using orthogonal polynomial basis functions," *J. Atmos. Sci.* **29**: 609–22.

Donner, L. (1988). "An initialization for cumulus convection in numerical weather prediction models," *Mon. Wea. Rev.* **116**: 377–85.

Dorney, C. (1975). *A vector space approach to models and optimization* (New York: Wiley and Sons).

Drozdov, O., and Shepelevskii, A. (1946). "The theory of interpolation in a stochastic field of meteorological elements and its application to meteorological map and network rationalization problems," *Trudy Niu Gugms Series* **1**, No. 13.

Duchon, J. (1976). "Interpolation des fonctions de deux variables suivant le principe de la flexion des plaques minces," *Rairo. Anal. Numer.* **10**: 5–12.

Eckart, C. (1960). *Hydrodynamics of oceans and atmospheres* (Oxford: Pergamon Press).

Eddy, A. (1964). "The objective analysis of horizontal wind divergence fields," *Quart. J. Roy. Meteor. Soc.* **90**: 424–40.

Eddy, A. (1967). "The statistical objective analysis of scalar data fields," *J. Appl. Meteor.* **4**: 597–609.

Eliasen, E., and Machenhauer, B. (1965). "A study of the fluctuations of the atmospheric flow patterns represented by spherical harmonics," *Tellus* **17**: 220–38.

Eliassen, A. (1954). *Provisional report on calculation of spatial covariance and autocorrelation of the pressure field. Rep. No. 5* (Oslo: Videnskaps-Akademiet Institut for Vaer og Klimaforskning).

Emanuel, K. (1983). "On the dynamical definitions of mesoscale." In: *Mesoscale meteorology: Theories, observations and models,* D. Lilly and T. Gal-Chen, eds. (Dordecht: Reidel).

Errico, R. (1981). "An analysis of interactions between geostrophic and ageostrophic modes in a simple model," *J. Atmos. Sci.* **38**: 544–53.

Errico, R. (1982). "Normal mode initialization and the generation of gravity waves by quasi-geostrophic forcing," *J. Atmos. Sci.* **39**: 573–86.

Errico, R. (1983). "Convergence properties of Machenhauer's initialization scheme," *Mon. Wea. Rev.* **111**: 2214–23.

Errico, R. (1984a). "Normal modes of a semi-implicit model," *Mon. Wea. Rev.* **112**: 1818–28.

Errico, R. (1984b). "The dynamical balance of a general circulation model," *Mon. Wea. Rev.* **112**: 2439–54.

Errico, R. (1984c). "The statistical equilibrium solution of a primitive-equation model," *Tellus* **36A**: 42–51.

Errico, R. (1989). "The forcing of gravitational normal modes by condensational heating," *Mon. Wea. Rev.* **117**: 2734–52.

Errico, R., and Rasch, P. (1988). "A comparison of various normal mode initialization schemes and the inclusion of diabatic processes," *Tellus* **40A**: 1–25.

Fadeev, D., and Fadeeva, V. (1963). *Computational methods of linear algebra* (San Francisco: W.H. Freeman).

Fillion, L., and Temperton, C. (1989). "Variational implicit normal mode initialization," *Mon. Wea. Rev.* **117**: 2219–29.

Fjortoft, R. (1952). "On a numerical method of integrating the barotropic vorticity equation," *Tellus* **4**: 179–94.

Flattery, T. (1971). "Spectral models for global analysis and forecasting." In: *Proc. Sixth AWS Exchange Conf. Air Weather Service Tech. Rep. 242*, NTIS ADA724093, pp. 42–54.

Forsythe, G., and Wasaw, W. (1960). *Finite difference methods for partial differential equations* (New York: Wiley and Sons).

Franke, R. (1985). "Sources of error in objective analysis," *Mon. Wea. Rev.* **113**: 260–70.

Franke, R. (1988). "Statistical interpolation by iteration," *Mon. Wea. Rev.* **116**: 961–3.

Gal-Chen, T. (1978). "A method for the initialization of the anelastic equations: Implications for matching models with observations," *Mon. Wea. Rev.* **106**: 587–606.

Gal-Chen, T. (1983). "Initialization of mesoscale models: The possible impact on remotely sensed data. In: *Mesoscale meteorology: Theories, observations and models,* D. Lilly and T. Gel-Chen (Dordrecht: Reidel), pp. 157–71.

Gal-Chen, T., and Kropfli, R. (1984). "Buoyancy and pressure perturbations derived from dual-Doppler radar observations of the planetary boundary layer: Applications for matching models with observations," *J. Atmos. Sci.* **41**: 3007–20.

Gandin, L. (1963). *Objective analysis of meteorological fields* (Leningrad: Gridromet). English translation (Jerusalem: Israel Program for Scientific Translation), 1965.

Gandin, L. (1974). "Contemporary problems of objective analysis of meteorological fields," *Trudy y vsesoyuznogo meteorologicheskogo s"yezda sektisiya prognoza pogody* **2** (1972). English translation by Foreign Technology Division, Wright-Patterson Air Force Base, FTD-HC-23-2100-74.

Gandin, L. (1988). "Complex quality control of meteorological observations," *Mon. Wea. Rev.* **116**: 1137–56.

Garbedian, P. (1964). *Partial differential equations* (New York: Wiley and Sons).

Gauss, K. (1963). *Theory of motion of the heavenly bodies* (translation of "Theoria Motus Corporum Coelestium," 1809) (New York: Dover).

Gelb, A. (1974). *Applied optimal estimation* (Cambridge: MIT Press).

Gel'fand, I., and Vilenkin, N. (1964). *Generalized functions, vol. 4. Applications of harmonic analysis* (New York: Academic Press).

Gent, P., and McWilliams, J. (1982). "Intermediate model solutions to the Lorenz equations: Strange attractors and other phenomena," *J. Atmos. Sci.* **39**: 3–13.

Ghil, M. (1980). "The compatible balancing approach to initialization and four dimensional data assimilation," *Tellus* **32**: 198–206.

Ghil, M., Cohn, S., Tavantzis, J., Bube, K., and Isaacson, E. (1981). "Applications of estimation theory to numerical weather prediction." In: *Dynamic meteorology: data assimilation methods,* L. Bengtsson, M. Ghil, and E. Kallen, eds. (New York: Springer-Verlag), pp. 139–224.

Gilchrist, B., and Cressman, G. (1954). "An experiment in objective analysis," *Tellus* **6**: 309–18.

Gill, A. (1982). *Atmosphere-ocean dynamics* (New York: Academic Press).

Gill, P., Murray, W., and Wright, M. (1981). *Practical optimization* (London: Academic Press).

Godske, C., Bergeron, T., Bjerknes, J., and Bundgard, R. (1957). *Dynamic meteorology and weather forecasting*, American Meteorological Society and the Carnegie Institute (Baltimore: The Waverly Press).

Goldman, S. (1953). *Information theory* (New York: Prentice-Hall).

Gordon, C., Umscheid, L., and Miyakoda, K. (1972). "Simulation experiments to determine wind data requirements in the tropics," *J. Atmos. Sci.* **29**: 1064–75.

Gradshteyn, I., and Ryzhik, I. (1965). *Tables of integrals, series and products* (New York: Academic Press).

Halberstam, I., and Tung, S. (1984). "Objective analysis using Hough vectors evaluated at irregularly spaced locations," *Mon. Wea. Rev.* **112**: 1804–17.

Hall, M. (1986). "Application of adjoint sensitivity theory to an atmospheric general circulation model," *J. Atmos. Sci.* **43**: 2644–51.

Haltiner, G., and Williams, R. (1980). *Numerical prediction and dynamic meteorology* (New York: Wiley and Sons).

Hane, C., Wilhelmson, R., and Gal-Chen, T. (1981). "Retrieval of thermodynamic variables within deep convective clouds: Experiments in three dimensions," *Mon. Wea. Rev.* **109**: 564–76.

Hauser, D., and Amayenc, P. (1986). "Retrieval of cloud water and water vapour contents from Doppler radar data in a tropical squall line," *J. Atmos. Sci.* **43**: 823–38.

Hayden, C. (1973). "Experiments in the four dimensional data assimilation of NIMBUS 4 SIRS data," *J. Appl. Meteor.* **12**: 425–36.

Hildebrand, F. (1956). *Introduction to numerical analysis* (New York: McGraw-Hill).

Hinkelmann, K. (1951). "Der mechanismus des meteorologischen larmes," *Tellus* **3**: 285–96.

Hinkelmann, K. (1959). "Ein numerisches experiment mit den primitiven gleichungen." In: *The atmosphere and the sea in motion* (New York: Rockefeller Institute Press), pp. 486–500.

Hoffman, R. (1984). "SASS wind ambiguity removal by direct minimization. Part II: Use of smoothness and dynamical constraints," *Mon. Wea. Rev.* **112**: 1829–52.

Hoffman, R. (1986). "A four-dimensional analysis exactly satisfying equations of motion," *Mon. Wea. Rev.* **114**: 388–97.

Hoke, J., and Anthes, R. (1976). "The initialization of numerical models by a dynamic relaxation technique," *Mon. Wea. Rev.* **104**: 1551–6.

Holbrook, J. (1959). *Laplace transforms for electronic engineers* (New York: Pergamon Press).

Hollingsworth, A. (1987). "Objective analysis for numerical weather prediction." In: *Short and medium range numerical weather prediction*, collected papers presented at WMO/IUGG NWP Symposium, Tokyo, August 1986, T. Matsuno, ed. Special volume of the *J. Meteor. Soc. Japan*: 11–59.

Hollingsworth, A. (1989). "The verification of objective analysis: diagnostics of analysis system performance," *Meteorology and Atmospheric Physics.* **40**: 3–27.

Hollingsworth, A., and Lonnberg, P. (1986). "The statistical structure of short-range forecast errors as determined from radiosonde data. Part I: The wind field," *Tellus* **38A**: 111–36.

Hollingsworth, A., Lorenc, A., Tracton, S., Arpe, K., Cats, G., Uppala, S., and Kallberg, P. (1985). "The response of numerical weather prediction systems to FGGE level IIb data. Part I: Analyses," *Quart. J. Roy. Meteor. Soc.* **111**: 1–66.

Hollingsworth, A., Shaw, D., Lonnberg, P., Illari, L., Arpe, K., and Simmons, A. (1986). "Monitoring of observation and analysis quality in a data assimilation system," *Mon. Wea. Rev.* **114**: 861–79.

Holloway, G., and West, B. (1984). *Predictability of fluid motions* (New York: American Institute of Physics).

Holton, J. (1972). *An introduction to dynamic meteorology* (New York: Academic Press).

Hoskins, B., Draghici, I., and Davies, H. (1978). "A new look at the omega equation," *Quart. J. Roy. Meteor. Soc.* **104**: 31–8.

Hoskins, B., McIntyre, M., and Robertson, A. (1985). "On the significance and use of isentropic potential vorticity maps," *Quart. J. Roy. Meteor. Soc.* **111**: 877–946.

Hough, S. (1898). "On the application of harmonic analysis to the dynamical theory of the tides. II, On the general integration of Laplace's dynamical equations," *Philos. Trans. R. Soc. London, Ser. A* **191**: 139–85.

Houghton, J., Taylor, F., and Rodgers, C. (1985). *Remote sounding of atmospheres* (Cambridge: Cambridge University Press).

Ikawa, M. (1984a). "Generalization of multivariate optimum interpolation and the roles of linear constraints and covariance matrix in the method. Part 1: In discrete form," *Papers in Meteorology and Geophysics* **35**: 81–102.

Ikawa, M. (1984b). "Generalization of multivariate optimum interpolation and the roles of linear constraints and covariance matrix in the method. Part 2: In continuous form," *Papers in Meteorology and Geophysics* **35**: 169–80.

Illari, L. (1986). "The quality of ECMWF humidity analysis." In: *Proceedings of ECMWF workshop on high resolution analysis* (Reading, England: European Centre for Medium Range Forecasting), pp. 41–68.

Isaacs, R., Hoffman, R., and Kaplan, L. (1986). "Satellite remote sensing of meteorological parameters for global numerical weather predictions," *Rev. Geophys.* **24**: 701–43.

Jazwinski, A. (1970). *Stochastic processes and filtering theory* (New York: Academic Press).

Jenkins, G., and Watts, D. (1968). *Spectral analysis and its applications* (San Francisco: Holden Day).

Jones, R. (1965). "An experiment in non-linear prediction," *J. Appl. Meteor.* **4**: 701–5.

Julian, P. (1984). "Objective analysis in the tropics: A proposed scheme," *Mon. Wea. Rev.* **112**: 1752–67.

Julian, P., and Thiebaux, H. J. (1975). "On some properties of correlation functions used in optimum interpolation schemes," *Mon. Wea. Rev.* **103**: 605–16.

Juvanon du Vachat, R. (1986). "A general formulation of normal modes for limited-area models: Applications to initialization," *Mon. Wea. Rev.* **114**: 2478–87.

Kailath, T. (1968). "An innovations approach to least square estimation. Part I: Linear filtering in additive white noise," *IEEE Trans. Automatic Control* **13**: 646–55.

Kalman, R. (1960). "A new approach to linear filtering and prediction problems," *Trans. ASME, Ser. D, J. Basic Eng.* **82**: 35–45.

Kalman, R., and Bucy, R. (1961). "New results in linear filtering and prediction theory," *Trans. ASME, Ser. D, J. Basic Eng.* **83**: 95–108.

Kasahara, A. (1976). "Normal modes of ultra-long waves in the atmosphere," *Mon. Wea. Rev.* **104**: 669–90.

Kasahara, A. (1978). "Further studies on a spectral model of the global barotropic primitive equations with Hough harmonic expansion," *J. Atmos. Sci.* **35**: 2043–51.

Kasahara, A. (1980). "Effect of zonal flows on the free oscillations of a barotropic atmosphere," *J. Atmos. Sci.* **37**: 917–29.

Kasahara, A. (1982a). "Non-linear normal mode initialization and the bounded derivative method," *Rev. Geophys. Space Phys.* **19**: 450–68.

Kasahara, A. (1982b). "Significance of non-elliptic regions in balanced flows of the tropical atmosphere," *Mon. Wea. Rev.* **110**: 1956–67.

Kasahara, A., Balgovind, R., and Katz, B. (1988). "Use of satellite radiometric imagery data for improvement in the analysis of divergent winds in the tropics," *Mon. Wea. Rev.* **116**: 866–83.

Kasahara, A., and Puri, K. (1981). "Spectral representation of three dimensional global data by expansion in normal mode functions," *Mon. Wea. Rev.* **109**: 37–51.

Khrgian, K. (1970). *Meteorology, a historical survey*. Translated by Israel Program for Scientific Translations (Jerusalem: Keter Press).

Kistler, R., and McPherson, R. (1975). "On the use of a local wind correction technique in four dimensional data assimilation," *Mon. Wea. Rev.* **103**: 445–9.

Kitade, T. (1983). "Non-linear normal mode initialization with physics," *Mon. Wea. Rev.* **111**: 2194–213.

Knox, J., Higuchi, K., Shabar, A., and Sargent, N. (1988). "Secular variation of Northern Hemisphere 50 kPa geopotential height," *J. of Climate* **1**: 500–11.

Koch, S., Desjardins, M., and Kocin, P. (1983). "An interactive Barnes objective map analysis scheme for use with satellite and conventional data," *J. Climate Appl. Meteor.* **22**: 1487–503.

Kolmogorov, A. (1941). "Interpolated and extrapolated stationary random sequences," *Izvestia an SSSR, seriya mathematicheskaya* **5**(2): 85–95.

Kontarev, G. (1980). *The adjoint equation technique applied to meteorological problems*, ECMWF Technical Report No. 21 (Reading, England: European Centre for Medium Range Forecasts).

Kopell, N. (1985). "Invariant manifolds and the initialization problem for some atmospheric equations," *Physica D* **14**: 203–15.

Kreiss, H. (1980). "Problems with different time scales for partial differential equations," *Comm. Pure Appl. Math.* **33**: 399–440.

Krishnamurti, T., Ingles, K., Cocke, S., Kitade, T., and Pasch, R. (1984). "Details of low latitude medium range numerical weather prediction using a global spectral model. Part II. Effects of orography and physical initialization," *J. Met. Soc. Japan* **62**: 613–49.

Krishnamurti, T., Bedi, H., Heckley, W., and Ingles, K. (1988). "On the reduction of spin up time for evaporation and precipitation in a global spectral model," *Mon. Wea. Rev.* **116**: 907–20.

Kruger, H. (1964). "A statistical-dynamical objective analysis scheme." In: *Canadian Meteorological Memoirs, No. 18* (Toronto: Meteorological Branch, Department of Transport), pp. 47–64.

Kruger, H. (1969). "General and special approaches to the problem of objective analysis of meteorological variables," *Quart. J. Roy. Meteor. Soc.* **95**: 21–39.

Kuhfittig, P. (1977). *Introduction to the Laplace transform* (New York: Plenum Press).

Kuo, Y.-H., and Anthes, R. (1984). "Accuracy of diagnostic heat and moisture budgets using SESAME-79 field data as revealed by observing system simulation experiments," *Mon. Wea. Rev.* **112**: 1465–81.

Kuo, Y.-H., Donall, E., and Shapiro, M. (1987). "Feasibility of short-range numerical weather prediction using observations from a network of profilers," *Mon. Wea. Rev.* **115**: 2402–27.

Lanczos, C. (1970). *The variational principles of mechanics, 4th ed.* Math. Expositions No. 4 (Toronto: University of Toronto Press).

Leary, C., and Thompson, R. (1973). "Shortcomings of an objective analysis scheme," *J. Appl. Meteor.* **12**: 589–94.

Le Dimet, F., and Talagrand, O. (1986). "Variational algorithms for analysis and assimilation of meteorological observations: theoretical aspects," *Tellus* **38A**: 97–110.

Leith, C. (1980). "Non-linear normal mode initialization and quasi-geostrophic theory," *J. Atmos. Sci.* **37**: 958–68.

Lewis, J. (1972). "An operational analysis using the variational method," *Tellus* **24**: 514–30.

Lewis, J. (1980). "Dynamical adjustment of 500 mb vorticity using P.D. Thompson's scheme. A case study," *Tellus* **32**: 511–24.

Lewis, J., and Bloom, S. (1978). "Incorporation of time continuity into sub-synoptic analysis by using dynamical constraints," *Tellus* **30**: 496–515.

Lewis, J., and Derber, J. (1985). "The use of adjoint equations to solve a variational adjustment problem with advective constraints," *Tellus* **37**: 309–27.

Lewis, J., and Grayson, T. (1972). "The adjustment of surface wind by Sasaki's variational matching technique," *J. Appl. Meteor.* **11**: 586–97.

Lewis, J., and Panetta, L. (1983). "The extension of P.D. Thompson's scheme to multiple time levels," *J. Clim Appl. Meteor.* **22**: 1649–53.

Lilly, D. (1983). "Mesoscale variability of the atmosphere." In: *Mesoscale meteorology: Theories, observations and models,* D. Lilly and T. Gal-Chen, eds. (Dordrecht: Reidel), pp. 13–24.

Lindzen, R., Batten, E., and Kim, J. (1968). "Oscillations in atmospheres with tops," *Mon. Wea. Rev.* **96**: 133–40.

Lions, J. (1971). *Optimal control of systems governed by partial differential equations* (Berlin: Springer-Verlag).

Longuet-Higgins, M. (1968). "The eigenfunctions of Laplace's tidal equations over a sphere," *Philos. Trans. R. Soc. London, Ser. A.* **262**: 511–607.
Lonnberg, P., and Hollingsworth, A. (1986). "The statistical structure of short-range forecast errors as determined from radiosonde data. Part II: The covariance of height and wind errors," *Tellus* **38A**: 137–61.
Loomis, E. (1885). *Contributions to meteorology* (New Haven: Tuttle, Moorehouse and Taylor).
Lorenc, A. (1981). "A global three-dimensional multivariate statistical interpolation scheme," *Mon. Wea. Rev.* **109**: 701–21.
Lorenc, A. (1986). "Analysis methods for numerical weather prediction. *Quart. J. Roy. Meteor. Soc.* **112**: 1177–94.
Lorenc, A. (1988a). "A practical approximation to optimal four-dimensional data assimilation," *Mon. Wea. Rev.* **116**: 730–45.
Lorenc, A. (1988b). "Optimal nonlinear objective analysis," *Quart. J. Roy. Meteor. Soc.* **114**: 205–40.
Lorenc, A., and Hammon, O. (1988). "Objective quality control of observations using Bayesian methods. Theory and a practical implementation," *Quart. J. Roy. Meteor. Soc.* **114**: 515–43.
Lorenz, E. (1980). "Attractor sets and quasi-geostrophic equilibrium," *J. Atmos. Sci.* **37**: 1685–99.
Lorenz, E., and Krishnamurthy, V. (1987). "On the nonexistence of a slow manifold," *J. Atmos. Sci.* **44**: 2940–50.
Lovejoy, S., Schertzer, D., and Ladoy, P. (1986). "Fractal characterization of inhomogeneous geophysical measuring networks." *Nature* **319**: 43–4.
Lynch, P. (1985a). "Initialization using Laplace transforms," *Quart. J. Roy. Meteor. Soc.* **111**: 243–58.
Lynch, P. (1985b). "Initialization of a barotropic limited-area model using the Laplace transform technique," *Mon. Wea. Rev.* **113**: 1338–44.
Lynch, P. (1986). *Numerical forecasting using Laplace transforms: Theory and application to data assimilation,* technical note 48 (Glasnevin Hill, Dublin 9: Irish Meteorological Service).
Lynch, P. (1987). *The slow equations*, technical note 50 (Glasnevin Hill, Dublin 9: Irish Meteorological Service).
Lyne, W., Swinbank, R., and Birch, N. (1982). "A data assimilation experiment and the global circulation during the FGGE special observing periods," *Quart. J. Roy. Meteor. Soc.* **108**: 575–94.
Machenhauer, B. (1977). "On the dynamics of gravity oscillations in a shallow water model with application to normal mode initialization," *Contrib. Atmos. Phys.* **50**: 253–71.
Mallonotte-Rizzoli, P., and Holland, W. (1986). "Data constraints applied to models of the ocean general circulation. Part I: The steady case," *J. Phys. Oceanogr.* **16**: 1665–82.
Marchuk, G. (1974). *Numerical solution of the problems of the dynamics of the atmosphere and ocean* (in Russian) (Leningrad: Gidrometeoizadat).
Mass, C., Edmon, H., Friedman, H., Cheney, N., and Reiter, E. (1987). "The use of compact discs for the storage of large meteorological and oceanographic data sets," *Bull. Amer. Met. Soc.* **68**: 1556–8.
Matsuno, T. (1966). "Numerical integrations of the primitive equations by a simulated backward difference method," *J. Meteor. Sco. Japan* **44**: 76–83.
McIntosh, P., and Bennett, A. (1984). "Open ocean modelling as an inverse problem: M2 tides in Bass Strait," *J. Phys. Oceanogr.* **14**: 601–14.
McPherson, R., Bergman, K., Kistler, R., Rasch, G., and Gordon, D. (1979). "The NMC operational global data assimilation system," *Mon. Wea. Rev.* **107**: 1445–61.
McWilliams, J. (1976). "Maps from the Mid-Ocean Dynamics Experiment: Part I. Geostrophic

streamfunction," *J. Phys. Oceanogr.* **6**: 810–27.

Merilees, P. (1968). "On the linear balance equation in terms of spherical harmonics," *Tellus* **20**: 200–2.

Miller, M., and Moncrieff, M. (1983). "The dynamics and simulation of organized deep convection." In: *Mesoscale meteorology: Theories, observations and models*, D. Lilly and T. Gal-Chen, eds. (Dordrecht: Reidel), pp. 451–95.

Miller, R. (1986). "Toward the application of the Kalman filter to regional open ocean modelling," *J. Phys. Oceanogr.* **16**: 72–86.

Miyakoda, K. (1956). "On a method of solving the balance equation," *J. Meteor. Soc. Japan.* **34**: 364–7.

Miyakoda, K. (1986). "Assessment of results from different analysis schemes." In: *International conference on the results of the Global Weather Experiment and their implications for the World Weather Watch*, GARP Publication Series No. 26, Vol. 1 (Geneva: World Meteorological Organization), pp. 217–53.

Miyakoda, K., and Moyer, R. (1968). "A method of initialization for dynamical weather forecasting," *Tellus* **20**: 115–28.

Miyakoda, K., Strickler, R., and Chludzinski, J. (1978). "Initialization with the data assimilation method," *Tellus* **30**: 32–54.

Miyakoda, K., Umscheid, L., Lee, D., Sirutis, J., Lusen, R., and Pratte, F. (1976). "The near real time, global, four dimensional analysis experiment during the GATE period, Part I," *J. Atmos. Sci.* **33**: 561–91.

Monin, A., and Yaglom, A. (1975). *Statistical fluid mechanics, Vol. 2* (translation of "Statisticheskaia gidromekhanika") (Cambridge: MIT Press).

Moore, A. (1990). "Normal mode initialization in a model of the tropical Pacific Ocean," submitted to *J. Phys. Oceanogr.*

Moore, A., Cooper, N., and Anderson, D. (1987). "Data assimilation in models of the Indian Ocean," *J. Phys. Oceanogr.* **17**: 1965–77.

Morel, P., and Talagrand, O. (1974). "Dynamic approach to meteorological data assimilation," *Tellus* **26**: 334–44.

Morse, P., and Feshbach, H. (1953). *Methods of theoretical physics. Part I* (New York: McGraw-Hill).

Munk, W., and Wunsch, C. (1982). "Observing the ocean in the 1990's," *Phil. Trans. Roy. Soc. London* **A307**: 439–64.

Navon, I., and Legler, D. (1987). "Conjugate-gradient methods for large-scale minimization in meteorology," *Mon. Wea. Rev.* **115**: 1479–502.

Navon, I., and de Villiers, R. (1983). "Combined penalty multiplier optimization methods to enforce integral invariant conservation," *Mon. Wea. Rev.* **111**: 1228–42.

Nitta, T., and Hovermale, J. (1969). "A technique of objective analysis and initialization for the primitive forecast equations," *Mon. Wea. Rev.* **97**: 652–8.

O'Brien, J. (1970). "Alternative solutions to the classical vertical velocity problem," *J. Appl. Meteor.* **9**: 197–203.

Ogura, Y., and Phillips, N. (1962). "Scale analysis of deep and shallow convection in the atmosphere," *J. Atmos. Sci.* **19**: 173–9.

Okland, H. (1970). "On the adjustment toward balance in primitive equation weather prediction models," *Mon. Wea. Rev.* **98**: 271–9.

Orlanski, I. (1975). "A rational subdivision of scales for atmospheric processes," *Bull. Amer. Meteor. Soc.* **56**: 527–30.

Paegle, J., and Paegle, J. N. (1976). "On geopotential data and ellipticity of the balance equation: A data study," *Mon. Wea. Rev.* **104**: 1279–88.

Panchev, S. (1971). *Random functions and turbulence* (Oxford: Pergamon Press).

Panofsky, H. (1949). "Objective weather-map analysis," *J. Appl. Meteor.* **6**: 386–92.

Papoulis, A. (1965). *Probability, random variables and stochastic processes* (New York: McGraw-Hill).

Parrish, D. (1988). "The introduction of Hough functions into optimal interpolation." In: *Proceedings of the eighth conference on numerical weather prediction* 1988 (Boston: American Meteorological Society).

Parrish, D., and Cohn, S. (1985). *A Kalman filter for a two dimensional shallow-water model: Formulation and preliminary experiments,* National Meteorological Center, Office Note 304 (Washington, DC: U.S. Dept. of Commerce, NOAA, National Weather Service).

Pedlosky, J. (1979). *Geophysical fluid dynamics* (New York: Springer-Verlag).

Peixoto, J., and Oort, A. (1984). "The physics of climate," *Rev. Mod. Phys.*, **56**: 365–429.

Penenko, V., and Obraztsov, N. (1976). "A variational initialization method for the fields of the meteorological elements," *Soviet Meteorology and Hydrology* **11**: 1–11.

Petersen, D. (1968). "On the concept and implementation of sequential analysis for linear random fields," *Tellus* **20**: 673–86.

Petersen, D. (1973). "A comparison of the performance of quasi-optimal and conventional objective analysis schemes," *J. Appl. Meteor.* **12**: 1093–1101.

Petersen, D., and Middleton, D. (1962). "Sampling and reconstruction of wavenumber limited functions in N dimensional Euclidean space," *Inform. Control.* **5**: 279–323.

Petersen, D., and Middleton, D. (1963). "On representative observations," *Tellus* **15**: 387–405.

Petersen, D., and Middleton, D. (1964). "Reconstruction of multidimensional stochastic fields from discrete measurements of amplitude and gradient," *Inform. Control.* **7**: 445–76.

Phillips, N. (1957). "A coordinate system having some special advantage for numerical forecasting," *J. Meteor.* **14**: 184–5.

Phillips, N. (1960). "On the problem of the initial data for the primitive equations," *Tellus* **12**: 121–6.

Phillips, N. (1963). "Geostrophic motion," *Rev. Geophys.* **1**: 123–76.

Phillips, N. (1977). *Variational analysis in pressure coordinates*, National Meteorological Center Office Note 134 (Washington, DC: U.S. Dept. of Commerce, NOAA, National Weather Service).

Phillips, N. (1981). "Variational analysis and the slow manifold," *Mon. Wea. Rev.* **109**: 2415–26.

Phillips, N. (1982). "On the completeness of multi-variate optimum interpolation for large-scale meteorological analysis," *Mon. Wea. Rev.* **110**: 1329–34.

Phillips, N. (1986). "The spatial statistics of random geostrophic modes and first-guess errors," *Tellus* **38A**: 314–22.

Platzman, G. (1967). "A retrospective view of Richardson's book on weather prediction," *Bull. Amer. Meteor. Soc.* **48**: 514–50.

Platzman, G. (1979). "The ENIAC computations of 1950: Gateway to numerical weather prediction," *Bull. Amer. Meteor. Soc.* **60**: 302–12.

Ploshay, J., White, R., and Miyakoda, K. (1983). *FGGE level II-b daily global analyses. Part I.* NOAA Data Report ERL-GFDL (Princeton: National Oceanic and Atmospheric Administration).

Pratt, R. (1979). "A space-time spectral comparison of the NCAR and GFDL general circulation models of the atmosphere," *J. Atmos. Sci.* **36**: 1681–91.

Provost, C., and Salmon, R. (1986). "A variational method for inverting hydrographic data," *J. Mar. Res.* **44**: 1–34.

Puri, K. (1981). "Local geostrophic wind correction in the assimilation of height data and its relationship to the slow manifold," *Mon. Wea. Rev.* **109**: 52–5.

Puri, K. (1983). "Some experiments in variational normal mode initialization in data assimilation," *Mon. Wea. Rev.* **111**: 1208–18.

Puri, K. (1985). "Sensitivity of low-latitude velocity potential fields in a numerical weather-prediction model to initial conditions, initialization and physical processes," *Mon. Wea.*

Rev. **113**: 449–66.

Puri, K. (1987). "Some experiments on the use of tropical diabatic heating information for initial state specification," *Mon. Wea. Rev.* **115**: 1394–1406.

Puri, K., and Bourke, W. (1982). "A scheme to retain the Hadley circulation during normal mode initialization," *Mon. Wea. Rev.* **110**: 327–35.

Rasch, R. (1985). "Developments in normal mode initialization. Part II: A new method and its comparison with currently used schemes," *Mon. Wea. Rev.* **113**: 1753–70.

Ray, P., Wagner, K., Johnson, K., Stephens, J., Bumgarner, W., and Mueller, E. (1978). "Triple-Doppler observations of a convective storm," *J. Appl. Meteor.* **17**: 1201–12.

Ray, P., Ziegler, C., Bumgarner, W., and Seraphin, R. (1980). "Single and multiple-Doppler radar observations of tornadic storms," *Mon. Wea. Rev.* **108**: 1607–25.

Reed, R. (1977). "Bjerknes memorial lecture," *Bull. Amer. Meteor. Soc.* **58**: 390–401.

Reinsch, C. (1967). "Smoothing by spline functions," *Numer. Math.* **10**: 177–183.

Richardson, L. (1922). *Weather prediction by numerical process* (Cambridge: Cambridge University Press).

Richtmeyer, R., and Morton, K. (1967). *Difference methods for initial value problems* (New York: Interscience).

Robinson, A., Carton, J., Mooers, C., Walstad, L., Carter, E., Reinecker, M., Smith, J., and Leslie, W. (1984). "A real time dynamical forecast of ocean synoptic/mesoscale eddies," *Nature* **309**: 781–3.

Robinson, A., and Leslie, W. (1985). "Estimation and prediction of oceanic eddy fields," *Prog. Oceanogr.* **14**: 485–510.

Rossby, C. (1938). "On the mutual adjustment of pressure and velocity distribution in certain simple current systems," *J. Marine Res.* (*Sears Foundation*) **2**: 239–63.

Roux, F. (1985). "Retrieval of thermodynamic fields from multiple-Doppler radar data using the equations of motion and the thermodynamic equation," *Mon. Wea. Rev.* **113**: 2142–57.

Rutherford, I. (1972). "Data assimilation by statistical interpolation of forecast error fields," *J. Atmos. Sci.* **29**: 809–15.

Rutherford, I., and Asselin, R. (1972). "Adjustment of the wind field to geopotential data in a primitive equations model," *J. Atmos. Sci.* **29**: 1059–63.

Sadokov, V., and Schteinbok, D. (1977). "Application of conjugate functions in analysis and forecasting of the temperature anomaly" (English translation), *Soviet Meteorology and Hydrology* **10**: 16–21.

Sasaki, Y. (1958). "An objective analysis based on the variational method. *J. Meteor. Soc. Japan* **36**: 77–88.

Sasaki, Y. (1960). *An objective analysis for determining initial conditions for the primitive equations*, Tech. Rep. (Ref. 60-16T) (College Station: Texas A&M University).

Sasaki, Y. (1969). "Proposed inclusion of time variation terms, observational and theoretical, in numerical variational analysis," *J. Meteor. Soc. Japan* **47**: 115–24.

Sasaki, Y. (1970a). "Some basic formalisms in numerical variational analysis," *Mon. Wea. Rev.* **98**: 875–83.

Sasaki, Y. (1970b). "Numerical variational analysis formulated under the constraints as determined by longwave equations and a low pass filter," *Mon. Wea. Rev.* **98**: 884–98.

Sasaki, Y. (1970c). "Numerical variational analysis with weak constraint and application to surface analysis of severe storm gust," *Mon. Wea. Rev.* **98**: 899–910.

Sasaki, Y., and McGinley, J. (1980). "Application of the inequality constraint in adjustment of superadiabatic layers," *Mon. Wea. Rev.* **109**: 194–6.

Satomura, T. (1988). "Dynamic normal model initialization for a limited area model," *J. Meteor. Soc. Japan* **66**: 261–76.

Saucier, W. (1955). *Principles of meteorological analysis* (Chicago: University of Chicago Press).

Savage, M., Weidner, G., and Stearns, C. (1988). "A diagnostic study of the influence of a gravity mode upon regional weather," *Mon. Wea. Rev.* **116**: 347–57.

Schlatter, T. (1975). "Some experiments with a multivariate statistical objective analysis scheme," *Mon. Wea. Rev.* **103**: 246–57.

Schlatter, T., and Branstator, G. (1979). "Estimation of errors in NIMBUS 6 temperature profiles and their spatial correlation," *Mon. Wea. Rev.* **107**: 1402–13.

Schroter, J., and Wunsch, C. (1986). "Solution of nonlinear finite difference ocean models by optimization methods with sensitivity and observational strategy analysis," *J. Phys. Oceanogr.* **16**: 1855–74.

Schulz, M. (1981). *Elliptic problem solvers* (New York: Academic Press).

Seaman, R. (1977). "Absolute and differential accuracy of analyses achievable with specified observational network characteristics," *Mon. Wea. Rev.* **105**: 1211–22.

Seaman, R. (1982). "A systematic description of the spatial variability of geopotential and temperature in the Australian region," *Aust. Met. Mag.* **30**: 133–41.

Seaman, R. (1983). "Objective analysis accuracies of statistical interpolation and successive correction schemes," *Aust. Met. Mag.* **31**: 225–40.

Seaman, R., and Gauntlett, F. (1980). "Directional dependence of zonal and meridional wind correlation coefficients," *Aust. Met. Mag.* **28**: 217–21.

Seaman, R., and Hutchinson, M. (1985). "Comparative real data tests of some objective analysis methods by withholding observations," *Aust. Met. Mag.* **33**: 37–46.

Semazzi, F., and Navon, I. (1986). "A comparison of the bounded derivative and the normal mode initialization methods using real data," *Mon. Wea. Rev.* **114**: 2106–21.

Shaw, D., Lonnberg, P., Hollingsworth, A., and Unden, P. (1987). "Data assimilation: The 1984/85 revisions of the ECMWF mass and wind analysis," *Quart. J. Roy. Meteor. Soc.* **113**: 533–66.

Shigehisa, Y. (1983). "Normal modes of the shallow water equations for zonal wavenumber zero," *J. Meteor. Soc. Japan* **61**: 479–93.

Shuman, F. (1957). "Numerical methods in weather prediction: I. The balance equation," *Mon. Wea. Rev.* **85**: 329–32.

Simmonds, I. (1976). "Data assimilation with a one-level primitive equation spectral model," *J. Atmos. Sci.* **33**: 1155–71.

Smagorinsky, J. (1956). "On the inclusion of moist adiabatic processes in numerical prediction models," *Ber. D. Deutsche. Wetterd.* **5**: 82–90.

Smith, W., Woolf, H., Hayden, C., Wark, D., and McMillin, L. (1979). "The TIROS-N operational vertical sounder," *Bull. Amer. Meteor. Soc.* **60**: 1177–97.

Sneddon, I. (1951). *Fourier transforms* (New York: McGraw-Hill)

Spilhaus, A. (1951). "World weather network." In: *Compendium of meteorology*, T. Malone, ed. (Baltimore: American Meteorological Society, Waverly Press), pp. 705–10.

Stephens, J. (1965). *A variational approach to numerical weather analysis and predictions*, Ph.D. dissertation (Austin: University of Texas).

Stephens, J. (1967). "Filtering response of selected distance-dependent weight functions," *Mon. Wea. Rev.* **95**: 45–6.

Stephens, J. (1970). "Variational initialization with the balance equation," *J. Appl. Meteor.* **9**: 732–9.

Stephens, J. (1971). "On definable scale reduction by simultaneous observations," *J. Appl. Meteor.* **10**: 3–25.

Stephens, J., and Polan, A. (1971). "Spectral modification by objective analysis," *Mon. Wea. Rev.* **99**: 374–8.

Stephens, J., and Stitt, J. (1970). "Optimum influence radii for interpolation with the method of successive corrections," *Mon. Wea. Rev.* **98**: 680–7.

Stern, W. (1974). "Updating experiments with a simple barotropic model," *J. Atmos. Sci.* **31**: 134–41.

Sugi, M. (1986). "Dynamic normal mode initialization," *J. Meteor. Soc. Japan* **64**: 623–36.
Tabjbakhsh, I. (1969). "Utilization of time dependent data in running solution of initial value problems," *J. Appl. Meteor.* **8**: 389–91.
Talagrand, O. (1981a). "A study of the dynamics of four dimensional data assimilation," *Tellus* **33**: 43–60.
Talagrand, O. (1981b). "On the mathematics of data assimilation," *Tellus* **33**: 321–39.
Talagrand, O., and Courtier, P. (1987). "Variational assimilation of meteorological observations with the adjoint vorticity equations. Part I. Theory," *Quart. J. Roy. Meteor. Soc.* **113**: 1311–28.
Tarantola, A. (1987). *Inverse problem theory* (Amsterdam: Elsevier).
Taylor, G. (1936). "The oscillations of the atmosphere," *Proc. R. Soc. London, Ser. A* **150**: 318–26.
Temperton, C. (1976). "Dynamic initialization for barotropic and multi-level models," *Quart. J. Roy. Meteor. Soc.* **102**: 297–311.
Temperton, C. (1977). *Normal modes of a barotropic version of the ECMWF gridpoint model.* ECMWF International Report No. 12 (Reading, England: European Centre for Medium Range Weather Forecasting).
Temperton, C. (1984). "Variational normal mode initialization for a multi-level model," *Mon. Wea. Rev.* **112**: 2303–16.
Temperton, C. (1988). "Implicit normal mode initialization," *Mon. Wea. Rev.* **116**: 1013–31.
Temperton, C. (1989). "Implicit normal mode initialization for spectral models," *Mon. Wea. Rev.* **117**: 436–51.
Temperton, C., and Williamson, D. (1981). "Normal mode initialization for a multi-level gridpoint model. Part I: Linear aspects," *Mon. Wea. Rev.* **109**: 729–43.
Testud, J. (1983). "Three-dimensional wind field analysis from Doppler radar data." In: *Mesoscale meteorology: Theories, observations and models,* D. Lilly and T. Gal-Chen, ed. (Dordrecht: Reidel), pp. 711–53.
Thacker, W., and Long, R. (1988). "Fitting dynamics to data," *J. Geophys. Res.* **93**: 1227–40.
Thiebaux, H. J. (1976). "Anisotrophic correlation functions for objective analysis," *Mon. Wea. Rev.* **104**: 994–1002.
Thiebaux, H. J. (1977). "Extending analysis accuracy with anisotropic interpolation," *Mon. Wea. Rev.* **105**: 691–99.
Thiebaux, H. J. (1981). "The kinetic energy spectrum vis-a-vis a statistical model for geopotential," *Tellus* **33**: 417–27.
Thiebaux, H. J. (1985). "On approximations to geopotential and wind-field correlation structures," *Tellus* **37A**: 126–31.
Thiebaux, H. J., Mitchell, H., and Shantz, D. (1986). "Horizontal structure of hemispheric forecast error correlations for geopotential and temperature," *Mon. Wea. Rev.* **114**: 1048–66.
Thompson, P. (1961). "A dynamical method of analyzing meteorological data," *Tellus* **13**: 334–49.
Thompson, P. (1969). "Reduction of analysis error through constraints of dynamical consistency," *J. Appl. Meteor.* **8**: 739–42.
Trenberth, K., and Olson, J. (1988). "An evaluation and intercomparison of global analyses from the National Meteorological Center and the European Centre for Medium Range Weather Forecasts," *Bull. Amer. Meteor. Soc.* **69**: 1047–57.
Tribbia, J. (1981). "Non-linear normal mode balancing and the ellipticity condition," *Mon. Wea. Rev.* **109**: 1751–61.
Tribbia, J. (1982). "On variational normal mode initialization," *Mon. Wea. Rev.* **110**: 455–70.
Tribbia, J. (1984). "A simple scheme for higher order non-linear normal mode initialization," *Mon. Wea. Rev.* **112**: 278–84.
Valentine, F. (1937). "The problem of Lagrange with differential inequalities as added side

conditions," In: *Calculus of variations* (Chicago: University of Chicago Press), pp. 407–48.

Van Isacker, J., and Struylaert, W. (1985). "Numerical forecasting using Laplace transforms," *Roy. Belgian Met. Inst. Publication Series A*, No. 115.

Vautard, R., and Legras, B. (1986). "Invariant manifolds, quasi-geostrophy and initialization," *J. Atmos. Sci.* **43**: 565–84.

Vederman, J. (1949). "The weather bureau–air force–navy analysis center," *Bull. Amer. Meteor. Soc.* **30**: 335–41.

Vinnichenko, N. (1970). "The kinetic energy spectrum in the free atmosphere: 1 second to 5 years," *Tellus* **22**: 158–66.

Wahba, G. (1982). "Vector splines on the sphere with applications to the estimation of vorticity and divergence from discrete noisy data," Technical Report no. 674, Department of Statistics, University of Wisconsin-Madison.

Wahba, G., and Wendelberger, J. (1980). "Some new mathematical methods for variational objective analysis using splines and cross validation," *Mon. Wea. Rev.* **108**: 1122–43.

Warn, T., and Menard, R. (1986). "Nonlinear balance and gravity-inertial wave saturation in a simple atmospheric model," *Tellus* **38A**: 285–94.

Washington, W., and Parkinson, C. (1986). *An introduction to three-dimensional climate modelling* (Mill Valley, California: University Science Books).

Watson, G. (1958). *A treatise on the theory of Bessel functions* (Cambridge: Cambridge University Press).

Wergen, W. (1983). "Initialization." In: *Interpretation of numerical weather products*, ECMWF Seminar/Workshop (Reading, England: European Centre for Medium Range Weather Forecasts), pp. 31–57.

Whittaker, E., and Robinson, G. (1924). *The calculus of observations* (London: Blackie and Sons).

Wiener, N. (1949). *Extrapolation, interpolation and smoothing of stationary time series* (New York: John Wiley).

Wilkinson, J. (1965). *The algebraic eigenvalue problem* (Oxford: Clarendon Press).

Williamson, D., and Daley, R. (1983). "A unified analysis-initialization technique," *Mon. Wea. Rev.* **111**: 1517–36.

Williamson, D., and Dickinson, R. (1972). "Periodic updating of meteorological variables," *J. Atmos. Sci.* **29**: 191–3.

Williamson, D., and Dickinson, R. (1976). "Free oscillations of the NCAR global circulation model," *Mon. Wea. Rev.* **104**: 1372–91.

Williamson, D., and Kasahara, A. (1971). "Adaptation of meteorological fields forced by updating," *J. Atmos. Sci.* **28**: 1313–24.

Williamson, D., and Temperton, C. (1981). "Normal model initialization for a multilevel grid-point model. Part II: Non-linear aspects," *Mon. Wea. Rev.* **109**: 745–57.

Winninghoff, F. (1973). "Note on a simple, restorative-iterative procedure for initialization of a global forecast model," *Mon. Wea. Rev.* **101**: 79–84.

Wolfsberg, D. (1987). *Retrieval of three-dimensional wind and temperature fields from single-Doppler radar data*, Ph.D. dissertation (Norman, Oklahoma: University of Oklahoma).

Wunsch, C. (1978). "The North Atlantic general circulation west of 50W determined by inverse methods," *Rev. Geophys. Space Phys.* **16**: 583–620.

Yaglom, A. (1962). *Stationary random functions*. English edition, 1973 (New York: Dover).

Yen, J. (1956). "On non-uniform sampling of band-width limited signals," *IRE Trans. Cir. Th.* **CT-3**(4): 251–7.

Ziegler, C. (1985). "Retrieval of thermal and microphysical variables in observed convective storms. Part I: Model development and preliminary testing," *J. Atmos. Sci.* **42**: 1487–1509.

Index